Hydrogen Energy

Hydrogen Energy

Production, Safety, Storage and Applications

L. M. Das

WILEY

This edition first published 2024

Registered Offices
John Wiley & Sons, Inc., 111 River Street, Hoboken, NJ 07030, USA
John Wiley & Sons Ltd, The Atrium, Southern Gate, Chichester, West Sussex, PO19 8SQ, UK

For details of our global editorial offices, customer services, and more information about Wiley products visit us at www.wiley.com.

Library of Congress Cataloging-in-Publication Data Applied for

Hardback ISBN 9781119900696

Cover Design: Wiley
Cover Images: © Scharfsinn/Shutterstock; Julija Vidjajeva/Shutterstock; Paulo Vilela/Shutterstock; art-siberia/Shutterstock; logoboom/Shutterstock; malp/Adobe Stock Photos; malp/Adobe Stock Photos; scharfsinn86/Adobe Stock Photos; malp/Adobe Stock Photos; Gyuszko/Getty Images; WillSelarep/Getty Images; RelaxFoto.de/Getty Images

Set in 9.5/12.5pt STIXTwoText by Straive, Chennai, India
Printed and bound by CPI Group (UK) Ltd, Croydon, CR0 4YY
C9781119900696_211223

This book is dedicated to my dear wife Bithika Das – my eternal life partner, my love and the fount of my unstinted inspiration—whose unwavering belief in my abilities and incessant words of encouragement spurred me to write this book. Though no longer with me, her words remain forever indelibly etched in my memory steering me through every step of my life.

Contents

Preface

First, a disclosure and a confession. It may even sound ironic, stemming from my childhood. Mutually conflicting images of hydrogen had, somehow or other, been treasured in my mind since childhood – for decades. As a child, when I heard about the hydrogen bomb (without an iota of knowledge of the thermonuclear reaction), I conceived a very devastating image of hydrogen in my mind. The horrifying image stayed in my head for years until I learned in high school that hydrogen could be obtained from water. That was a relief; it soothed my mind. I convinced myself that something derived from as common a substance as water cannot be that malicious and destructive. This self-conviction triggered my enthusiasm to study further. On and on it has continued, pulling me into a world that far from threatening seemed a distinct boon for mankind and the Planet Earth. At a later stage when I read more and more and started working on hydrogen energy, I learned through my progressive research that many of its unique distinctive characteristics could immensely benefit mankind if only the lab knowledge/experiments could be translated into everyday life reality and practice. It is my privilege to share with readers my thoughts and experience from the little I have learnt from my academic pursuits and research on hydrogen energy.

Hydrogen is currently viewed from two perspectives: the proponents emphasising its ability to solve energy–environment challenges and the opponents emphasising its explosive properties, backfires, hydrogen bombs and Hindenburg. Safety-related issues have been more prominent in common minds. As the cliché suggests: 'A known devil is better than an unknown angel'. In fact, enormous benefits emerge from hydrogen in terms of efficiency and low emission features, if some of its temperamental combustion characteristics are properly handled. However, mishandling of the same properties can result in disasters if they are not handled correctly. Chapters 2 and 5 attempt to discuss these aspects.

Rapid shift to energy transition and strategies for its implementation with an integrated hydrogen technology have been formulated in several countries. Hydrogen has accelerated its penetration worldwide. Thirty-two countries and the European Union, at the United Nations Climate Change Conference (Conference of Parties [COP26]) were unanimous in their view to work together for the development of clean hydrogen to ensure that 'affordable renewable and low-carbon hydrogen is globally available by 2030' (UNFCCC, 2021). At the 27th annual United Nations Conference on Climate Change (COP27) in 2022, green hydrogen was a massive push. There was plenty of discussion about 'how communities can benefit from local green hydrogen value chain'. Now greater focus is on green hydrogen as that can help achieve zero emissions. Global concern for energy security coupled with climate change has been the major driver towards focusing on renewable hydrogen energy and its role towards decarbonisation of so-called hard-to-abate sectors. Several stringent measures are being taken to prevent global warming, greenhouse gas emissions and achieve net-zero emissions by 2050. Research and development activities in many parts of the world have initiated efforts and programmes for hydrogen's rapid penetration into numerous energy sectors including power distribution, road, water and air transportation as well as for domestic applications. Economic viability of an adequate amount of green hydrogen is a concern.

It is likely that some feasible solutions will come out soon since several countries have initiated projects with time-bound goals. Material for use in hydrogen applications, safe storage and delivery at the point of use are improving. IC-engine-based hydrogen-fuelled cars, buses and three-wheelers have already been designed and tested for road transport. A number of hydrogen-powered trains have been operating on railway tracks. Hydrogen-fuelling stations have been built in many countries. Boeing and Airbus are developing passenger planes that are hydrogen fuel compliant. An array of fuel cells has been designed and is being used in the transport and power sectors across various countries. Hydrogen is also used for heating homes. Sets for distributed power generation, irrigation pump sets, lawn mowers and stoves for domestic use, too, have been developed.

This book aims at providing information on a relatively wide range of topics on hydrogen energy technology, aimed at the needs of a broad spectrum of readers. Special attention has been paid to defining the overall energy transition to total hydrogen energy. An integrated hydrogen energy system consists of a wide range of topics such as production, transfer, storage and delivery, safety, combustion, emissions, life cycle analysis and application in a variety of energy sectors. Separate independent books with more elaborate descriptions can be written on each of these aspects.

Mine is a modest effort to provide information to students, researchers and professionals in academia and industry (updated and referenced, up until early-2023), in a manner such that students of different disciplines with an interest in hydrogen energy would get basic information to read further. For an advanced reader, it is aimed at providing background knowledge of the subject and triggering ideas for further study and research in their own narrow specialised and niche areas of interest. Recently, several chemical industries, automobile companies, oil companies and power corporations around the world have developed hydrogen-based programs. I hope this book will also be useful for professionals in these industries. A caveat though would be in order: in a rapidly changing field of hydrogen energy where advancements are being made literally daily, technical information needs to be updated regularly by both researchers and stakeholders alike to bring them up to speed and help leverage the advancements made by all interested parties.

Windsor, Ontario,
Canada

Dr. L. M. Das
Former Professor and Emeritus Professor
Centre for Energy Studies
Indian Institute of Technology, Delhi,
New Delhi, India

Acknowledgement

It would be unfair to limit my gratitude to only those who have had a significant impact on me during the writing of this book. My four-decade journey to unravel hydrogen energy potential has been steered by countless individuals—in the classroom, research laboratories and industries or stages of national and international conferences. I have derived immense benefits from these invaluable interactions.

I take pleasure in extending my gratitude to Professor T N Veziroglu, the Founding President of the International Association of Hydrogen Energy, who also serves as the Founding Editor-in-Chief of the *International Journal of Hydrogen Energy*. My first meeting with him in 1982 followed by a series of subsequent personal meetings and discussions has significantly influenced and inspired me to strengthen my research pursuits in the area hydrogen energy. Furthermore, I am indebted to my doctoral thesis advisor, Prof. H. B. Mathur at the Indian Institute of Technology (IIT) Delhi, for his guidance and encouragement. His unwavering support ignited my zeal to delve into the study of hydrogen energy. Mr G P Singh and Mr Attar Singh, both from IIT, Delhi, offered invaluable expertise and assistance in effectively managing the complexities of hydrogen in laboratory setting.

Over the course of the ensuing four decades, continuous learning and progress have marked my journey. This journey has not only advanced my understanding but also has helped me in erasing from my mind the undeserved stigma attached to hydrogen as an unsafe fuel. I wish to express my sincere thanks to the Ministry of New and Renewable Energy (MNRE) of Government of India and the United Nations Industrial Development Organisation (UNIDO). Their sponsorship on projects centred on hydrogen utilisation in engines and vehicles has been instrumental in driving this research forward.

My interactions, either in personal meetings or through correspondence with some of the esteemed international researchers on hydrogen energy like Prof Wm. D. Van Vorst, Prof P. C. T. De Boer, Dr. W. J. McLean, Prof M. R. Swain, Prof G. A. Karim, Mr. Frank Lynch, Mr. C. A. Mac Carley, Prof Roger Sierens, Dr. Giuseppe Spazzafumo, Prof K. S. Varde, Prof C. J. Winter, Mr. J. J. Fagelson, Prof Makato Ikegami, Prof S. Furuhama at different phases of my professional career have proven invaluable. They have kept alive in me the passion for continued research on this emerging area of technology.

My list of inspirational people would be incomplete without acknowledging the contribution of some individuals whose combined efforts has smoothened out the wrinkles and bumps and brought the book to its present shape. These include Sudhansu Mohanty-my wife Bithika's younger brother, my son-in-law Himanshu Bhankar and my daughter Ishani

Das.Sudhansu meticulously edited and rephrased text throughout all chapters of the book to enhance its readability. Both Himanshu and Ishani provided invaluable technical reference materials and data. They precisely refined the graphs and upgraded the resolution of several figures, visibly enhancing the quality of the book's contents. The credit for the front cover, which depicts hydrogen energy's versatility, goes to both Himanshu and Ishani, not me. Special thanks to Dr. Divesh Bhatia, Dr. Pradeepta Sahoo, Dr. M. R. Nouni, Dr. Avinash Agarwal, and Dr. Saket Verma for their invaluable and diverse input in refining the text materials.

I want to extend my heartfelt appreciation to all the Wiley executives who I have had the privilege of interacting with on numerous occasions. From the moment I submitted the initial book proposal to its evolution into the present form, I have been genuinely thankful for their consistent courtesy, patience and prompt responsiveness to all inquiries.

If unintentionally I have missed the names of some who inspired me throughout this journey – either with their words of encouragement, technical guidance or emotional support , I want to convey my profound gratitude – more than that can be expressed within the confines of this acknowledgement page.

Now to the ultimate people for whom this book was written —you, my dear readers—for your time and interest in this book. You are the ultimate judge and I must confess I greatly value your input. I shall feel gratified if my humble endeavour has helped you appreciate the theme and character of discussion in the book. I will consider it a source of genuine satisfaction if the book has been as beneficial to you as it has been a delight for me writing it.

List of Figures

Author Biography

Dr. L. M. Das is a former Professor at the Indian Institute of Technology Delhi (IIT Delhi), India, where he later became an Emeritus Professor. He was a Visiting Professor at the University of California, Riverside, in 2001–2002. He has about forty-five years of teaching and research experience – a lifetime interest he continues to pursue today. His primary research focus has been on developing alternative-fuelled low-emission engines/vehicles. He has supervised a number of Ph.D. theses and master's degree projects besides publishing many papers in international journals in this area.

Dr. Das has had various international research assignments in the USA, France, UK and Portugal. He has delivered invited talks in India as well as in the USA, Canada, England, France, Germany, Japan, Switzerland, Australia, Belgium, Netherlands, Thailand, Moscow, Turkey, Singapore, Philippines, Nepal and Sri Lanka. He was the Principal Investigator (PI) of diverse sponsored projects funded by the Government of India's Ministry of New and Renewable Energy and Ministry of Science and Technology; General Motors (USA); European Union, Brussels; United Nations Industrial Development Organization (UNIDO); and Shell, United Kingdom. In 2012, the Indian Institute of Technology (IIT), Delhi, under his stewardship was successful in launching the first hydrogen-powered three-wheeler in the world in collaboration with Mahindra and Mahindra (India) and Air Products (USA). This project was sponsored by the UNIDO. As part of another sponsored project funded by General Motors, USA, the road testing of a long-distance biodiesel-fuelled bus from Delhi to Kanyakumari (the southernmost tip of India) turned out to be a winner.

Dr. Das is the recipient of several awards, including winner at the innovator's competition for DST-Lockheed Martin Indian Innovation Growth Programme (2012); 'Outstanding Service Award for his altruistic contribution to the cause of hydrogen economy conferred by the International Association of Hydrogen Energy' (2013); T.N. Veziroğlu Award presented by the Hydrogen Association of India (2013); Biju Patnaik Award for Excellence in Science (2014) bestowed by the Odisha Bigyan Academy; Lifetime Achievement Award accorded by the International Society of Energy, Environment and Sustainability (ISEES) (2015), among others.

As a researcher author, he has contributed varied chapters to books that have been published internationally. His forthcoming book Hydrogen as an Alternative Fuel for Internal Combustion Engines, also published by Wiley, is likely to be released in spring 2025.

1

Overall Energy Perspective

We can and should seize upon the energy crisis as a good excuse and great opportunity for making some very fundamental changes that we should be making anyhow for other reasons.

– Russell E. Train

CHAPTER MENU

1.1 Introduction

The present state of energy transition coupled with targeted strategies for total decarbonisation has put major economic and logistical challenges before all countries. A critical phase was reached when renewable energy and planet-friendly technology became our topmost priority for safe survival. A journey through time, to the edge of our civilisation, shows that energy in some form or other has been indispensable to human life. Since the dawn of

Hydrogen Energy: Production, Safety, Storage and Applications, First Edition. L. M. Das.

civilisation, man has adopted several methods to light the fire and provide heat for cooking. Survival of mankind has been very closely linked with an ever-increasing demand for energy. In all these processes of evolution, man has been solely dependent on Mother Earth. A closer look at the history of ancient living patterns of mankind shows that man extracted, stored, transmitted and used the primary energy available in nature as for his needs.

1.2 Energy Overview

A broad schematic diagram of the energy conversion chain shows how the primary energy has passed through the secondary energy stage and finally gives out useful energy as shown in Figure 1.1.

Starting from the first stage of extraction till the stage of utilisation, each stage remains broadly the same in character with implementation of modern technological advances at each stage to optimise the process. In the present context, it is extremely important to understand that the residue and the various forms of pollutants coming out at the last stage cause immense harm to human beings and the entire ecosystem. A recent report by International Energy Agency discusses the background that has led to the recent phase of growing energy crisis. Energy scenario in the transition phase and the roadmap for Net Zero Emission (NZE) by 2050 have been discussed with pertinent data and analysis [1].

During the agrarian era, thousands of years ago, people used wood as a means to generate fire, a resource of immense value for maintaining warmth during periods of cold weather. Besides, wood was useful for cooking as well as for building construction. In a way, wood was the first source of energy for mankind and remained the main energy-releasing material for centuries. Even now, wood is used for specific applications in the form of firewood, chips, charcoal and saw dust, and subsequently man discovered that coal was a better source of heat and light. Efforts were made to substitute coal in place of wood in several applications. Coal mining was commercially adopted, and coal was used in several energy sectors. It was observed that half a tonne of coal was able to produce four times the energy that the same amount of wood would produce. Thus, coal gradually displaced wood and entered into several energy sectors. Steam engines were successfully operated with coal. Railways were built for extensive inland transportation, which became faster and cheaper. People and cargo were moved

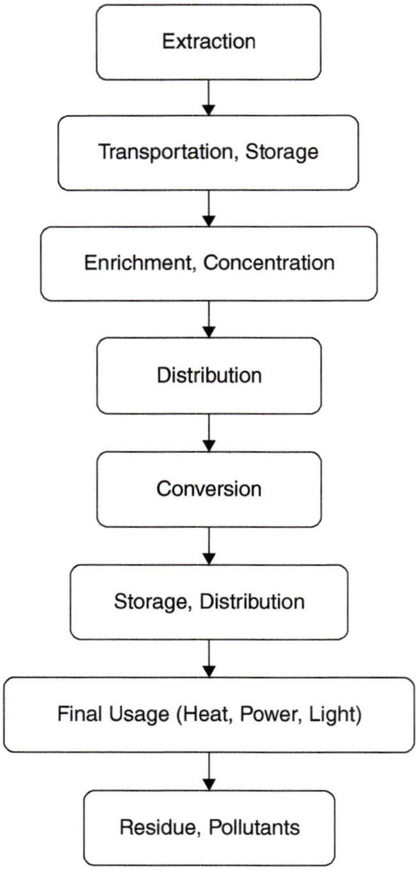

Figure 1.1 Energy conversion chain.

conveniently across the land. Use of coal in several industries substantially enhanced productivity. Even now, after such a long time, coal is still being used in several industrial sectors and contributing to the economic growth of nations and societies.

International Energy Agency has made some studies on the global primary energy demand and its projected pattern till 2070 in two scenarios –stated policy scenario (STEPS) and sustainable development scenario. STEPS provides 'a more conservative benchmark' [2]. The data have been projected in Figure 1.2 for the situations on existing energy policies without any major modification in the announced targets. As evident in the graph, while there is a drastic fall in fossil fuels, the demand for renewables rises from 15% in 2019 to about 60% in 2070. Widespread implementation of carbon capture, usage and storage (CCUS) in power plants and the several measures taken in the transport sector to reduce pollution have resulted in changes in the share of energy demand from fuels. Such a step is useful in decarbonisation by drastically reducing the high levels of carbon dioxide (CO_2) emitted by several industrial and transport sectors.

Figure 1.3 shows the change in the global energy demand by sustainable development scenario during the period 2019–2070. There is a rapid decline in the demand for fossil

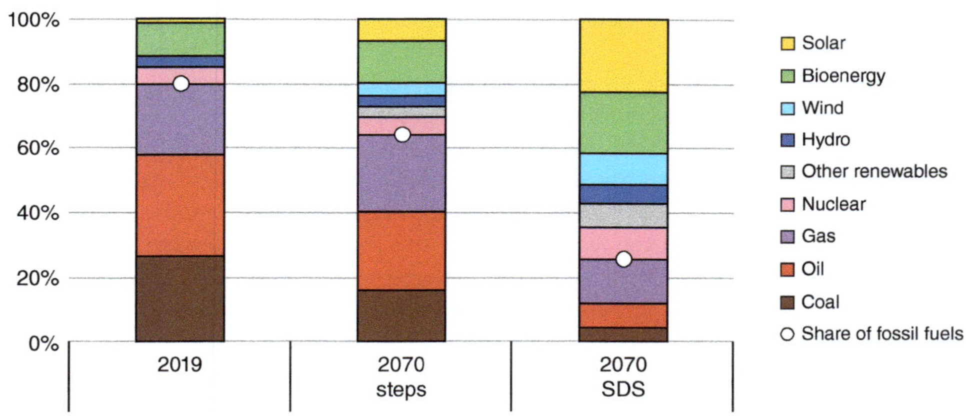

Figure 1.2 Global energy demand by fuel share in two scenarios [2]/IEA/Licensed under CC BY 4.0: stated policy scenario (STEPS) and sustainable development scenario (SDS).

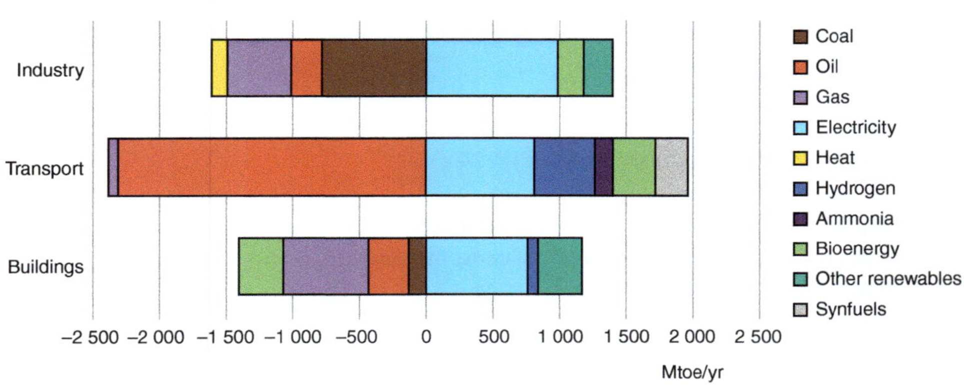

Figure 1.3 Demand for energy for sustainable development scenarios 2019 and 2070. [2]/IEA/ Licensed under CC BY 4.0.

fuels, not only in the transport sector but also in industries and buildings. There is a large demand for electricity as a dominant source. Several steps are being adopted in different countries to ensure conditions of sustainable environment. The transport sector now happens to be a major offender in environment degradation. Several low-emission alternative vehicular fuels are being tried and used in many countries. Hydrogen, being the 'cleanest burning' fuel, is receiving a lot of global attention. The surge of demand for hydrogen-based fuels is evident in the graph. The technology for producing hydrogen for the power and transport sectors has already been developed. It is feasible to produce hydrogen for onsite use in industries.

These developments in the eighteenth century literally brought an end to agrarian economy and heralded the advent of industrial economy. Subsequently, in the nineteenth century, a lot of research activities were carried out on synthesised hydrocarbons. In the course of time, it was observed that petroleum was more flexible and had a wider field of application as compared to coal. Energy consumption has been rather a history of energy transition. Now, even after the living pattern being radically changed, the world energy consumption using different fuels goes on increasing over the years in different sectors. Energy from renewable sources has started being used. With the renewed spate of interest and emphasis on renewables, demand for renewables is likely to go up at a much faster rate soon. This has been primarily because of the enrichment in quality of human life, multifold increase in population, rapid industrialisation, widespread transportation and several other factors leading to economic growth for which energy consumption is essential.

The period when petroleum was used for several fields of applications was a very crucial period in the history of energy transition. By the year 1919, gasoline had already been used as the fuel for trucks and tanks In fact, the use of petroleum fuels for military application in World War I established the feasibility of petroleum as a transportation fuel. In the meantime, while petroleum was dominating the energy sector, in the 1920s natural gas received the attention of being used in the transportation sector. Till then, natural gas that was being produced along with oil was flared off because the combustible features of natural gas were not still known. After the combustion characteristics of natural gas were recognised, it started being used for heating and also for several industrial applications. Thus, hydrocarbon fuels such as petroleum and natural gas dominated the energy scene for several decades. These fuels very aggressively entered domestic, transportation and many more industrial sectors. It is only recently that the emphasis has been shifted from dwindling fossil fuels and concentrated on renewables because of their source potential and non-polluting characteristics.

1.3 Sun as the Source of All Energy

Sun has been the main source of all energy to Earth which is received by our planet in the form of heat and light. The wavelength of most intense radiation is inversely proportional to the temperature. As shown in Figure 1.4, Sun has a surface temperature of about 6000 K and it emits radiation at short wavelengths of 0.5 μm, whereas Earth's atmosphere (average temperature of 278 K) emits peak radiation at longer wavelengths of 10 μm – infrared part of the spectrum.

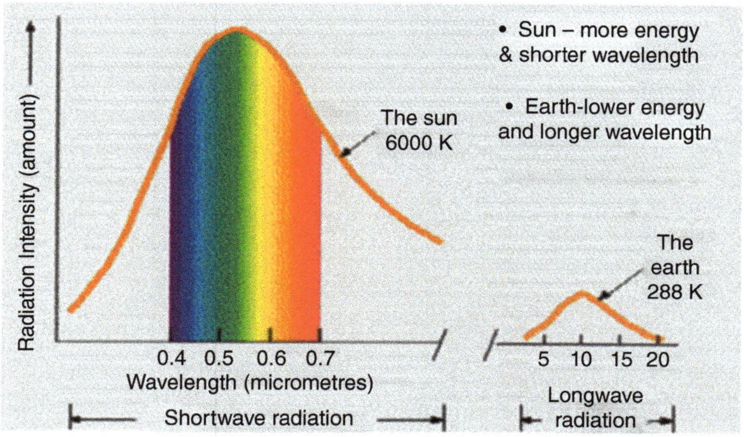

Figure 1.4 Radiation from Sun and Earth.

The peak of solar output is in the visible (light) part of the electromagnetic spectrum, while Earth emits most of its energy in the infrared (heat) portion of the electromagnetic spectrum. Most of Sun's energy is emitted in a spectrum from 0.15 to 4 μm, 41% of it is visible, 9% is ultraviolet (UV) and 50% infrared, whereas Earth's radiant energy stretches from 4 to 100 μm, with maximum energy falling at about 10.1 μm (infrared).

Earth's energy budget (Figure 1.5) represents Earth's solar energy flux. From among the solar radiant energy reaching Earth, about 34% is reflected back to the space, clouds, chemicals and dust in the atmosphere and Earth's surface, and only a tiny fraction is captured by the green plants.

Climate change is the greatest challenge of our time, so there's a lot at stake. From rising global temperatures to shifting weather patterns, the impacts of climate change are already being felt around the globe. Overcoming this challenge will require radical changes to how we generate and consume energy.

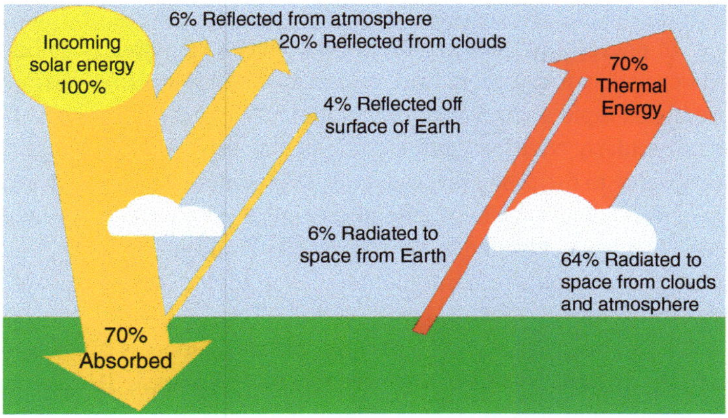

Figure 1.5 Solar energy flux of Earth.

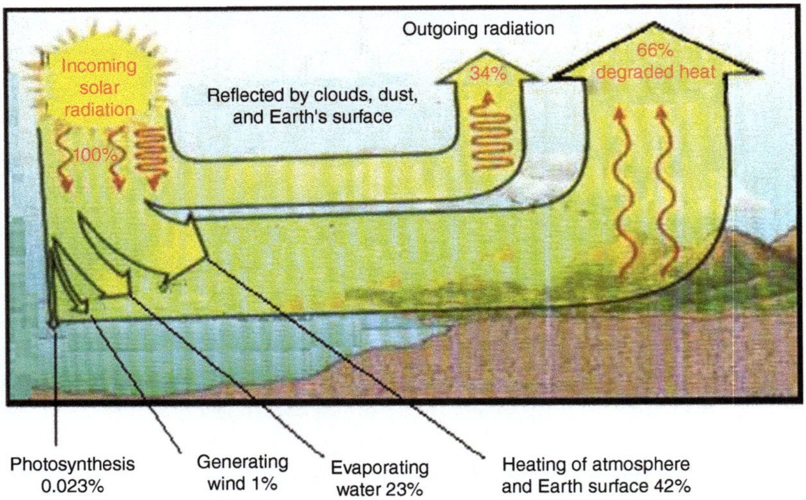

Figure 1.6 Energy flow from Earth.

Renewable energy technologies have developed drastically in recent years, with improved efficiency and lower cost. But one major barrier to their uptake is the fact that they do not provide a constant supply of energy. Wind and solar farms only produce energy when the wind is blowing or Sun is shining – but we need electricity around the clock, every day. Ultimately, we need a way to store energy on a large scale before we can make the switch to renewable sources. Figure 1.6 shows how part of the incoming solar radiation is used to different activities. Major part of this energy is used for heating the surface of Earth as well as for evaporating water.

1.4 Energy Consumption in Transport, Agriculture and Domestic Sectors

As discussed, earlier types of energy and the consumption have been changing from time to time. All through the twentieth century, there were remarkable and significant progress in energy sectors because of the rapid technological developments and their implementation. The energy sector, particularly transport and electricity, has been a major cause of environment issues, contributing to the problem caused due to greenhouse gases. It is thus essential that integration of affordable and clean technologies into the current energy landscape must be established. With this objective in view, strategies focused on energy security, improved efficiency and ensuring clean environment are being adopted with innovation to this phase of the energy transition.

The energy sector (electricity, transport, etc.) is one of the principal sources of environmental degradation, being the highest contributor to global greenhouse gas (GHG) emissions. In this context, significant concerns linked to the sustainability of energy arise. It is thus of paramount importance to undertake actions fostering the integration of affordable and clean technologies into the current energy landscape. To that end, strategies focused on enhancing the security of the energy supply, increasing the efficiency, reducing the emissions and expanding the energy market while encouraging the innovation and the competitiveness of the energy transition are set out in different countries of the world. Widespread electrification has been implemented in several countries of the world. Historically, for many years, the agricultural sector has been making use of resources such as land and biomass [3]. An effective interaction between agriculture and electricity sectors can be very useful to renewable electricity production.

Development of transportation engines resulted in the development of a large number of vehicles in several developing countries. Research and development activities on nuclear power were initiated in several countries of the world. While the destructive features of nuclear power were demonstrated in Hiroshima and Nagasaki, the potential of nuclear energy, as a source of electricity, was established without generating CO_2 or any other pollutant. The energy consumption has changed in almost all industries starting from small industry to very large ones. It has also affected the transportation, agriculture and domestic sectors. Canan G. Corlu, Rocio de la Torre and co-investigators [4] carried out some investigation on some of the energy-related optimisation challenges in road, train, maritime and air transportation. In the analysis, they were able to take into consideration several factors depending on the type of transport (land, water or air). Parameters such as energy-efficient driving and speed optimisation for road transport and weather-oriented monitoring along with fuel economy conditions for maritime systems were analysed. Aircraft trajectory optimisation and monitoring airport ground operation are very significant to energy-related air transportation systems.

Paramati Sudharshan Reddy and a team of researchers [5] analysed several robust panel econometric models in their study considering cross-sectional dependence and heterogeneity. This study had been carried out to assess the energy consumption in the agricultural sector in G20 nations. The study also includes the overall economic impact in these countries in both renewable and non-renewable energy consumption in industrial sectors. Domestic energy consumption chiefly depends on the living standard. It also depends upon a lot of factors such as the type of residence, age group of the inhabitants and the climate of the locality. There are different types of fuels such as solid fuels, liquid fuels, electricity and biomass which are being used in different parts of the world. The average household energy consumption for modest living consists of energy for cooking, lighting, washing, drying, cooling and refrigeration and several miscellaneous loads required at different periods.

Ali Najmi, Hamed Shakouri G. and Abbas Keramati [6] carried out research in critically identifying the factors that are responsible for residential energy consumption pattern. Practical test data were collected for a more rational analysis. In this study, a regression model was developed to identify the most dominant parameters using natural gas and the amount of electricity consumed. The results of such an analysis were useful in identifying which are the parameters that need to be regulated for reducing the domestic energy consumption.

It has been observed that one-third of global energy consumption and one-fourth of CO_2 emissions are due to buildings [7]. Population, urbanisation, present living style, etc., contribute to this level of energy consumption. Figure 1.7, based on International Energy

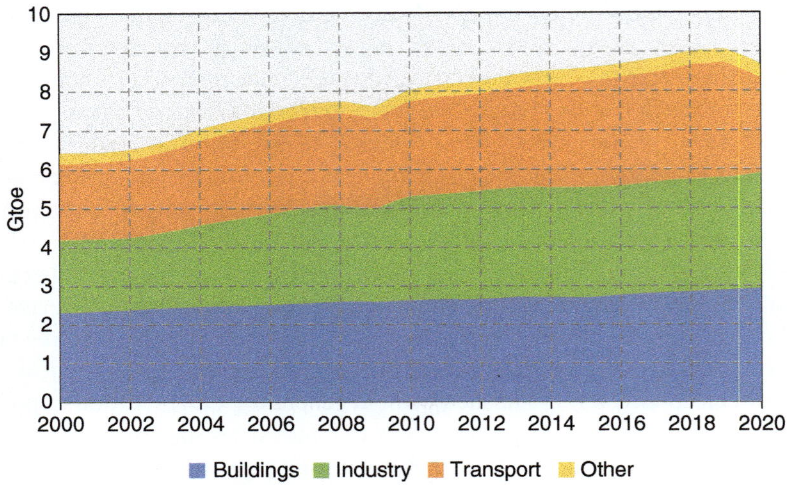

Figure 1.7 Global energy consumption by sector [7]/with permission of Elsevier.

Agency (IEA) data, gives a comparative picture of global energy consumption in buildings with respect to transport and industrial sectors [7]. Therefore, design, construction and maintenance of buildings have to be considered from the point of view of energy efficiency and climatic conditions.

1.5 Energy Crisis: Starvation of Fossil Fuels

There are several reasons leading to shortage of petroleum-based fuels. The factors leading to this fuel starvation conditions have been mainly due to geographical and geopolitical reasons. Extreme dependence on energy imports and lack of effort to exploit the local renewable resources have led to the present situation. The character of this problem could be sometimes from local to regional, but the high magnitude of the present problem is universal. Localised shortages and market manipulation are responsible for local problems. The nature of the problem of energy crisis in a particular region can emerge temporarily because of organised union strike, sudden tax hikes, over consumption and ageing infrastructure. These local and regional problems are very temporary and can be sorted out easily. But the real energy crisis is essentially universal in character.

In the modern industrial society with rapid economic growth, world's energy demand on the natural resources is very rapidly increasing. At the same time, the limited reserves are depleting fast. Situations are likely to worsen in future because of the expected world population and pattern of life style being more dependent on the energy supply. Inadequate oil and gas supply to meet the surge in global energy demand would affect our living pattern. Such a situation would disturb the planetary ecosystem. It would affect several areas such as transportation, power, agriculture, chemical and a host of other industrial activities in the entire world. As evident in Figure 1.8, the data have been collected for every quarter of

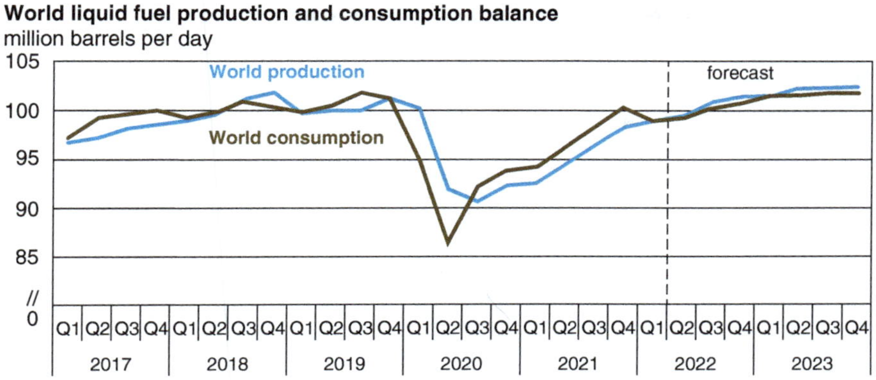

Figure 1.8 Global oil consumption [8]/US Energy Information Administration/Public Domain.

the year since 2017 and, on the basis of trend of variation in liquid fuel production and consumption, the possible consumption and production in 2023 have been indicated [8].

The consequences of fossil fuel starvation have affected the transport and industry sectors, and even the everyday life of a common citizen. In fact, it has very seriously affected the vulnerable population who have been depending on traditional sources for their livelihood. Last but not the least, it has caused massive global economic disruption. But the positive effects ultimately far outweigh these disadvantages. Reduction in fossil fuel consumption leads to substantial reduction of greenhouse gases, contributing to mitigating climate change and global warming. This has accelerated the transition to renewable energy sources with infinite source potential – such as solar, wind, hydro and geothermal. Every nation, depending upon its renewable resources, can implement strategies of energy mix as suitable to the country. Availability of plenty of energy can help formulate country-specific plans to develop innovation technologies, fostering advancements in various industries to enrich the quality of life of a common citizen. Pollution causes biodiversity loss with the destruction of habitats. It also leads to decline and extinction of plant and animal species, which results in total disruption of ecosystems and food chains.

1.6 Environmental Degradation Due to Fossil Fuel Combustion

Fossil fuel starvation is closely linked to environmental pollution. The extent of the effects of the pollutants depends on the concentration and duration of exposure to pollutants, as well as the resilience of the affected species or structures. Several industries, power plants and automobiles use fossil fuels and release a lot of noxious pollutants into the atmosphere. Inhalation of pollutants like particulate matter, volatile organic compounds (VOCs) and nitrogen oxides (NO_X) can cause respiratory issues, cardiovascular problems and even premature death. Prolonged exposure to certain pollutants, such as heavy metals and persistent organic pollutants (POPs), can lead to chronic health conditions, including cancer and neurological disorders. High levels of sulphur dioxide (SO_2) and NO_X can lead to the degradation of buildings, monuments and other structures. These pollutants react with the materials in the structures, causing corrosion, discoloration and weakening over time. Depletion of air, water, soil, etc., has resulted in rapid destruction of the ecosystem. Petroleum-based fuels have had lots of influence over human civilisation over centuries in several sectors. The problem of rapid environmental degradation had been one of the major concerns, and its damaging effects have been glaringly felt. Environmental pollution has threatened the very safe survival of this planet through destruction of natural habitats and depletion of natural resources. The degradation of the environment due to disturbances of significant magnitude, caused by both natural factors and the loss of biodiversity, has greatly impacted the condition of the surroundings, including aquatic animals, wildlife and structures.

Thus, the wide spectrum of problems ranges from flood, cyclone, heavy rain, deforestation, soil erosion, etc. The intensity of some such events are often location-specific. A closer look at the cause of all such disturbances in the environment shows that some of these phenomena are predominantly natural in origin, even though the influence of anthropogenic activities cannot be ignored. On the other hand, global warming, climate change, greenhouse gas emission, acid rain, ozone layer depletion, smog, urban heat island, vehicular pollution, etc., are some of the important phenomena of environmental degradation which are caused or accelerated by anthropogenic activities.

1.6.1 Green House Effect: Greenhouse Gas and Global Warming

Solar radiation from Sun reaches Earth's atmosphere. The surface of Earth absorbs most of the short-rayed insolation and later releases this heat in the form of infrared radiation into the atmosphere. Radioactively active gases that absorb wavelengths longer than 4 μm are called 'greenhouse gases'. This absorption heats the atmosphere, which in turn radiates energy back to Earth as well as out of space.

Most of the long-wavelength energy radiated by Earth is absorbed by a combination of radioactively active gases like greenhouse gases. Some of the heat is absorbed by particles known as 'greenhouse gases', such as CO_2, methane (CH_4), nitrous oxide (N_2O), ozone, chlorofluorocarbons and water vapour. The heat retained by the gases acts as a thermal blanket. The impact of greenhouse gases on global warming and the ultimate effect on climate change are too well known to be elaborated. Figure 1.9 shows how global warming occurs and which are the greenhouse gases that contribute to global warming.

Greenhouse gases absorb infrared radiation in the form of heat which is released into the atmosphere. This is referred to as the greenhouse effect. CO_2 contributes maximum (about 60% of radioactive forcing), CH_4 around 15–30%, NO_X up to 15% of warming, chlorofluorocarbons (CFCs), around 12% of warming and the rest is contributed by the trace gases. Figure 1.10 shows the greenhouse effect on Earth.

To compare the effect of different greenhouse gases on the global energy balance, global warming potential (GWP) has been adapted. It is a weighing factor that enables comparisons to be made between the global warming impact of 1 kg of any greenhouse gas and 1 kg of CO_2. It is a dimensionless quantity and includes a time horizon during which the impact is felt. By including the time horizon, the GWP accounts for the greater impact that a gas

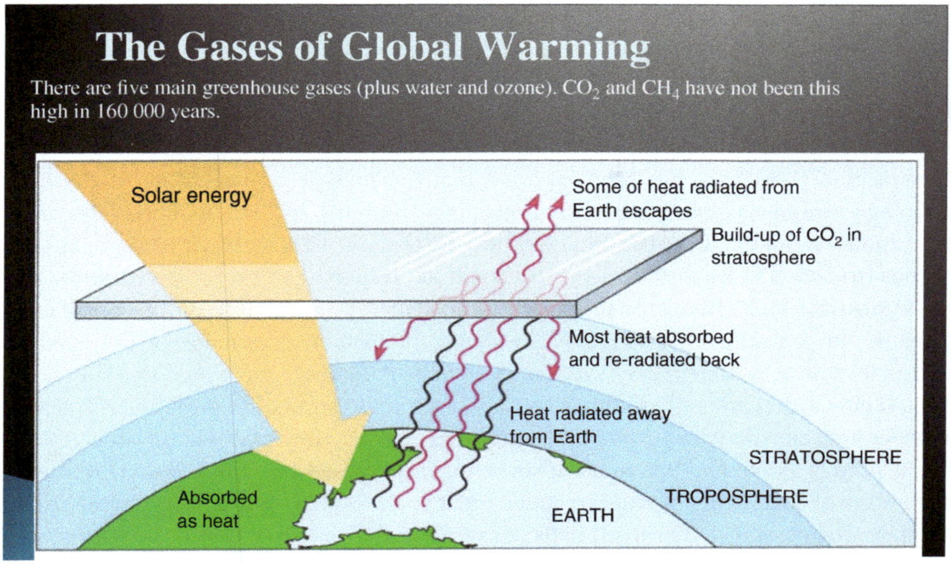

Figure 1.9 Phenomenon of global warming.

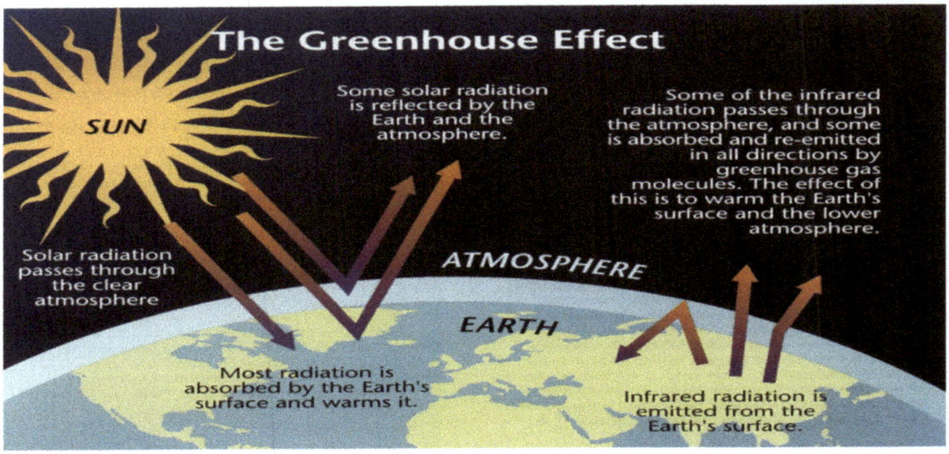

Figure 1.10 Greenhouse effect.

with a long atmospheric lifetime will have compared with one that quickly disappears once it has been emitted. GWP is a ratio of the cumulative radiative forcing (which defines the warming effect on the climate) for 1 kg of a greenhouse gas over the same period of time to the cumulative radiative forcing for 1 kg of a reference gas (chosen to be CO_2). Mathematically,

$$\mathrm{GWP} = \left[{}_0\!\int^{T} \Delta F_{\mathrm{g}} \cdot R_{\mathrm{g}}(t)\mathrm{d}t \right] / \left[{}_0\!\int^{T} \Delta F_{\mathrm{CO2}} \cdot R_{\mathrm{CO2}}(t) \cdot \mathrm{d}t \right]$$

where ΔF_g is the radiative forcing of greenhouse gas in question (W m^{-2} kg^{-1}), ΔF_{CO2} is the radiative forcing of CO_2 (W m^2 kg^{-1}), $R_g(t)$ is the fraction of 1 kg of greenhouse gas remains in the atmosphere at time 't', $R_{CO2}(t)$ is the fraction of 1 kg of CO_2 remaining at time 't' and T is the time period of cumulative effects (years).

In a recent greenhouse gas bulletin on Global Atmosphere Watch of World Meteorological Organisation (WMO), it has been reported that the present concentration of the heat-trapping greenhouse gases has become exceedingly high [9]. In 2021, 'carbon dioxide concentrations were 415.7 parts per million (ppm), methane at 1908 parts per billion (ppb) and nitrous oxide at 334.5 ppb. These values constitute, respectively, 149%, 262% and 124% of pre-industrial levels before human activities started disrupting natural equilibrium of these gases in the atmosphere' [9]. Figure 1.11 shows the history of variation of CO_2 concentration level in atmosphere over thousands of years [10]. Fossil fuels such as coal, oil and natural gas (hydrocarbons) produce CO_2 and water when burned for any application such as generating electricity in power plants, running a vehicle, heating or for any other domestic purpose at home and use in aeroplanes. CO_2 is released into the atmosphere during respiration of consumers. It is also released into the atmosphere through decomposition of animals. The combustion of fossil fuels adds CO_2 to the air along with other natural processes like volcano eruptions. Decomposition of detritus also gives out CO_2 which happens to be major greenhouse responsible for global warming.

Concentration of CO_2 emission level started taking a sharp increasing trend with the onset of the Industrial Revolution. The situation got further worsened with the increase of automobiles. Increase of CO_2 in the atmosphere has resulted in global warming and greenhouse effect. It has also been observed that CO_2 reaches a peak in winter and a trough in summer. This is so because green plants are most active in summer and the total amount of photosynthesis exceed the total amount of respiration; as a result, CO_2 is removed from atmosphere. In winter, the total amount of photosynthesis decreases and becomes less than that of the amount of respiration, so CO_2 concentration increases.

Figure 1.12a,b shows the recent data of carbon dioxide in atmosphere.

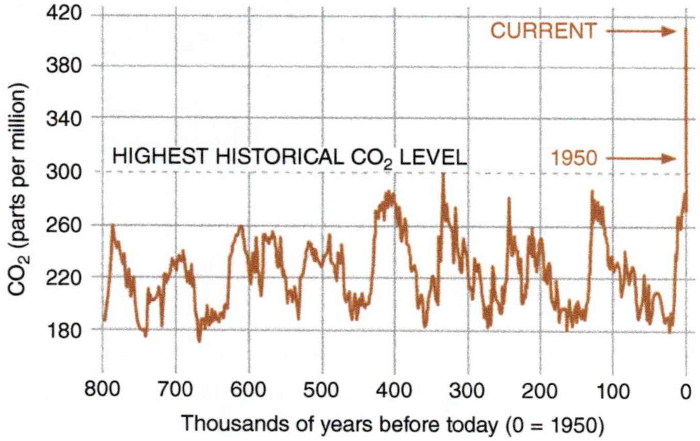

Figure 1.11 Historical record of CO_2 in concentration in atmosphere [10]/NASA/Public Domain.

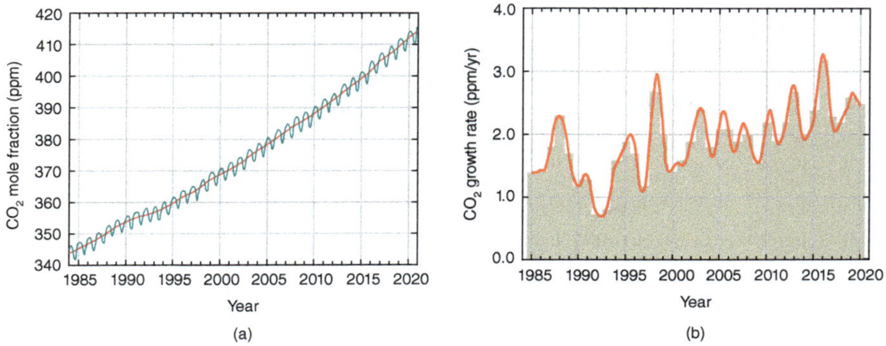

Figure 1.12 Atmospheric concentration of CO_2 takes up an increasing trend [9]/World Meteorological Organisation.

Concentration of CO_2 in the atmosphere breached the milestone of 400 ppm in 2015. Just five years later, it exceeded 413 ppm. Even the pandemic 'COVID-19', with absolute lockdown in several countries, did not have any overall discernible impact on the atmospheric levels of greenhouse emissions, even though sometimes it temporarily declined. It has been reported that, during COVID-19, a large number of countries adopted stringent steps with carbon-neutral targets. The present level of atmospheric CO_2 emission, which has reached 149% of the pre-industrial level in 2021, is mainly due to anthropogenic activities and combustion of fossil fuels. At present, there is a very serious concern that in future the land ecosystems and the oceans would lose their characteristic role of acting as 'sinks' and would not be able to absorb CO_2.

It is essential that policies and implementation strategies for the industrial sector and transportation system need to be formulated giving priority to environmental degradation and climate change.

Figure 1.13 shows the growth pattern of other important greenhouse gas, CH_4, from 1985 to 2020. It is clear from the graph that the global average CH_4 concentration level has been taking up a higher rate of increasing trend since 2007. The data recorded since 1983 show

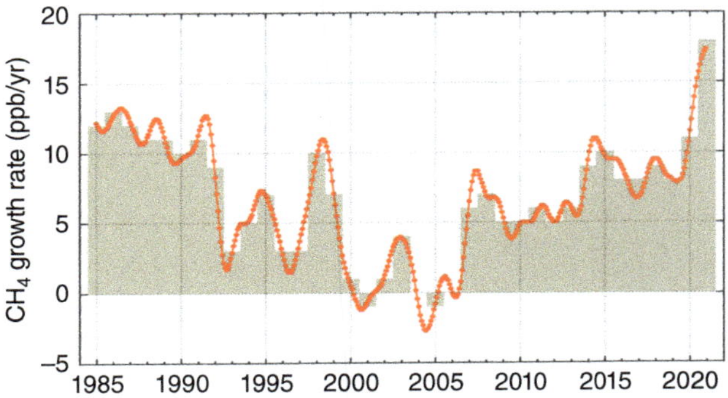

Figure 1.13 Atmospheric CH_4 concentration level [9]/World Meteorological Organisation.

that the annual increase in 2020 was 15 ppb and in 2021 it has been 18 ppb. Increase in CH_4 has also been due to anthropogenic activities and natural sources. The atmospheric CH_4 emission has been taking an upward trend since 2011. CH_4 has an atmospheric lifetime of about 12 years compared to the lifetime of CO_2 which is in centuries.

The emission of N_2O has been estimated from natural sources and also from anthropogenic activities. Apart from a host of industrial activities, the major sources have been oceans, soils, combustion of biomass burning, use of fertiliser, etc.

There is a slow yet continuously increasing trend in its growth rate over past several years. Each GHG has a different GWP and each GHG has got different length of lifetime in the atmosphere. There are two criteria that determine the impact of a greenhouse gas on the climate. These are the active residence time of the GHG in the atmosphere and also the amount of energy absorbed by it when it remains in the atmosphere. Figure 1.14 shows the variation of N_2O concentration over the period of last several years.

Chlorofluorocarbons, apart from being a potent greenhouse, have also been responsible for damaging the ozone layer and thus leading to global warming. Figure 1.15 shows the

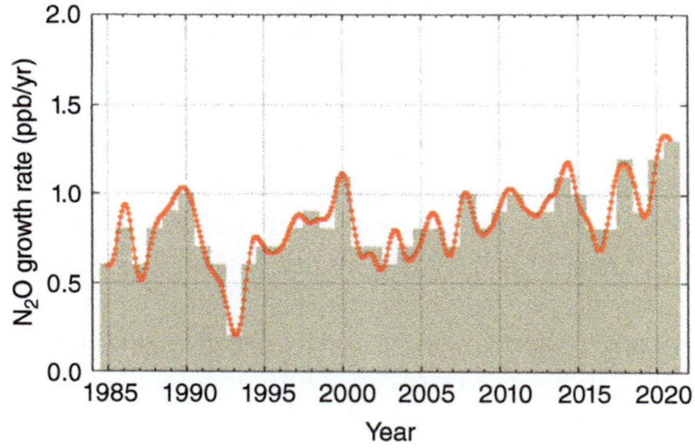

Figure 1.14 Atmospheric N_2O concentration level [9]/World Meteorological Organisation.

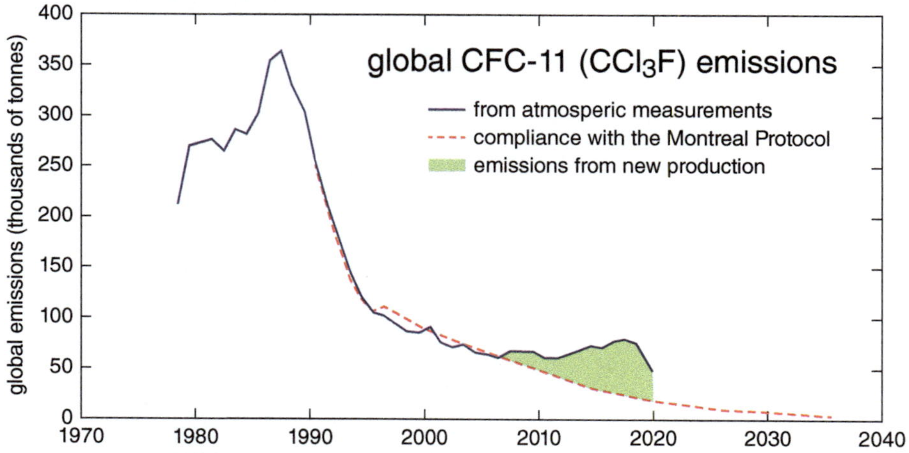

Figure 1.15 Atmospheric global CFC-11 emission [11].

global CFC measurements [11]. Both hydrofluorocarbons (HFCs) and hydrochlorofluorocarbons (HCFCs) are considered as substitutes to CFCs on the basis of their lesser damage to the ozone layer, but they significantly contribute to climate change. Such an abrupt turnaround has been detected by the global monitoring network of National Oceanic and Atmospheric Administration (NOOA) that 'five years after an unexpected spike in emissions of the banned ozone-depleting chemical chlorofluorocarbon CFC-11, emissions dropped sharply between 2018 and 2019' [12]. There are several other factors causing environment degradation and disruptions in climate. The present discussions are restricted to phenomena smog, acid rain, pollution in transport and industrial sectors where hydrogen has got the ability to play a significant role.

1.6.2 Smog

There are two types of smog: industrial smog or sulphur smog, and photochemical smog. Sulphurous smog is commonly referred to as 'London Smog' because in London the combined polluting effects of smoke and fog claimed 4000 fatalities in 1952. This was due to inefficient combustion of high-sulphur coal that produced high concentration of unburned carbon soot and other particulates, acidic sulphate aerosols (such as sulphuric acid, H_2SO_4) and high levels of SO_2. Sulphuric acid is corrosive and attacks bronchial passageways. It is of special concern to asthma/emphysema sufferers. There had been several stringent steps over the period of time and sulphurous smog does not occur now.

Photochemical smog is a very common pollution phenomenon in the urban areas with the massive population of vehicles and other industrial activities. Exhaust emissions from vehicles and pollutants from industries release hydrocarbons (HCs) and NOx into the atmosphere which come in the presence of sunlight and form ozone at the ground level. The primary responsibility for causing smog lies with this 'bad ozone'. The phenomenon of smog formation is shown in Figure 1.16. Photochemical smog was earlier referred to as 'Los Angeles Smog' because its damaging effects were observed in Los Angeles and also to differentiate its characteristics from sulphurous smog. Now, it has become a global phenomenon affecting the environment in several countries and is presently referred to simply as 'smog'.

At first, nitrogen dioxide (NO_2) is broken down by sunlight to form nitric oxide (NO) and an oxygen radical (O) as shown in Eq. (1.1):

$$NO_2 + \text{sunlight} \rightarrow NO + O \tag{1.1}$$

Sunlight is necessary to supply enough energy to break one of the nitrogen–oxygen bonds. Atomic oxygen also generates the hydroxyl radical by reacting with atmospheric water vapour. The oxygen radicals formed react with atmospheric oxygen (O_2) to form ozone (O_3) as shown in Eq. (1.2):

$$O + O_2 \rightarrow O_3 \tag{1.2}$$

Ozone (oxidant) concentration does not increase until most of NO is converted into NO_2.

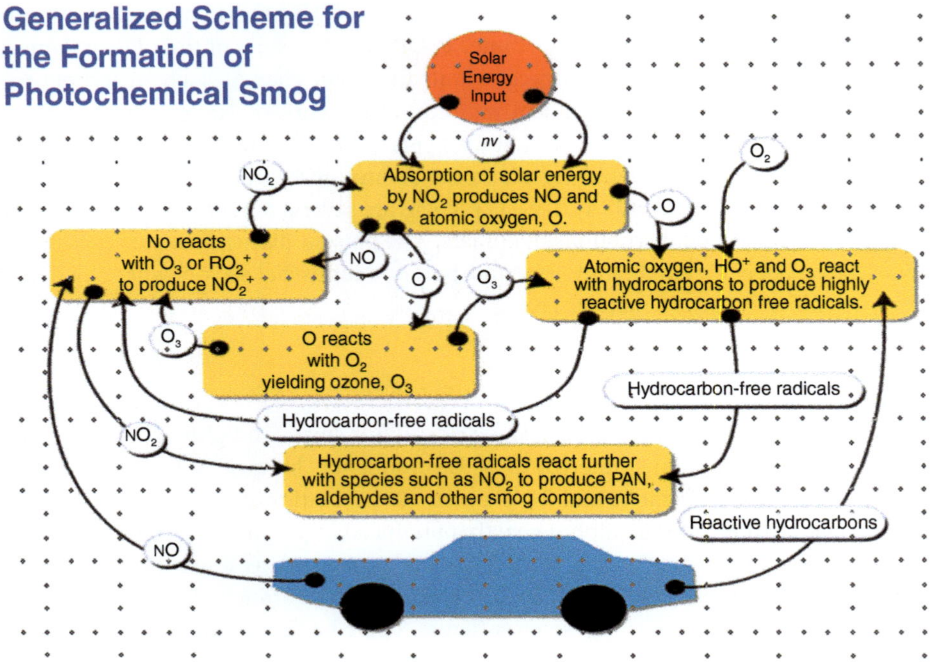

Figure 1.16 Formation of photochemical smog.

This is due to following competing atmospheric chemical reaction and this is why sunlight is necessary for smog formation. Reaction of ozone with NO produces NO_2 and oxygen as indicated in Eq. (1.3):

$$O_3 + NO \rightarrow NO_2 + O_2 \tag{1.3}$$

Harmful products, such as peroxyacetyl nitrate (PAN), are produced by reactions of NO_2 with various hydrocarbons (R), which are compounds made from carbon, hydrogen and other substances:

$$NO_2 + R \rightarrow \text{products such as PAN} \tag{1.4}$$

The primary origin of these hydrocarbons can be attributed to VOCs. Similarly, oxygenated organic and inorganic compounds (ROx) react with NO to produce more NO_X:

$$NO + ROx \rightarrow NO_2 + \text{other products} \tag{1.5}$$

The significance of the presence of the VOCs in these last two reactions is paramount. Ozone is normally consumed by NO, as in reaction (1.3). However, when VOCs are present, NO and NO_2 are consumed as in reactions (1.4) and (1.5) allowing the build-up of ground level ozone. In the context of smog formation, ultraviolet (UV) radiation from Sun plays a crucial role in the photochemical reactions. In a simplified manner, the overall reaction can be expressed as:

$$NO_2 + VOC + O_2 + UV \rightarrow NO + O_3 + PAN + \text{other oxidants}$$

1.6.3 Acid Rain

Acid rain by definition is rain that is more acidic than normal water. This is so because it contains sulphuric acid and nitric acid originating from oxides of sulphur (SO_X) and oxides of nitrogen (NOx), respectively, in atmosphere. 'Acid deposition' is a better term than 'acid rain' because acidity can reach Earth's surface not only in rain but also as gases and solid particles. Acidity of substances in water is commonly expressed in terms of pH. Solution with pH less than 7 is acidic. Solution with pH more than 7 is base. Natural precipitation is slightly acidic with a pH of 5.0–5.6. Acid rain degrades building materials such as limestone, marble, galvanised steel and certain paints. In fact, weathering and erosion of monuments were identified much earlier to indicate the adverse effect of acid deposition. Similarly, sulphuric acid and nitric acid are formed when SO_X and NO_X react with water and other chemicals in the atmosphere (Figure 1.17).

Sources of acid deposition, such as SO_2 and NO_X from the combustion of coal, oil and natural gas, are mainly responsible for causing acid rain. The chemical reactions resulting in the formation of sulphuric acid and nitric acid are shown below:

$$\text{Sulphuric acid: } 2SO_2 + O_2 + 2H_2O \rightarrow 2H_2SO_4$$

$$\text{Nitric acid: } NO + NO_2 + O_2 + H_2O \rightarrow 2HNO_3$$

$$\text{Hydrochloric acid: } H_2SO_4 + 2NaCl \rightarrow 2HCl + Na_2SO_4$$

$$\text{Carbonic acid: } CO_2 + H_2O \rightarrow H_2CO_3$$

Hydrochloric and carbonic acids also contribute to the problem of acid deposition. Combustion of petroleum fuels and coal, which are normally used to generate electricity, happens to be the source of SO_2 and NO_X. Similarly, the combustion of fossil fuels in the transport

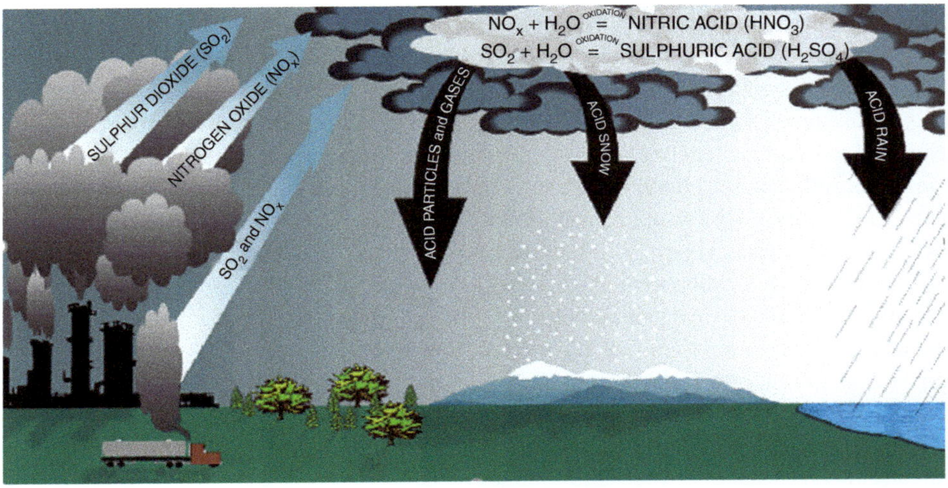

Figure 1.17 Sulphuric acid and nitric acid formation.

sector (cars, buses and trucks aeroplanes) also contributes to the formation of NO_X and SO_X, which in turn are responsible for causing acid rain. Figure 1.17 shows the process of acid formation in the atmosphere.

It has been broadly observed that 70% of the sulphur emissions and 88% of the NO_X emissions are caused by human activities. The rest are formed due to biosphere, marine sources and volcanoes.

At one point in time in past, very tall smokestacks were constructed in belief that if the exhausts from power plants were dispersed high into the air it would no longer create a problem with pollution. This has proved fallacious. Sending exhausts with pollutants high into the sky increases the time the pollutants stay in the air. It enhances the chances of acid rain as the exhaust come into contact with water droplets from rain or other sources. Moreover, depending upon the speed of wind, sometimes the pollutants get carried away to some locations where they come into contact with water to cause local acid rainwater.

1.6.4 Vehicular Pollution

Increase in world population and improved living standard the entire world over have resulted in the widespread use of petroleum fuels. It was realised that petroleum fuels were dwindling fast, and it is virtually impossible to maintain our existing lifestyles in the absence of petroleum fuels. It wasn't possible to go back in time and adopt the earlier style of living. The entire world encountered the problem of energy crisis which is essentially the crisis emerging out of depletion of petroleum fuels. Efforts around the world in academic and research institutes, industries and government levels were accelerated to identify appropriate substitute to petroleum fuels. Transportation sector throughout the world was very badly affected in an era of fast mobility. Energy conservation and alternative fuels were identified, studied and applied as substitutes for petroleum fuels. No universally acceptable alternatives have emerged so far. Several countries are carrying out extensive research and development (R&D) activities to utilise the energy source available in their country. In the context of energy and environment, it would perhaps be appropriate to discuss the constraints and damaging role of petroleum fuels in the transportation sector. In an effort to conserve energy and protect the environment, presently with the use of gasoline, diesel, natural gas, liquefied natural gas (LNG), liquefied petroleum gas (LPG), propane and several other alternatives, such as ethanol, methanol and bio fuels (bio diesel, bio gas, bio oils, bio ethanol, etc.), dimethyl ether, hydrogen, etc., are being studied extensively in several parts of the world. However, detailed discussion on these aspects is beyond the scope of this chapter. Figure 1.18 shows the combustion of petroleum fuels in an automobile.

The major pollutants coming out of tail pipe of an automobile are carbon monoxide (CO), HCs and NOx. Hydrogen combustion does not produce any CO or HCs. It has already been discussed that automobiles are one of the major culprits for causing environmental degradation. They have been

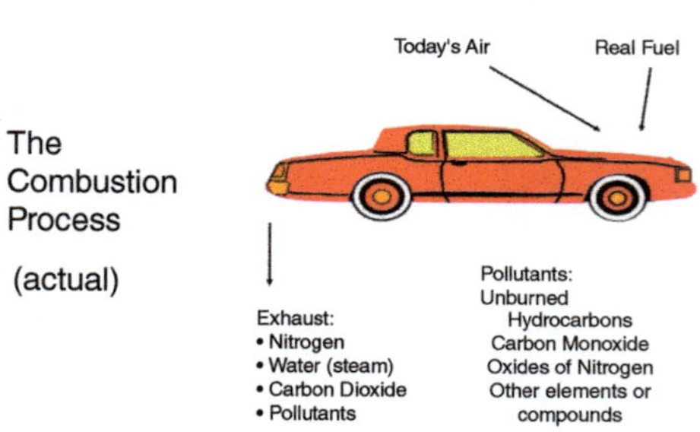

Figure 1.18 Combustion process in an internal combustion engine.

responsible for causing conditions leading to smog, acid rain and a host of other problems. As would be discussed later, the use of hydrogen in engines/vehicles can have a very clean exhaust with practically no noxious pollutants.

Chemical industries that utilise conventional fuels emit the same pollutants such as CO_2, SO_X, NO_X, particulate matter and volatile compounds. There could also be heavy metals such as mercury, lead and cadmium besides benzene and polycyclic aromatic hydrocarbons (PAHs). Power plants also emit similar type of pollutants. Most of these would be almost eliminated if hydrogen can be used in these industrial applications.

1.7 Energy Transition Towards Sustainability

It is now evident that energy transition has been occurring since long. Transition from wood to coal, subsequently from coal to natural gas and oil, has taken place at different stages across different parts of the world and in various industries. Now the thrust is on renewables. It is worthwhile to appreciate that these transitions have not been abrupt and radical. All stages of transitions have taken some time to effectively come into play. Moreover, these transitions have not taken place simultaneously all over the world. It has taken place at different countries at different times depending upon the resources, socio-economic conditions and several other factors such as geographical location, political system, transportation infrastructure, etc.

Situations have drastically changed over the years. The sole objective of present global energy transition aims at identifying an infinite source of energy that can be used without destroying the environment. Wide diverse arrays of alternatives are being considered which can confront the twin crises of fossil fuel depletion and combustion-generated pollution. Solar, wind, geothermal, hydro and tidal and nuclear energy have been and are being effectively used for a variety of applications that were earlier monopolised by conventional sources of energy. One of the most distinctive sustainable characteristics of hydrogen is that it is produced from water and, upon combustion, it gives out water (Figure 1.19). Thus, it is a part of the water cycle.

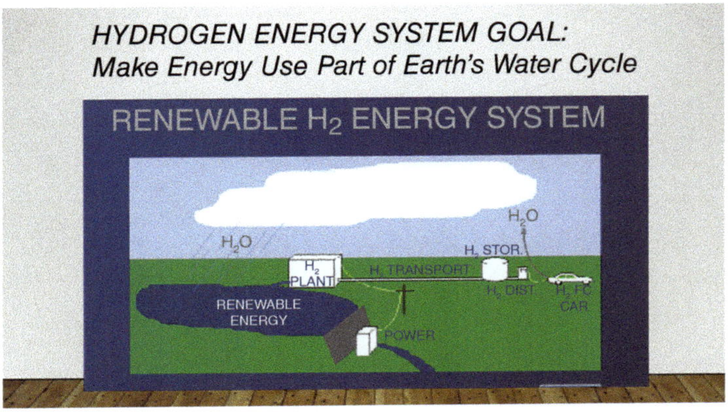

Figure 1.19 Sustainable characteristics of hydrogen energy.

The potential of hydrogen has been realised, and many countries across the world have initiated hydrogen energy programmes and several others have strengthened their existing hydrogen energy activities. Figure 1.20, from the US Department of Energy, shows a wider spectrum of the possible generation and utilisation potential of hydrogen [13].

The trend of energy transition is very closely linked to the growing threats of climatic disruption. Scientists and technologists are pondering over the complexity of the problem, and feasible options are being worked out. Generation of electricity from renewable sources, such as solar, wind and geothermal, etc., can facilitate widespread electrification. This might be carried out more conveniently for some applications such as buildings (including hospitals, shopping centres, several other structures, etc.) as well as some vehicles (cars, etc.). However, in many other important industrial sectors, such as steel, cement and several other pharmaceutical companies, this would be a relatively difficult problem now. A switchover from the existing systems to electricity to transport such as long-haul road transport, aviation and maritime industry is also not an easy task. In some applications where electrification is not the ready answer, green hydrogen can be of great use. Figure 1.21 shows the versatility of an integrated hydrogen system.

Several groups of researchers have been carrying out R&D activities to capture carbon from air to lower atmospheric CO_2 levels. Carbon capture and storage (CCS) or carbon capture and sequestration are very important techniques being adopted to control the environment. CO_2 is captured from industries such as coal-fired power plants and other large point sources and stored underground. The objective of this process is to prevent the CO_2 emitted from these industries from reaching the atmosphere, thereby mitigating the effects of climate change. Recently, research activities have begun to develop specially designed electrodes on combining the approaches from bio-electrochemistry, enzyme biology and synthetic biology. This electrode is capable of using renewable energy to drive enzymes that can produce valuable organic molecules from greenhouse gases.

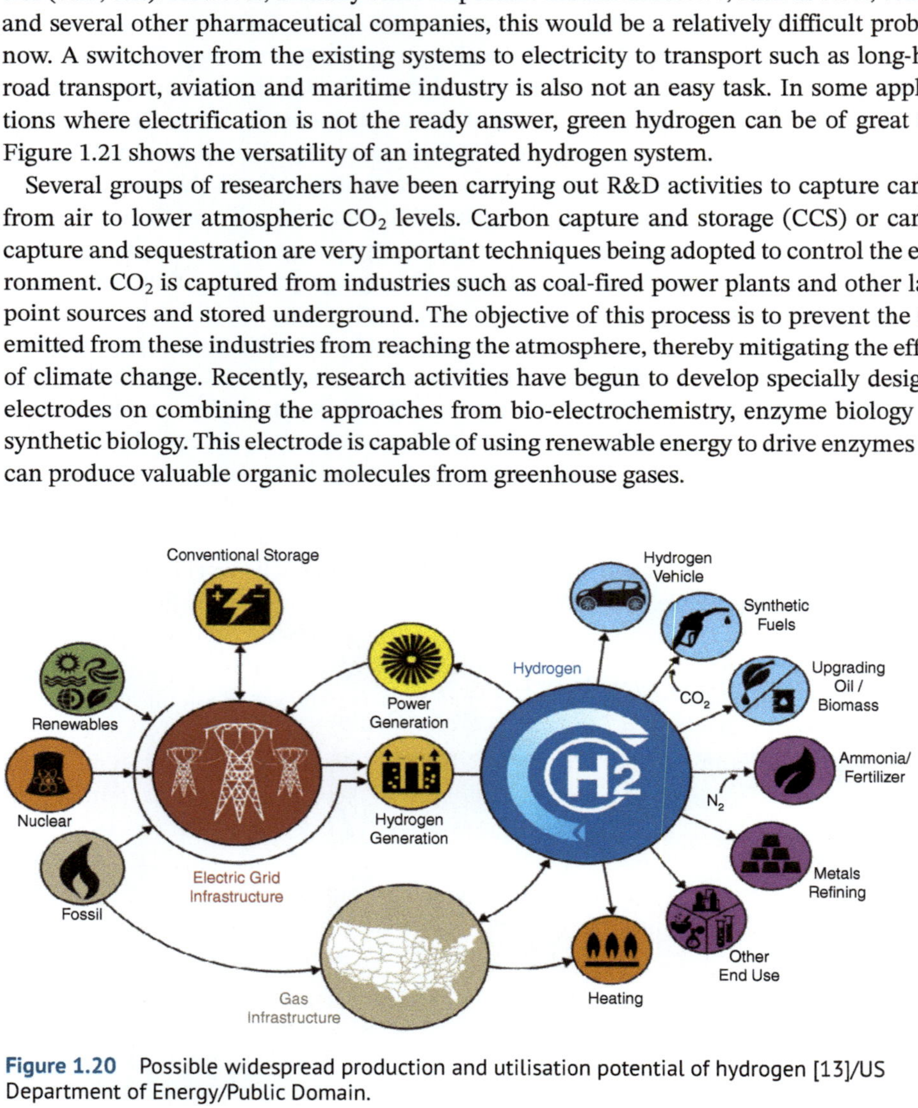

Figure 1.20 Possible widespread production and utilisation potential of hydrogen [13]/US Department of Energy/Public Domain.

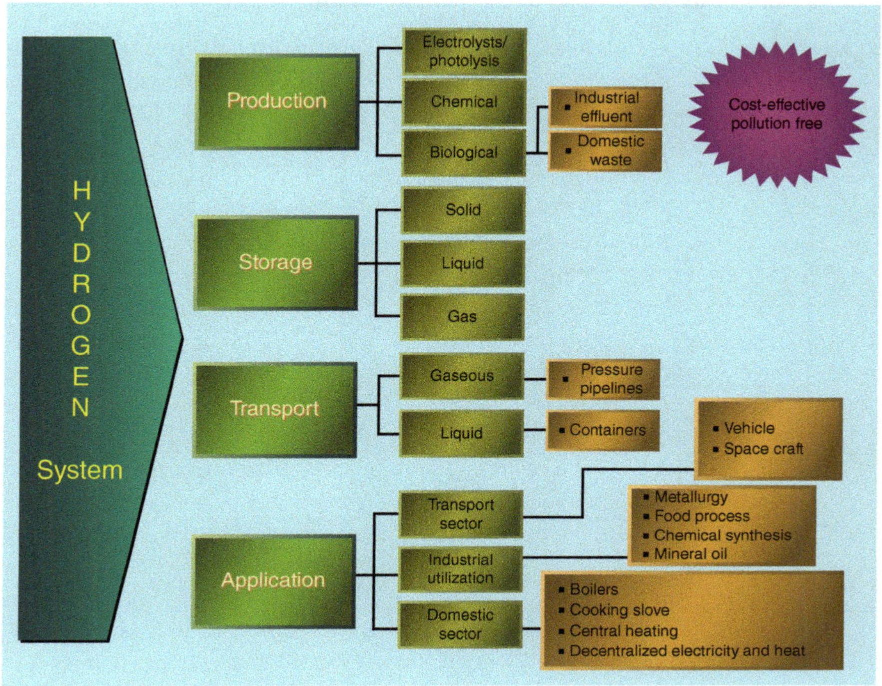

Figure 1.21 An integrated hydrogen energy system.

1.8 Role of Hydrogen in Present Energy-environment Context

Hydrogen stands as the most abundant element in the cosmos, comprising 75% of the universe's composition. Alongside hydrogen, helium constitutes 23%, while the remaining 2% encompasses various other gases. In Sun, 400 million tonnes of H_2 is burned every second to produce helium and 3.8×10^{26} W of energy, out of which only 1.75×10^{17} W reaches Earth's surface. Approximately 70% of our planet is covered by water, which comprises the element hydrogen. A hydrogen energy system consists of the production, transmission, storage, delivery and utilisation of hydrogen. Hydrogen is not available freely in nature, and it is essentially considered to be an energy carrier. Hydrogen has infinite source potential. It can be produced from water, as well as from fossil and a host of non-fossil sources. It can be used in rockets, internal combustion engines, fuel cells and several space applications. A schematic diagram in Figure 1.22 shows various primary energy sources from which hydrogen is produced, indicating the immediate, midterm and long-term supply routes for specific applications. However, in the context of the present thrust on environment, useful energy obtained from hydrogen has to be initially produced from renewable energy sources.

It may be relevant to recall here that most countries have realised the damaging effects of environmental pollution and have felt it essential to take immediate and aggressive steps to

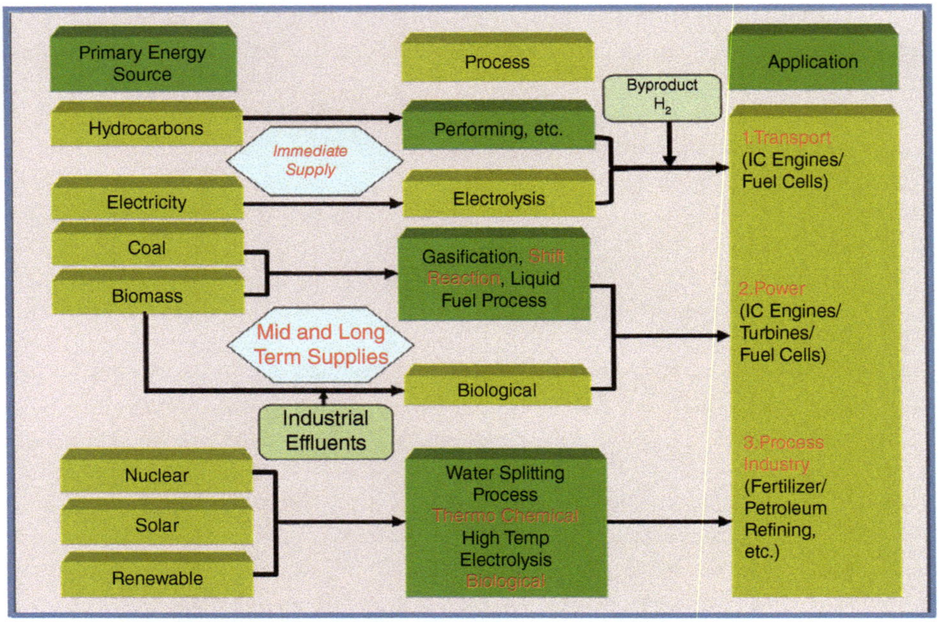

Figure 1.22 Methods of hydrogen production.

reduce greenhouse gas emissions to a net-zero level. In the Paris Agreement of 2015, the 197 participating countries agreed to 'achieve a balance between anthropogenic emissions by sources and removals by sinks of greenhouse gases in the second half of the century'. The Intergovernmental Panel on Climate Change has also categorically stressed that achieving net-zero CO_2 emissions is extremely important to avoid the worst encounter with climate disruption. Hydrogen energy is the most ideal candidate for such a situation because it is possible to avoid greenhouse as emission in a well-designed integrated hydrogen system. Green hydrogen can be produced by the use of renewables or solar energy. Greenhouse gas emission does not come into picture in the subsequent stage of transmission, distribution and delivery. Last but not the least, hydrogen combustion does not produce any greenhouse gases. It has been discussed later in the book how NO_X can be drastically reduced in combustion of hydrogen with air in vehicles. Recently, in the 26th UN Climate Change Conference of the Parties (COP26), it was observed that the world is not on the track to maintain global warming below 1.5 °C.

1.9 Demand for Hydrogen

The demand for hydrogen stood at around 94 Mt in 2021. The pre-pandemic value in 2019 was about 91 Mt [14]. However, the increase to such a large quantity was caused due to its wide use in conventional applications such as in refineries and chemical industries. Lockdown and slow economic activities in 2021 in many countries had drastically brought down the use of hydrogen as the demand in the industrial, transport and power sectors was extremely sluggish. The overall demand has taken up an increasing trend in the refining, industry, transport, buildings and power generation sectors. Refineries use hydrogen to remove impurities, such as sulphur, and also to upgrade heavy oil fractions into lighter products. Sharp and abrupt fall in the mobility during 2020/2021 literally stalled the refining activities. It is expected that due to the recovery of global crude processing and the recent regulations in several countries for lower sulphur emissions would help improve the situations for hydrogen demand. Deployment of fuel cell electric vehicles (FCEVs), heavy-duty trucks in particular in several countries, and improvement in refining and in steel production in some countries have also increased the demand for hydrogen.

Hydrogen is used in meeting the demands of refineries and industries. Industries use hydrogen as a clean fuel for heating purposes, such as in furnaces or other industrial processes requiring high-temperature applications. The major demand for hydrogen has been for the production of ammonia. The next in order of demand is for the production of methanol, which is almost a quarter of the demand for ammonia production. The demand for methanol and its derivatives is essentially owing to their fuel applications. In fact, almost one-third of global methanol and its derivatives is used for fuel applications. A very small amount is in demand for steelmaking by using the direct reduced iron (DRI) process. In the direct reduction process, hydrogen is used as a reducing agent to convert iron ore (usually in the form of iron oxide pellets or lump ore) into DRI or sponge iron. This is typically done in a shaft furnace or a direct reduction reactor. The chemical reaction involves the removal of oxygen from iron oxide, resulting in the production of metallic iron and water vapour. By using hydrogen as the reducing agent, the process becomes much cleaner and more environmentally friendly compared to traditional methods that use carbon-based reducing agents like coal or coke. The traditional methods produce CO_2 emissions, while the hydrogen-based process only generates water vapour, making it a potential pathway for reducing carbon emissions in the steel industry. DRI usage is highly concentrated in the Middle East and India. The Middle East has about 40% and India has about 30% of global production. India's steel production suffered a very serious setback during the pandemic period, resulting in drastic fall in the output. However, recently, after the dark phase of the pandemic, there has been a burgeoning demand for hydrogen in all these sectors, such as in ammonia production, methanol production and the DRI process. The other potential application of hydrogen in the steel industry, which is still in early stage of development, is that hydrogen can also be used as a reductant in electric arc furnaces (EAFs), which is another common method of steelmaking. EAFs use electric arcs to melt recycled scrap steel, and the process can be enhanced by injecting hydrogen gas into the furnace. The injected hydrogen reacts with any remaining oxygen and other impurities in the molten metal, leading to a cleaner product. This is a promising technology to improve the quality of steel while reducing energy consumption and to enhance sustainability by decreasing greenhouse gases. Thus, hydrogen is likely to play a significant role in the future of steel industries. Novel techniques and processes are being adopted for low-emission routes for ammonia, methanol and steel production. Carbon capture, utilisation and storage techniques are adopted for such production processes. In refineries, hydrogen is used in the hydrocracking and hydrotreating processes. Hydrocracking involves breaking down heavy hydrocarbon molecules into lighter ones using hydrogen, which is essential for producing high-octane gasoline and other valuable products. Hydrotreating uses hydrogen to remove impurities from refined products, such as sulphur, nitrogen and other contaminants, to meet environmental and product quality specifications. Refineries are increasingly exploring hydrogen-related technologies, such as hydrogen production through steam methane reforming (SMR) with CCS, to reduce carbon emissions. Hydrogen is also used in the desulphurisation of fuels, such as gasoline and diesel, for ensuring compliance with environmental regulations and reducing harmful emissions when these fuels are burned.

The demand for hydrogen in the transport sector has recently gone up. Even though there is a steady increase in the number of fuel cell electric cars, the demand for hydrogen is more for trucks and buses mainly due to their heavy weight compared to cars.

Heavy-duty hydrogen trucks have increased significantly in number, and the demand for commercial vehicles, such as vans and trucks, is also rising significantly, for the first time since 2021, compared to buses. It is estimated that this demand was up to 45% of the total hydrogen demand in the transport sector. The demand for hydrogen is also expected to rise in decarbonising diesel rail lines, particularly in locations where either electrification is difficult or where it is difficult to cover a very large distance by battery electric trains. Thus, with the ambitious climate goals set towards clean energy transition, it is expected that hydrogen energy technologies would diffuse very rapidly to road transport, shipping and aviation sectors apart from its regular application in chemical industries.

Yusuf and a team of researchers [15] have carried out a study using the system dynamics approach and building a conceptual model to assess the possible growth of hydrogen energy by 2050. The study considered the routes of hydrogen production and had also taken into account the possible factors that could intervene in reaching the targeted goals. As far as the demand for hydrogen is concerned, this study indicated that the demand for hydrogen would continue to move upward. The global hydrogen demand would rise from 73 to 158 Mt in 2030, to 300 Mt in 2040, and climb to 568 Mt in 2050. The other interesting result from their analysis shows that NOx is a major pollutant in combustion systems. Effective steps are necessary to control this pollutant.

A number of projects are being taken up to produce hydrogen from fossil fuels adopting CCUS. Recently, a global spate of interest in many countries has initiated and strengthened their efforts to achieve total decarbonisation through green hydrogen. Innovation technologies for economically viable routes of adequate amount of green hydrogen are necessary for putting the world on the track of sustainable and 'net-zero' path. In a recent report, IEA [14] has indicated the quantity of hydrogen demanded by the entire globe and expected by different sectors in the net-zero scenarios, 2020–2030 (Figure 1.23). Demand from some hydrogen-consuming countries and from the rest of the world (RoW) has also been estimated.

Many countries have already made time-bound climate commitment goals aimed at long-term net-zero targets. It is expected that, by the year 2030, about 25% of hydrogen demand would be for new emerging applications and the remaining 75% would continue to be used in the existing conventional route. In light of the growing awareness about environmental pollution, a study conducted at the International Centre for Hydrogen Energy Technology, UNIDO, has projected that hydrogen energy in its diverse forms is progressively becoming integrated into everyday lifestyles. Furthermore, the study suggests that the hydrogen economy might surpass the dominance of petroleum-based economies sooner than anticipated, as depicted in Figure 1.24.

As far as the environment aspects of hydrogen are concerned, hydrogen can be obtained in 'green' form with techniques to produce it without CO_2 emissions. Additionally, hydrogen

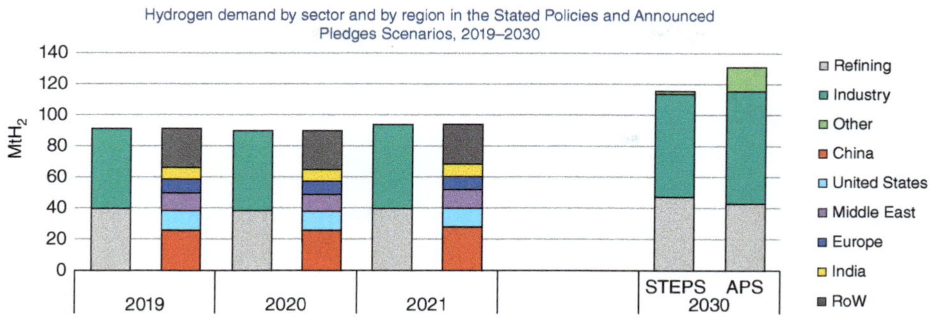

Figure 1.23 Demand for hydrogen [14]/IEA/Licensed under CC BY 4.0.

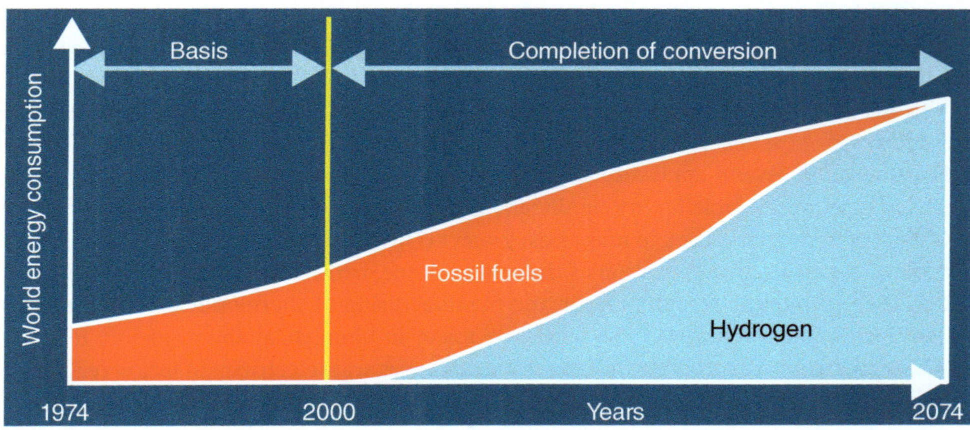

Figure 1.24 Transition to hydrogen economy. *Source:* T. Nejat Veziroglu, Hydrogen Energy Technologies, UNIDO.

combustion is environmentally friendly as it does not emit harmful pollutants such as CO, CO_2, hydrocarbons, SO_X, particulate matter, etc. Hydrogen combustion in internal combustion engines produces NO_X. It has been reported in the literature that there are several effective techniques to control the NO_X emission level. Thus, hydrogen, in view of its infinite source potential and renewable carbon-neutral characteristics, is the appropriate candidate to ensure eventual freedom from the twin problems arising out of fossil fuel usage – starvation and environmental degradation.

1.10 Structure and Phases of Hydrogen

There are three isotopes of hydrogen, as shown in Figure 1.25. These are (i) protium – mass 1, makes up 99.98% of the natural element; (ii) deuterium – mass 2, makes up about 0.02%; and (iii) tritium – mass 3, occurs in extremely small amounts in nature but may be produced artificially by various nuclear reactions. Normal hydrogen is a mixture of 25% and 75% of ortho and para forms of hydrogen under normal conditions of temperature and pressure. The ortho–para hydrogen conversion is accompanied by a release of heat.

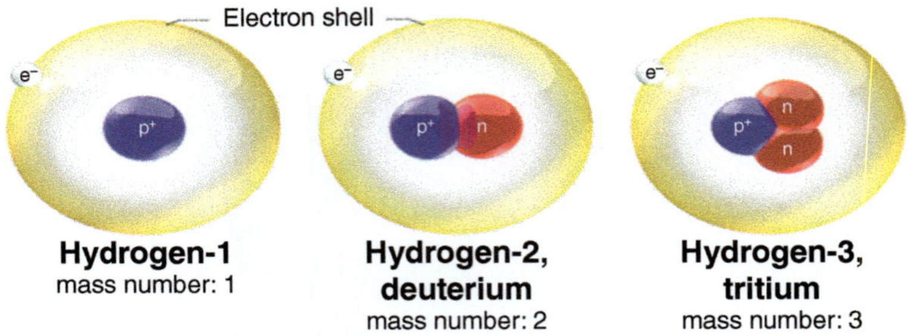

Figure 1.25 Hydrogen isotopes.

The heat is 703 kJ kg^{-1} at 20 K for ortho- to para-hydrogen conversion and 527 kJ kg^{-1} for normal to para-hydrogen conversion. It has also been observed that liquefaction is a very slow exothermic process that can take several days to complete. However, the process can be accelerated with the use of a paramagnetic catalyst.

In chemistry and biology, deuterium compounds have a very relevant application in studies of isotope effects on reaction rates. In the deuterium form (hydrogen-2), hydrogen is also used in several applications, particularly in nuclear fission and fusion. In nuclear fission reactions, its role is very significant as a moderator to slow down the reaction. The isotope of hydrogen is used to make heavy water (D_2O) that is used in nuclear reactors as a coolant. As far as the application of the other isotope, tritium (hydrogen-3), is concerned, it is used in nuclear reactors and in the manufacturing of thermonuclear weapons and hydrogen bombs. Tritium is also used as a tracer in biomedical and academic research. R&D activities are in progress in some countries to generate electricity in fusion reactors by using tritium (hydrogen-3).

Figure 1.26 shows the phase diagram of hydrogen. Slush hydrogen (melting point −252 °C) happens to be the mixture of solid and liquid hydrogen at the triple point temperature. Liquid hydrogen is a clear liquid with a blue tint. The critical point (the end point

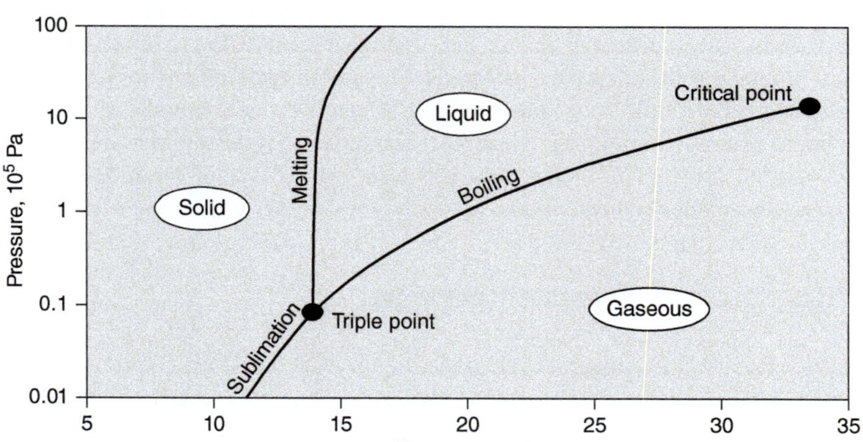

Figure 1.26 Phase diagram of hydrogen.

of the pressure–temperature curve under which both liquid and vapour coexist) is defined by $T = 33.2$ K (-239.5 °C) and $p = 1.3$ Mpa, and the triple point (where all the three phases coexist) is defined by $T = 13.8$ K (-259 °C) and $p = 7.2$ kPa.

1.11 Discovery and Occurrence of Hydrogen

Hydrogen was discovered almost four centuries ago. Valuable scientific information was acquired at different stages from the exhaustive tests carried out by Robert Boyle, Cavendish, Antoine Lavoisier, William Nicholson and Sir Anthony Carlisle and a few others. Properties of hydrogen could be evaluated from such repeated tests by the group of scientists. The British physicist Cavendish carried out some experiments using metals with strong acids. However, Robert Boyle had also earlier carried out experiments in which he was able to collect a unique gas by dissolving iron in dilute hydrochloric acid. An interesting observation from the experiments of Cavendish was that the gas that was formed was found to be combustible. Hence, he had named this gas 'inflammable air'. In subsequent experiments, he also discovered that water vapour (steam) was produced while burning this gas. On the basis of this property of producing water vapour, French chemist Lavoisier, in 1783, called this gas 'hydrogen'. *Hydrogenium* in Greek language is a combination of '*hydro*' (meaning 'water') and '*genium*' (meaning 'to give birth to'). The very nomenclature indicates that hydrogen can be obtained from water. With progress in the scientific discipline of chemistry, such a process has been defined as *electrolysis*.

Earth's atmosphere does not have hydrogen because our planet's weak gravitational field is not able to retain such extremely light molecules. Therefore, because of its light weight, it rapidly escapes into space. It is commonly observed in some volcanic gases as well. Most of the hydrogen on Earth is found in water, hydrocarbons and several other compounds. However, hydrogen is the most abundant element in the universe and contains about 75% of its mass in different forms. If all the atoms in the observable universe are taken into account and their masses measured, approximately 75% of the total mass would be from hydrogen atoms, and around 23% of the total mass would be from helium atoms. The remaining mass is made up of trace amounts of other elements, with heavier elements like carbon, oxygen, and iron comprising only a tiny fraction of the universe's mass. It is postulated that, at the initial stage of formation, the temperature of the universe was extremely high, approximately about a billion degrees. This hydrogen- and helium-rich composition is a result of the Big Bang, the prevailing cosmological model for the origin of the universe. Subsequently, the universe went through phases of heating and cooling, and after a period of several millions of years, hydrogen molecules were formed. Over billions of years, these primordial elements of hydrogen and helium served as the building blocks for the formation of the galaxies, stars and planets that we observe today. Hydrogen, helium, oxygen and carbon are the most common elements among the 118 elements in the universe. Interestingly, most of these elements have gases such as hydrogen and helium.

Hydrogen is the most plentiful element and makes up most of the universe. Almost nine out of every 10 atoms in the universe are hydrogen atoms. Sun, at present, is approximately about 4.5 billion years old, and Sun, stars and planets have a large amount of hydrogen in them. Hydrogen and helium are the two major elements that constitute Sun. The radius of

Sun's radius is approximately 700,000 km, and about three-fourths of Sun is constituted by hydrogen. Sun contains about 92.1% hydrogen, 7.9% helium and several other metals with less than 0.1% of the total mass. Hydrogen in Sun gets converted into helium by nuclear fusion. It happens over an extended period, as long as there is hydrogen fuel available in the Sun's core. The process is very gradual, with immense amounts of hydrogen atoms undergoing fusion every second. Other stars too have hydrogen and helium as their major constituents. Four planets in the solar system – Jupiter, Saturn, Uranus and Neptune – are mostly composed of hydrogen that exists under extremely high pressure. However, it is almost absent in free form on Earth. It is present in the atmosphere in extremely negligible amounts such as less than one part per million by volume. As explained earlier, this is because hydrogen is so light that it escapes the influence of Earth's gravity. However, water, which consists of almost two-thirds of Earth, is composed of hydrogen and oxygen. About 70% of human body also contains water. Hydrogen occurs in very large amounts in the form of a number of chemical compounds. Indeed, hydrogen ranks as the Earth's tenth most abundant element. It manifests in diverse forms such as numerous petroleum derivatives, natural gas and minerals, and serves as a constituent in a wide array of chemical compounds.

1.12 Uses of Hydrogen

Some of the uses of hydrogen have already been discussed earlier, in the context of demand for hydrogen. Today, hydrogen is one of the twenty-first century's key vectors of clean energy. Figure 1.27 shows several energy-consuming sectors where hydrogen can be used.

Hydrogen is being used in several sectors right now and has the potential to be adopted for much wider application in many more areas such as agriculture, chemical industry, aviation, power sector, road transportation, space exploration, petroleum refinery, metal working, welding, electronic industry, glass manufacturing, plastics food industry, gas welding, ultraviolet lamps, gas chromatography and several other applications.

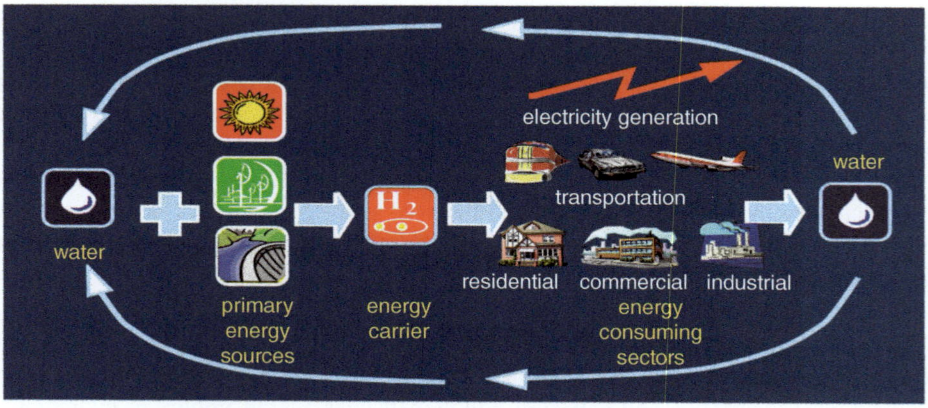

Figure 1.27 Hydrogen in a wide spectrum of energy-consuming sectors.

Hydrogen has also been used in potentiometric titration that uses two electrodes: an indicator electrode and a reference electrode. Hydrogen gas is passed into the reference electrode under a pressure of one atmosphere. However, the purity of hydrogen affects this analysis.

In the gas welding process, the temperature level reaches about 4000 °C. Metals melt at this high temperature and thus can be used to join the broken surfaces. Atomic hydrogen welding (AHW) is an effective way of arc welding that prevents oxidation and manufacturing defects in a hydrogen environment in which hydrogen is used as a shielding gas: a mixture of hydrogen and nitrogen is used to avoid oxidation and thus defects during glass manufacturing. In this welding process, hydrogen, besides generating heat, also helps prevent the highly reactive metals from reacting with other elements such as nitrogen and carbon during the process. Presently, about 55% of hydrogen the world over is being used for the production of ammonia (NH_3) and about 10% for methanol production. Almost 90% of ammonia produced the world over is used as a component in fertiliser that is very significant for food production. Hydrogen has also been used as a food additive (E 949) to identify leakages in food package testing. Besides, it has an anti-oxidising property. There are a lot of industrial and laboratory activities in which hydrogen plays a very important role, particularly in chemical analysis. In the petroleum industry, hydrogen is used to remove sulphur content. In several other processes as well, hydrogen has a lot of application potential. In the process of hydro cracking, long-chain hydrocarbons are broken into shorter ones. Hydroisomerisation is adopted to convert normal paraffin into iso-paraffin and also for de-aromatisation, which is the process of converting aromatic compounds into cycloalkanes.

According to basic chemistry definition, the addition of hydrogen implies *reduction*, whereas the removal of hydrogen results in *oxidation*. Hydrogen is also used as a reducing agent for both the electro-negative and electro-positive elements. This electro-positive nature is used in redox reactions. Hydrogen has got another very useful application in atomic absorption spectroscopy and also in gas chromatography. Atomic absorption spectroscopy uses hydrogen to generate heat, and in gas chromatography, hydrogen is used as a mobile phase to separate volatile substances from a mixture. Hydrogen peroxide (H_2O_2), which is formed by the combination of hydrogen with oxygen, is used as a sterilising agent, mostly used in clinics and hospitals. It is also used for the cleaning of wounds and cuts and for several other pharmaceutical applications. Its application is also being studied in research activities for testing the antioxidant potential of enzymes like catalase. Hydrogen has also been used for structural identification of many compounds using the nuclear magnetic resonance (NMR) spectroscopy. This technique is quite useful particularly in compounds where it is difficult to study the nature of bonds between atoms and molecules.

Pure hydrogen or a mixture of hydrogen and nitrogen has been used as a tracer gas to detect very minute leakages. This leak detection technique has widespread application in industries such as chemical power generation aerospace, telecommunication and automotive industry. It is also known that ammonia is being used as a very environment-friendly refrigerant (R-717). As far as the utility aspects are concerned, hydrogen is used to turn unsaturated fats into saturated oils and fats along with hydrogenated vegetable oils such as margarine and butter spreads. It is also used for metal working. Hydrogen is used in the petroleum refining industry in hydro cracking to create petroleum products such as gasoline and diesel. It is also used to remove contaminants such as nitrogen, sulphur and metals. Hydrogen also plays a very important role in electronics manufacturing of semiconductors, LEDs, photovoltaic segments, etc., where it is used as a reducing agent. Recently, hydrogen is being investigated as a therapeutic agent for several ailments. Hydrogen has been used in the aviation industry for many years and is also used for space applications. In the present situation, the prospect of the rapid penetration of hydrogen-based technologies into personal and public transport sectors is receiving a lot of attention.

Hydrogen has only about 7% of the density of air; it was used as a lifting gas in balloons. This is one of the earliest uses of hydrogen which, at that point in time, had attracted a lot of attention and interest. However, in view of its explosive characteristics, hydrogen was withdrawn and replaced by helium at a later stage. Figure 1.28 shows a hydrogen-operated flying balloon.

Some of the specific properties of hydrogen such as low molecular weight, high flame speed and minimum ignition energy have been very useful in some specific applications. Combustion of hydrogen with oxygen provides the highest specific impulse and thus has been used as rocket fuel. National Aeronautics and Space Administration (NASA) uses hydrogen for its space shuttle. Hydrogen has been used in several projects as signature fuel in the space programme of the United States of America. In modern times, there are also several experimental projects in the aviation industry such as unmanned aircrafts Pathfinder and Helios aircrafts in which fuel cell has been used. Recent 'ZEROe' aircrafts are modified gas turbines that utilise liquid hydrogen to provide power to the engine. NASA's Apollo projects of space exploration have been very successfully operated by liquid hydrogen (LH_2) fuel. NASA has also carried out research using rocket propellants made from atomic hydrogen, carbon or boron. It might be relevant to mention here that the use of hydrogen for transportation dates back to the eighteenth century when Charles and Robert made 'hydrogen gas aerostat' as the first manned flight indicating man's desire to

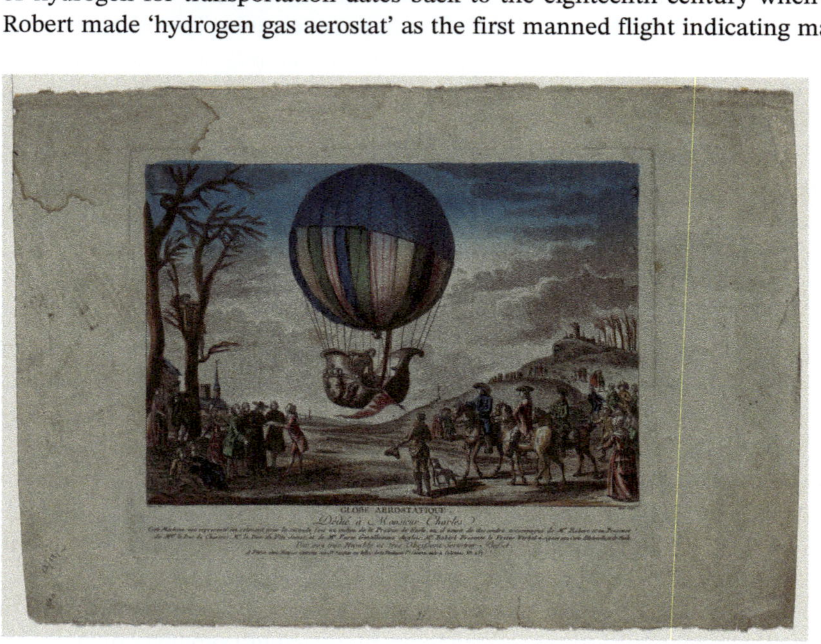

Figure 1.28 Hydrogen-operated balloon.

explore into the sky. Use of hydrogen has rapidly expanded from the initial stage of Saturn Rockets to several government projects and other commercial sectors such as United Launch Alliance, Boeing and Blue Origin. Studies of Bacon on fuel cell started sometimes in 1932. Ultimately, these investigations gave rise to the prototype fuel cell development in 1953. R&D activities in this area continued with full vigour and the prototype thus developed was a model for a series of Apollo space missions, and the final outcome of these evolutionary designs resulted in the moon landing.

Hydrogen possesses the potential to contribute to the creation of highly potent weapons, including hydrogen bombs. These bombs are designed on the basis of thermonuclear reaction built on the principle of nuclear fusion reaction. Liquid H_2 has received a very wide application in superconductivity studies and many other areas of cryogenic research. As indicated above, hydrogen is aggressively entering into wide-scale application in several types of transportation systems.

The utilisation of hydrogen as a fuel has a rich and extensive historical background. Town gas, which was used as a regular fuel in the early part of the twentieth century, had a substantial amount of hydrogen (to the tune of 50%) in its composition. Today, hydrogen is considered one of the most promising vectors of the twenty-first century for clean energy in transportation. Many global companies have started adopting hydrogen fuel cells in their power trucks, lifts, etc. Reputed companies such as Toyota, Hyundai, Nikola Motors, Kenworth Truck, etc., have started switching over to hydrogen-powered trucks and vans. Hydrogen fuel cells have already started being used and also being seriously considered for use in public buses in several major cities such as Chicago, Vancouver, London, New Delhi and Beijing. Hydrogen-powered trains have been introduced in Germany, and they might soon appear in the United States, Great Britain, Japan, France, South Korea, Italy, India and possibly several other countries. Active R&D projects for the development of prototype vehicles for personal use with fuel cells as well as internal combustion engines are in progress in many developed and developing countries. Several such projects are the outcome of effective collaborations between academic institutes/research organisations and industries.

Hydrogen is also being considered for use in power generation. It already has application in cooling power plant generators and also has promising application for electrical grid stabilisation. Uninterrupted power supply plays a very critical role in hospitals and data centres. R&D activities have already resulted in technology development to produce hydrogen and store it for subsequent end-use application. As a means of backup power generation, stationary fuel cells can be used for uninterruptible power supply (UPS) systems. With the existing transmission and distribution infrastructure and keeping the NZE scenario in mind, efforts could be made to use low-carbon hydrogen in place of fossil fuels, particularly in several industrial sectors such as refineries and power plants. It has been reported that the global use of hydrogen would go beyond 200 Mt in 2030 from a figure less than 90 Mt in 2020 [13]. Similarly, the use of low-carbon hydrogen will rise from 10% in 2020 to 70% in 2030 (Figure 1.29). As per the present trend, it is expected that a substantial portion (around 50%) of low-carbon hydrogen will be produced from conventional routes such as electrolysis and the rest from natural gas and coal adopting CCUS. However, the situation varies at different locations and regions. Since natural gas is being used as a regular fuel in the transportation sector in several countries, hydrogen could be blended in some percentages

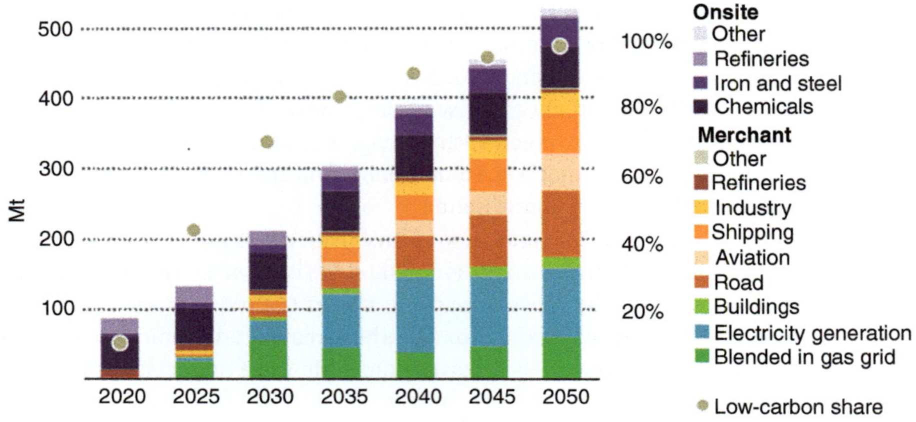

Figure 1.29 Global use of hydrogen and hydrogen-based fuel: net-zero emission scenario. *Source:* Adapted from [14].

(about 20% by volume) to natural gas. This is expected to bring down the corresponding level in terms of about 6% of CO_2. A study conducted by the IEA has examined and forecasted the favourable scenario required to attain an NZE state by the year 2050. CO_2 emissions from energy-related activities and industrial processes need to be reduced by about 40% in the period from 2020 to 2030 so that NZE is reached in 2050. It is expected that, during the period 2020–2030, the total energy supply would fall by 7%. By the year 2030, solar photovoltaic (PV) and wind power are poised to emerge as the primary electricity sources. Their prominence is projected to increase, with estimates suggesting that they could account for approximately 70% of global electricity generation by 2050. Coal, oil and natural gas are likely to have a drastic fall in demand. Application of the CCUS process would have a much wider application, particularly in sectors such as cement production where it is difficult to adopt low-emission technologies [16].

Concluding Remarks

We are now seeing an unprecedented global shift towards decarbonisation, and hydrogen is gathering momentum as a key energy transition pillar. More than 30 countries have initiated hydrogen energy programmes under different titles, such as hydrogen law, hydrogen strategy, hydrogen vision, hydrogen policy, etc., and country-specific, time-bound goals have been prepared through well-defined routes. Bilateral international cooperation on different aspects of hydrogen energy in the form of projects, technology collaboration and trade is increasing very rapidly. Industries have announced more than 200 projects and ambitious investment plans. Different governments worldwide have committed more than $700 billion in public funding. The process of initiating bilateral interactions through the establishment of memoranda of understanding (MoUs) has commenced among various countries. Nations that were already engaged in hydrogen-energy-related endeavours have re-evaluated their targets and bolstered their aims.

Hydrogen's entry into energy infrastructure depends on several aspects of the related technology such as production, storage, transmission, delivery and utilisation. Hundreds of research papers highlighting the experimental descriptions on these aspects have been discussed later in different chapters of this book. With the present level of technology, hydrogen production costs are high relative to conventional fuels. It is essential that improved technology for green hydrogen production be developed, and appropriate delivery infrastructure be built for hydrogen to be available to the consumer at a relatively lower price. In the initial phase, until an adequate quantity of green hydrogen has been produced to replace the existing commercial hydrogen production methods (such as SMR, multi-fuel gasification, electrolysis, etc.), efforts should be made to improve the efficiency and purity to have inexpensive and high-purity hydrogen with extremely low carbon or preferably no carbon emissions. At present, there exists a healthy marketability for hydrogen in chemical industries and refineries. Cheaper production rates will encourage the expansion of transmission and storage systems. This would enhance the possibility of hydrogen's demand to be implemented in transport and power sectors. Hydrogen could be produced in centralised or decentralised mode. Centralised facilities should have a distribution system to carry hydrogen to the point of use. In decentralised facilities, hydrogen produced could be stored and used onsite. Efforts need to be directed towards creating a storage system that is lightweight for easy mobility and is also cost-effective. Rapid penetration of hydrogen energy technology needs a well-coordinated and multifaceted approach involving various stakeholders, including governments, industries, researchers and consumers. Policy support and incentives will accelerate the adoption of technology. Development of hydrogen infrastructure, scaling up hydrogen production, public awareness, safety regulation and demonstration projects based on collaborations among governments, industries and research institutions are some of the steps which can ensure rapid penetration of hydrogen technologies, leading to a more sustainable and carbon-free future.

Abbreviations

STEPS	stated policy scenario
SDS	sustainable development scenario
GHG	Greenhouse gas
IEA	International Energy Agency
CCUS	carbon capture usage and storage
CFC	Chlorofluorocarbon
GWP	global warming potential
WMO	World Meteorological Organisation
ppm	parts per million
ppb	parts per billion
HFC	hydrofluorocarbon
NOOA	National oceanic and atmospheric administration
HCs	hydrocarbons
PAN	peroxyacetyl nitrate

VOC	volatile organic compounds
LNG	liquified natural gas
LPG	liquified petroleum gas
COP	Conference of the Parties
FCEV	fuel cell electric vehicle
DRI	direct reduced iron
RoW	rest of the world
ICHET	International Centre for Hydrogen Energy Technologies
UNIDO	United Nations Industrial Development Organization
D_2O	heavy water (deuterium)
AHW	atomic hydrogen welding
NMR	nuclear magnetic resonance
UPS	uninterruptible power supply
MoU	Memorandum of Understanding
NZE	Net Zero Emission

References

1 World Energy Outlook (2022). *International Energy Agency (IEA).* Revised Version November 2022.

2 Energy Technology Perspectives (2020). *International Energy Agency (IEA).*

3 Sutherland, L.-A., Peter, S., and Zagata, L. (2015). Conceptualising multi-regime interactions: the role of the agriculture sector in renewable energy transitions. *Research Policy* 44 (8): 1543–1554.

4 Corlu, C.G., de la Torre, R., Serrano-Hernandez, A. et al. (2020). Optimizing energy consumption in transportation: literature review, insights, and research opportunities. *Energies* 13: 1115. https://doi.org/10.3390/en13051115.

5 Paramati, S.R., Apergis, N., and Ummalla, M. (2018). Dynamics of renewable energy consumption and economic activities across the agriculture, industry, and service sectors: evidence in the perspective of sustainable development. *Environmental Science and Pollution Research* 25: 1375–1387.

6 Najmi, A., Hamed, S.G., and Keramati, A. (2016). Energy consumption in the residential sector: a study on critical factors. *International Journal of Sustainable Energy* 35 (7): 645–663, https://doi.org/10.1080/14786451.2014.930466.

7 González-Torres, M., Pérez-Lombard, L., Coronel, J.F. et al. (2022). A review on buildings energy information: trends, end-uses, fuels and drivers. *Energy Reports* 8: 626–637.

8 US Energy Information Administration (EIA) (2021, March). *Short Term Energy Outlook.*

9 World Meteorological Organization (WMO) (2022). *WMO State of the Global Climate* (Publication date 08 Nov 2022).

10 https://climate.nasa.gov/vital-signs/carbon-dioxide.

11 News. (2021, February). Global CFC emissions now declining again as expected under the Montreal Protocol CSIRO Atmospheric composition and Chemistry.

12 NOAA Global Monitoring Laboratory and Climate.gov.

13 Pivovar, B. (2019). H2@Scale Overview, 2019 DOE Hydrogen and Fuel Cells Program Review.

14 Global Hydrogen Review. (2022). *International Energy Agency (IEA)*.

15 Yusaf, T., Laimon, M., Alrefae, W. et al. (2022). Hydrogen energy demand growth prediction and assessment (2021–2050) using a system thinking and system dynamics approach. *Applied Sciences* 12: 781. https://doi.org/10.3390/app12020781.

16 International Energy Agency (IEA). (2021, October). Net Zero by 2050: A Roadmap for the Global Energy Sector, United Nations Climate Change.

2

Hydrogen Energy: Properties and Quality

Hydrogen is an investment that will inevitably provide benefit, but at an unknown time in the future, and to someone other than the original investor.

– Derek P. Gregory (Institute of Gas Technology, 1980)

CHAPTER MENU

2.1 Introduction

Needless to say, at this point in time, hydrogen continues to draw global attention as a critical enabler of energy transition and decarbonisation. The hydrogen economy broadly

Hydrogen Energy: Production, Safety, Storage and Applications, First Edition. L. M. Das.

defines the situations where hydrogen is the main energy carrier. In such situations, hydrogen would be produced from available primary energy sources in adequate quantities to replace the areas where fossil fuels are being used. Currently, the additional dimension of the problem hinges upon the environmental compatibility of all stages, including production, storage, transmission, delivery and utilisation. Transitioning to sustainable energy requires identifying the challenges and proposing solutions.

The significance and relevance of hydrogen energy in the context of energy–environment crises are briefly highlighted in Chapter 1. An introduction to different stages of hydrogen energy, such as production, storage, transmission and delivery, along with the application shown in Figure 2.1, provides an overview of the hydrogen economy from production to use. These topics are discussed in detail later in the other chapters of the book.

Each of these stages already has an operating mechanism. There are some difficulties in improving the efficiency or effectiveness of any of these stages. Challenges have been identified, and suggested methods are being proposed. A brief discussion on these aspects is provided here. As far as commercial methods of hydrogen production are concerned, it is essential that efforts should be made to improve the existing methods of Steam methane reforming (SMR). SMR is the most dominant commercial method that accounts for the majority of global hydrogen production, especially for industrial applications and large-scale hydrogen plants. Although SMR is a mature and cost-effective technology, other hydrogen production methods are gaining momentum due to increased interest in low-carbon and renewable hydrogen production. Advanced technology through biological routes and other methods using nuclear or solar energy should be widely implemented. However, all these steps need to be followed during the intermediary stage for achieving adequate production with improved efficiency. An optimum integration of supportive policies, technological advancements, market incentives, and collaborative efforts among

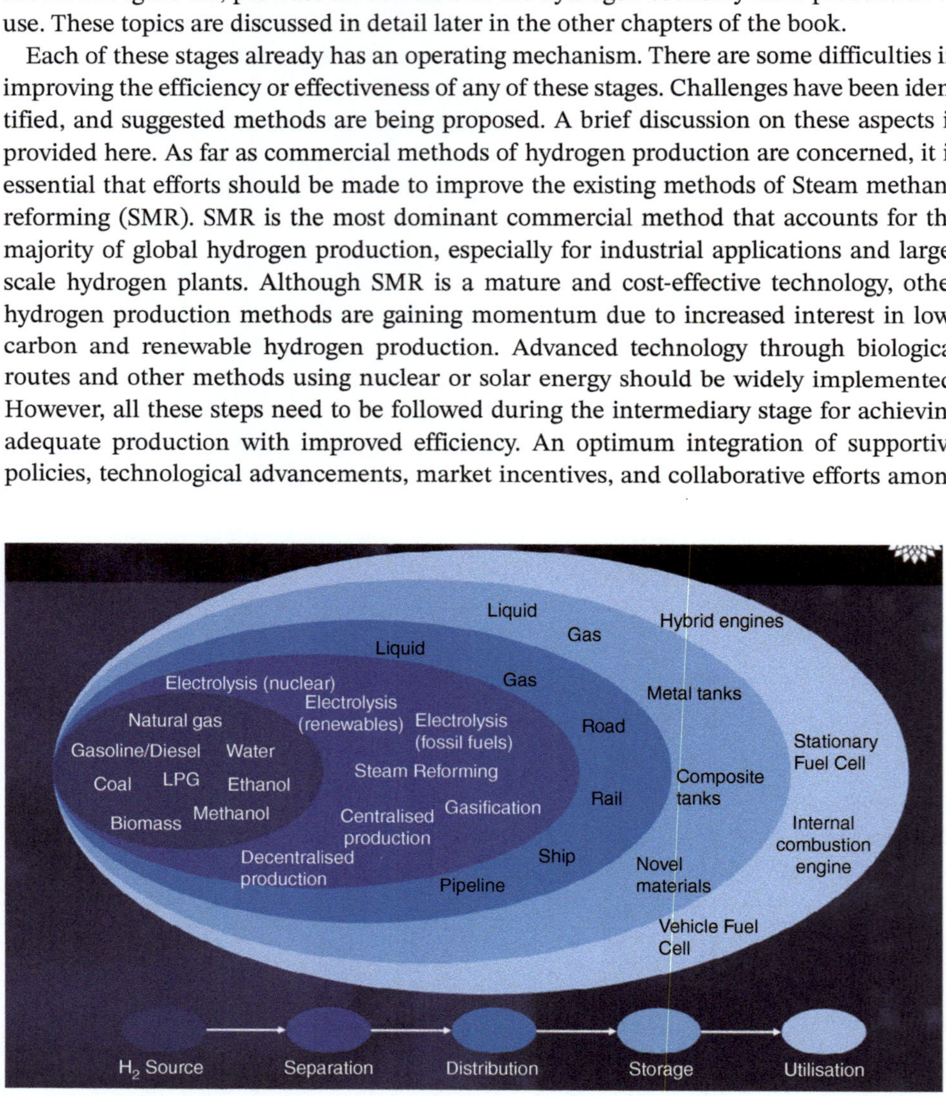

Figure 2.1 Pathways to hydrogen economy from production to application.

governments, industries and research institutions can contribute to the early entry of green hydrogen into global energy infrastructure.

It is well known that with the present level of technology, hydrogen production costs are higher than those of producing conventional fuels. However, there is no comparison between the cost of delivering one unit of hydrogen and that of delivering one unit of conventional fuels to the consumer. Over a period of time, well-developed infrastructure has been established for oil, coal, electricity and now also for natural gas. There is no doubt that hydrogen has already received market demand in refineries and some chemical industries. But, a lot is yet to be achieved because of the global strategy for high-quality hydrogen implementation in the transport, power, industrial and domestic sectors. Thus, hydrogen production from renewable energy technologies needs to be cost-competitive.

Apart from production, hydrogen delivery systems should also be improved and made available at the location of application, and it is always advisable that hydrogen can be produced onsite and utilised in specific stationary and portable applications. However, if the application area is farther from the point of production, then the hydrogen produced can be transmitted or distributed through pipelines or via rails and trucks. A detailed discussion on the storage and delivery of hydrogen is provided in a separate chapter in the book. Younggeun, UngLee and Kyeongsuin [1] carried out a techno-economic analysis and a quantitative risk analysis (QRA) of hydrogen delivery infrastructure. In their analysis, they compared delivery option using both gaseous and liquid hydrogen. Their research results indicated that gaseous hydrogen delivery was cheaper than liquid hydrogen, whereas liquid hydrogen delivery was observed to be safer than that in the gaseous mode.

Thus, an ideal 'hydrogen economy could evolve with all those steps, such as production, transmission, storage, delivery and utilisation, optimised at each stage and interacted into an environmentally friendly system'. Thus, it is appropriate at this point to start the description with the physico-chemical properties of hydrogen.

2.2 Properties of Hydrogen

Figure 2.2 shows a sketch of a hydrogen element. The physical and chemical properties of hydrogen are vastly different from those of conventional fuels. Precisely, this is the reason that the effect of hydrogen in several combustion systems is characteristically different from that of conventional fuels. Therefore, a liberal use of hydrogen in a particular combustion system might lead to problems of safety. Since there is a common problem of safety closely associated with hydrogen, a detailed discussion on some of the properties, such as density, buoyancy, diffusivity, quenching distance, ignition energy, flammability limit, burning velocity, etc., is provided in Chapter 5, from the safety point of view, and details are not discussed here. Safety-related details are imperative mainly in safe handling and combustion processes.

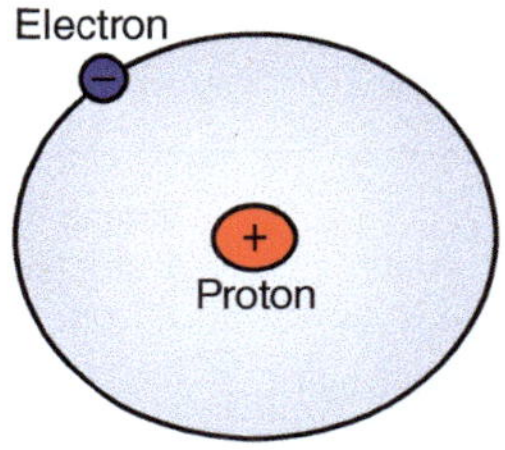

Figure 2.2 Hydrogen element.

2.3 Physical Properties

The ionisation potential of hydrogen atoms is 13.54 V. Hydrogen is a mixture of ortho- and para-hydrogen in equilibrium, distinguished by the relative rotation of the nuclear spin of the individual atoms in the molecule. Molecules with spins in the same direction (parallel) are termed 'ortho-hydrogen', and those in the opposite direction are termed 'para-hydrogen'. These two molecular forms have slightly different physical properties but have equivalent chemical properties. At ambient temperature, normal hydrogen contains 75% ortho-hydrogen and 25% para-hydrogen. The ortho-to-para conversion is associated with the release of heat. For example, at 20 K, 703 kJ kg^{-1} of heat is released during the ortho-to-para conversion. The conversion is slow but occurs at a finite rate (takes several days to complete) and continues even in the solid state.

Hydrogen is the smallest and lightest molecule in the universe. Therefore, it is the lightest element in the periodic table and consists of one proton and one electron. Catalysts are useful for accelerating the production of liquid hydrogen, more than 95% of which is para-hydrogen. There are three different types of hydrogen: (i) proton, mass 1; (ii) deuterium, mass 2; and (iii) tritium, mass 3. Among these elements, protium constitutes 99.98% of the natural elements, whereas deuterium constitutes about 0.02%. Tritium occurs in nature in extremely small quantities and is a rare radioactive isotope of hydrogen. Hydrogen is not present in its free state in the atmosphere, although free hydrogen exists in Sun and the stars. It is approximately 50 000 times smaller than the thickness of human hair.

Since hydrogen is the lightest element and contains no heavy carbon atoms, it has the highest energy-to-weight ratio for all fuels. As hydrogen has the lowest atomic weight, it has a very low density with respect to both gas and liquid. *Density*, by definition, means mass per unit volume. Density is always referred to with respect to temperatures and pressures, particularly for gases, because the compactness of the molecular arrangement in gases is very sensitive to temperatures and pressures. Usually, the density of gas is referred to as the 'vapour density', and the density of liquid is called the 'liquid density'. Hydrogen has a vapour density of 0.08376 kg m^{-3} (Table 2.1).

The thermodynamic properties of hydrogen are compared to the corresponding properties of methane and gasoline as shown in Table 2.2. The comparison shows that there is a growing surge of interest in using hydrogen as a substitute for gasoline and methane in several energy applications, including the transport sector. It has the lowest atomic weight and hence has a very low density both as a gas and as a liquid. Hydrogen is termed 'liquid' when it is below its boiling point of 20 K (−253 °C) and termed 'solid' when it is below its melting point of 14 K

Table 2.1 Vapour density and liquid density of hydrogen.

Fuel	Vapour density at 20 °C and 1 atm (kg m^{-3})	Liquid density at normal boiling point and 1 atm (kg m^{-3})
Hydrogen	0.08376	70.8
Methane	0.65	422.8
Gasoline	4.4	700

Table 2.2 Thermodynamic properties of hydrogen methane and gasoline (generally accepted values from literature).

Properties	Hydrogen	Methane	Methane
Molecular weight	2.016	16.043	107
Density of gas at NTP (kg m^{-3})	0.08376	0.651	4.4
Liquid density (at normal boiling point 1 atm) (kg m^{-3})	70.8	422.8	700
Heat of combustion (low) (kJ g^{-1})	119.93	50.02	44.5
Heat of combustion (high) (kJ g^{-1})	141.86	55.53	48
Specific heat (C_P) of NTP gas (J g^{-1} K^{-1})	14.86	2.22	1.62
Viscosity of NTP gas (g cm^{-1} s^{-1})	0.0000875	0.00011	0.00005
Specific heat ratio (γ) of NTP gas	1.383	1.308	1.05
Gas constant (R) (cm^2 atm g^{-1} K^{-1})	40.703	5.11477	0.77
Diffusion coefficient in NTP air (cm^2 s^{-1})	0.61	0.16	0.005

(259 °C) and atmospheric pressure. This characteristic is more prominent in gas. When used as a vehicular fuel, the low density of hydrogen shows that a large volume of hydrogen is required to provide an adequate driving range. The specific volume of hydrogen gas is 11.9 m^3 kg^{-1} at 20 °C and 1 atm, and the specific volume of liquid hydrogen is 0.014 m^3 kg^{-1} at −253 °C and 1 atm. Even as liquid, hydrogen is not very dense. Ironically, every cubic metre of water (constituted of hydrogen and oxygen) contains 111 kg of hydrogen, whereas a cubic metre of liquid hydrogen contains only 71 kg of hydrogen. Under normal conditions of temperatures and pressures, the weight of 1 l of hydrogen gas is approximately 90 mg. The viscosity of gaseous hydrogen is 89.48 μPoise at normal temperature and pressure (NTP) and 11.28 μPoise at normal boiling point (NBP), and the viscosity of liquid hydrogen is 132.0 μPoise at NBP. The specific heat of gaseous hydrogen at constant pressure (C_P, J kg^{-1} K^{-1}) is 14.85 at NTP, 14.304 at standard temperature and pressure (STP) and 12.15 at NBP. The specific heat of liquid hydrogen is 9.66 at NBP. *The specific heat ratio* (γ) of hydrogen gas is 1.39 at NTP and 1.405 at STP. The thermal conductivity of hydrogen is much higher than that of other gases. The thermal conductivity of hydrogen is 0.187 W m^{-1} K^{-1} at NTP and 0.0169 W m^{-1} K^{-1} at NBP, whereas the thermal conductivity of liquid hydrogen is 0.09892 W m^{-1} K^{-1} at NBP. The speed of sound in gaseous hydrogen is 1294 m s^{-1} at NTP and 355 m s^{-1} at NBP, whereas the speed of sound in liquid hydrogen is 1093 m s^{-1} at NBP.

Usually, gases cool down when they expand from high to low pressure. As a result of the Joule–Thomson effect, the temperature of hydrogen rises upon depressurization and this may lead to ignition. However, this increase in temperature is not sufficient to ignite the hydrogen–air mixture. Hydrogen energy is typically measured by weight in kilograms (kg) instead of volume because gaseous hydrogen has a low volumetric energy density. In electrical terms, 1 kg of hydrogen effectively holds 33.3 kWh of usable energy. Hydrogen occupies about 75% of the mass of the universe. Its buoyancy in the air is 1.2 kg m^{-3}, and it is about 14 times lighter than air. At STP, hydrogen is a colourless, odourless, non-metallic, tasteless, highly flammable diatomic gas with the molecular formula H_2. It has an atomic number of 1 and an atomic weight of 1.00797. On cooling, hydrogen condenses to liquid at 253 °C and to solid at 259 °C. Solid metallic hydrogen conducts electricity more efficiently than any other solid elements, and in the gaseous form, hydrogen shows the highest heat capacities. This element is a major constituent of water and all organic matter and is widely distributed not only on Earth but also throughout the universe. Some of the significant thermodynamic properties of hydrogen are given in Table 2.2. To avoid repetition in description, several properties of hydrogen are discussed in Chapter 5, mainly considering the implication of such properties with respect to safety.

An extremely low volume of hydrogen has been one of the major constraints in its application for storage as a fuel. Hydrogen storage in the liquid form at a very low temperature (20 K) and in the gaseous form at different ranges of pressures up to 700 bar consumes a substantial amount of energy. It has been estimated that liquid storage consumes about 30% of the energy content of hydrogen, whereas the gaseous storage consumes about 10% of the energy content of hydrogen. Figures 2.3 and 2.4 provide a clear understanding of the behaviour of hydrogen from a mass and a volume basis. As evident from Figure 2.3, on a mass basis, hydrogen energy (120 MJ kg^{-1}) is almost 3 times that of gasoline (44 MJ kg^{-1}) and 2.5 times that of methane. However, the results are very different from the perspective of volume. Gasoline has a density of 32 MJ l^{-1}, whereas liquid hydrogen has a density of only 8 MJ l^{-1} (Figure 2.4).

In Chapter 5, toxicity and asphyxiation are discussed, and it is shown that hydrogen, even if itself is non-toxic, can act as a simple asphyxiant by dislodging oxygen in the air. It has been observed that some gases, such as carbon dioxide, carbon monoxide, nitrogen and traces of some other gases, also displace oxygen, thereby acting as asphyxiates. Carbon monoxide is a poisonous gas, and its affinity for haemoglobin is 240 times that of oxygen. Thus, the inhalation of carbon monoxide prevents oxygen from reaching the bloodstream, leading to asphyxiation conditions. This can cause a wide range of problems, ranging from mild headaches and dizziness to convulsions, unconsciousness and even death. At exposure concentrations of around 1200 ppm, carbon monoxide poses a health hazard without any warning symptoms. Moreover, carbon monoxide has almost

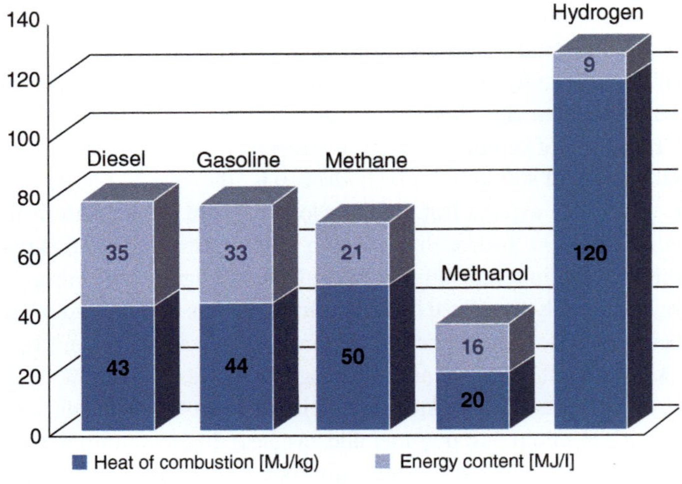

Figure 2.3 Hydrogen's energy on a mass and a volume basis compared to other fuels.

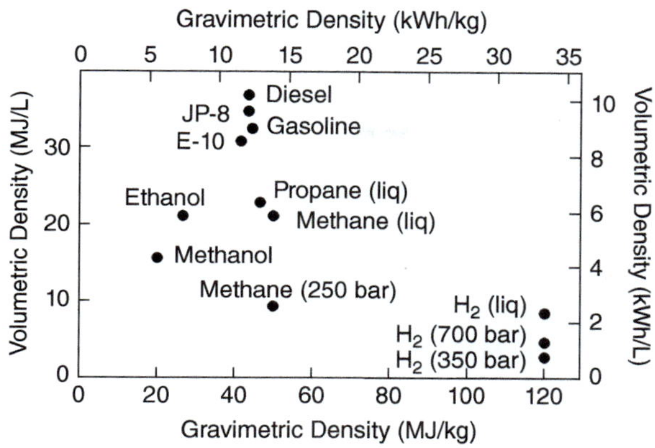

Figure 2.4 Volumetric energy density and gravimetric energy density of hydrogen and other fuels. *Source:* https://www.energy.gov/eere/fuelcells/hydrogen-storage

the same density as that of air. Thus, in the event of leakage, it is unlikely to diffuse by rising. Therefore, the anticipated presence of carbon monoxide should be very meticulously assessed to avoid any possible hazard.

The molecules of hydrogen gas are smaller than all other gases, and they can diffuse through many materials considered airtight or impermeable to other gases. Hydrogen has two very significant properties in terms of safety: diffusion and buoyancy. It has a diffusion coefficient of 61 $cm^2\,s^{-1}$ in the air. This property makes hydrogen more difficult to contain than other gases. Considering its high diffusivity and rapid dispersion rate, hydrogen diffuses much faster than any other gases. As indicated earlier, the relevance of these properties with respect to safety is discussed in detail in Chapter 5. When discussing the physical properties of hydrogen, it is important to consider the expansion ratio because hydrogen is widely used in both the liquid and gaseous forms. The expansion ratio is the ratio of the volume at which hydrogen (in the gas or liquid form) is stored compared to the volume of the gas or liquid at atmospheric pressure and temperature.

Stored liquid hydrogen vaporises on expansion to atmospheric pressure, thus resulting in an increase in volume – hydrogen has an expansion ratio of 1:848. Therefore, gaseous hydrogen occupies 848 times more volume than its liquid state as shown in Figure 2.5. Considering its non-polluting properties, hydrogen is being identified as an alternative to

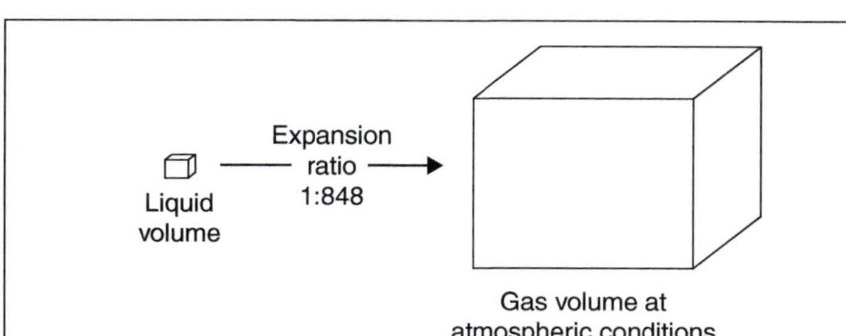

Figure 2.5 Expansion ratio of hydrogen (liquid to gas).

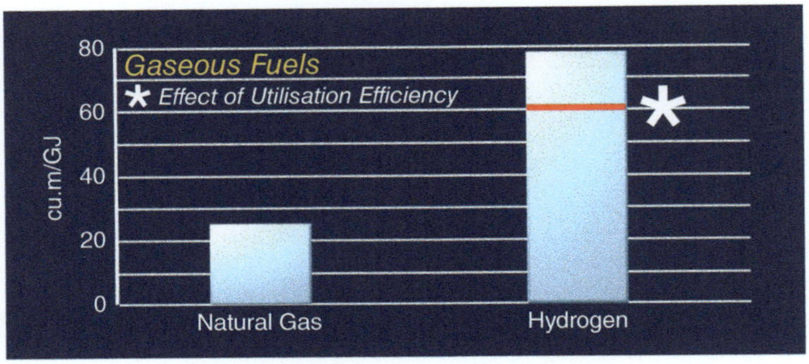

Figure 2.6 Comparative depiction of utilisation efficiency of hydrogen and natural gas.

other gaseous fuels such as natural gas in several energy sectors, particularly for vehicular applications. Figure 2.6 shows a comparative depiction of the bulkiness of hydrogen with that of natural gas, both in the gaseous form. The improved effect of utilisation efficiency is distinctly reflected in this figure. Since one cubic metre of liquid hydrogen consists of 71 kg of hydrogen, whereas one cubic metre of water (which consists of hydrogen and oxygen) contains 111 kg of hydrogen, therefore, water contains more hydrogen per unit volume than the corresponding amount of hydrogen itself. Such a characteristic is very distinct even from other compounds, such as methanol and heptane, which also contain liquid hydrogen.

The other characteristic is the leakage of hydrogen, which has been a point of concern in various applications. Safety is discussed in detail in Chapter 5. Hydrogen molecules of smaller sizes diffuse through several other materials that are otherwise impermeable to many other gases. In the low boiling point of liquid hydrogen, it evaporates much faster.

2.4 Chemical Properties

At ordinary temperatures, hydrogen is comparatively non-reactive unless it has been activated in a specific manner. On the contrary, a hydrogen atom is chemically very reactive, which is the reason that it is not chemically free in nature. In fact, very high temperatures are needed to dissociate molecular hydrogen from atomic hydrogen. Even at 5000 K, about 5% of hydrogen remains dissociated. Figure 2.7 shows the chemical structure of some common hydrocarbon fuels along with that of hydrogen.

As elemental hydrogen is not found on Earth in free form, it is most often bound to either oxygen or carbon atoms. Thus, hydrogen is an energy carrier that can store and transmit energy that is extracted from a primary energy source by adopting some industrial processes. In order to derive hydrogen from these compounds, some amount of energy should be spent. Hydrogen along with helium constitutes most of Sun. However, due to a very high temperature, the atoms are almost completely ionised into hydrogen ions and helium ions. Such an ionised state of the gas is referred to as plasma. Similar to stars, hydrogen is often found in its plasma state in stars. A very significant chemical characteristic of atomic hydrogen is that it is a very strong reducing agent even at room temperature. It can react with a host of compounds such as chlorides and oxides of other metals, such as silver,

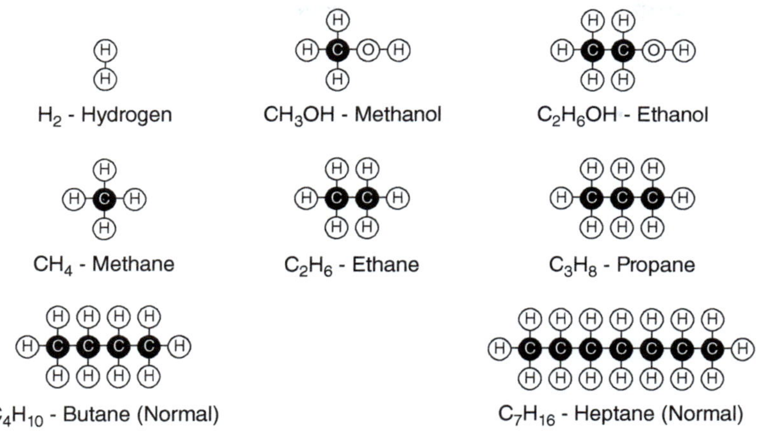

Figure 2.7 Chemical structures of some common fuels.

copper, lead, mercury, etc. As a result, free metal is produced. It can also reduce many salt compounds, such as nitrites and nitrates, to the corresponding metals. In addition, the compound reacts with a host of metals and non-metals to form hydrides. It additionally combines with sulphur to create a number of sulphur hydrides (the most common being H_2S). H_2O_2 is also formed when it reacts with oxygen. Hydrogen reacts with a variety of organic compounds and oxidisers under different conditions. Reaction with ethylene produces C_2H_6 and C_4H_{10}. It also reacts with nitrous oxide, fluorine and chlorine (halogens), and acetylene (unsaturated hydrocarbons).

It is well known that the reaction of hydrogen with oxygen results in the formation of water vapour. The reaction could be either combustion or electrochemical. This reaction, which is usually very slow, can be accelerated by an electric spark or by using a catalyst such as platinum. Some of the most remarkable properties of hydrogen combustion are its wide range of flammability, low ignition energy, small quenching distance, high auto-ignition temperature and high flame speed at stoichiometric ratios. The chemical properties of hydrogen, along with methane and gasoline, are provided in Table 2.3.

Ignition energy, flammability limit, burning speed, quenching distance and explosion phenomena, such as deflagration, detonation, etc., are some of the properties that need to be monitored not only to benefit from using hydrogen but also to eliminate some undesirable combustion phenomena. In addition, some other properties are discussed here, including auto-ignition temperature, the heat of combustion and electro-conductivity.

Figure 2.8 shows the energy state in a chemical reaction. It is imperative to understand the reactivity of fuels from the perspective of chemical reactions. In the event of a chemical reaction, it is necessary to know that initially the unreacted fuel molecules are at a certain level of energy. This level decreases as the reacted molecules form

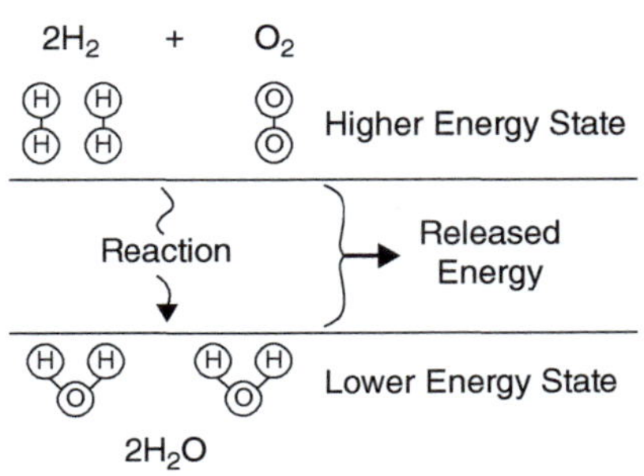

Figure 2.8 Energy states illustrating a chemical reaction.

Table 2.3 Combustion properties of hydrogen, methane and gasoline (generally accepted values from the literature).

Properties	Hydrogen	Methane	Gasoline
Limits of flammability in air (volume %)	4.0–75.0	5.3–15.0	1.0–7.6
Stoichiometric composition in air (volume %)	29.53	9.48	1.76
Minimum energy for ignition in air (mJ)	0.02	0.29	0.24
Auto-ignition temperature (K)	858	813	501–744
Flame temperature in air (K)	2318	2148	2470
Burning velocity in NTP air (cm s^{-1})	265–325	37–45	37–43
Quenching gap in NTP air (cm)	0.064	0.203	0.2
Percentage of thermal energy radiated from flame to surrounding (%)	17–25	23–32	30–42
Diffusivity in air (cm^2 s^{-1})	0.63	0.2	0.08
Normalised flame emissivity (2000 K, 1 atm)	1.00	1.7	1.7
Limits of flammability (equivalence ratio)	0.1–7.1	0.53–1.7	0.7–3.8

bonds with oxygen from the air in a chemical reaction. As molecules react with changes in chemical energy, energy is released. This energy could be diverted to more useful work. The reactions leading to the formation of water from hydrogen and oxygen are reversible. Thus, water existing at a lower level of energy could be converted to hydrogen and oxygen at a higher state of energy by some additional energy.

2.4.1 Flammability Limit

Hydrogen possesses a very wide flammability range of 4–75% by volume as shown in Figure 2.9. Hydrogen has a lower minimum ignition energy (MIE) than gasoline, which means that it requires less energy to ignite. The MIE of hydrogen in the air is typically

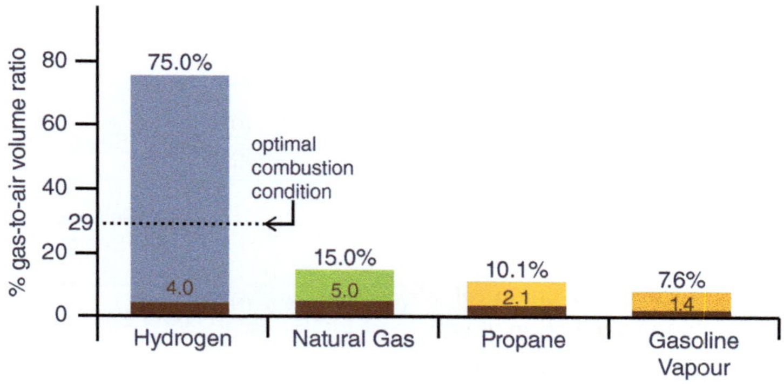

Figure 2.9 Flammability limit of hydrogen with other fuels.

around 0.02 mJ. Therefore, under optimal combustion conditions, the energy required to initiate hydrogen combustion is much lower than that of gasoline.

However, it has been observed that the flammability range linearly expands as the temperature rises (Figure 2.10). The lower flammability limit has been observed to decrease by about 2.5% by volume with an increase in temperature from 27 to 427 °C. On the other hand, the upper flammability limit increases to a much higher range of about 12.5% by volume for a similar temperature range.

2.4.2 Minimum Ignition Energy

Minimum ignition energy is the energy required to ignite a fuel–air mixture under specified conditions. Figure 2.11 shows comparative data of the MIE of hydrogen with other prospective gaseous fuels, such as methane and propane, along with conventional gasoline

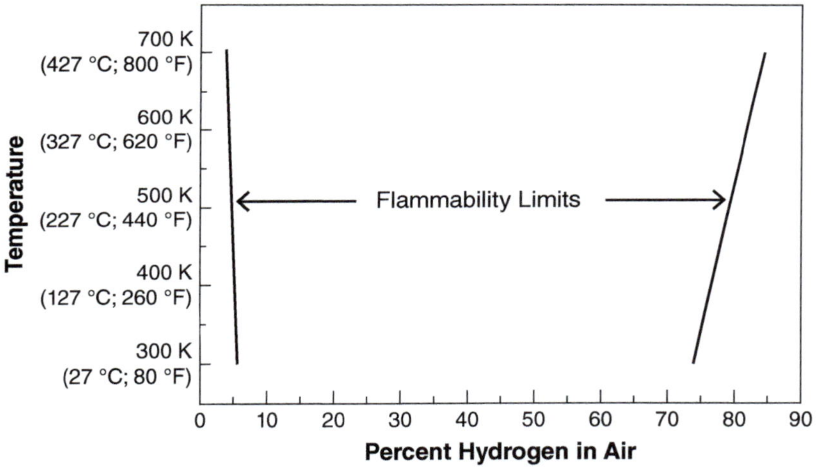

Figure 2.10 Variation in temperature with the flammability limit of hydrogen.

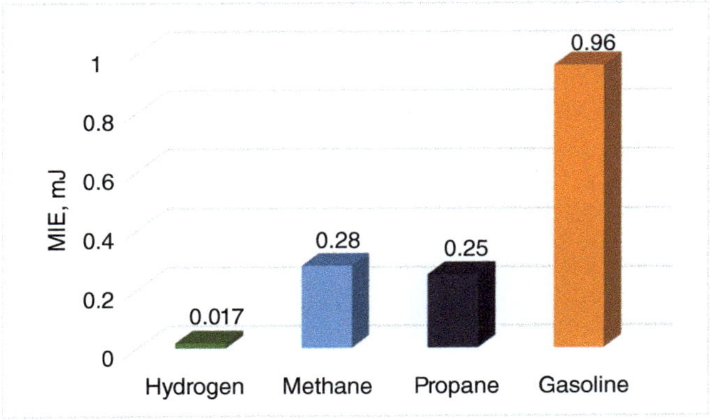

Figure 2.11 Minimum ignition energy for hydrogen and methane, and propane and gasoline.

(petrol). As a combustion-related property, hydrogen's low MIE is advantageous for engine applications that use a less-sophisticated ignition system. Hence, hydrogen can be ignited in a conventional spark ignition (SI) engine with a very low energy spark. At the same time, the low ignition energy makes the system susceptible to surface ignition. This is a very crucial property responsible for causing the uncontrolled combustion phenomenon, which is known as 'backfire' in an engine.

2.4.3 Flashpoint

The flashpoint is the lowest temperature at which the fuel forms enough vapour in the air near its surface. This means that it will flash when exposed to a brief flame under certain conditions. The flashpoint indicates how easily a fuel can be converted to vapour. Thus, by definition, the *flashpoint* is the temperature at which the fuel produces enough vapour to form an ignitable mixture with air at its surface. The fuels burn only in a gaseous or vapour state. Liquid fuels, such as gasoline and diesel, need to be converted to a vapour state so that they burn. On the other hand, hydrogen is already in the gaseous form. In this case, the flashpoint is far below the ambient temperature. Figure 2.12 illustrates the flashpoint of hydrogen compared with other fuels that are being used in several energy sectors.

2.4.4 Auto-ignition Temperature

The auto-ignition temperature of a fuel defines the minimum temperature to initiate self-sustained combustion in a combustible fuel mixture in the absence of a source of ignition. In conventional systems, the fuel is heated until it bursts into a flame. Hydrogen has a relatively high auto-ignition temperature as shown in Figure 2.13. It is a very relevant property of internal combustion engines. Some definite guidelines related to combustion could be generated from engine tests and then applied to vehicles operating on an internal combustion engine. It allows larger compression ratios to be used in a hydrogen engine than in a gasoline engine. On the other hand, hydrogen is difficult to ignite in a compression ignition engine with a diesel configuration because the temperatures needed for these types of ignition are relatively high. Hydrogen has a very high diffusivity and facilitates the formation of a uniform mixture of fuel and air. When leakage occurs, hydrogen rapidly disperses. Hydrogen has a very low density. This results in two principal problems: a very large volume is required to store enough hydrogen to supply an adequate driving range and the energy density of a hydrogen–air mixture, and hence the power output, is reduced.

	Hydrogen	Methane	Propane	Petrol	Diesel
Flashpoint, °C	−253	−188	−96	−(11–45)	37–110

Figure 2.12 Flashpoint of hydrogen along with other fuels, such as methane, propane, petrol and diesel.

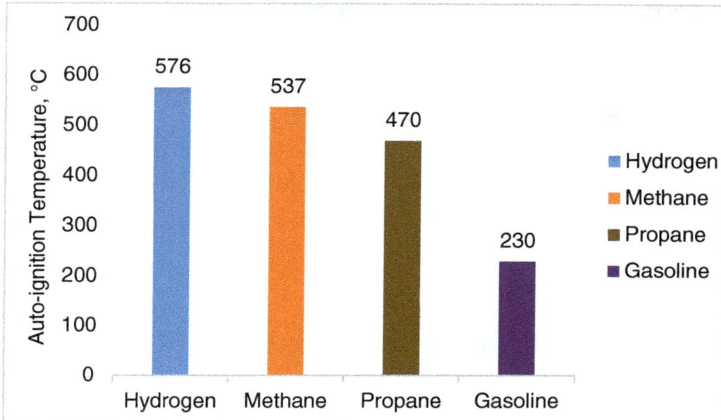

Figure 2.13 Auto-ignition temperature of hydrogen, methane, propane and gasoline.

2.4.5 Octane Number

The octane number is an important property because it indicates the anti-knock property of the fuel. Knock is a century-old undesirable combustion phenomenon in an SI engine. Knock takes place when the conditions ahead of the flame front auto-ignite due to heat build-up in the combustion chamber where the local temperature exceeds the auto-ignition temperature. Knock very seriously affects the efficiency. The anti-knock characteristic of fuel is indicated by a relative octane number rating of 100. The higher the octane number is, the better the knock resistance capacity. Hydrogen has a very high octane number of above 130. Therefore, a hydrogen engine can run in a lean mode.

2.4.6 Heat of Combustion

The lower heating value (heat of combustion) of hydrogen is 241.7 kJ mol^{-1}, and its higher heating value is 286.1 kJ mol^{-1} (Table 2.4). About 16% of this difference is the heat of condensation of water vapour, which is relatively higher compared to the corresponding gases.

Table 2.4 HHV (higher heating value) and LHV (lower heating value).

Component	HHV (MJ kg^{-1})	LHV (MJ kg^{-1})
Hydrogen	141.8	121
Methane	55.53	50
Ethane	51.9	47.8
Propane	50.35	46.35
Butane	49.51	45.75
Pentane		45.35
Gasoline	47.30	44.40

2.5 Electro-conductivity and the Joule–Thomson Effect

Hydrogen has the added property of low electro-conductivity so that the flow or agitation of hydrogen gas or liquid may generate electrostatic charges that result in sparks. Thus, all hydrogen-carrying equipment must be thoroughly grounded. Gases cool down when they are expanded from high pressure to low pressure. As the Joule–Thomson process begins, the temperature of hydrogen continues to rise from its ambient temperature. Nevertheless, the temperature level does not reach the point where the hydrogen–air mixture ignites.

2.6 Emissivity of Hydrogen Flame and Adiabatic Flame Temperature

The maximum temperature of a hydrogen flame is approximately 2380 °C with air as an oxidiser. However, the radiant heat produced is extremely low and negligible. On the other hand, gasoline, when burned, can emit radiant heat that can be felt from a much longer distance. Ethanol emits almost twice the radiant heat for the same volume of gasoline. The emissivity of hydrogen has been estimated to be very low and even may be less than 0.3, and even less than 0.1 under some conditions. A hydrocarbon flame, however, has an emissivity close to 1.

During the combustion process, some amount of energy is released to the products of combustion. The temperature of the combustion products is termed 'adiabatic flame temperature' if no heat is lost during the combustion reaction, and all the heat released is utilised to heat the combustion products. Figure 2.14 shows the pattern of variation of adiabatic temperature with respect to the equivalence ratio. Under atmospheric conditions, the adiabatic flame temperature of methane is 2328 K or 2055 °C, whereas for hydrogen it is 2400 K or 2127 °C.

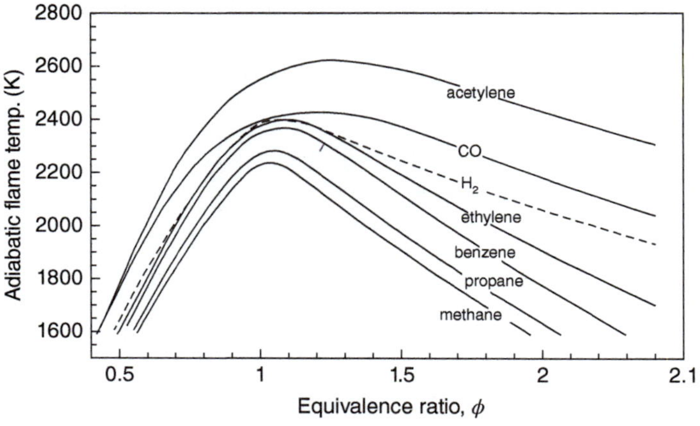

Figure 2.14 Adiabatic flame temperature of hydrogen and several other fuels.

2.7 Laminar Burning Velocity

The *laminar burning velocity* (often referred to simply as the 'burning velocity') of a combustible mixture is defined as the relative velocity at which a flame propagates with respect to the unburned gas ahead of it under specific thermodynamic conditions of temperatures and pressures of the unburned gas. The laminar burning velocity is a crucial parameter in the process of hydrogen combustion. Several studies have been conducted to determine the laminar burning velocity of hydrogen–air at various equivalence ratios and under various mixture conditions. Figure 2.15 demonstrates a comparative depiction of the laminar burning velocity of hydrogen with that of methane under certain experimental conditions with excess air in different proportions. The laminar burning velocity of a stoichiometric hydrogen–air mixture has been estimated to be about 1.91 m s^{-1}. In some experimental investigations, it has also been found that the laminar flame speed of the hydrogen–air mixture reaches a value of 2.933 m s^{-1} at an equivalence ratio of 1.7. It has also been observed that the laminar flame speed rises along with an increase in the volumetric fraction of hydrogen in the mixture.

2.8 Hydrogen–Oxygen Reaction Mechanism

Hydrogen is a flexible energy carrier that has tremendous potential for use in several sectors and ensures an improved lifestyle. A detailed description of hydrogen's penetration into the transportation, power, industrial and commercial sectors is presented later in the book. In order to identify the specific properties that need close observation, it might be appropriate here to discuss the mechanism of the hydrogen–oxygen reaction mechanism in combustion in a broader perspective. Hydrogen has many properties, such as low MIE, wide ignition limits, high flame propagation rate, etc., which are extremely different from those of conventional fuels. In this context, it is essential to have a clear idea about the

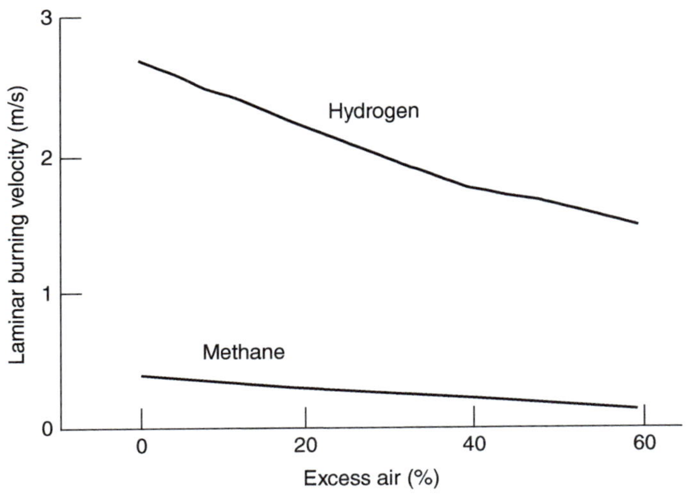

Figure 2.15 Laminar burning velocity as a function of excess air.

hydrogen–oxygen reaction mechanism and its application to several systems. The reaction of oxygen from water is represented by the following equation:

$$\text{Hydrogen} + \text{Oxygen} \rightarrow \text{Water} + \text{Energy}$$

$$H_{2(g)} + \frac{1}{2}O_{2(g)} \rightarrow H_2O + E$$

A review of the literature shows that hydrogen combustion with oxygen has been extensively studied and successfully applied in several systems. Hydrogen is not very reactive under NTP conditions. As is well known, hydrogen has a very pale blue flame that is practically invisible. This process of combustion is extremely slow under ambient conditions. However, the reaction rate could be accelerated by a catalyst. It provides very little radiant heat, and the limiting oxygen index (LOI). LOI is the minimum concentration of oxygen, expressed in percentages, which supports flame propagation in a mixture of fuel, air and nitrogen. It has been reported that there would not be any flame propagation in a mixture of hydrogen, air and nitrogen under NTP conditions if the mixture contained less than 5% by volume of oxygen [2].

Under ambient temperature conditions, no direct reaction takes place between hydrogen and oxygen in the absence of a catalyst. It has also been observed that when hydrogen and oxygen are exposed to light, oxygen is usually activated through dissociation. Sensitisers of Cl, N_2O and NH_3, a set of secondary reactions, take place. These reactions form H atoms that subsequently enter into a reaction with activated oxygen, resulting in the formation of H_2O. A catalytic reaction occurs with the metal platinum, where H_2O forms as a by-product. Schlosser's [3] branching-chain reaction discusses that each cycle of the chain results in a net increase in the number of chain carriers such as radicals or atoms. A typical example of a hydrogen–oxygen branching-chain explosion is provided below [4]:

$$H + O_2 \rightarrow 20H(\text{Initiating})$$

$$H + 0 \rightarrow OH + 0(\text{Branching})$$

$$O + H_2 \rightarrow OH + H(\text{Branching})$$

$$OH + H, \rightarrow H, O + H(\text{Propagating})$$

$$H + H + M \rightarrow H, +M(\text{Terminating})$$

One cycle of such a reaction produces a net increase of two hydrogen atoms in the reactive system. These should be used in the same length of time or else the reaction rate continues to increase exponentially, causing an explosion. Figure 2.16 shows the explosion-limit diagram for a chain branch in the reaction of a hydrogen–oxygen system [5]. This figure very clearly shows the occurrence of an explosion peninsula and the boundary of three well-defined pressure limits over a relatively short temperature range.

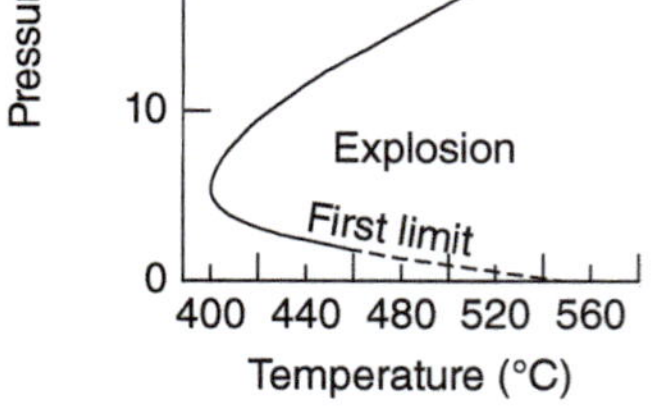

Figure 2.16 Explosion limit of a hydrogen–oxygen system.

The literature reports that one of the greatest concerns with hydrogen engines is pre-ignition. A closer look at the map can provide related information about the type of pre-ignition. Various limits as evident in the graph are referred to as the first limit (due to the wall), the second limit (due to gas phase recombination) and the third limit (due to high temperature). The second explosion limit varies with the geometry, size and surface material of the system. The concentration of any inert species (if present) also has some definite influence on the second explosion limit. Here, we do not provide a detailed analysis of the combustion chemistry of hydrogen. As a result, we limit the discussion of undesirable combustion phenomena, such as pre-ignition and backfire, which are closely associated with hydrogen combustion. Some pre-ignition occurs by heat transfer from the wall or from hot particles during the intake stroke, as shown by the second explosion limit of 50 mmHg. Parameters such as temperature, pressure, induction time and duration period have been studied by several researchers.

An internal combustion engine usually experiences pre-ignition during the compression stroke, which is explained by the third explosion limit. Due to an increase in pressure, the third explosion limit (the temperature level) decreases. Pre-ignition can occur during the intake stroke specified by the second explosion limit of 50 mmHg. This usually occurs during heat transfer from the wall or from hot particles. To find the conditions that lead to pre-ignition, the temperature of the charge should be constant for a minimum duration specified as the induction time. This holds true for temperature ranges much above the explosion limit. Therefore, pre-ignition occurs if the charge remains at a particular level of high temperature for the minimum period specified as the induction time.

2.9 Hydrogen Colours and Carbon Footprint

The fact that a colourless gas, such as hydrogen, can be described with adjectives of different colours to reveal a rainbow of hydrogen colours is very interesting. These colours are adjectives to specify how and where production takes place. A rainbow of hydrogen colours is useful because not only do they show various types of technologies used to produce hydrogen, but they also show the carbon footprint of each process, which is very different. In order to identify the type of hydrogen production, it is imperative to consider the feedstock used for hydrogen production. It is pertinent to consider how much energy is required for hydrogen production. Finally, it is necessary to decide the treatment of the carbon emissions that might release during the process. There are various codes adopted for different colours of hydrogen. In certain situations, it is classified into the following nine colours: green, blue, grey, brown or black, turquoise, purple, pink, red and white. In this method of classification, there is no strict line of demarcation between black and brown hydrogen. Both black and brown hydrogen are produced from coal. The colours black or brown are assigned to the type of coal such as bituminous (black) or lignite (brown) coal, respectively. In some literature, hydrogen that is naturally available (even though in negligible quantities) is referred to as 'white' hydrogen. The other method is a broad classification of hydrogen colour spectra. Hydrogen produced from black (bituminous) or brown coal (lignite) by the method of gasification is usually referred to as black or brown. There are several methods of gasification, and all of them operate at temperatures higher than 900 °C. Hydrogen and carbon dioxide are the two major products of gasification. Recently, several improved techniques have been adopted to reduce emissions caused by this process. However, this method of production has one of the highest carbon footprints compared to other methods.

COLOUR	DESCRIPTION: FEEDSTOCK
	Grey: natural gas reforming without CCUS
	Brown: brown coal (lignite) as feedstock
	Blue: natural gas reforming with CCUS
	Green: electrolysis powered through renewable electricity
	Pink: electrolysis powered through nuclear energy
	Turquoise: methane pyrolysis
	Yellow: electrolysis powered through electricity from solar
	Orange: electrolysis powered through electricity from wind

Figure 2.17 Hydrogen colour spectrum.

Irrespective of the colour assigned to the type of hydrogen, it can always be identified by way of its production from a definite source through a definite route, as shown in Figure 2.17.

Since the colours are assigned depending on the source/feedstock, this is not universally accepted. Orange hydrogen is not commonly mentioned in the literature. Since active research is being pursued worldwide, it is quite likely that developing fresh feedstocks and novel methods of hydrogen production will change the rainbow of hydrogen colours. Grey/black hydrogen, blue hydrogen and green hydrogen are currently identified as the most significant colours of hydrogen. Grey/black hydrogen is the most common colour of hydrogen, which is commercially produced in the maximum quantity. Blue hydrogen is currently receiving a lot of attention because it is associated with carbon capture, usage and storage (CCUS) technology. Thus, blue hydrogen has the potential to significantly contribute to the problem of decarbonisation. Green hydrogen is the most environmentally friendly colour of hydrogen because it emits no gases during production. We discuss grey, blue and green hydrogen in this chapter.

2.10 Grey, Blue and Green Hydrogen

Hydrogen can be produced from different sources by adopting different technologies, and the properties and utilisation characteristics of various types of hydrogen correspondingly vary. Thus, a colour code nomenclature is used. However, there are some cases that do not fall under a single category of colour and can have mixed sources. The factors leading to several systems involving such mixed colour codes need to be carefully studied from the point of view of the life cycle of greenhouse gas (GHG) emissions.

2.10.1 Grey Hydrogen

In terms of the hydrogen colour spectrum, grey hydrogen, which has been used for decades, is produced from fossil fuel sources through SMR or coal gasification. During the SMR process, a large quantity of CO_2 is produced. In the present context, where decarbonisation is the ultimate goal, grey hydrogen is not an acceptable option. Since the GHG emissions from grey hydrogen are considerable, there is an upsurge of interest in the natural gas industries to shift to blue hydrogen and ultimately adopt green hydrogen. Presently, grey hydrogen creates almost nine parts of carbon dioxide for each part of hydrogen. In an economic analysis, it is pertinent to ascertain whether the cost of converting an existing plant is a preferred option for constructing a new plant for green hydrogen production. Other key design features to be evaluated include the type of renewable energy that would be used for electrolysis and whether hydrogen should be stored and distributed, and also whether it is in the gaseous form or liquid form. Location intelligence is a very useful tool because it adopts the technology necessary to visualise spatial data. For green hydrogen companies, this information is crucial for making strategic business decisions. An appropriate Geographic Information System (GIS) would assist in assessing this situation by using the renewable energy database. Sustainable operation of the green hydrogen industry is crucial for the industry's success. Improvements in efficiency, clean energy certifications and appropriate safety regulations are very critical for the sustained survival and growth of the industry. It is very important to follow very stringent procedures while dealing with the current decarbonisation thrust. Maps, dashboards and real-time data are essential to achieve a sustainable hydrogen economy.

In this context, it is vital to understand the amount of major greenhouse gas emissions, such as carbon dioxide and methane, resulting from grey hydrogen production. The GHG emissions vary in the range of 10.9–18.4 kg CO_2eq/kg H_2. Oxides of nitrogen are much lower and thus not considered significant. The life cycle analysis of methane includes the main components such as extraction (mining), transport, storage and utilisation. In addition to natural gas needed to produce hydrogen, it is also required to power the carbon capture process. The carbon capture process itself may require additional energy from different sources, including natural gas or other fossil fuels. As indicated in the following reaction, one mole of carbon dioxide and four moles of hydrogen gas (H_2) are produced per mole of methane consumed:

$$CH_4 + 2H_2O \rightarrow CO_2 + 4H_2$$

It has been estimated that the sum of emissions of carbon dioxide (75.6.0 g CO_2 per MJ) and unburned methane (77.4 g CO_2eq per MJ) to produce grey hydrogen is 153 g CO_2eq per MJ [6].

2.10.2 Blue Hydrogen

Grey hydrogen turns blue through an effective implementation of the carbon capture and storage (CCS) technique. Functionally, it is a bridge to green hydrogen, which aims at total decarbonisation. It is well known that grey hydrogen is the most common type of hydrogen currently. Under such circumstances, retrofitting carbon capture to the well-established processes, such as SMR, auto-thermal reformers (ATRs) and gasification reactors, can be adapted to produce blue hydrogen. It is imperative to decide whether CO_2 is captured and utilised (CCUS) or captured and stored (CCS) in underground storage systems in such cases. These steps depend on the likelihood of commercialisation over time. At present, this can be adopted as a short-term transition until adequate infrastructure is built up with renewable energy capacity and facilities exist for green hydrogen to move ahead on the path to net-zero emissions. Blue hydrogen is characteristically grey hydrogen with CCS. It has relatively better market potential than grey hydrogen. The existing grey hydrogen system could be retrofitted with the CCS mechanism, and blue hydrogen can be produced with much lower levels of greenhouse gases. Blue hydrogen can be useful in some industrial processes, such as the production of steel that requires a continuous flow of hydrogen. Sometimes, blue hydrogen is loosely referred to as having extremely low emissions. However, it is not feasible to capture all carbon dioxide. Moreover, some amount of carbon dioxide is emitted while producing blue hydrogen from natural gas, in which methane is a major constituent. It is well known that methane is a more powerful greenhouse gas than carbon dioxide and has a much higher global warming potential. It has been estimated that, in recent decades, almost 25% of the net global warming has been caused by methane. Recently, at the 26th Session of the Conference of the Parties (COP26) in Glasgow, United Kingdom, in 2022, 104 countries promised to reduce their methane emissions by at least 30% by 2030.

Therefore, during the transition phase, blue hydrogen could be the initial solution until appropriate infrastructure is built and adequate green hydrogen becomes available. This will enable a continuous flow of hydrogen. However, blue hydrogen does not offer a long-term feasible solution as it depends on fossil fuels that have limited reserves, and their prices continue to fluctuate. The efficiency of CCS could at best be in the range of 85–95%. Therefore, 5–15% of CO_2 would still be emitted during the production of blue hydrogen. A cost would be involved in transporting and storing CO_2 as well as monitoring it closely. Additionally, methane, a potent greenhouse gas, is used in the process of blue hydrogen production. During several stages of the production of blue hydrogen, even small amounts of methane can leak, which can have a negative effect on the environment. Thus, blue hydrogen is not capable of creating a situation of net-zero mission conditions. However, several production techniques, which adopt different levels of CCS efficiencies, have been useful in generating blue hydrogen. Figure 2.18 shows the colour-coded hydrogen spectrum with the source and method of production.

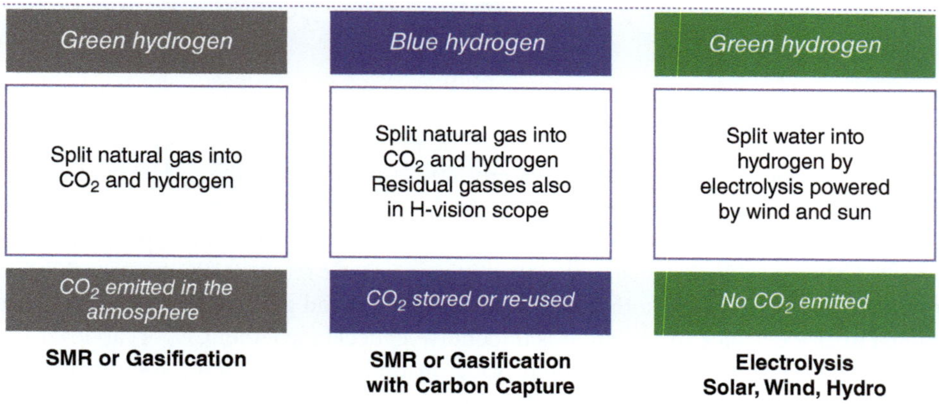

Figure 2.18 Hydrogen colour spectrum (source and route of production).

A number of industries, including the energy and transportation technology sectors, have set net-zero emissions targets and implemented the CCS process. In the CCS process, atmospheric carbon dioxide is captured and stored in carbon 'sinks'. Carbon sequestration has always existed in natural sinks such as forests and soils. Anthropogenic activities have largely destroyed these natural sinks. Right now, because of the increasing awareness of protecting the environment, many countries are putting efforts to strengthen their terrestrial carbon sink by replanting forests, revegetating the cleared land and implementing the cultivation of crops that capture high levels of carbon. It is also possible to transform carbon dioxide gas into the liquid form and store it underground. Figure 2.19 shows a schematic diagram of hydrogen production with CCS.

Howarth et al. [6] provided an analysis of the life cycle of greenhouse gas emissions of blue hydrogen accounting for emissions of both carbon dioxide and unburned fugitive methane. They studied the greenhouse effect of grey hydrogen (with the SMR process) and compared it with blue hydrogen with carbon capture. This study compared greenhouse gases of grey hydrogen (produced by SMR) with blue hydrogen (produced after carbon capture). The authors observed that far from being low carbon, 'greenhouse gas emissions from the production of blue hydrogen are quite high, particularly due to the release of fugitive methane'. The researchers reported that due to the 'default assumption (3.5% emission rate of methane from natural gas and a 20-year global warming potential), total carbon dioxide equivalent emissions for blue hydrogen are only 9–12% less than for grey hydrogen'. Additionally, the study observed that fugitive methane emissions from blue hydrogen are higher due to the higher amount of natural gas used to capture carbon. The sensitivity analysis carried out by the authors leads to the conclusion that 'the greenhouse gas footprint of blue hydrogen, even with the capture of carbon dioxide from exhaust flue gases, is as large as or larger than that of natural gas' [6].

This study stresses that 'blue hydrogen as a strategy only works to the extent it is possible to store carbon dioxide indefinitely for a long term in the future without leakage back to the atmosphere' [6]. Figure 2.20 shows the greenhouse gas footprint of grey hydrogen and blue hydrogen with those of other fuels such as natural gas, coal and diesel oil per unit of energy produced when burned.

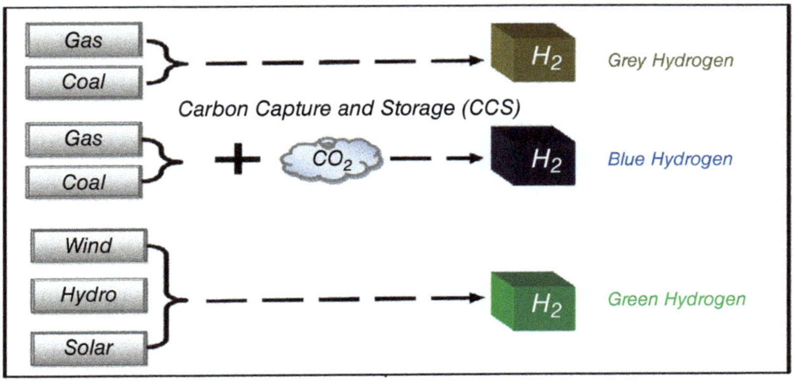

Figure 2.19 Different colours of hydrogen production with carbon capture and storage.

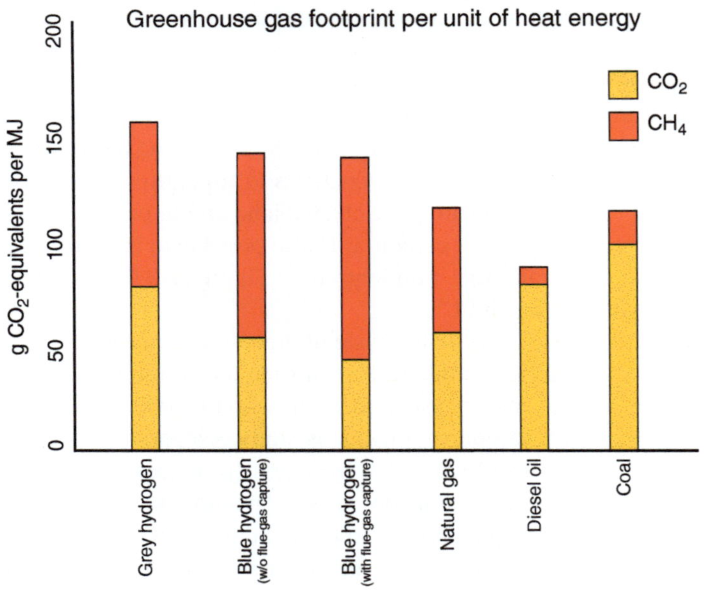

Figure 2.20 Greenhouse gas footprint for per unit of heat energy [6]/John Wiley & Sons/CC BY 4.0.

2.10.3 Turquoise, Brown, Black, Pink, Red, Yellow and White Hydrogen

Turquoise hydrogen is a relatively new entrant to the colour chart, which is still in the pilot stage [7, 8]. This method of hydrogen production uses methane and natural gas as feedstock and yet does not produce any CO_2. Turquoise hydrogen is obtained through pyrolysis. In most cases, this method involves three processes: thermal, plasma (Kvaerner process) and catalytic. These processes produce hydrogen and solid carbon black as by-products. Therefore, in this method, there is no emission of CO_2. It offers some specific advantages. Solid carbon black is easy to store compared to CO_2. Additionally, it has already gained some commercial value and contributes to additional revenue due to the existing market for carbon black already. Active research is being carried out to explore the low-emission characteristics of turquoise hydrogen by adopting the appropriate thermal process and using renewable energy. Efforts are also being made to store carbon permanently or utilise it. Depending upon the colour of the coal that is used to produce hydrogen, sometimes the term 'black' or 'brown' is used for the colour of hydrogen that essentially falls into the broad category of grey hydrogen because of their production source. Brown hydrogen is produced by gasifying solid fossil fuels such as coal or lignite and subsequently collecting hydrogen from the resulting gas by SMR. Approximately 23% of all hydrogen generation is brown hydrogen currently.

Black hydrogen refers to the hydrogen generated from black coal and brown hydrogen from brown coal or lignite coal that is brown in colour. There is a tendency to use black and brown quite liberally for hydrogen produced from fossil fuels through gasification. Before discussing in detail about green hydrogen, it may be relevant to discuss some upcoming entries to the hydrogen colour charts. Hydrogen generated by electrolysis with nuclear energy is known as 'pink', 'purple' or 'red' hydrogen. The high temperature of nuclear reactors can also play an instrumental role in steam methane reforming by producing steam that can be utilised as a fuel. Moreover, the steam generated at the high temperature of nuclear reactors could be used for more efficient electrolysis. Similarly, hydrogen made through electrolysis with solar power is referred to as yellow hydrogen, and as such it is also a distinct phase in the chart. Sometimes yellow hydrogen is referred to as hydrogen produced using electricity supplied from the grid. In that case, the total GHG emissions of yellow hydrogen completely depend on the mode of electricity generation.

2.11 Green Hydrogen

The world strives to reduce greenhouse gas emissions and reach carbon neutrality by 2050. Energy-intensive industries and transportation have a difficult time controlling emissions using available fuels and energy. In the midst of such challenges, hydrogen provides the solution. Green hydrogen, among all hydrogen shades, stands out due to its capability to provide a fully sustainable and environmentally friendly energy transition. The flexibility of the system can be increased by accommodating location-specific seasonal storage depending on the availability of solar and wind energy. Solar photovoltaic (PV) and wind power costs are steadily declining. Green hydrogen can be produced by the well-established process of electrolysis achieved by renewable electricity. Electrolysis has only a limited number of applications. As discussed earlier, most of the hydrogen produced now is obtained from SMR. A very scant amount of up to 5% of global hydrogen is obtained through electrolysis as a by-product during the production of chlorine. It could be derived from solar, wind or nuclear energy. Although green hydrogen has received international attention, production technology has not yet matured to a commercial scale. Rapid R&D activities are in progress in several countries to improve technology and decrease the cost of production of green hydrogen. As a result, its applications will start operating quickly. Research activities are being pursued to produce green hydrogen by several novel routes. In their research paper, Nelson Bion and co-researchers described the design of nano-catalysts to produce green hydrogen from bioethanol [9]. At present, hydrogen with an estimate of about 120 million tonnes is being produced worldwide. Almost two-thirds of this amount is obtained in the form of pure hydrogen, and the rest is obtained in the form of a mixture of other gases [10]. The relevance of green hydrogen can be judged in a particular energy scenario depending on the system's greenhouse production target. More than 120 countries have committed to programmes to achieve net-zero emissions. The net-zero emission goal has already been achieved by seven countries [11]. Many more countries have already initiated and implemented net-zero goals.

Green hydrogen is one of the most ideal methods to generate hydrogen from renewable sources and has the potential to provide solutions to the problems related to GHG emissions. Renewable sources of energy, such as solar, wind, small sustainable hydropower, geothermal, tidal waves, etc., do not use fossil energy sources.

Figure 2.21 shows a broader scheme for green hydrogen production pathways and their potential areas of applications [12]. Demand for hydrogen has already reached an estimated 87 million metric tonnes (MT) in 2020. It has been predicted that this demand will increase to 500–680 million MT by 2050. As is well known, currently about 95% of

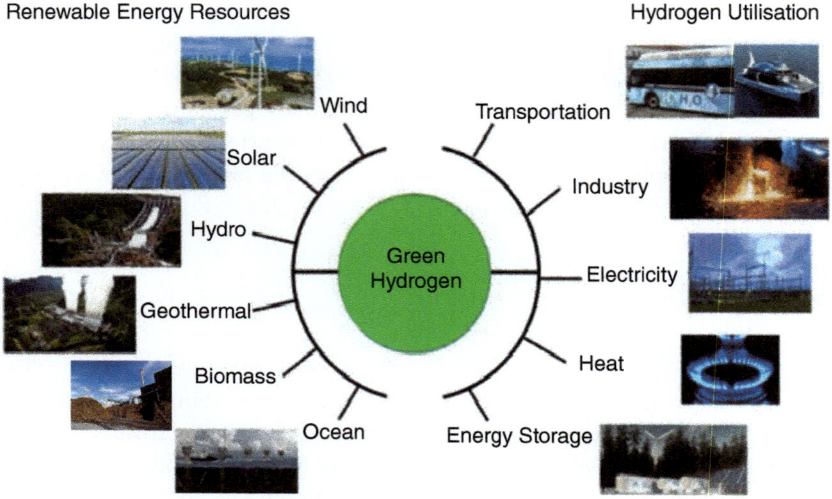

Figure 2.21 Pathways of green hydrogen production and utilisation (photo: [12]. C B Agaton, et al. (2022)/Elsevier/licensed under CC BY 4.0). (a) Andrei Merkulov/Adobe Stock.

hydrogen is produced from fossil fuels. Although the importance of green hydrogen has been acknowledged, production capacities have not substantially improved. However, there is a renewed interest in green hydrogen as the environmental compatibility of green hydrogen is of paramount significance currently. The use of hydrogen is rapidly expanding in multiple sectors, such as power generation, the transport sector, steel making, cement production, refrigeration, environmentally friendly ammonia production for fertilisers, etc., as shown in the figure, and green hydrogen is the obvious choice. Presently, most green hydrogen production developments are under process or have been operating at a pre-commercial stage. The government, the private sector and multilateral organisations are actively collaborating in several countries to achieve the industrial scale-up of green hydrogen. The hydrogen generated can be used as a fuel for transport, to heat buildings or to power various industries, as shown in Figure 2.22. It can also be used as a feedstock for chemicals such as fertilisers and for several other products.

The technologies of producing hydrogen from non-fossil sources are being increasingly used to generate electricity that can be further used for carrying out electrolysis to split water into its components: hydrogen and oxygen. A closer look at Figure 2.17 on the hydrogen colours shows that hydrogen thus produced from renewables is green hydrogen, which, on production, does not release harmful greenhouse gases. It has been estimated that the efficiency of electrolysis can be around 80%. This is a very positive point in favour of the energy value of electricity being stored as hydrogen. The electrolysis process produces hydrogen that can be used to store intermittent renewable energy sources. It is worthwhile to recall that during the recent UN climate conference COP26, the role of green hydrogen in decarbonising heavy industries and transport (land, shipping and aviation) was elaborated, and measures were suggested towards achieving a net-zero economy.

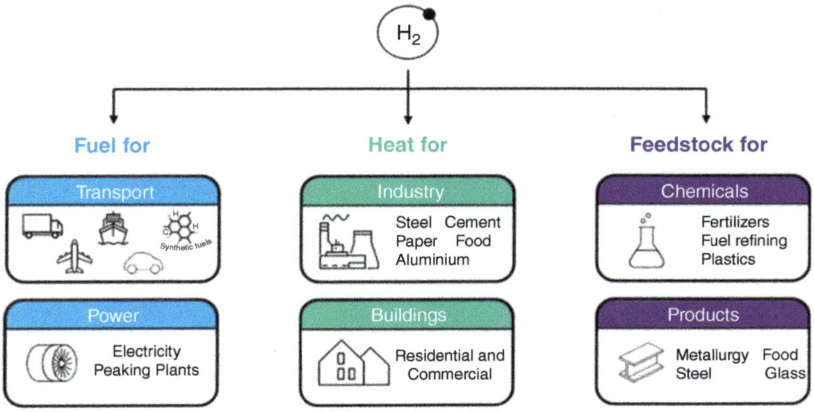

Figure 2.22 Routes and areas of utilisation of hydrogen. *Source:* Bloomberg NEF, Hydrogen energy outlook March 2020.

It is certain that the significance of the role played by green hydrogen substantially varies in the emerging energy transition scenario. In the transition scenario, it is essential to prioritise the systems according to the requirement of green hydrogen for each system. When the goal is total decarbonisation, a large amount of green hydrogen becomes essential. By the time adequate amounts of green hydrogen are available, renewable power and electrification may be adopted in certain applications. In the meantime, more hydrogen pathways would be identified, which could enable a faster deployment of technology along with the economic viability of large-scale production of green hydrogen. The most prominent advantage of green hydrogen is that it is 100% sustainable and does not emit any pollutants, either during production or combustion. It also has other beneficial features that are very closely related to hydrogen. It can be stored and used for several applications later on if necessary. It can be transformed into electricity or synthetic gas and can be used in several domestic and industrial sectors.

It can also be used in several commercial sectors for mobility purposes. Hydrogen is receiving a lot of attention for use in the transport sector either as neat hydrogen or blended with natural gas. Several studies and field applications have been conducted in this area. Green hydrogen is an ideal form for such practical applications. The production of green hydrogen is very expensive currently as it is generated from renewable sources by electrolysis. Compared to other fuels, hydrogen requires more energy to produce. The production of green hydrogen consumes even more energy. Hydrogen has a very low MIE and a much wider flammability range. Thus, adequate safety measures are necessary to ensure the safe handling of hydrogen.

Grey hydrogen is currently being produced in large quantities, which emits high levels of CO_2 and methane emissions. A step towards decarbonisation could be the replacement of ammonia from natural gas with environmentally friendly ammonia. Blue hydrogen is being considered an intermediate step towards the implementation of green hydrogen.

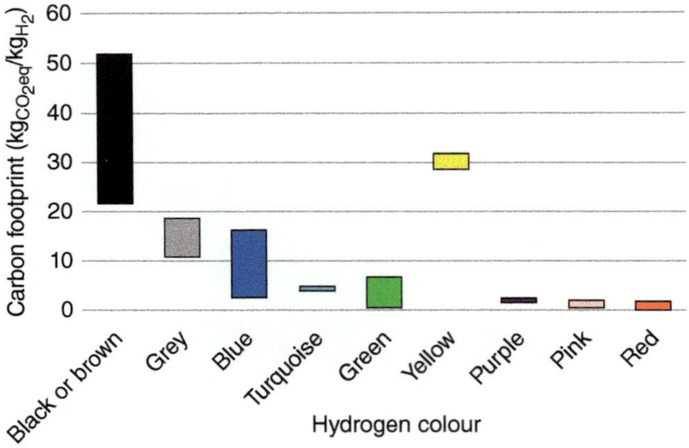

Figure 2.23 Carbon footprints of various hydrogen colours [13]/MDPI/CC BY 4.0.

Existing pilot-scale units could be upgraded for commercialisation, and appropriate infrastructure may be built up with CCS and geological storage of CO_2. Last but not least, it is necessary that this process be cost-competitive with the existing grey hydrogen production process. The net-zero world strategy of replacing grey hydrogen with blue hydrogen will likely accelerate the transition to green hydrogen. As a result, green hydrogen would be economically stable by scaling up the manufacturing process of renewable technology. Needless to say, these renewable energy technologies have already matured to a stage where competitive green hydrogen production is achievable. Electrolysers that have already been developed must be scaled up to reduce their cost. Green hydrogen's potential contribution to lowering greenhouse gases has not yet been adequately appreciated in several key industrial sectors. Currently, there is no well-defined international standard or prescription for differentiating between grey hydrogen and green hydrogen. Due to the lack of incentives, green hydrogen has been unable to penetrate several markets where its products could be of immense benefit. Figure 2.23 shows the carbon footprints of various hydrogen colours. A comparative depiction of the effect on the environment is provided in this figure [13]. As discussed above, it has additionally been shown in this graph that black or brown tends to have the highest carbon footprint. It is also evident from the figure that electrolysis is not always the cleanest method. It is also seen that higher levels of emissions than the grey hydrogen are generated by yellow hydrogen that utilises electricity from the grid to feed the electrolyser.

An overall analysis of the present situation shows that black and brown hydrogen have the highest carbon footprints. Blue and turquoise hydrogen have lower GHG emissions, whereas green hydrogen and nuclear-based hydrogen have the lowest carbon footprint. The available literature also indicates that blue hydrogen produced from biomethane could also be considered a renewable source along with green and white hydrogen. Hydrogen energy should be scrutinised at each stage, such as production, transmission, storage and utilisation. Figure 2.24 indicates the emission intensity associated with different methods of hydrogen production [14].

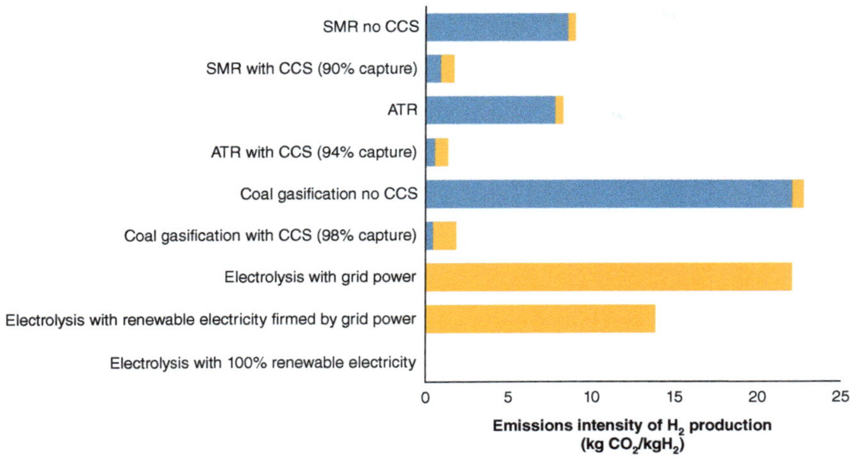

Figure 2.24 Emission intensity of hydrogen production through various processes [14]/Global CCS Institute.

2.12 Benefits of Green Hydrogen

Hydrogen is an infinite resource, and its supply potential is virtually limitless. Its advantage over batteries is that it can either be used where it is produced or can be transported to a different location as needed. Batteries do not have the ability to store large amounts of electricity for a longer period. However, it is possible to produce sufficient amounts of hydrogen from renewable energy and store it for a much longer time. Hydrogen has been the subject of interest at different points in history. There have been waves of interest at different times for different reasons, including sudden increases in oil prices, insufficient supply during periods of peak demand, etc. Research activities on alternative fuels have also been a topic of interest for hydrogen. However, apart from energy security, the present wave is driven by the thrust on total decarbonisation, and this is the situation when green hydrogen provides the solution.

One of the most attractive features of green hydrogen lies in its capability to be produced from variable renewable energy (VRE). Electrolysis from renewable resources is the preferred route to hydrogen production. Investigations have been carried out using a polymer electrolyte membrane (PEM) and an alkaline electrolyser. These investigations were carried out under varying operating conditions, such as settling time, response time, turn down, ramp rate, part load operation, etc. These studies provided very useful information on the effect of frequency disturbances and start and shutdown capabilities. Recently, many countries have strengthened their R&D efforts, and in several countries, the cost of solar photovoltaic and wind electricity has decreased rapidly. It has been estimated that the capital cost of electrolysers has decreased considerably. Since the technology has reached a mature stage, these trends are likely to gain momentum in several other countries. In addition, the cost might drop even further, leading to green hydrogen production in large quantities. Stöck et al. [15] observed that the role of green hydrogen in decarbonising the transport sector is not clearly understood. However, a correct analysis of the power sector is yet to develop. Their study has indicated that more energy-efficient and less flexible small-scale onsite electrolysis is appropriate in the case of lower shares of renewables and hydrogen, whereas large-scale hydrogen supply chains are more flexible but less energy-efficient. Researchers have suggested that distinct flexibility characteristics of hydrogen supply chains should be clearly studied to obtain a more distinct depiction of the role of green hydrogen in future energy scenarios.

Green hydrogen can also be applied to fuel cells that power electric vehicles and a wide range of electronic devices, another promising and highly prospective area of application. In this context, fuel cells have a clear advantage over batteries. This is because the fuel cell will continue to work as long as it has hydrogen fuel and does not need to be recharged. Fuel cell vehicles offer some definite advantages. These vehicles are much more efficient than vehicles running on internal combustion engines. A fuel cell vehicle can be refuelled in less than four minutes. All other basic advantages of hydrogen are as such applicable to green hydrogen. Fuel cells do not depend upon the grid and hence can be of substantial use in military applications as well as in disaster zones. However, in certain applications, it is also possible to connect them to the grid to obtain power in a more reliable and consistent manner. Several collaborative projects of different magnitudes are being carried out to derive the benefits of green hydrogen. In Germany, the 'GreenHydroChem' project encompasses linking of a series of subprojects. The large-scale production of green hydrogen using electrolysis powered by renewable energy can be transported through pipelines and can be stored for a longer duration in a hydrogen cavern. This green hydrogen can be used in several ways, including chemical processes and also during feeding into the regional gas grid. In another project called 'H2Med' with Portugal, Spain, France and Germany at a later phase, the pipeline under the Mediterranean Sea will carry green hydrogen, made from water through electrolysis using renewable energy. It is expected that about 10% of the European Union's hydrogen demand will be met by 2030 from this project.

Fukushima Hydrogen Energy Research Field (FH2R) was developed by the New Energy and Industrial Technology Development Organization (NEDO), a Japanese governmental agency. It is a pioneering project in Japan which involves a large-scale hydrogen production and the supply facility of a 10 MW electrolysis plant located at Namie, Fukushima prefecture, Japan. The project was designed to integrate multiple renewable energy sources, such as solar and wind power, to generate electricity, which would be utilised to power electrolysers and split water into hydrogen and oxygen through electrolysis. Facilities for hydrogen storage as well as for utilisation in transport and power generation were planned in this project. The Green Energy H2 Valley project in South Korea aims at developing a 20 MW electrolysis plant that produces hydrogen from renewables for use in transport and several other industrial applications. HYstock is a collaborative project between companies in the Netherlands and Belgium, which involves a green hydrogen production and storage facility in the port of Antwerp. The facility utilises renewable electricity to produce hydrogen through electrolysis. The Zhangjiakkou green hydrogen demonstration project is a major initiative in China and Chile, aimed at using the production of green hydrogen from solar and wind energy through electrolysis.

Apart from ongoing projects, a number of projects are on the table in the United States of America, India, Australia, China, Canada, the Netherlands, Norway, European Union, the United Kingdom, France, Japan, South Korea, Germany and many other countries. In fact, there is hardly any country now which does not use hydrogen-related activities.

In the context of the present chapter, a few isolated projects are highlighted here on green hydrogen. The hydrogen landscape is constantly evolving at a rapid rate. It is quite likely that by the time this book is completed, the global picture of hydrogen energy will be substantially changed.

2.13 Obstacles and Challenges to Green Hydrogen

There are several obstacles that must be overcome to ensure the rapid and effective implementation of green hydrogen. In order to produce green hydrogen, the technologies used to generate a sufficient amount of hydrogen are dependent on strategies and policies related to sources of VRE. One of the major problems is its high production costs. With the present level of technology, green hydrogen is about three to four times more expensive than grey hydrogen. It was only recently that there was a surge of interest in green hydrogen. There is still a lack of value recognition. Hydrogen has not even acquired the official status of being counted towards the estimated energy consumption statistics. There is no well-defined green hydrogen market. The capability to deliver lower or no greenhouse gases has not received any international recognition. To date, there is no definite procedure or any international practice to differentiate between green and commercially available grey hydrogen. There have been no incentives until now to either produce or use green hydrogen. Appropriate dedicated infrastructure should be built for transport and storage. Here, it is relevant to discuss some existing challenges and constraints that create difficulties in producing adequate amounts of green hydrogen at comparable prices. Although the indispensable role of green hydrogen in the context of environmental pollution has been realised by all, the production of green hydrogen through electrolysis is currently more expensive compared to conventional hydrogen produced from fossil fuels. The widespread adoption of green hydrogen requires significant investments in renewable energy infrastructure, such as wind farms and solar power plants. Such renewable sources are subject to intermittency and seasonal variability, affecting hydrogen production. The efficiency of electrolysis needs to be further improved. Scaling up its production sometimes raises financial and logistic challenges. Although green hydrogen is produced from renewable energy, its overall impact depends upon the entire supply chain. The life cycle assessment should take into account all emissions starting with production, transport and utilisation. At this point in time, when an adequate amount of green hydrogen is not available, similar greenhouse gas emission targets cannot be uniform in all sectors and all places. Thus, the transition technology options widely vary among different scenarios due to a number of factors that are region specific.

There are several silver linings that are driving green hydrogen fast towards becoming a sustainable energy option. It is well known that the cost of electricity directly impacts the production of green hydrogen. Within the past three to four years, the price of electricity from solar PV and wind plants has substantially reduced. Countries such as the United Arab Emirates (UAE), India, South Africa, Brazil, China, the United States of America, Germany, Australia, Spain, Chile and several others are rapidly expanding their solar capacities and have implemented appropriate strategies in large projects. Most of these countries have driven down the solar PV cost, and some countries have adopted technological innovation and have started mass production. This cost reduction in solar PV electricity is a very favourable condition for green hydrogen production. Many countries, such as Denmark, China, the United States of America, Germany, India, Brazil, the United Kingdom, France, Canada, Sweden, Australia, Turkey, Mexico, Spain and Italy, have onshore wind plants, and these countries have adopted these plants as part of their renewable energy mix. Many other counties are likely to implement onshore wind farm activities to reduce their dependence on fossil fuels and mitigate climate change.

The landscape of wind energy installations is changing as new projects are being developed. This activity will be useful for green hydrogen production. The cost of electricity from solar PV and onshore wind plants has substantially decreased in the past few years, and the trend is still continuing. The production of green hydrogen becomes economically more attractive as the costs of solar PV and wind electricity continue to reduce.

Of course, some of the problems are common to all shades of hydrogen, and green hydrogen is no exception. The transport of hydrogen is quite difficult on large scales even though the method of transfer has been well established. It should be cooled to −253 °C to liquefy or should be compressed to 700 bar for being transferred as a compressed gas. It has been discussed elsewhere in the book that at present hydrogen is transferred through dedicated pipelines, low-temperature liquid tanker trucks, tube trailers, rails, etc.

It is possible to transport hydrogen over long distances either by pipelines or by shipping. At present, only a few of these two methods are deployed. They are either pure hydrogen or blended hydrogen. A rational comparison shows that hydrogen infrastructure has fallen behind natural gas infrastructure. Presently, there are only about 5000 km of hydrogen transmission pipelines around the world [16], whereas natural gas has spread more than three million kilometres. Similarly, there are approximately 470 hydrogen refuelling stations around the world, whereas natural gas refuelling stations are far more numerous. In many countries, hydrogen and natural gas are blended for some applications, such as vehicles. Perhaps some upgrades or modifications could be made to existing natural gas pipelines for blending purposes, and then dedicated hydrogen pipelines could be built. The UN's International Maritime Organisation (IMO) has already suggested that the international shipping industry should reduce greenhouse gas emissions by 50% by 2050. As far as road transport and aviation are concerned, hydrogen tanks for vehicular applications are much more expensive than their fossil fuel counterparts. Similarly, if used for aviation, it might be several times more expensive than conventional jet fuel. There are significant energy losses associated with green hydrogen at several stages. In the process of electrolysis, about 30–35% of the energy is lost in producing hydrogen. In addition, transporting hydrogen requires additional energy inputs, which are in the range of 10–12% of the energy of hydrogen itself.

Hydrogen generated from renewable energy is green in character. Electricity produced by a renewable energy plant also results in the production of green hydrogen. Grid-connected electrolysers, however, perform for longer periods of time and are relatively cheaper, but they do not provide a sustainable system since grid electricity is generated by fossil fuel plants. The global renewable energy potential is much higher than the level of hydrogen demand. Therefore, it is essential that appropriate locations with abundant renewable energy resources be identified where green hydrogen could be produced.

It is difficult to distinguish different colours of hydrogen from a safety point of view. There is no difference between the most common hydrogen (grey) and the most desirable hydrogen (green). Both can be considered as hydrogen in general. It is well known that hydrogen gas is safe and non-toxic. Green hydrogen has the additional positive feature of being free of greenhouse gases.

2.14 Cost of Green Hydrogen

In the preceding discussions, it has been made clear that hydrogen of all hues does not meet the criteria for sustainable and zero-emission energy. Hydrogen is the only energy source that can meet such needs. Therefore, an adequate amount of green hydrogen should be available for effective utilisation in several energy sectors including power.

An integrated approach to the overall system depends on the development of the optimum design features to minimise the cost, and at the same time it can function efficiently under fluctuating operating conditions. Several factors dictate the system design. The cost of electricity is a major driver to ensure an accelerated implementation of green hydrogen. There is already an encouraging trend in this direction: the cost of electricity from solar PV and onshore wind plants has substantially reduced during the past few years. If the electricity supply is from a grid, the consumption pattern would differ widely for the supply from a solar or wind source. In the former situation, consumption would be more or less constant. It also depends upon the type of system adopted for the stack. It is well known that alkaline, alkaline electrolyte membrane (AEM) and PEM are relatively more flexible than solid oxides. The other important design factor depends upon the nature of the hydrogen demand. In certain chemical processes, there is a constant demand for hydrogen supply, whereas in some other applications, the demand is variable. As discussed earlier, about 95% of hydrogen, at present, is produced from fossil fuels. Green hydrogen is produced from renewable energy resources, such as wind, solar and biomass, to obtain clean, renewable fuel. It has been reported that, at the end of 2021, only about 4% of hydrogen was produced by electrolysis. Looking at the global average of electricity share, it is observed that only about 1% of global hydrogen output is produced with renewable energy.

Green hydrogen is very expensive currently with an approximate cost of US$6 kg^{-1} compared to US$1–3 kg^{-1} of conventional hydrogen. It is therefore essential that an adequate amount of green hydrogen should be produced, and it must be more cost-efficient. The location intelligence technology can form an integral part of a business model. It can be adopted using virtual 3D models and real-time dashboards as well as several other techniques to obtain answers to many spatial questions. Such an exercise can provide guidelines for several societal problems. Energy companies need to reduce costs and at the same time increase production to facilitate the wider application of hydrogen. Access to renewable resources, local demand for hydrogen and existing infrastructure (piping, shipping, etc.) for the transport of hydrogen are some of the factors that also affect the cost of green hydrogen. Keeping in mind the rate at which green hydrogen penetrates the regular life stream, it is important to have a correct assessment of the amount of renewable energy required for the estimated production rate. Other factors that also influence the cost of green hydrogen are local weather patterns and local market needs. The pollution-free characteristics of green hydrogen are clearly known, and people would like to use it. However, definite strategies need to be adopted to reduce the cost of green hydrogen so that more people can buy it for their specific use. Apart from regulatory standards to limit emissions, policy priorities and incentives to support green hydrogen production could be very useful.

It has been estimated that the demand for hydrogen will increase 10-fold by 2050 (Figure 2.25) due to the rapid shift of the global market towards a sustainable future with the decarbonisation of the industry, energy and transportation sectors [17].

It has been estimated by a study in NRDC that the cost of green hydrogen is gradually reducing and will be substantially lower compared to conventional hydrogen as shown in Figure 2.26 [18].

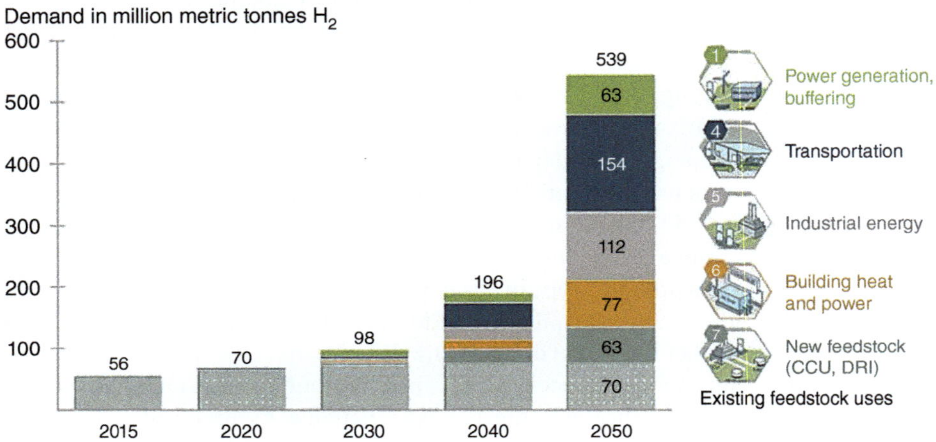

Figure 2.25 Growing trend of green hydrogen demand. *Source:* Data from [17].

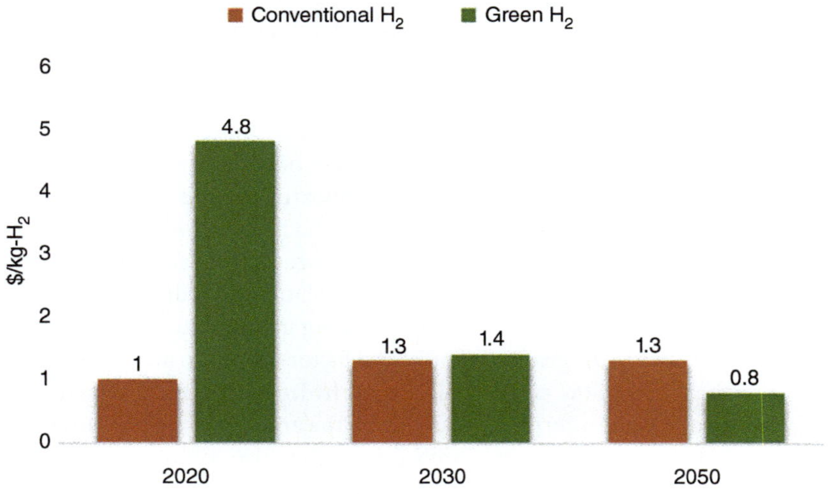

Figure 2.26 Cost of green hydrogen over a period of time.

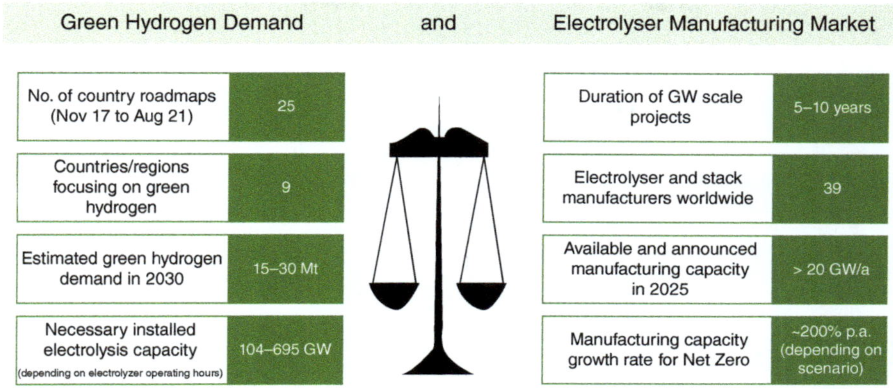

Figure 2.27 Building up of green hydrogen market [18]/Hydrogen Council.

According to Wapler and a team of researchers [18], electrolysers will be required by 2030 to ensure the effective and widespread distribution of green hydrogen on the market. The study analysed 25 road maps on hydrogen from different countries and their views on green hydrogen. In 18 studies, the future demand for green hydrogen in 2030 and 2050 was taken into account. The production capacities of some electrolyser companies were considered along with their identified production capacities. On the basis of this analysis, it has been estimated that more than 700 GW of water electrolysis units need to be installed to achieve the net-zero goal. It has also been observed that the manufacturing capacity of electrolysers must be increased by 200%. Figure 2.27 shows the current demand for hydrogen and the strategies for building up the green hydrogen market with emphasis on the manufacture of electrolysers.

The problem of large-scale production of hydrogen depends on the reduction of the cost of the electrolyser, which in turn depends upon a number of factors. In order to ensure improved system efficiency, it is necessary to evaluate optimal design criteria before building and constructing an appropriate system of the desired size. The low price of electricity is a critical factor for cost reductions of electrolysers. The selection and procurement of materials and the design and installation of the plant contribute to the overall economics. It has been estimated [20] that under such desirable conditions, green hydrogen production costs can be reduced by 85% as shown in Figure 2.28.

One of the most decisive factors in reducing the cost of green hydrogen is the availability of low-cost electricity. Many projects and activities are already underway to deploy renewable energy sources around the world, and the cost is on the decline. Locally available renewables at a particular site can produce electricity at a relatively cheaper rate than producing green hydrogen. The availability of local resources will also reduce the cost of procurement and construction.

'Several strategic reports on green hydrogen have highlighted its merit to enhance the system flexibility, particularly by seasonal storage and making efforts to integrate major contributions from solar, wind and renewables. The situation is expected to be cost-competitive as the cost of electricity from renewables is rapidly reducing. The most

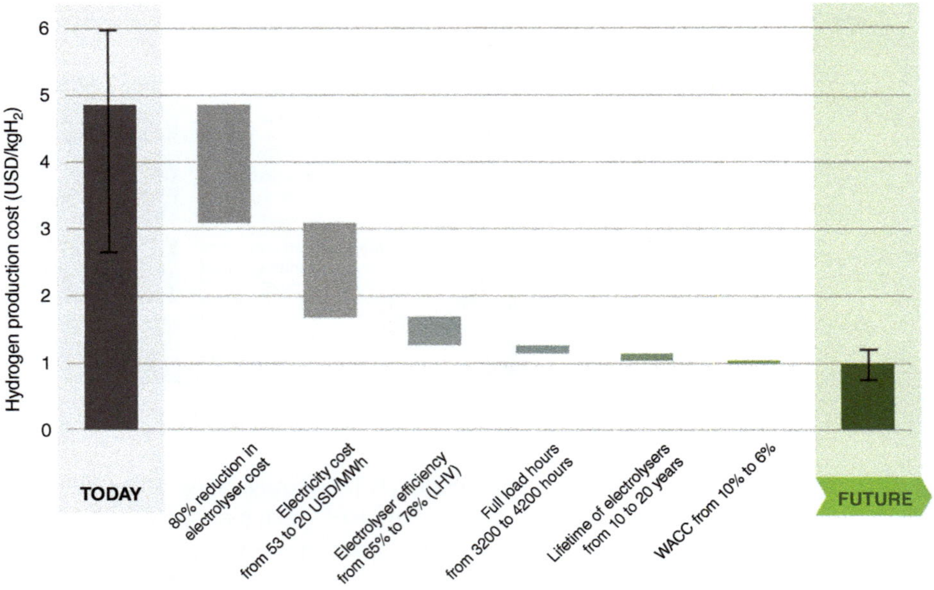

Figure 2.28 Situation leading to a lower hydrogen production cost by 80% [19]/IRENA.

significant cost driver to ensure adequate availability of green hydrogen is renewable electricity, which is gradually becoming cheaper every year. During the past decade, solar photovoltaic and wind power costs have reduced by 80% and 40%, respectively, and the trend continues [19]. Today' captures best and average conditions. 'Average' signifies an investment of US\$770 kW^{-1}, an efficiency of 65% (lower heating value [LHV]), an electricity price of US\$53 MWh^{-1}, full load hours of 3200 (onshore wind) and a weighted average cost of capital (WACC) of 10% (relatively high risk). 'Best' signifies an investment of US\$130 kW^{-1}, an efficiency of 76% (LHV) and an electricity price of US\$20 MWh^{-1}, full load hours of 4200 (onshore wind) and a WACC of 6%. These data are based on the IRENA analysis [19].

Research and development activities are continuing to establish the most effective performance and long-term durability of electrolysers. This would further reduce the cost of green hydrogen production. In their report, the International Energy Agency (IEA) [20] provided a very informative overview of several parameters showing the trade-offs between efficiency, durability and cost of electrolysers (Figure 2.29).

Concluding Remarks

It is essential to understand the physico-chemical properties of hydrogen in order to understand the role of hydrogen energy in a decarbonised world. Hydrogen has a range of combustion characteristics that set it apart from conventional fuels, and therefore the

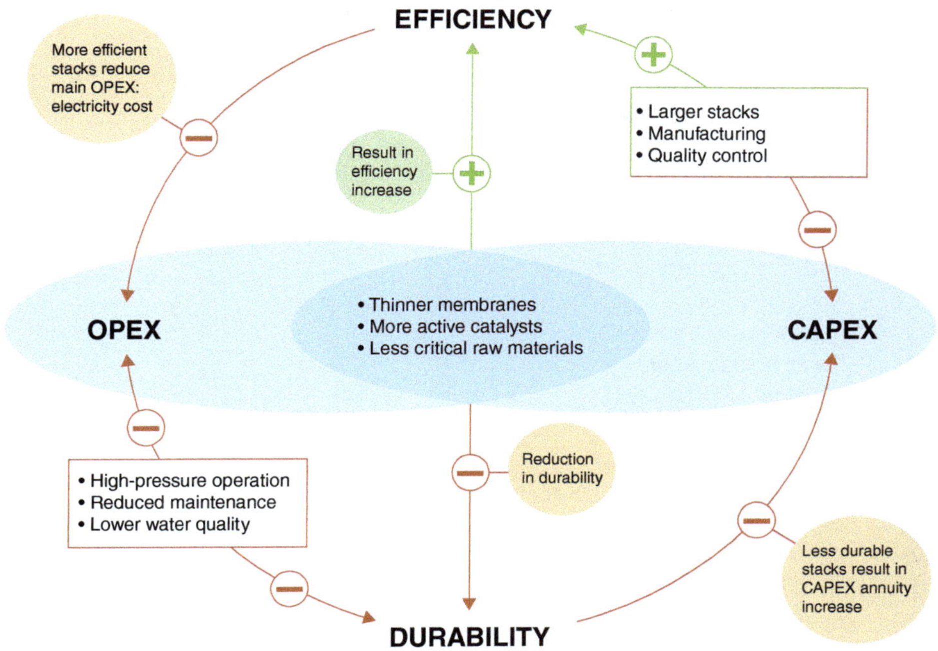

Figure 2.29 Trade-offs between efficiency, durability and cost of electrolysers [19]/IRENA.

hydrogen–oxygen reaction mechanism plays a critical role in a specific combustion system. Properties such as flammability limits, flame speed, MIE, auto-ignition temperature, etc., need to be closely monitored during the combustion process. Hydrogen is one of the most environmentally compatible fuels. The possibility of achieving net-zero goals has been widely recognised. Hydrogen's classification along with its colours provides valuable information about its production and possible applications. The demand for hydrogen is growing worldwide with unprecedented momentum. Several countries have strengthened their ongoing activities, and many new projects have also been initiated in some other countries to implement sustainable hydrogen economy to decarbonise a range of sectors, including long-haul transport, power and several chemical industries along with iron and steel. A series of mission-oriented projects are being formulated in many countries with technological advances throughout the value chain, from generation and storage to distribution and until the end use of hydrogen energy. There are hundreds of projects of different magnitudes in the areas of hydrogen production, such as renewables, infrastructure development to ensure safe and efficient transfer of hydrogen, deployment of hydrogen-fuelling stations to promote hydrogen transportation, hydrogen-powered vehicles, hydrogen for ammonia production and steel making, storage solutions including underground storage, hydrogen-based cogeneration, etc. There are many international collaborations and large-scale demonstration projects showing the potential of hydrogen as a 'forever fuel' and a versatile energy carrier.

Abbreviations

SMR	Steam methane reforming
QRA	quantitative risk analysis
NTP	normal temperature and pressure
NBP	normal boiling point
STP	standard temperature and pressure
MIE	minimum ignition energy
SI	spark ignition
HHV	higher heating value
LHV	lower heating value
LOI	limiting oxygen index
CCUS	carbon capture, usage and storage
GHG	greenhouse gas
GIS	Geographic Information System
CCS	carbon capture and storage
ATRs	Auto-thermal Reformers
COP	Conference of Parties
MT	metric tonnes
VRE	variable renewable energy
PEM	polymer electrolyte membrane
IMO	International Maritime Organisation
PV	Photovoltaic
AEM	alkaline electrolyte membrane
NRDC	National Resources Defense Council
IEA	International Energy Agency

References

1 Lee, Y., Lee, U., and Kim, K. (2021). A comparative techno-economic and quantitative risk analysis of hydrogen delivery infrastructure options. *International Journal of Hydrogen Energy* 46 (27): 14857–14870.

2 NASA. (1997). Safety standard for hydrogen and hydrogen systems. Guidelines for hydrogen system design, materials selection, operations, storage, and transportation. *Technical report NSS1740.16*. Office of Safety and Mission Assurance, Washington. Available from: http://www.hq.nasa.gov/office/codeq/doctree/canceled/871916.pdfwascancelledonJuly252005 [accessed 06.11.20].

3 Schlosser, E.G. (1967). Catalytic hydrogen combustion on nickel oxide. *Berichte der Bunsengesellschaft für physikalische Chemie* 71.

4 Strehlow, R.A. (1984). *Combustion Fundamentals*. McGraw Hill Book Company.

5 I. N. Momtchioff, E. D. Taback and R. F. Buswell. (1963). Kinetics in hydrogen–air flow system. I. Calculation of ignition delays for hypersonic ramjets. *NASA Technical Report 63N18522*.

6 Howarth, R.W. and Jacobson, M.Z. (2021). How green is blue hydrogen. *Energy Science and Engineering* 12: https://doi.org/10.1002/ese3.956.

7 Philibert, C. (2020). Methane Splitting and Turquoise Ammonia, Ammonia Energy Association, New York. http://www.ammoniaenergy.org/articles/methane-splitting-and-turquoise-ammonia.

8 Monolith. (2020). Monolith Materials Plans to Build Country's First Large Scale Carbon-Free Ammonia Plant, Monolith Materials, Nebraska. http://www.monolithmaterials.com/news/monolithmaterials-carbon-free-ammonia-plant.

9 Bion, N., Duprez, D., and Epron, F. (2012). Design of nano catalysts for green hydrogen production. *ChemSusChem* 5: 76–84. https://doi.org/10.1002/cssc.201100400.

10 IRENA (2019). *Hydrogen: A Renewable Energy Perspective*. Abu Dhabi: International Renewable Energy Agency.

11 WEF (World Economic Forum) (2020). Insight Report: The Net-Zero Challenge: Fast-Forward to Decisive Climate Action, World Economic Forum in collaboration with Boston Consulting Group. http://www3.weforum.org/docs/WEF_The_Net_Zero_Challenge.pdf.

12 Agaton, C.B., Batac, K.I.T., and Reyes, E.M. Jr. (2022). Reye prospects and challenges for green hydrogen production and utilization in the Philippines. *International Journal of Hydrogen Energy* 47 (41): 17859–17870. https://doi.org/10.1016/j.ijhydene.2022.04.101.

13 Ustolin, F., Campari, A., and Taccani, R. (2022). An extensive review of liquid hydrogen in transportation with focus on the maritime sector. 10 (9): 1222. https://doi.org/10.3390/jmse10091222.

14 Naschert, C. (2021, September). Blue hydrogen under microscope over emissions, role as transition fuel, Hydrogen production; Different methods different emission (Theme Energy, S&P Global Market Intelligence).

15 Stöckl, F., Schill, W.-P., and Zerrahn, A. (2021). Optimal supply chains and power sector benefits of green hydrogen. *Scientific Reports* 11: 14191. https://doi.org/10.1038/s41598-021-92511-6. http://www.nature.com/scientificreports.

16 IEA (2019, June) The Future of Hydrogen – Seizing Today's Opportunities, IEA, Paris. https://www.iea.org/reports/the-future-of-hydrogen

17 Hydrogen Analysis Resource Center (2016), *Hydrogen Pipelines September 2016*, Hydrogen Tools.

18 Hydrogen Council. (2017). Hydrogen scaling up, A sustainable pathway for the global energy transition. Hydrogen Roadmap Europe.

19 Wappler, M., Unguder, D., Lu, X. et al. (2022). Building the green hydrogen market – current state and outlook on green hydrogen demand and electrolyzer manufacturing. *International Journal of Hydrogen Energy* 47 (79): 33551–33357.

20 IRENA. (2021). *Making the Breakthrough: Green Hydrogen Policies and Technology Costs* Abu Dhabi: International Renewable Energy Agency.

3

Production of Hydrogen

I believe that water will one day be employed as fuel, that hydrogen and oxygen which constitute it, used singly or together will furnish an inexhaustible source of heat and light, of an intensity of which coal is not capable . . . water will be the coal of the future.
– Jules Verne (*The Mysterious Island*, 1874)

CHAPTER MENU

(Continued)

Hydrogen Energy: Production, Safety, Storage and Applications, First Edition. L. M. Das.

3.1 Introduction

Rapid urbanisation along with an increase in population and improved lifestyles has significantly increased the demand for energy. At the same time, the depletion of fossil fuels and environmental degradation due to anthropogenic activities have shown the urgency of alternative sources of energy based on renewable sources. Hydrogen's enormous potential in providing eventual freedom from energy-related environmental problems has been discussed in Chapters 1 and 2. However, as is well known, hydrogen is not available in free form on Earth although it is the most abundant element in the universe. Hydrogen, an energy carrier rather than an energy source, is found in combination with water, hydrocarbons and alcohol. It is also found in natural biomass that includes plants and animals. In this chapter, an overview of hydrogen production technologies from fossil fuels, water and biomass has been discussed. The existing established commercial methods of hydrogen production have been dominated by fossil fuels, such as coal and natural gas, which have a significant impact on the environment. Hydrogen production technology with electrolysis involving renewable sources needs to be expedited to generate environmentally friendly green hydrogen. Several other methods, which have been tested on a small or modest scale from different sources, some of which have the technology to be scaled up, have also been highlighted.

3.2 Routes of Hydrogen Production

Presently, hydrogen is produced using various processes and technologies. Global hydrogen production, at present, is estimated at 75 million tonnes, and consumption increases annually by around 6%. It is essential that technologically sound and economically viable routes for producing large quantities of hydrogen are established. This is to cater to the growing demand for hydrogen in several sectors.

While it is possible to produce enough hydrogen from both renewable and non-renewable sources, all methods of global hydrogen production (such as steam reforming [SR], partial oxidation [POx], etc.) are derived from petroleum. Hydrogen production from water is a well-proven technology, restricted to laboratories (which has not been extensively scaled up for large quantities of production). The inherent constraints of electrolysis in producing large quantities of hydrogen are discussed later in the chapter. It is only recently that a lot of attention has been drawn to the production of green hydrogen, considering a rapid transition to a zero-carbon future.

The existing methods of hydrogen production through different routes (such as Steam methane reforming, partial oxidation, auto-thermal reforming, gasification, electrolysis, haemolysis, etc.) from different feedstocks (such as natural gas, biomass, coal, oil and water) using various primary energy sources are discussed here. Purification methods, such as desulphurisation and water–gas shift (WGS) reactions, are also briefly discussed. Table 3.1 and Figure 3.1 illustrate various sources from which hydrogen can be produced and highlight the methods of production.

Table 3.1 Hydrogen production: process and raw material.

	Process	Raw materials	Source of energy
Electrochemical	Electrolysis	Water	• Electricity from renewable energy sources (e.g. wind, geothermal, solar and hydro) • Electricity from non-renewable energy (e.g. fossils fuels and nuclear)
Thermochemical	Reforming	• Natural gas • Hydrocarbons • +Water	• Combustion of natural gas/syngas • Concentrating solar thermal
	Gasification	• Coal • Carbonaceous materials • Biomass • +Water	• Combustion of coal/biomass/carbonaceous materials/syngas • Concentrating solar thermal
	Decomposition	• Natural gas • Fossil fuel hydrocarbons • Biomethane • Biohydrocarbons	• Natural gas combustion • Concentrating solar thermal
	Thermolysis	• Water	• Concentrating solar thermal
	Thermochemical cycle	• Water	• Concentrating solar thermal • Nuclear heat
Photochemical	Photosynthesis	• Water	• Solar radiation and artificial light
	Photobiological	• Microbial (e.g. algae) • Water	• Solar radiation

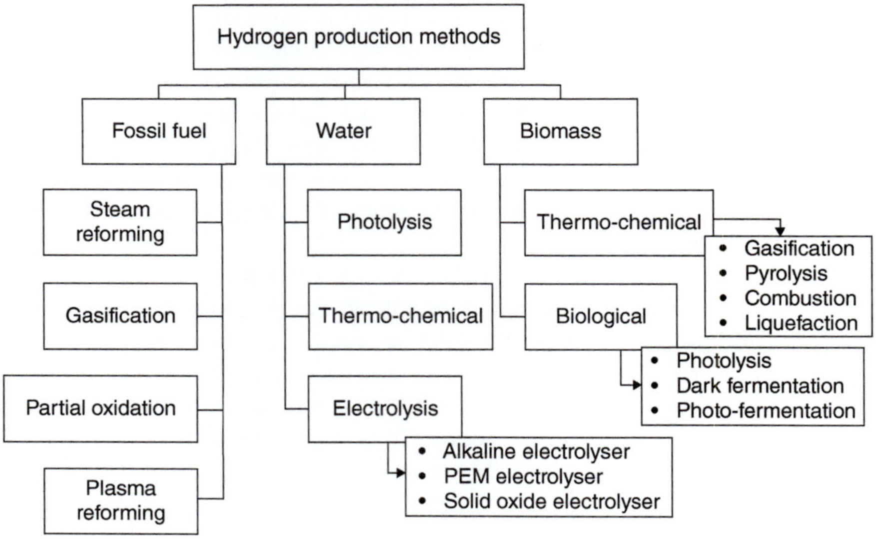

Figure 3.1 Hydrogen production from several sources through various routes.

The above information essentially lists out the broad well-established and technologically mature routes of hydrogen production, mainly from fossil fuels, water and biomass. Further discussions later in the text describe and highlight some additional sources, such as wind, ocean thermal energy conversion (OTEC), geothermal, nuclear energy, etc., some of which are still in the stage of research and development. A brief discussion is also provided on some emerging technologies, such as the production of hydrogen using anaerobic bacteria, plasma formation, ammonia dissociation, hydrogen from hydrates, etc.

Almost 48% of the present global hydrogen demand is met by steam methane reforming, a process that releases an enormous amount of greenhouse gases (GHGs). Therefore, although this is a commercially well-established process, it is not an acceptable long-term solution from an environmental point of view. Figure 3.2 shows that 30% of energy is generated from coal, while 18% of energy is generated from oil-based sources, 3.9% from water electrolysis and 0.1% from other sources [1–4].

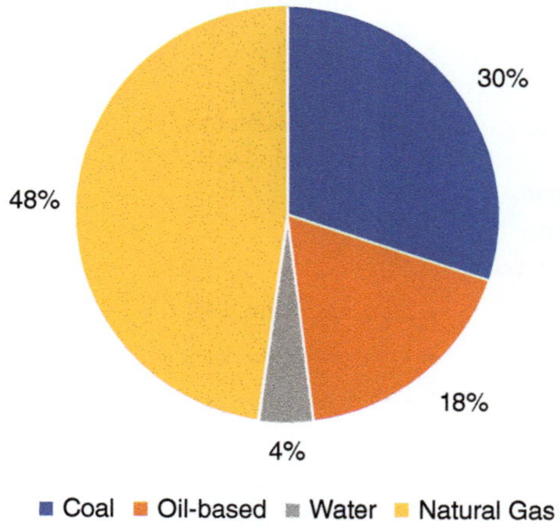

Figure 3.2 Global hydrogen production.

The present scenario of global hydrogen production is expected to change radically in the coming years. It has been trending upwards as shown in Figure 3.3.

Direct reduced iron (DRI) steel production: Methanol and 'other mixed' represent the demand for applications that use hydrogen as part of a mixture of gases, such as synthesis gas, for fuel or feedstock [4]. Figures 3.4 and 3.5 show 'global hydrogen production

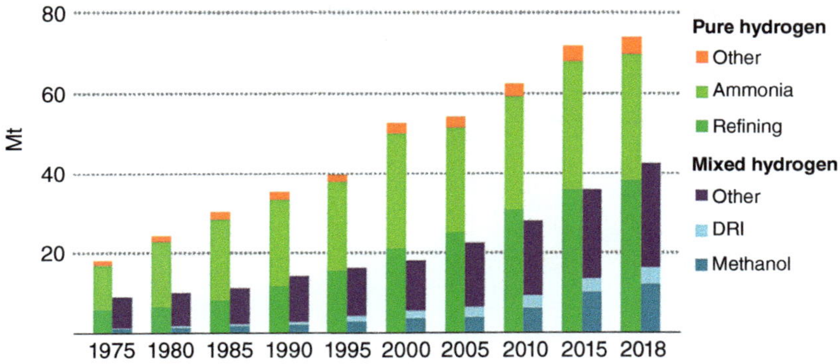

Figure 3.3 Global annual demand for hydrogen by application [4]/IEA/Licensed under CC BY 4.0.

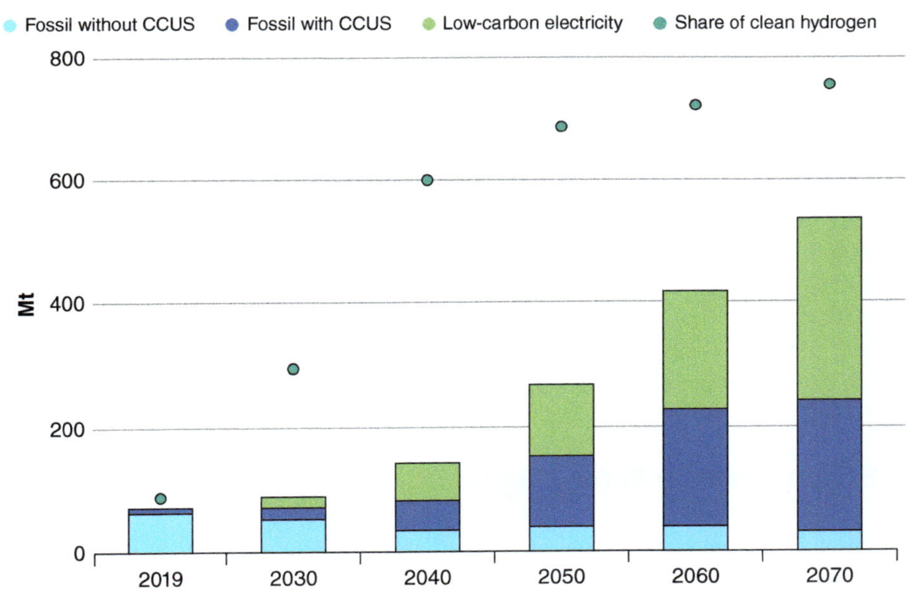

Figure 3.4 Global hydrogen production for sustainable development. *Source:* Adapted from [5].

by fuel and hydrogen demand by sectors in the Sustainable Development Scenario, 2019–70' [5].

Refining catalytic naphtha reformer (CNR) refers to the production of hydrogen as a by-product of catalytic naphtha reforming in refineries. It is obvious that green hydrogen is the ultimate solution. The existing process should be improved, however, as long as an adequate production of green hydrogen is not achieved to meet demand. This will reduce environmental pollution.

It is not possible to evaluate the production process exclusively from an ecological perspective. However, there are other significant criteria that dictate the overall implementation of a particular technology. Electrolytic and plasma processes, despite their high

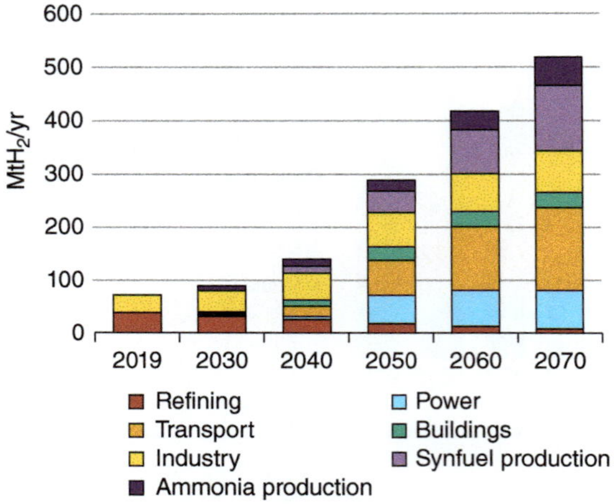

Figure 3.5 Global hydrogen use for sustainable development [5]/IEA/Licensed under CC BY 4.0.

efficiency, cannot be adopted as they are extremely energy intensive [6]. Similarly, processes such as photolytic water cracking are environmentally friendly. However, this process cannot be adopted for large-scale industrial applications because of its low efficiency. Therefore, it is necessary to produce energy-efficient, environmentally friendly and economically feasible hydrogen production from fossil fuels [6]. As discussed earlier, there are three broad types of technologies: SR, partial oxidation and automatic thermal reforming. However, all three technologies produce a large amount of carbon monoxide (CO) during production. CO to CO_2 conversion is sometimes accomplished using methane action reactions, WGS or preferential oxidation (PrOx) reactions.

3.3 Steam Methane Reforming (SMR)

Fossil fuels such as hydrocarbons, gasoline, methanol and ethanol produce hydrogen. At present, methane is the most common source of hydrogen production by SMR, which is relatively less expensive. A mixture of methane and water vapour undergoes a strongly endothermic reaction as shown below in the equation:

$$\mathrm{CH_4 + H_2O + heat = CO + 3H_2}$$

Broadly speaking, reforming is the process by which hydrocarbon molecules are broken from the compound to be removed from:

$$\mathrm{C}_n\mathrm{H}_m + n\mathrm{H}_2 \rightarrow n\mathrm{CO} + \frac{(m+n)}{2}\Delta H(298K) > 0$$

SR is the most effective and least expensive method of industrial hydrogen production. Figure 3.6 shows the present level of hydrogen production through different routes.

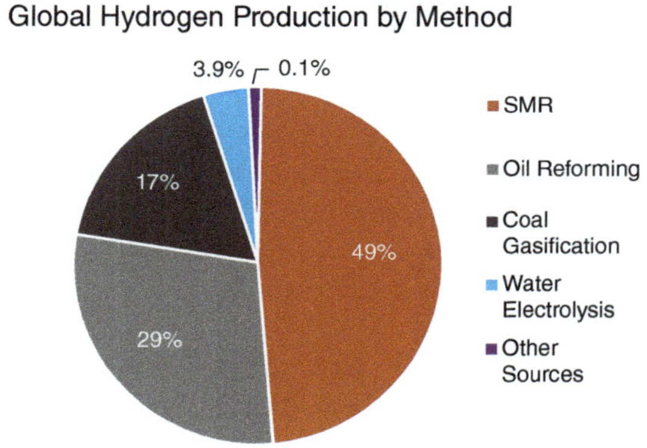

Figure 3.6 Global hydrogen production by different methods and processes.

As shown in the equation, methane and water vapour undergo an endothermic reaction in the first stage:

$$CH_4 + H_2O \rightarrow CO + 3H_2 - \Delta H^\circ$$

where ΔH° is the enthalpy change (= 252.3 kJ mol^{-1}) at ambient temperature and pressure (298 K and 0.1 MPa). If water (H_2O) is in the gaseous/vapour form, the value of ΔH^0 becomes equal to 206.2 kJ mol^{-1}. The process typically takes place at temperatures of 700–850 °C and at a pressure of 2.5×10^6 Pa. The product of reaction (i), which is a mixture of CO and hydrogen, is called 'synthesis gas'. This product gas contains about 12% CO. In the penal stage, CO is converted to CO_2 and H_2 through a water–gas shift reaction:

$$CO + H_2O = CO_2 + H_2 - \Delta H^0$$

Here, $\Delta H^\circ = -41.1$ kJ mol^{-1} where reactants are in the gaseous form, and under ambient temperature and pressure conditions, $\Delta H^\circ = -5$ kJ mol^{-1} if water is in the liquid form. In the final stage, H_2 is separated from CO_2 using chemical adsorption, and the stream of H_2 is purified in a pressure swing adsorption (PSA) unit. Figure 3.7 shows a schematic diagram of hydrogen production from natural gas. As is well known, methane is a major constituent of natural gas [7].

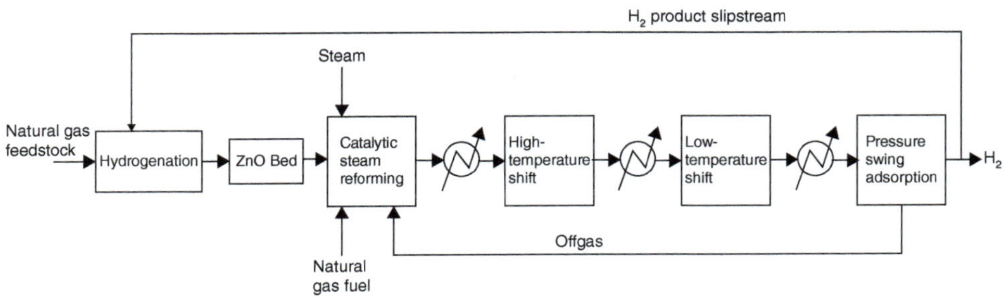

Figure 3.7 Hydrogen production from natural gas [7]/US Department of Energy/Public Domain.

Before SR, it is essential to remove any sulphur present in the feedstock. To achieve this objective, the feedstock (natural gas) is subjected to pre-treatment in a hydrogenation vessel with a small amount of hydrogen recycled from the product stream. In this way, the sulphur compounds present in the feedstock are converted to H_2S, which is subsequently removed in the ZnO bed. As evident from the figure, natural gas and steam are put in the steam reformer. The synthesis gas, thus formed, is then allowed to pass through the high-temperature shift (HTS) and low-temperature shift (LTS) reactors, and the water–gas shift reaction converts 92% of the CO into H_2. Hydrogen, thus formed, is purified in a pressure swing absorption unit [7].

Natural gas reforming is the most widely used technology and contributes around 45% to hydrogen production. Biomass is actively produced through several research and development programmes worldwide. With some technical problems still existing, its economic viability has been established. Hydrogen is also generated by electrolysis and thermolysis, which decompose water to produce hydrogen. However, despite its efficiency in small-scale units, the transfer of the technology to large commercial applications is still to be demonstrated. While microbiological processes are highly promising, they have only been demonstrated in a laboratory setting. With the present level of technological advancement, microorganisms in the lab produce extremely low amounts of hydrogen. The process is far from being adopted for commercial industrial production. Methane reformation has achieved an efficiency of 70–85%. Research and development activities are also conducted on solid waste, food industry waste and agricultural biomass. These materials also produce the same level of effusing. The main shortcoming is a higher CO_2 production level. The other design factor is temperature. Although the steam reformation process needs a relatively low temperature of about 180 °C for methanol and oxygenated hydrocarbons, a temperature beyond 500 °C is usually required for most conventional hydrocarbons.

3.3.1 Water–Gas Shift Reactor

Usually, the product gas mixture obtained by the reforming process contains carbon monoxide up to about 5% by volume. Often the product gas is passed through a WGS reactor at high temperatures to increase hydrogen content and reduce carbon monoxide levels. As shown in Figure 3.8, the water shift reaction for carbon monoxide reduction can occur in two steps, namely, HTS and LTS reactors.

An Fe_3O_4/Cr_2O_3 catalyst is used in the first step reaction with a temperature range of 310–450 °C, reducing carbon monoxide levels by up to 10%. Using $Cu/ZnO/Al_2O_3$, the second step of the reaction is performed in a temperature range of 180–250 °C, resulting in a further reduction in the CO concentration of about 500 ppm. As indicated in the figure, the

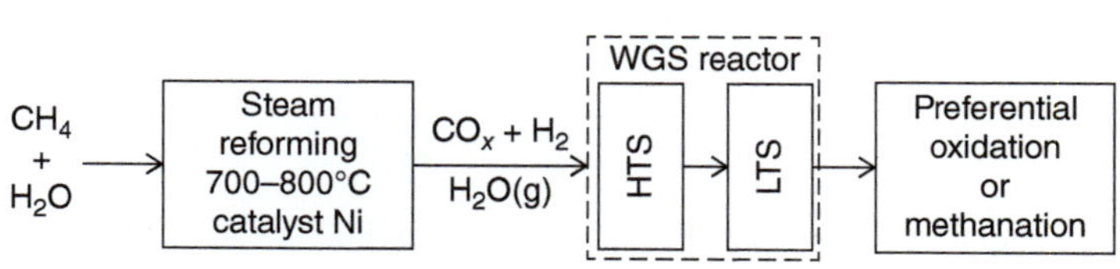

Figure 3.8 Steam reforming in two-step water shifts as reactors.

use of a PrOx reactor or a carbon monoxide selective methanation reactor is effective in further lowering the CO level.

3.3.2 Selection of Catalysts

In addition to natural gas/methane, other petroleum-based fuels are also used. All these fuels contain sulphur. The sulphur content depends on the type of fuel and the source. Desulphurisation is essential for this fuel-processing mechanism. There are several desulphurisation methods. The most common and commercially available methods for removing sulphur are those in which organosulphur compounds are converted using catalysts. This reaction for SR requires a catalyst. The catalyst is usually nickel or aluminium oxide, cobalt alkali and rare earth mixtures. Sulphur-containing compounds are avoided to prevent catalyst deactivation. To achieve higher conversion efficiency, heat inputs are often taken from cooling reactants in the reaction area. This is carried out to control the reforming process, which has a steam-to-methane ratio of 2–3. A palletised catalyst is usually less than 5% effective in conventional SR. SMR is a very old and established hydrogen production technique. In the past century, research and development have been performed to identify more reliable and improved catalysts. Improvements have also been made to the reactor design. In a review research paper, Advis et al. [8] analysed existing systems and discussed the performance improvement aspects of several configurations for different reforming applications.

3.3.3 Ethanol and Methanol Steam Reforming

As discussed above, methane reforming is the most widely accepted method of producing hydrogen. Several other hydrocarbons are being tested and are in the experimental stages.

Ethanol is a renewable alcohol that can be produced using crops such as corn, sugarcane or cellulosic biomass. It can be used as a feedstock for hydrogen production through SR. During SR, ethanol is reacted with high-temperature steam (water vapour) over a catalyst to produce hydrogen gas (H_2), carbon monoxide (CO) and carbon dioxide (CO_2). The resulting gas mixture is known as 'synthesis gas' or 'syngas'. Syngas can be further processed to separate and purify hydrogen:

$$CH_3CH_2OH\,(\text{ethanol}) + 3H_2O\,(\text{steam}) \rightarrow 3H_2\,(\text{hydrogen}) + 2CO\,(\text{carbon monoxide})$$

Goula and co-researchers [9] investigated ethanol steam-reforming reactions and observed that a 5 wt.% Pd/Al_2O_3 catalyst performs very well and shows high catalytic activity and stability for the reaction. Haryanto [10] and his team tested steam-reforming ethanol for hydrogen production. It has been reported that ethanol conversion and hydrogen production greatly vary depending upon several factors, such as the type of catalyst, catalyst preparation method and reaction conditions. Catalysts such as Co/ZnO, ZnO, Rh/Al_2O_3, Rh/CeO_2 and $Ni/La_2O_3–Al_2O_3$ showed promising performance. Techno-economically, steam ethanol reforming remains a research area that need to be further investigated.

Steam methanol reforming for hydrogen production is being investigated [11–15] Similar to ethanol, it can also undergo reforming to produce hydrogen-rich gas. In this process, methanol is first converted into synthesis gas (syngas) through SR, a reaction involving methane (CH_4), steam (H_2O) and a catalyst. The syngas consists of hydrogen (H_2), carbon monoxide (CO) and carbon dioxide (CO_2). Hydrogen can then be separated and purified.

$$CH_3OH\ \text{(methanol)} + H_2O\ \text{(steam)} \rightarrow 3H_2\ \text{(hydrogen)} + CO\ \text{(carbon monoxide)}$$

Technological prospects and economic challenges for steam methanol reforming have been identified, and solutions have also been discussed to improve the technology. Methanol decomposition, methanol SR and water–gas shift reaction pathways over the Ni2P catalyst have also been investigated using density functional theory (DFT), and it has been found that Ni2P is a promising new catalyst for methanol SR [13]. The results of studies on the analysis of a series of A-site strontium-doped perovskite catalysts synthesised via the ethylene diamine tetraacetic acid (EDTA) sol–gel method for hydrogen production by methanol SR have also been reported [14]. Deqing Me and the researchers, in their paper [15], discuss the methanol-reforming process and its potential role in hydrogen production. They identified four specific areas, highlighted as 1, 2, 3 and 4 in Figure 3.9, which need to be considered and improved. The purpose is to ensure an efficient methanol-reforming process.

3.3.4 Fuel Processing for Fuel Cell Application

Studies are being conducted on the thermodynamic analysis of parameters, improving catalyst performance and analysing several hydrocarbons and oxygenates. Farrauto et al. [16] discuss the application potential of catalyst and adsorbent technologies for hydrogen production from natural gas to be applied in a proton exchange membrane (PEM) fuel cell. Polymer electrolyte membrane (PEM) and Proton exchange membrane are often used interchangeably. Both refer to a critical component of proton exchange membrane fuel cell which use a polymer electrolyte membrane. Pipeline natural gas could be used for residential use. However, processing natural gas with sulphur requires adsorbents for desulphurisation. Additionally, SR or autothermal reforming requires heterogeneous catalysts. This chapter also discusses hydrogen

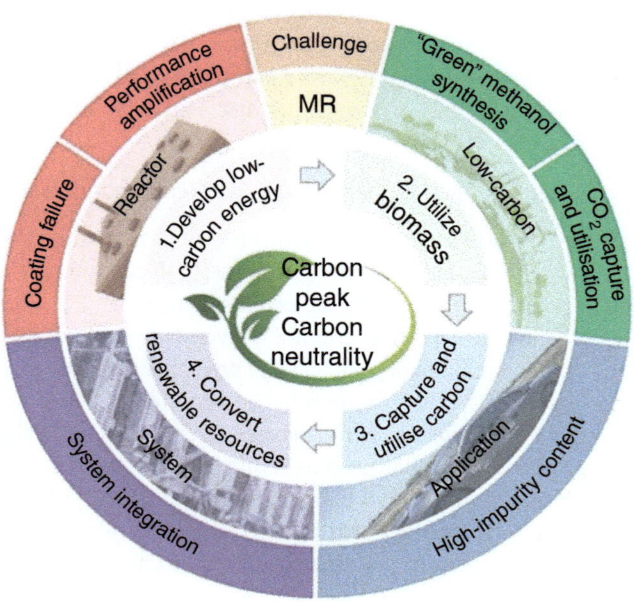

Figure 3.9 Methanol reforming for hydrogen production [15]/with permission of Elsevier.

production by reforming methanol. This can be used in a number of portable power applications such as laptop computers, cellular phones and several other digital equipment.

Song and co-researchers [17] investigated the possibility of fuel processing for low-temperature and high-temperature fuel cells. At present, infrastructure exists for hydrocarbon fuel production and distribution. However, for use in fuel cells, fuel processing needs to be improved to achieve higher energy efficiency. Furthermore, the fuel processor size should be reduced. Alcohol fuels are free of sulphur and can be formed at lower temperatures. Hydrocarbon reforming is different. For fuel cells to work, sulphur must be removed before and after reforming.

3.4 Partial Oxidation of POx

Partial oxidation has been studied as an effective hydrogen production method. Catalytic partial oxidation (CPOx) of *n*-decane and *n*-hexadecane with air over an Rh-coated monolith can be carried out to produce synthesis gas with H_2 and CO [18]. The estimated conversion efficiency and contact time of an effective reaction were evaluated. Researchers have studied partial oxidation from a different technological viewpoint. Lutz et al. [19] analysed the process from two different approaches to chemistry: (i) global reaction balances and (ii) chemical equilibrium. A comparative analysis of both processes' outcomes provides very useful information. While the global reaction analysis is performed considering the steam-to-carbon ratio and the oxygen-to-carbon ratio as process parameters, the analysis from chemical equilibrium gives a realistic idea about the composition of the product and the thermal efficiency of the process. In the POx process, the raw material is usually methane. The process can also be used with biogas and heavy oil fractions. It is primarily a moderate and partial oxidation process in which heat is generated so that the reaction is exothermic:

$$CH_4 + \frac{1}{2}O_2 \rightarrow CO + 2H_2 + \Delta H$$

where $\Delta H = -36\ \text{kJ mol}^{-1}$.

If the oxygen-to-fuel ratio is increased, the reaction becomes more exothermic, such as

$$CH_4 + O_2 \rightarrow CO_2 + 2H_2, + \Delta H \ \text{where}\ \Delta H = -319$$

A close look at the above two equations shows that during the POx process of methane, the amount of heat generated can be increased from a lower value of −36 to −319 kJ mol^{-1} just by raising the oxygen in the reaction. The implication of adding more oxygen is that by adding more air (hence oxygen) the additional fuel is also combusted. The practical application of this could lower reformer start-up time by increasing air quality. In a more general way, the equation can be expressed as

$$C_nH_m + 1/2nO_2 \rightarrow nCO + 1/2mH_2 - \Delta H°$$

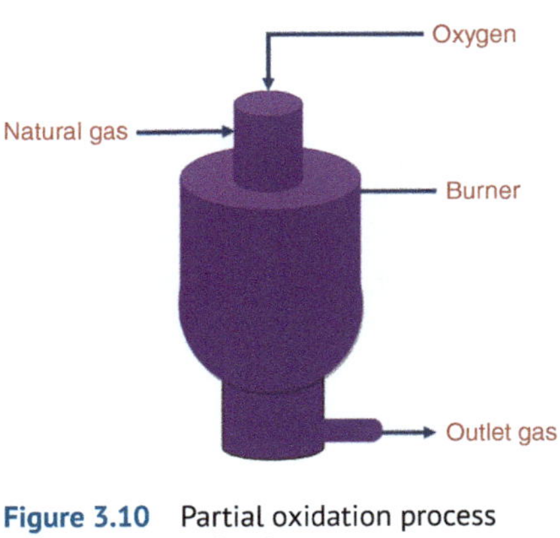

Figure 3.10 Partial oxidation process of hydrogen production.

The POx process, as shown in Figure 3.10, involves gasifying the raw material at a temperature range of 1300–1500 °C in a non-catalytic manner. In this non-catalytic process, it is not required to remove sulphur from natural gas (methane) as it is mostly (about 95%) converted to hydrogen sulphide and the remaining 5% is converted to carbon oxysulphides (COS). In partial oxidation systems, catalysts are sometimes recommended when operating temperatures are lower, between 700 and 1000 °C. Catalysts based on Ni or Rh have been tested. However, they have performance limitations. Ni has a strong tendency to coke. It has been demonstrated in the literature that exothermic reactions often cause coke to form, leading to difficult temperature management situations. Rh, on the other hand, is very expensive and cannot be widely used. Catalytic partial oxidation has been a very useful process for producing hydrogen for fuel cell applications. POx and CPOx undergo complex reactions involving intermediate species, and radicals are formed. The gaseous mixture consists of CO, CO_2, H_2, CH_4, H_2O_2, H_2S and COS. However, a portion of this gaseous mixture provides the heat for the endothermic process to occur.

3.5 Partial Oxidation of Heavy Oils and Naphtha

Certain oil products are used for hydrogen production by partial oxidation. This is essentially a gasification process suitable for feedstocks, such as heavy oil and naphtha, with a high carbon-to-hydrogen ratio. Here, similar to the conventional process of partial oxidation of methane, the process starts with an input of pure oxygen. It has been reported in the literature that a mixture of CO and H_2 is formed when the feedstock reacts with oxygen below the stoichiometric ratio of oxidation at a temperature in the range of 1200–1350 °C [20, 21] as shown in the following equation:

$$\mathrm{C}_n\mathrm{H}_m + \frac{1}{2}n\mathrm{O}_2 \rightarrow n\mathrm{CO} + \frac{1}{2}m\mathrm{H}$$

3.6 Auto-thermic Reaction (ATR)

ATR is a process that combines SR and catalytic partial oxidation. SR is an endothermic process, whereas CPOx is an exothermic process. ATR has some distinct advantages over SR and POx. It does not require external heat, and the process is relatively cheaper. As shown below, the CO produced is converted to H_2 through the water–gas shift reaction:

$$\mathrm{CH}_4 + \mathrm{H}_2\mathrm{O} + \mathrm{Heat} = \mathrm{CO} + 3\mathrm{H}_2$$

$$\mathrm{CH}_4 + 1/2\mathrm{O}_2 = \mathrm{CO} + 2\mathrm{H}_2 +$$

$$\mathrm{CO} + \mathrm{H}_2\mathrm{O} = \mathrm{CO}_2 + \mathrm{H}_2$$

The ATR system can quickly turn on and off, and at the same time, it can produce more hydrogen than POx [22]. With less capital cost, the ATR system could be used for on-board hydrogen production from hydrocarbons. The following equation shows the ATR process for a hydrocarbon:

$$\mathrm{C}_n\mathrm{H}_m + 0.5m\mathrm{O}_2 + 0.5m\mathrm{H}_2\mathrm{O}\, m\mathrm{CO} + (0.5m + 0.5n)\mathrm{H}_2$$

where 298 K $= 0$ kJ mol^{-1}.

An ATR reactor has higher heat and power efficiency than a steam reformer reactor. A small ATR processor design is much simpler as shown in Figure 3.11.

ATR is also useful as it can produce hydrogen on-board from gasoline and other higher hydrocarbons by using suitable catalysts [23]. It is essential that an ATR system is designed considering its limitations of mass and heat transfer, which can create hotspots. The mass transfer rate of the reactant to the catalyst site controls the exothermic reaction rate on the surface of the catalysts. Moreover, while reforming methanol on Cu-based catalysts, the exothermic reaction rate is much faster than the endothermic reaction.

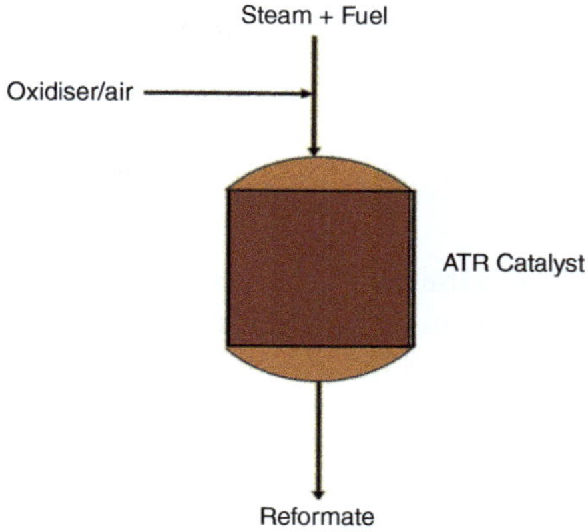

Figure 3.11 Auto-thermal reformer process for hydrogen production.

Figure 3.12 shows a comparative depiction of these major hydrogen production routes. This figure illustrates that the choice of process and operating conditions depends on specific requirements. Almost certainly, the most desirable target is to produce the maximum amount of hydrogen while reducing carbon monoxide emissions. SR creates such situations. However, SR is an endothermic process. It demands a lot of energy that should be supplied from an outside source.

It is evident from the previous discussion that natural gas is one of the potential sources of hydrogen production by using any of the three above-mentioned processes. Table 3.2

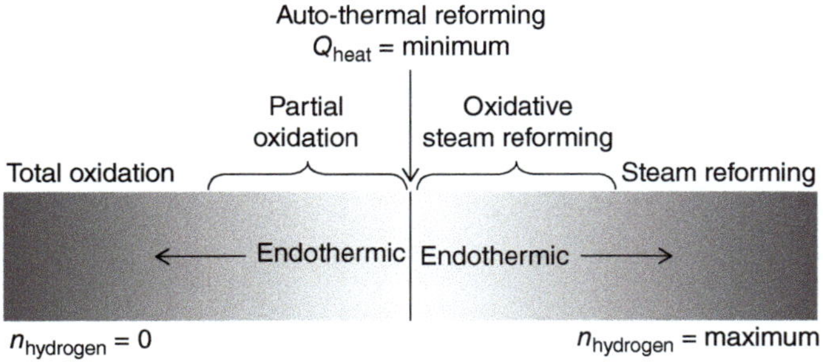

Figure 3.12 Range of operation for fuel processors for hydrogen production.

Table 3.2 Comparison of hydrogen production process.

Process of hydrogen production	Chemical process	Energy	Material	Temperature range (°C)	Efficiency range (%)
Steam methane reforming (SMR)	X	External heat	Catalyst	800–1100	70–85
Partial oxidation (POx)	Pure oxidation	X	X	950–1500	55–75
Auto-thermic reaction (ATR)	Pure oxidation	X	Catalyst	700–1000	60–75

shows comparative evidence of these merits, indicating requirements and outcomes in terms of efficiency and operating parameters.

3.7 Hydrogen from Coal Gasification

Coal is the most abundant fossil fuel with hydrogen production potential, and gasification is the route to this conversion process. Gasification is the process in which coal or biomass is converted into gaseous components by applying heat under pressure and steam. It also produces synthetic gas by using air, oxygen or water vapour. It is possible to use several types of raw materials, such as coal, sawdust, wood, car tyres, sewage, a host of biomass and plastic waste, for gasification. This will generate a product gas that includes some or all of the constituents. As such, waste-to-wealth, which is the new grail for energy industries, has been an old-fashioned concept for centuries, while gasification is a modern methodical technological process to optimise this old-age concept. This mixture of gases needs to be cleaned up to separate out useful gases, such as H_2, CH_4 and CO. However, the amount of hydrogen obtainable from gasification depends upon the characteristics of the feedstock and the gasification temperature mode. In addition, it depends on the efficiency of the clean-up mechanism. There are several modes of gasification, such as fixed bed gasification, moving bed gasification, fluidised bed gasification, entrained flow gasification and plasma gasification. In these modes, system designs and operating conditions are different. The most common feedstocks are coal and biomass. Figure 3.13 shows a general schematic of the gasification process. Gasification achieves a higher conversion efficiency than other modes of solid fuel conversion technologies [24].

3.7.1 Types of Coal Gasification

A high-temperature entrained flow is preferred because the conversion of carbon to gas is maximum, preventing the formation of tar, char and phenol. It has been found to be the cleanest and most efficient gasification process among existing systems. There are two types of coal gasification: surface coal gasification (SCG) and underground coal gasification (UCG). SCG has been around for a long time, whereas UCG is a relatively new technology for producing hydrogen.

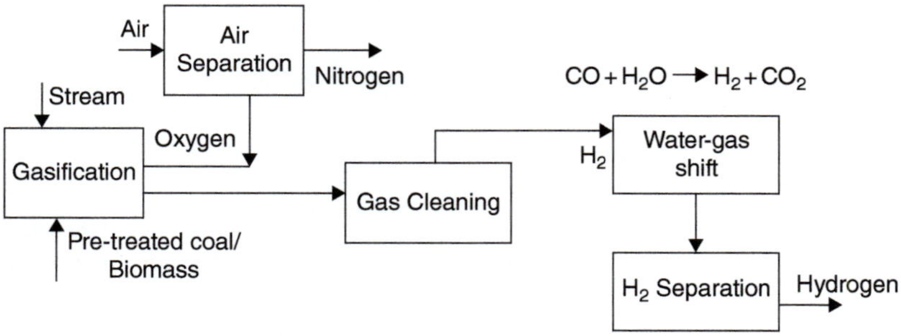

Figure 3.13 Schematic diagram of the gasification process. Adapted from [24].

3.7.2 Mechanism of Hydrogen Production by Gasification

The basic mechanism for hydrogen production through gasification is the same as that described earlier. In this process, coal is fed into a gasifier where it reacts with oxygen and steam at a relatively high temperature. The following equation shows the formation of carbon monoxide and hydrogen from coal:

$$C(s) + H_2O + heat \rightarrow CO + H_2$$

This is an endothermic reaction, and additional heat is required for methane reforming. As discussed earlier, the water–gas shift reaction converts CO into CO_2 and H_2:

$$CO + H_2O \rightarrow CO_2 + H_2 + heat$$

Finally, CO_2 is removed from the mixture. The process is usually carried out in a PSA unit. PSA facilitates calcium carbonate formation by combining CO_2 with calcium oxide (CaO). Thus, hydrogen of very high purity is obtained. In many countries, the coal available has a high sulphur content, and it is imperative to remove the sulphur from the coal. The hydrogen production efficiency by the coal gasification process is about 60–75% compared to 70–85% achievable by the SMR method.

One of the major concerns with coal gasification is its damaging effect on the environment. In an entire life cycle analysis of hydrogen production, it has been observed that coal gasification produces about twice the amount of CO that is produced through the SMR process. Of all hydrogen production processes, coal gasification has the highest global warming potential (GWP). GWP is calculated in terms of CO_2 equivalent as the common scale to evaluate the climate effect. Moreover, due to sulphur in the coal feedstock, it has a high acidification potential (AP). The amount of CO_2 produced during the hydrogen production process is dependent upon the characteristics of the feedstock. To control CO_2 emissions, some decarburisation methods need to be adopted. Researchers have investigated the carbon capture and storage (CCS) technology to reduce GHG emissions. Ash agglomeration with fluidised bed gasification and appropriate CCS has shown very encouraging results where GHG emissions have been substantially reduced. Effective coal gasification is environmentally better than conventional coal-based thermal power plants with almost 20% less CO_2 emissions.

CO_2 can be used in an energy conversion process for electricity. The other option is to store CO_2 in oil and gas fields. The present trend of CO_2 storage in depleted/depleting oil fields is a very promising technique as it is relatively less expensive and risky.

3.8 Underground Coal Gasification

The use of UCG seems to be a very useful technology and a very promising future option for 'unworked coal' at present. In this process, the conversion of coal to syngas takes place (which consists of hydrogen, carbon monoxide and methane). UCG has the distinctive advantage of on-site coal gasification. Ignition is achieved by introducing the required amount of steam, and oxygen is introduced into the steam. Thus, situations leading to partial combustion are created by mixing coal with oxygen, air or steam at very high temperatures. This mode of gasification can be used for several purposes, including the production of hydrogen. Apart from the biggest advantage of its potential to link carbon capture and

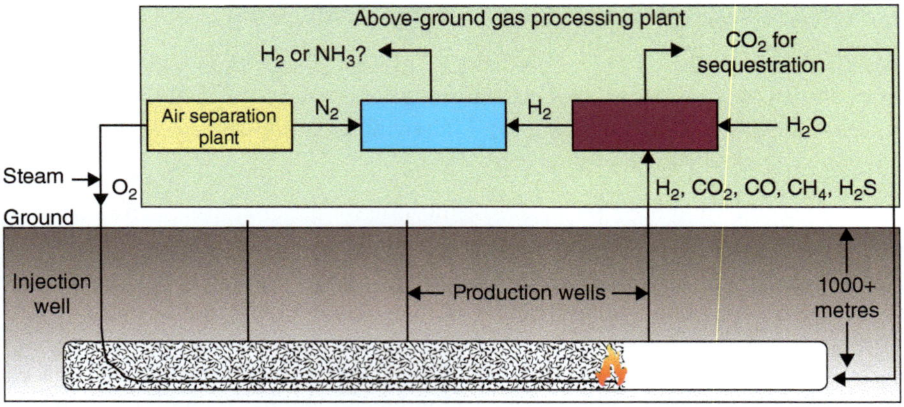

Figure 3.14 Underground gasification.

sequestration, it is relatively cheaper. There is a rapid surge of interest in the UCG technology in view of the depletion of oil and gas coupled with global climate change. In view of these strong positive features, R&D activities are being pursued to assess the possibility of generating commercially competitive syngas using low-grade coal from inaccessible locations. The characteristics of the feedstock, the design of the gasification system and the operating conditions are closely studied to optimise the process. Figure 3.14 shows a schematic diagram of the UCG processes in sequence.

In certain locations, particularly near coal seams, UCG could be a very attractive option from the point of view of economic viability. The available high pressure deep in the underground will require a small plant that is not expensive. Moreover, syngas captures CO_2 at a higher level. Researchers have suggested a process of converting deep unminable coal to syngas by the UCG method and then obtaining hydrogen from it by adopting well-known conventional technologies [25].

In their data-intensive model analysis, the life cycle of GHG emissions was assessed for hydrogen production by the UCG method. CCS was not considered in this analysis. The results showed that the life cycles of GHG were 0.91 and 18.00 kg CO_2eq/kg H_2 in H_2 production from UCG with and without CCS, respectively [25], thereby illustrating the advantages of UCG with CCS.

3.9 Hydrogen Production from Biomass

Solid waste and various types of biomass are responsible for the highest anthropogenic methane emissions into the atmosphere. Reformation of landfill gas (LFG), produced from municipal solid waste (MSW), is a promising route for hydrogen production. There are several other methods to generate hydrogen from different types of biomasses. Apart from MSW, biomasses of different kinds are available from a wide range of sources. These include wood, grass, agricultural products, crops and residues, wastes from plants and animals, forestry crops and residues, the biodegradable fraction of industrial residues, and a host of others. However, hardly any major commercial plants produce hydrogen from biomass on a large scale. With the development of appropriate technology, it is possible to obtain an adequate amount of hydrogen from biomass and fuels derived from biomass.

3.9.1 Thermochemical Conversion of Biomass to Hydrogen

The thermochemical and biochemical processes of microorganisms are the main routes of hydrogen production from biomass. Steam gasification (direct or indirect), entrained flow gasification, gasification in supercritical water, application of thermochemical cycles or the conversion of intermediates (e.g. ethanol and bio-oil) are the prospective routes to generate hydrogen from biomass. At present, most biomass feedstocks are heterogeneous. The quality and quantity of fuel hydrogen produced depend on a number of factors, such as the consistency and homogeneity of the biomass feedstock, geographical location and climatic conditions. It also depends upon transport and handling logistics. An appropriate understanding of all these factors provides guidelines for identifying the method to adopt for producing fuel from hydrogen. Figure 3.15 shows the thermochemical route to hydrogen production.

3.9.2 Gasification of Biomass

Although coal gasification is much more established and has been adopted for a long time, biomass gasification is considered a route for hydrogen production with high potential. There are some specific crops, along with agricultural and forest product residues, that can be used in the gasification process coupled with the water–gas shift reaction. The designs of gasifiers with different configurations and factors affecting gasifier performance are very significant parameters. As far as the production of hydrogen is concerned, it is necessary that the syngas exiting the biomass gasifier should be able to be used in a thermal

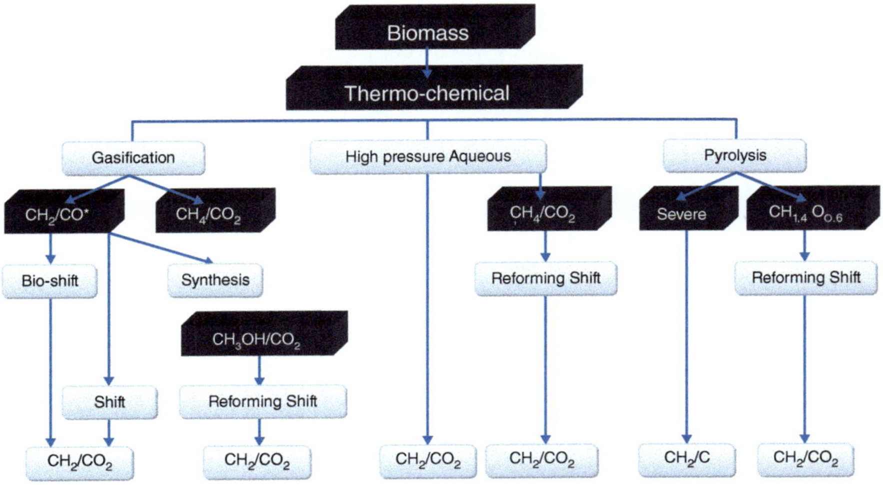

Figure 3.15 Thermochemical conversion of biomass for hydrogen production [15], Deqing et al., 2022/with permission from Elsevier.

environment with a temperature of about 800–900 °C and a pressure level of about 20–30 kPa. The production of high-quality gas and a techno-economic analysis are needed. As the technology is gradually evolving towards a mature stage, it is paramount that the system is designed for a drastic reduction in tar formation. Operational conditions must be optimised for improved performance depending on the characteristics of biomass. Life cycle sustainability, techno-economic analysis and, last but not least, socio-economic impacts must be carefully reviewed to pave the way for commercialisation [26]. Investigations have been carried out to assess the environmental features and efficiency of producing hydrogen from biomass via two processes: biomass gasification followed by reforming of the syngas and gasification followed by electrolysis to generate electricity [27]. It has been observed that gasification by SR in conjunction with the PSA route is the most energy-efficient system. However, further study of life cycle analysis shows that biomass gasification followed by electricity generation through electrolysis has better environmental characteristics.

Wang and his team [28] have already reported that the thermochemical route of renewable hydrogen production is economically viable. A schematic diagram of the gasification of biomass is shown in Figure 3.16. Several investigators working in this area have already achieved about 50% efficiency from biomass gasification. It is observed that ample scope still exists for further improvement in the technology, leading to higher efficiency and wider application. As shown in the equation below, steam is used to gasify by mixing with oxygen and air. Steam has been found to be an excellent agent for ensuring a better reforming reaction. As a gasification agent, it ensures the highest H_2/CO ratio and least impurities:

$$\text{Biomass} + H_2O \rightarrow H_2 + CH_4 + CO + CO_2 + \text{Tar} + \text{Char}$$

Several types of biomass-based feedstocks, such as rice husk, sugarcane press-mud bark, sorghum, algae biomass, etc., have been used in gasification. Parameters, such as the nature of biomass, mixing ratio, etc., are critical to the gasification process depending upon the type of biomass. Researchers have adopted several ways to improve product composition by increasing hydrogen. The exact nature of the problem is highly dependent upon the particular type of biomass. There are times when very high levels of tar are formed due to a complex biomass composition, and under certain circumstances, the catalysts do not work and some biomass impurities are much higher than others. Sometimes, tar does not

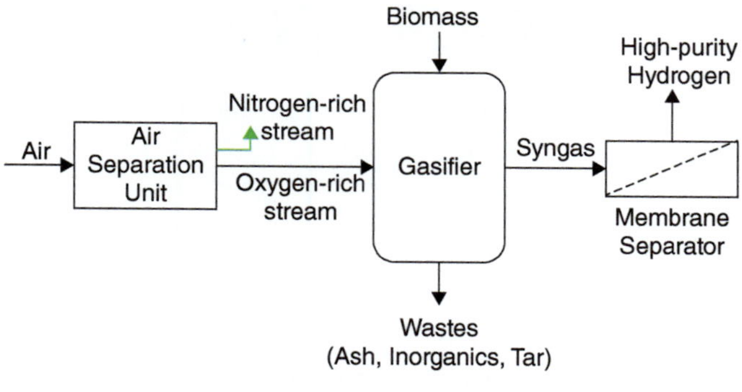

Figure 3.16 Hydrogen production by biomass gasification.

decompose at that temperature and causes clogging. Extended operation leads to wear and tear problems. Tars decompose only at temperatures higher than 1250 °C. Generating energy from trash is a very beneficial technology. However, when applied at a massive scale, noxious pollutants and toxic chemicals will be emitted, causing grave environmental problems.

3.9.3 Plasma Gasification Process

The plasma gasification process (PGP) is the latest technology to address these environment-related problems. PGP is sometimes called 'plasma arc gasification' because it works on the principle of creating a high-energy electrical arc between two electrodes. Hot gases from the plasma torch or arc interact with the gasifier at a high temperature of between 1500 and 5000 °C depending upon the characteristics of the feedstock. Plasma gasification can be applied to a wide range of feedstocks regardless of composition, size and moisture content. In fact, almost any material subjected to PGP is broken down to its elemental components in the gaseous form. During biomass plasma gasification, organic components are converted into organic gases, whereas inorganic components form a vitreous slag. Because of the high temperature, no tar or furans are produced. Some studies have been carried out in which a process has been developed to simulate the biomass PGP considering the effect of some relevant parameters, such as gasification temperature, equivalence ratio (ER) and steam-to-biomass ratio (SBR), on fuel gas composition [29]. The results of the analysis are in close agreement with the experimental data, which show that 'low temperatures are more favourable for hydrogen production', whereas the ER 'shows a negative effect on hydrogen production' [29].

3.9.4 Pyrolysis of Biomass

Biomass pyrolysis for hydrogen production usually takes place without oxygen except in situations where partial combustion needs thermal energy to perform the process. In a recent paper, Lopez et al. [30] reviewed the process of hydrogen generation by pyrolysis. This publication also highlights limitations and discusses ways to overcome them. As discussed later in detail, the choice of an appropriate catalyst is very critical for producing hydrogen of high purity.

An effective pyrolysis process for biomass depends upon the characteristics of the biomass, pyrolysis temperature and heating rate. It also depends upon the types of pre-treatment, such as physical, chemical, biological, etc. During the process, temperature and the associated thermal environment produce hydrogen, methane, CO and CO_2. Biochar and bio-oil are also produced under some circumstances. Bio-oil and biochar reduce the efficiency of the pyrolysis process. Gases are mostly formed at temperatures greater than 800 °C, and biochars are formed at temperatures less than 450 °C. Bio-oil is a major product in intermediate temperature ranges with high heating rates. To produce hydrogen using biomass pyrolysis, the thermal environment must be closely monitored. A large number of research results are available on bio-oil production and pyrolysis of char. Technical discussions related to biomass pyrolysis kinetics and the effects of temperature, particle size, catalyst effect, etc. have been studied. The present descriptions only cover hydrogen production.

Wang and co-investigators have reported interesting research results on bio-hydrogen production from organic waste [28]. Methods of sustainable production of hydrogen and chemicals from biomass-derived polyols have been reported by Xu et al. [31].

This paper describes the processing of biomass-derived C6 and C5 polyols, such as sorbitol and xylitol, for producing several chemicals, fuels and hydrogen. Relevant information is provided for the development of appropriate catalysts and greener chemical processes that can be useful for the valorisation of sorbitol and xylitol derived from biomass resources [31].

3.9.5 Supercritical Water Gasification of Biomass (SWGB)

A supercritical fluid is referred to as a fluid whose temperature and pressure are beyond its critical point and where it exists as neither in a liquid state nor a gaseous state but as a distinct state with unique properties. Water temperature and pressure conditions are 374 °C and 22.1 MPa at the critical point. Biomasses are of various types. However, it has been found that major constituents of the specific biomass, such as carbohydrates, lignin, ash, proteins and lipids, greatly influence the chemical reaction. Apart from the reaction conditions, heterogeneous catalysts have also been observed to be very relevant to hydrogen production. The supercritical water gasification (SCWG) process involves feeding biomass, such as agricultural residues, forestry wastes, etc., under stringent conditions in a high-pressure reactor containing supercritical water. Water acts both as a reaction medium and as a reactant. SWGB is a very promising technology for hydrogen production because it does not need drying pre-treatments, which means algae and sewage sludge; even with their high moisture contents, it can be used for SCWG [32].

Supercritical water gasification is often carried out with biomass, and the process can be referred to as SWGB in the in abbreviated form. Biomass contains cellulose, hemicellulose and lignin, so their degradation must be carefully performed. Several researchers have reported their results on SWGB [31, 33, 34]. It has been observed that both pressure and temperature significantly affect hydrogen production. At higher reaction temperatures, gasification becomes easier and more effective. A potential future technology for the acid hydrolysis treatment of feedstock could be hydrothermal gasification as part of a biorefinery that uses hydrogen produced internally. Features such as diffusivity, high reactivity and solubility of supercritical water help in easily mixing organics and gases. Versatile feedstock consisting of a wide range of biomass makes it more flexible than some other biomass conversion processes. Therefore, the homogeneous phase facilitates reactant conversion. Figure 3.17 shows a schematic diagram of the major pathways leading to conversion.

It is evident from the above figure that biomass in supercritical water undergoes a series of complex reactions, such as pyrolysis, hydrolysis, condensation and dehydrogenation [35].

In this step, the large molecules are hydrolysed into small organic molecules. The overall chemical reaction is given by:

$$CH_xO_y + (2-y)H_2O \rightarrow CO_2 + (2-y+0.5x)H_2$$

The process has a higher conversion efficiency and can be performed at a lower temperature than steam gasification. Although the process has been technically well proven on the laboratory scale, it has not yet been commercialised. This is mainly because of the high cost

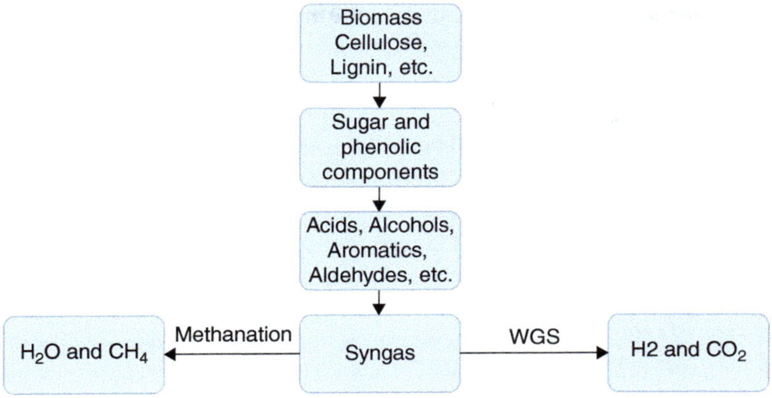

Figure 3.17 Stages of SWGB.

involved in developing industrial conditions. To scale up the technology and ensure early commercialisation, it is critical that parameters such as pressure, temperature, feedstock characteristics and concentration of reactants, residence time, and choice of appropriate catalysts need to be optimally integrated.

The optimum reactant concentration is crucial for effective SWGB. When the reactant concentration with large amounts of biomass is increased, the water content is reduced, affecting the steam-reforming reaction. On the other hand, a high reactant concentration leads to several undesirable intermediate products, such as coke, that often clog the reactor.

The critical residence time also plays a significant role. In certain cases, the extended residence time helps in completing the reaction, which helps in higher conversion. However, if the residence time is unduly extended beyond the critical residence time, it leads to the methanation reaction. In the same way, selecting suitable catalysts is very crucial as different catalysts produce different effects. Moreover, high-pressure and high-temperature conditions require robust reactor materials and safety measures. In certain cases, the process could be energy intensive depending upon the feedstock and operating conditions.

3.10 Biological Production of Hydrogen

Khan et al. [36] discussed the potential of different anaerobic systems for hydrogen production. Further research is needed to identify long-term-feasible solutions for sustainable production from various substrates.

The process and technology of producing biogas (with methane as a major constituent) from anaerobic systems is very well known. It has been reported that it is possible to produce increased hydrogen concentrations by 'inhibition of microorganisms that lower hydrogen concentrations and the constant removal of hydrogen to promote hydrogen-producing bacteria' [37]. Studies have demonstrated that the process is environmentally friendly and economically viable. Further studies may provide relevant guidelines for the development of large-scale systems. The production of hydrogen through the biological route has been known for a long time. Researchers have investigated the performance parameters necessary for an optimum system design depending on the feedstock and operating conditions. The limiting factors preventing process improvement have also been identified [38].

One of the major obstacles is the low energy content of solar irradiation, which governs photosynthetic processes. In addition, the cost of the photobioreactor is very high, and the conversion efficiencies for direct bio photolysis are as low as about 1% only. Figure 3.18 provides a general overview of the several biological routes of hydrogen production from biomass. Broadly speaking, these processes involve organic compounds and microorganisms. The principal biological processes are (i) anaerobic digestion and (ii) fermentation.

Microorganisms such as microalgae and cyanobacteria contain certain hydrogen-producing enzymes with very different catalytic activities, which cause biological conversions such as biophotolysis, photo-fermentation and dark fermentation [28, 39]. Biotransformation can broadly be divided into three categories: (i) biophotolysis, (ii) photo-fermentation and (iii) dark fermentation.

3.10.1 Biophotolysis

Biophotolysis is an environmentally friendly process as it uses water as a raw material. There are two types of biophotolysis: direct and indirect. In the direct photolysis process, hydrogen is produced by water decomposition. Microorganisms use light to facilitate this decomposition, whereby light energy is converted into storable chemical energy. Direct photolysis occurs in organisms that can perform photosynthesis, such as algae, which have chlorophyll and the process requires enough light. However, indirect photolysis is carried out in two stages. In the first stage, CO_2 is fixed as carbohydrates that are broken down by hydrogenising enzymes to produce hydrogen [40]:

$$12H_2O + 6CO_2 + \text{light} \rightarrow C_6H_{12}O_6 + 6O_2$$

$$C_6H_{12}O_6 + 12H_2O + \text{light} \rightarrow 12H_2 + 6O_2$$

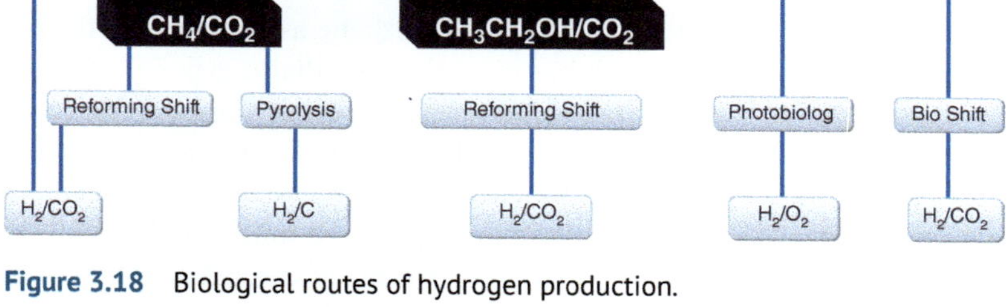

Figure 3.18 Biological routes of hydrogen production.

Excessive oxygen sometimes deters this process. Biophotolysis slows down as it produces hydrogen and oxygen, and the oxygen concentration increases as the reaction proceeds. Research efforts are being carried out to ensure sustained hydrogen production by providing cells with an intermittent sulphur-deficient environment to partially inactivate oxygen [41, 42].

Further studies are being carried out to improve the conversion efficiency for direct biophotolysis, which is very low currently. Hydrogen production by biophotolysis based on microalgal photosynthesis and research on the hydrogen metabolism of marine microalgae and photosynthetic bacteria are being actively investigated. Figure 3.19 illustrates photobiological hydrogen production.

3.10.2 Photo-fermentation

Photo-fermentation and dark fermentation are suitable processes and can be preferred to biophotolysis because they produce hydrogen and consume waste. During the photo-fermentation process, organic substrates are converted into hydrogen and CO_2 by microorganisms with light energy. This is shown in Figure 3.20.

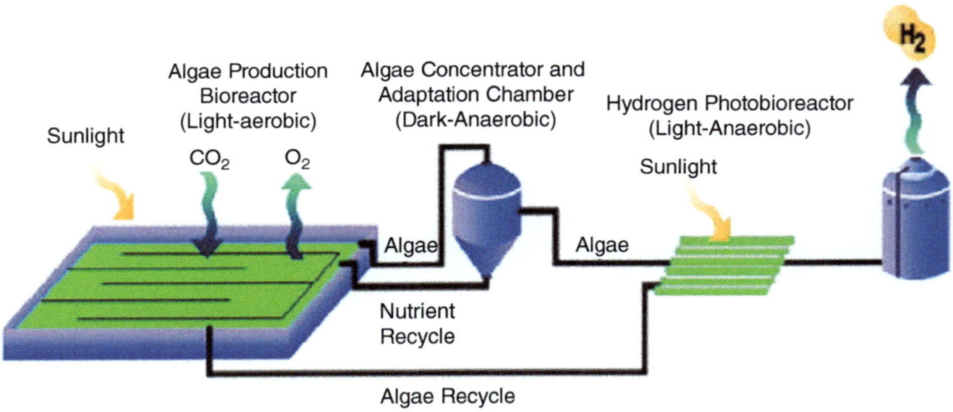

Figure 3.19 Photobiological hydrogen production.

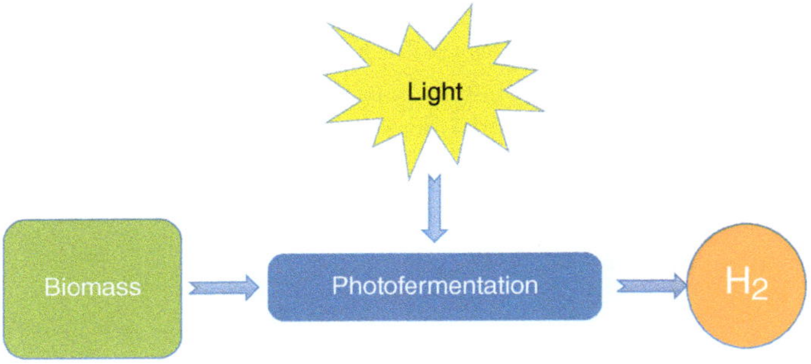

Figure 3.20 Photo-fermentation process.

Two enzymes, namely nitrogenase and hydrogenase, are used for bio-hydrogen production. It has been observed that purple non-sulphur bacteria (PNS) can significantly contribute to biological hydrogen production potential. PNS bacteria depend upon nitrogenase to catalyse the following reaction in a nitrogen-efficient environment [43]:

$$CH_2O + H_2O \rightarrow 2H_2 + CO_2$$

External conditions show significant impacts on the biochemical process, so the culture environment for bacteria and light conditions need to be strictly controlled to ensure smooth hydrogen production. It has been reported that the suitable temperature and light wavelengths were 30–36 °C and 400–1000 nm, respectively. The light should be spatially uniformly distributed throughout the bioreactor. In addition, a strictly anaerobic environment and a near-neutral pH of 6.8–7 are also necessary conditions.

It has been observed that hydrogen production through this route can be effectively conducted only if the appropriate culture environment for bacteria and illumination conditions are properly monitored. Stringent conditions need to be maintained for an effective biochemical process. The temperature should be within 30–36 °C. The light wavelength must be in the range of 400–1000 nm, and the light distribution must be spatially uniform. The entire process needs to be carried out in an almost neutral pH condition of around 6.8–7 and in an anaerobic environment [44, 45].

Nevertheless, there is a lot of scope for further research to identify the optimum environmental conditions and the problem of low energy conversion efficiency.

3.10.3 Dark Fermentation

Dark fermentation, as its name indicates, does not require light for hydrogen production. In this method, microorganisms such as bacteria and microalgae break down organic matter into small molecules under anaerobic conditions. They produce hydrogen without light. It is a complex process that utilises a variety of carbon sources (including biomass waste), and biochemical reactions are carried out in three stages. In the first stage, organic matter is decomposed into soluble carbohydrates. In the next stage, they are hydrolysed into simple carbohydrates such as monosaccharides. In the final stage, organic acids, alcohols, H_2 and CO_2 are produced. Acetic acid and butyric acid are also produced as by-products of these reactions [46].

Dark fermentation only uses a small fraction of energy in the material, resulting in a low conversion efficiency. Most energy is transferred to intermediate products of metabolic reactions. Thus, it can be further improved if these by-products are used [47]. Unfortunately, some of these by-products deter hydrogen production because they interfere with microorganism metabolism. Research and development activities are being pursued to explore the possibility of modifying microorganisms genetically [48].

3.10.4 Combined Dark–Photo Co-fermentation

Hydrogen production is improved by combining the dark fermentation and photo-fermentation techniques. The basic approach to this mode is to ensure that both processes mutually complement each other. The shortcomings of one process can be useful for the benefit of the other. In this way, it is possible to enhance hydrogen production efficiency and use a wider range of substrates for hydrogen production.

The photo-fermenting mechanism handles shorter organic molecules, whereas larger complex organic molecules are handled by the dark fermentation process. It may be relevant to mention that short organic molecules cannot be decomposed by dark fermentation. Microbial co-culture and two-step hydrogen production are the significant stages of this combined process of deep photo-fermentation. Two more microorganism species have been successfully cultured in the laboratory. Detailed studies on co-culture related to bio-hydrogen production have been reported in the literature [49–51].

Co-culture has inherent limitations. It is not possible to inoculate microorganisms with widely different growth patterns into one system and expect them to efficiently contribute to hydrogen production. Perhaps it is easier and more feasible to adopt a two-step hydrogen production process by alternating dark fermentation and photo-fermentation.

3.11 Hydrogen Production Based on Electrolysis

Some of the major routes of hydrogen production have been described earlier. Sources such as natural gas, biomass and coal have been discussed. It is perhaps worthwhile to remember that water electrolysis is one of the most well-known hydrogen production processes. Figure 3.21 [52] provides an idea about different sources of hydrogen production. It is clear that electrolysis has a major contribution to hydrogen production.

Electrolysis of water is the first and most common hydrogen production method. In the course of time, as the wider application potential of hydrogen was gradually realised, other routes of hydrogen production, such as SMR, partial oxidation, coal gasification, etc., were tested in the lab and gradually entered the field. The well-known method of hydrogen production by electrolysis received renewed interest during the energy crisis in the 1970s. Numerous research and development activities have been conducted on the electrolysis

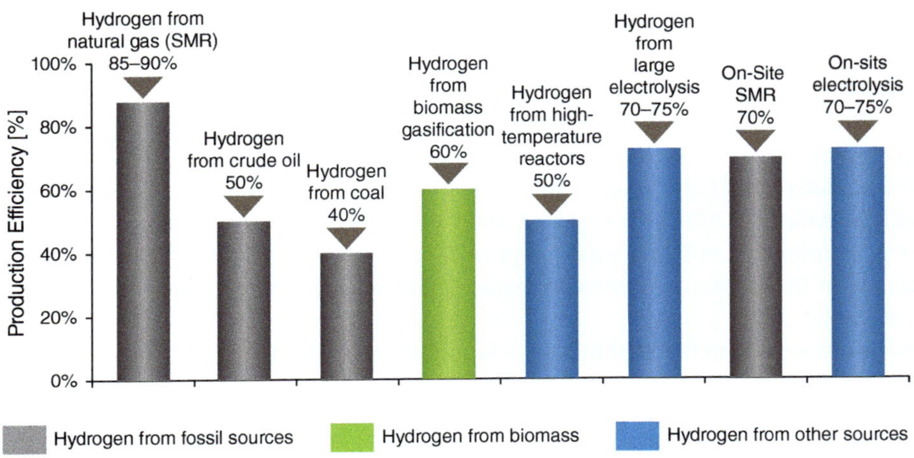

Figure 3.21 Energy chain efficiency of sources to hydrogen fuel [52]/European Union/Licensed under CC BY 4.0.

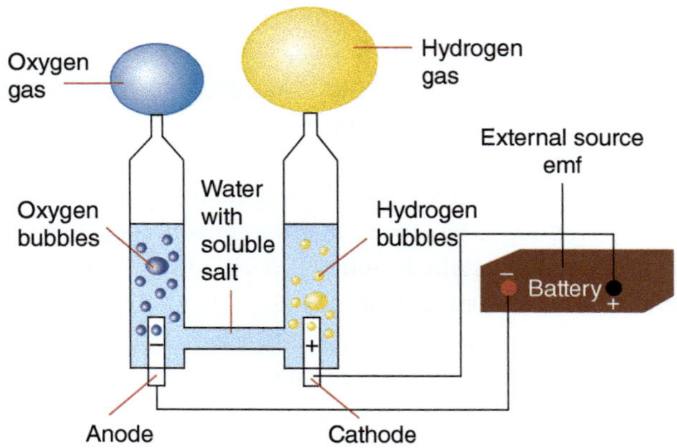

Figure 3.22 Electrolysis of water.

process using fossil fuels and local biomass sources. Figure 3.22 shows a schematic flow diagram of a well-known conventional electrolysis process.

$$2H_2O \rightarrow 2H_2 + O_2$$

In the meantime, several routes of water electrolysis have been developed, which are carried out in electrolysers using different types of electrolytes. The electrolysis process can be divided into the following types: alkaline electrolysis (AEL), polymer electrolyte membrane (PEM) electrolysis, solid oxide electrolysis (SOEL) and alkaline anion exchange membrane (AEM) electrolysis.

3.11.1 AEL and PEM Electrolysis

AEL and PEM electrolysis are mature and well-proven technologies with a long industry history.

Braun et al. review AEL and renewable energy systems [53]. This analysis has focused on very prospective sources such as solar and wind energy. A comparative evaluation of the effectiveness of water electrolysis by renewable energy has been made with conventional fossil fuel electrolysis, and it has been suggested that each component of the hydrogen energy system based on renewables needs to be optimised to match the system efficiency of fossil fuel electrolysis [53]. Changqing Li and Jong-Beom Bae discuss the environmental benefits of AEM and PEM electrolysis. AEM electrolysis can be integrated with intermittent energy resources and produces high-purity hydrogen gas used in fuel cells. The AEMs developed in the laboratory show comparable performance with commercially available systems [54].

Test results have been reported by researchers after producing high current density AELs based on Pt/C cathodes and stainless-steel anodes [55]. Tests with a catalyst coating of the Pt/C cathode showed that it was possible to achieve 'optimal AEL performance and nearly zero average voltage stability' [55] with 25 wt.% Nafion binder. Apart from binder content optimisation, studies were also carried out by investigators in other areas, such as AELs using nanostructured catalysts coating, the economic viability of the AEL technology and several others. Tests have been reported on highly hydrophilic separators for alkaline water electrolysis. These separators were 'prepared by plasma-initiated grafting of acrylic acid on porous polypropylene (PP) membranes' [56]. In this system, membranes were activated, and the results of the analysis favour acrylic-acid-grafted microporous membranes as excellent separators for alkaline water electrolysis. AEL is characterised by two electrodes immersed in a high-concentrate liquid alkaline solution, which is either KOH or NaOH. These electrodes are separated by a selective porous diaphragm, thereby separating the product gases and transporting the hydroxide ions from one electrode to another. It is relevant to mention here that the porous diaphragms should be a highly wet table in the AEL or else bubbles would be produced and increase the resistance. Thus, it plays a critical role in separating the hydrogen and oxygen produced and hence affects the purity of the gas produced. Earlier chrysotile asbestos was used. However, now it has been banned as it is carcinogenic in character. Another performance limitation of chrysotile asbestos is that it is not resistant to alkaline solutions above 80 °C. A novel porous separator material called 'Zirfon' [57, 58], which has been synthesised from zirconium oxide nanoparticles and polysulphone, has been observed to be extremely effective. Studies show that, on the basis of diaphragm resistance, resistances in zero-gap alkaline electrolysers with the Zirfon separator show higher values [57]. On the basis of the simulation analysis by researchers, it is believed that one of the causes of the increased resistance could be due to the uneven current distribution. Studies have shown that the Zirfon separator remains stable in concentrated KOH solutions even at relatively high temperatures [59]. The characteristics of Zirfon separators have also been studied for AELs with respect to the diffusivity of hydrogen and electrolyte permeability [60]. Specific Zirfon separators with reduced average pore sizes have been developed and commercialised to ensure more controlled load operation of the alkaline electrolyser system [61]. Further improvements in its performance have been found by changing the raw material ratio. Tests carried out by Burnat et al. [62] on a composite membrane of mineral filler and polysulphone have shown that it is possible to achieve higher purity in hydrogen thus obtained than that produced using a Zirfon membrane, and tests are being conducted on several polymeric materials with high stability in alkaline liquids.

The basic electrochemical reactions that guide PEM electrolysis and AEL are similar. However, PEM electrolysis uses an acidic polymer membrane as a solid electrolyte, and thus the process is carried out in an acidic environment. The terms 'proton exchange membrane' and 'polymer electrolyte membrane' are used interchangeably in electrolysis. The terms refer to the same type of membrane used in the same type of function (electrolysis processes), specifically in PEM electrolysis or polymer electrolyte membrane electrolysis. In both electrolysis processes, the membrane serves as a critical component that separates the hydrogen and oxygen gases produced during the electrolysis of water. The membrane is a thin, selectively permeable layer made of a special polymer material that allows only protons (positively charged hydrogen ions) to pass through while blocking the migration of other species such as oxygen and electrons. As far as sustainability and environmental impacts are concerned, PEM is considered one of the most promising technologies. However, AEL has some advantages of economy and flexibility. PEM electrolysers use expensive noble metals such as iridium, ruthenium and platinum as electrodes, whereas nickel- or iron-based electrodes are used in AELs. Moreover, PEM electrolysis equipment has a relatively shorter service life than AELs because of degradation of PEM electrolytes during operation. All these factors contribute to substantial cost reductions [63–65].

Research and development efforts are being made for non-noble materials and catalysts to reduce costs. Non-fibre membranes have been tested in electrode assemblies [66, 67]. The effectiveness of several other catalysts, such as cobalt chelate and oxide-based catalysts, Pt-black, phosphides and carbides of transition metals, is also being studied, and the effect of molybdenum disulphide (MoS_2) as a catalyst has also been reported [68–70].

3.11.2 Alkaline Electrolysis

In the meantime, several routes of water electrolysis have been developed, which are carried out in electrolysers using different types of electrolytes. The process of electrolysis can be divided into several types, such as AEL, PEM electrolysis, SOEL and alkaline AEM electrolysis.

These processes of AEL and PEM electrolysis have been commercialised. Alkaline electrolysers have been used for a long time in the chemical industry. This electrolyser uses an aqueous KOH solution (caustic) as an electrolyte that circulates through the electrolytic cells:

$$\text{Cathode: } 2H_2O + 2e^- \rightarrow H_2 + 2OH^-$$

$$\text{Cathode: } 2H_2O + 2e^- \rightarrow H_2 + 2OH^-$$

$$\text{Sum: } 2H_2O = O_2 + 2H_2$$

AEL is characterised by two electrodes immersed in a high-concentrate liquid alkaline solution, KOH or NaOH. These electrodes are separated by a selective porous diaphragm, thereby distinguishing the product gases and transporting the hydroxide ions from one electrode to another. Here, it is relevant to mention that the porous diaphragms should be highly wettable in AELs or else bubbles would be formed and increase the resistance. Thus, it plays a critical role in separating the hydrogen and oxygen produced and thus affects the purity of the gas produced.

Previously, chrysotile asbestos was used, but now it has been banned as it is carcinogenic in some way. Another performance limitation of chrysotile asbestos is that it is not resistant to alkaline solutions above 80 °C. Laboratory tests are being conducted on several polymeric materials with high stability in alkaline liquids [71].

3.11.3 Polymer Electrolyte Membrane Electrolysis

Based on sound technological features, PEM electrolysis can be used for both mobile and stationary applications. The chemical principle of PEM electrolysis is provided in the following equations:

$$\text{Anode: } H_2O \rightarrow 1/2O_2 + 2H^+ + 2e^-$$

$$\text{Cathode: } 2H^+ + 2e^- \rightarrow H_2$$

Problems related to low current density and low-pressure operation and issues of partial load, which are the major shortcomings of alkaline electrolysers, are of no concern for PEM electrolysers. However, the limited lifespan of membranes is one of its major constraints. The system becomes more expensive because critical design considerations due to higher density and higher operating pressures add to the cost. Thus, relatively higher cost, poor efficiency and short lifespan of alkaline electrolysers have proven to be the primary obstacles to the commercialisation of this technology. Currently, research is being carried out to sort out these problems and explore the potential for wider applications of PEM fuel cell technology. This is performed using better materials and improved efficiency.

Although both PEM electrolysis and AEL are basically characterised by the same mechanism, PEM electrolysis occurs in an acidic environment. It uses an acidic polymer membrane as the solid electrolyte. The dominant features of PEM electrolysis are that it has a compact system along with a rapid response and a higher current density. Furthermore, PEMs are based on superior technology to AELs as far as reaction dynamics are concerned. PEMs also demonstrate slightly higher efficiency than AEL.

On-site electrolysis is a very effective hydrogen production method. Usually, in this process, 30% KOH is added to water to improve the conductivity of the electrolyte. Although this electrolyser has a lower efficiency, it is widely used as it is a very old and well-proven technology. Moreover, it is relatively less expensive than other modes. As far as hydrogen purity is concerned, it is possible to obtain extremely pure (99.99%+) or better hydrogen. However, hydrogen purity depends on the user's requirements and for which application it has been made. Figure 3.23 shows a schematic diagram of on-site electrolysis.

Solid oxide electrolyser has a distinctive characteristic so that the ceramic material used in the cells can function both as an electrolyser and as a separate membrane which separates hydrogen and oxygen. Ni et al. [72] reviewed the technological development of hydrogen production by solid oxide electrolyser cells (SOECs). It has been observed that the present level of research activities needs to be strengthened on 'modelling analyses of the

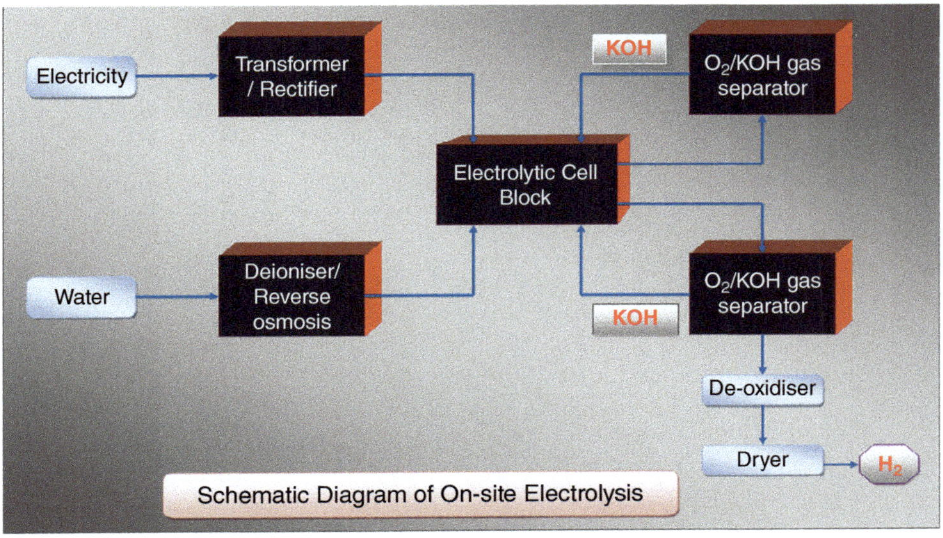

Figure 3.23 Schematic diagram of on-site electrolysis.

coupled heat/mass transfer and electrochemical reaction phenomena of the SOEC' [72]. Experimental and theoretical modelling analyses, carried out simultaneously, would be useful for efficient and economically viable processes for hydrogen production.

In SOEC, water is fed into the cathode and undergoes a water reduction reaction (WRR), thereby being converted into hydrogen gas and oxide ions. Hydrogen gas is purified and separated. Solid oxide fuel cells (SOFCs) operate in the temperature range of 800–1000 °C. Such high temperatures are essential to thermally activate oxide ion migration and ensure effective electrochemical reactions on both electrodes.

Hydrogen production by electrolysis from seawater has been extensively researched and developed due to the requirement of high-purity water for electrolysis. Khan and a team of researchers [73] have focused their research efforts on the seawater reverse osmosis (SWRO) process coupled to PEM electrolysis. They analysed a relatively unexplored area. They observed that since commercial water electrolysis consumes a lot of energy compared to SWRO, it is a very viable technology as both capital and operating costs are significantly lower. Thus, it is necessary to pursue R&D activities in the areas of a 'nascent direct seawater electrolysis technology'. In this context, it is perhaps appropriate to emphasise that the level of sodium chloride (NaCl) in seawater is an environmental threat.

Photo-electrolysis is the process where the splitting of water molecules into hydrogen and oxygen occurs in the presence of light. Photovoltaic (PV) systems coupled to electrolysers are used and offer the flexibility of obtaining electricity from photovoltaic cells or hydrogen from the electrolyser. Figure 3.24 shows the photolysis process.

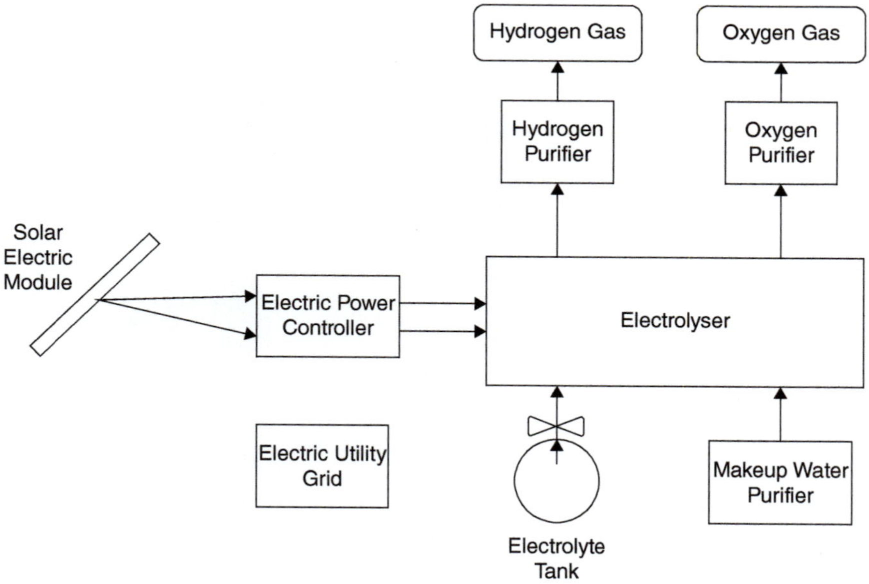

Figure 3.24 Principle of photolysis.

3.12 Hydrogen Production Using Solar Energy

Solar energy may be the single most abundant renewable primary energy source capable of providing infinite clean and sustainable hydrogen. For hydrogen production, there are three main sources of solar energy: thermal, photovoltaic and photoelectrochemical.

The present level of technology clearly shows that there are several options to produce hydrogen using concentrated solar thermal energy sources. Some of the most significant are solar thrombolysis, solar thermochemical cycle, solar gasification, solar electrolysis, photoelectrolysis and biophotolysis. Figure 3.25 shows hydrogen production thermal paths.

Figure 3.26 shows the routes of hydrogen production from solar heat by thermochemical water splitting, thermal cracking, reforming and gasification. As indicated in the figure, the range of feedstock for hydrogen generation could be fossil fuels, biofuels, biomass, chemicals and water depending upon the process used for production. In another way of producing hydrogen, solar photovoltaics generate electrical power for electrolysis. Solar gasification and solar ammonia reforming are also hydrogen production methods.

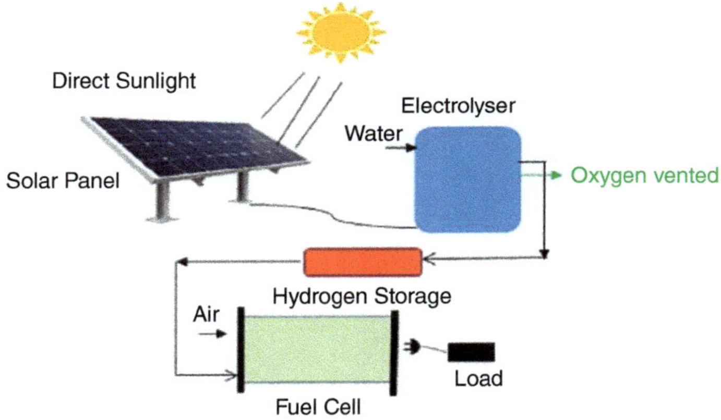

Figure 3.25 Hydrogen production through solar panels.

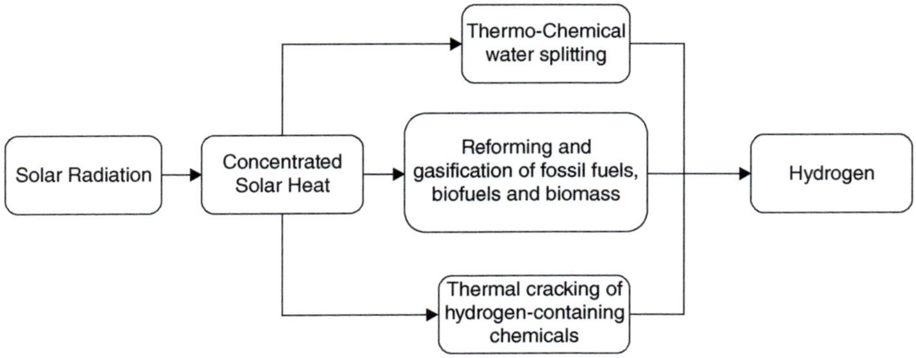

Figure 3.26 Solar thermal paths for hydrogen production.

3.12.1 Solar Thermal Methane Splitting

Abuseda et al. [74] studied the feasibility of producing high-yield hydrogen and graphite from methane using direct, concentrated solar thermal energy. Concentrated irradiation from a simulated solar source was used to convert methane into graphitic carbon and hydrogen. Methane was allowed to flow within a photo-thermal reactor under steady-state conditions, and the temperature (room temperature) was maintained. It was observed that methane conversion up to 70% was achieved even with large flow areas and low irradiation levels. In this process of hydrogen production with solar thermal splitting, the high-temperature level of concentrated solar energy is used to split methane into hydrogen and carbon. This is essentially an environmentally friendly thermochemical reaction that occurs quickly. Solar energy is also used for hydrogen production by thermal dissociation of ZnO/Zn in which concentrated solar radiation is used for hydrogen production. Rahul [75] studied solar hydrogen production by a ZnO/Zn-based thermochemical cycle. In their studies, they found that increasing temperatures from 1040 to 1340 K through several steps always showed an increasing trend in ZnO-reduction (ZnO-R). This shows a very significant effect on increasing hydrogen production by the water-splitting step. In such routes of hydrogen production by the thermal dissociation of ZnO/Z, a serious concern is that the system needs a large amount of argon for quenching Zn and O_2. An alternative method is that hydrolysis is not required to be carried out at the solar plant itself. On the other hand, Zn can conveniently be taken to the place where electrolysis is carried out. Another alternative method of hydrogen production is using the solar carbo-thermal reduction of the Zn/ZnO technique; argon could be avoided for quenching the resulting gases. By adding carbonaceous materials, such as coal or charcoal, to the reaction chamber, the solar carbothermic reduction of ZnO (SCR) is facilitated.

3.13 Solid Oxide Electrolyser

SOEL cells adopt the reverse concept of SOFCs. Kamlungua et al. [76] present an overview of this hydrogen generation method. This chapter briefly describes low-temperature electrolytic cells and solid oxide electrolytic cells. A number of other aspects of material selection, as well as the research gaps, challenges and future prospects, are discussed.

This system has the unique feature that the cells use ceramic materials that can act both as an electrolyser and as a separate membrane, that separates hydrogen and oxygen. This chapter also discusses challenges and future prospects. Reversible SOEC functions as a fuel cell for electricity generation and as an electrolysis cell for hydrogen production. Therefore, it is essential that the materials used for reversible SOECs have some specific properties such as high chemical stability in both oxidising and reducing environments. Furthermore, the electrocatalytic activity involved in fuel/electrolysis cell reactions is of extreme practical significance. However, adequate materials have not yet been identified to cater to such essential needs.

3.14 Seawater Electrolyser

In certain locations, particularly in coastal areas, abundant water available in the oceans is an everlasting source of hydrogen production. Hydrogen is produced through electrolysis.

However, the high level of sodium chloride (NaCl) in seawater creates environmental problems [77]. This study analysed the electrolysis of pure water and seawater from a hydrogen production perspective. It was observed from the current–voltage curves and the impedance spectra that the electrolysis of simulated seawater at 800 °C and the electrolysis of pure water had nearly the same performance features. Very useful observations were made in this study such that both contaminated and uncontaminated cells exhibited almost similar initial performance. Moreover, the degradation rates of the contaminated cell and the uncontaminated cell were almost the same.

3.14.1 Photo-electrolysis (Photolysis)

Photo-electrolysis is the process whereby light splits water directly into hydrogen and oxygen. At present, many research activities are being pursued in this area as a very promising route for producing hydrogen from renewable energy. Photovoltaic systems coupled to electrolysers provide the option of obtaining electricity from photovoltaic cells or hydrogen from the electrolyser. Figure 3.27 shows the photolysis process.

Studies have been conducted on optimising photovoltaic electrolyser coupling. It has been observed that although the technology of both photovoltaic and coupling topics has been well studied, the high variability of solar radiation has been the main cause of deciding an efficient relative sizing [78].

Relative sizing is very critical when coupling renewable electricity sources to water electrolysers. Modelling and simulation are used to extract a cloud of maximum power points in this context.

Solar hydrogen production systems based on water electrolysis require detailed information and guidelines on radiation and temperature conditions for PV generators. Researchers have conducted investigations [79] on hydrogen production by photolysis. Newly synthesised solid solutions of rare earth, gallium and indium oxides were used as catalysts in this study. From their photolysis tests, solid solution $Y_{0.8}Ga_{0.2}InO_3$ doped with sulphur atoms was the most effective catalyst. However, the study also included sonolysis and sonophotolysis.

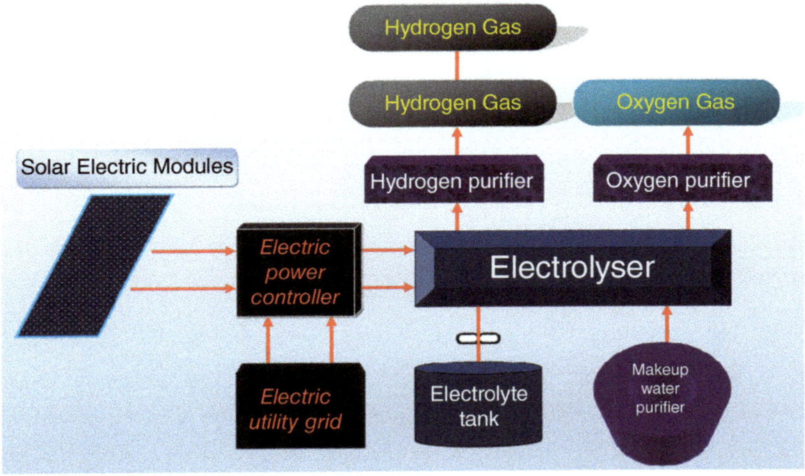

Figure 3.27 Process of photolysis.

3.15 Hydrogen Generation Using Wind Energy

Wind is also available in plenty and perpetually. The most significant part is the choice of an appropriate location for wind farms. Many countries have initiated efforts, and wind power has been successfully used to produce hydrogen. It is well known that wind turbines extract wind energy by electrolysing water to produce hydrogen in an electrolyser. There are two types of wind turbines: offshore and onshore wind turbines. As the name suggests, onshore wind turbines are located on land and use wind to generate electricity. On the other hand, offshore wind turbines are installed on open water (typically the ocean) to generate electricity. Onshore wind farms have certain advantages over offshore farms.

Although they are easy to install, the infrastructure required to transmit electricity from these turbines is relatively less than that required for offshore farms. In addition, these systems do not have any erosion problems as the moisture level in the area is low. However, one of the major shortcomings of this system is that wind directions change very frequently, and it is difficult to predict this sudden change. Such a situation results in reduced efficiency as the turbines are designed and optimised for a specific speed. Offshore wind turbines offer more reliable and efficient energy as wind speeds are higher. Offshore systems also generate more energy because of higher wind speeds. In addition, there are practically no restrictions on tall buildings as these wind farms are located far from the sea. Figure 3.28 shows the schematic diagram of a hydrogen production unit based on wind.

As indicated in Figure 3.28, electrical energy extracted from wind can be used for electrolysis and to produce hydrogen. Although adequate research and development activities have not been reported in this area, it is a promising technology for clean, efficient and sustainable energy production.

In a typical case study [80], six key criteria were adopted for the selection of locations for wind farm construction in Fars Province in Iran. Wind power density was the most critical selection parameter. This was determined from three-hour wind speed data between 2004 and 2013. In this research, the ranking order of the cities was fixed using data envelopment analysis (DEA). Based on the results of this analysis, Izadkhast city was found to be a suitable location for construction of a wind farm. It was estimated that installing a 900 kW wind turbine could provide 21.9 tonnes of hydrogen per year.

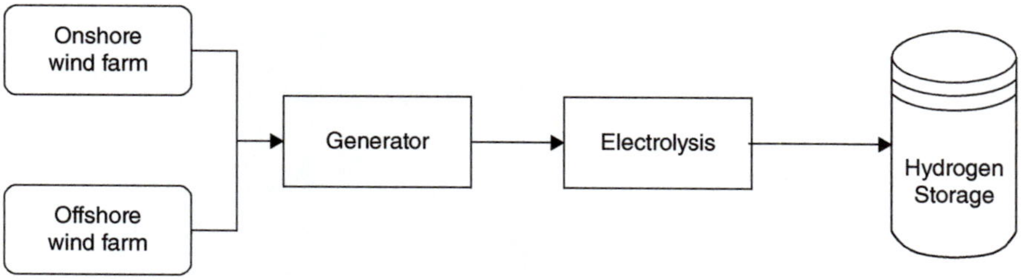

Figure 3.28 Wind-energy-based hydrogen production unit.

3.16 Ocean Thermal Energy Conversion for Hydrogen Production

Oceans, which cover 85% of the global surface, can be an attractive source of hydrogen production. A large-scale electrolysis process can be adopted for converting the power produced in an OTEC system into hydrogen. Studies are being conducted on hydrogen production by OTEC. Hydrogen energy generated through OTEC in a distant far-off location could be used for several purposes, including power generation through fuel cells.

Energy conversion from an OTEC system into hydrogen energy is an extremely promising technology. Such technology can be used for generating power through fuel cells. Researchers [81] analysed hydrogen production through an OTEC system coupled to a polymer electrolyte membrane electrolyser. The optimum temperature drop between the working fluid and seawater (dTp) was considered the reference for the analysis. Data on various temperature differences between surface and deep seawater (DT) were analysed. It was observed from the analysis that both the temperature drop and temperature difference had a definite impact on the electrical power output and conversion efficiency. It was also found that the actual rate of hydrogen production varied from 2.5 to 60 $Nm^3\ h^{-1}$ as DT increased from 5 to 25 °C, respectively.

3.17 Geothermal Energy for Hydrogen Production

Geothermal energy is heat generated inside Earth's subsurface. This can be used by transmitting it to Earth's surface using steam and hot water. Geothermal energy can be used for cooling, heating and generating power. The literature reports that geothermal power can be applied to hydrogen production. Figure 3.29 shows a schematic diagram of hydrogen production through geothermal energy [82].

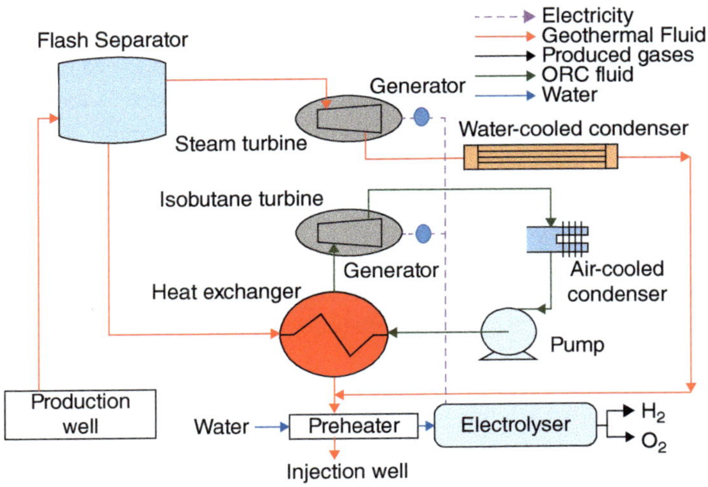

Figure 3.29 Hydrogen production through geothermal energy [82]/with permission of Elsevier.

Dincer and Acar [83] assessed various energy sources and ranked them based on technical, economic and environmental criteria. It was observed that geothermal was the most suitable option when viewed in the context of 'non-air pollution criteria'. Under conditions of heating and cooling, geothermal ranked high and had the same rank as biomass from a technical, environmental and cost performance point of view. As is well known, geothermal energy is weather independent. It can generate power throughout the year. Soltani and co-authors [84] reviewed their paper on geothermal heating and cooling systems and discussed several aspects of the technology, such as geothermal energy distribution systems, ground source heat pumps and ground heat exchangers. Several prospects for the technology's application have been discussed. The relevant technical information reported in the paper can provide applicable practical guidelines for geothermal heating or cooling system installation.

There are several other applications of geothermal energy for improved and more efficient water electrolysis for use in heat pump systems [85]. The major obstacle to adopting a geothermal heat pump (GHP) is mainly the cost of the geothermal heat exchanger (GHE). Cost reduction through optimisation has been discussed for residential buildings. This includes significant parameters, such as length, radius, well numbers, the flow discharge inside the pipe and the pipe's external radius. In this study, aiming at optimisation and energy-economy evaluation of a GHP, a 10-year simulation study was conducted. The objective of this investigation was to estimate the decrease in energy consumption based on varying levels of fluids, such as ethylene glycol, methanol, potassium acetate, sodium chloride, etc., compared to water in the GHP. Based on a comprehensive analysis of a decade-long operational performance assessment and the associated loss metrics, ethylene glycol emerges as the most optimal selection for utilization within a Ground-Source Heat Pump (GHP) system.

Tolga Balta et al. [86] and a team of researchers studied the following four methods of hydrogen production using geothermal energy: (i) direct production of hydrogen from geothermal steam, (ii) conventional water electrolysis using the electricity generated through a geothermal power plant, (iii) using both geothermal heat and electricity for high-temperature steam electrolysis (HTSE) and/or hybrid processes and (iv) using the heat available from the geothermal resource in thermochemical processes. Geothermal heat for hydrogen production is a very promising and environmentally friendly technology. In their thermodynamic analysis of performance evaluation by energy and exergy efficiency, it was observed that the energetic and exergetic efficiencies of the copper–chlorine (Cu–Cl) cycle were 51 and 65%, respectively [86]. Moreover, the Cu–Cl cycle was totally environmentally friendly as there was no greenhouse gas emission into the atmosphere. Thus, the Cu–Cl cycle has been found to be a very prospective route for geothermal-based hydrogen production. One of the most attractive features of hydrogen production using the Cu–Cl cycle is the efficiency of the cycle without greenhouse gas emission into the atmosphere. It has been suggested by researchers that the thermodynamic feasibility of low-temperature thermochemical and hybrid cycles for hydrogen production coupled with geothermal sources requires to be investigated [86].

3.18 Hydrogen from H_2S in Black Sea Waters

Hydrogen sulphide, an industrial by-product, poses a threat to the ecological balance. H_2S combines with water in the atmosphere and forms sulphuric acid, which causes acid rain. Acid rain harms agriculture, buildings and monuments and disrupts the ecological balance. Hydrogen sulphide naturally occurs in many gas wells, gas hydrates and gas-saturated sediments. The Black Sea is the largest water body in the world that contains H_2S. It has been estimated that 90% of Black Sea water is anaerobic in character, and the hydrogen sulphide layer in the Black Sea begins about 200 m below the surface. The total sulphide pool in the Black Sea is approximately 4.6×10^9 tonnes. H_2S is formed by the decay of organic matter over the past thousands of years. Thus, the concentration of H_2S differs according to the depth of the water layers: deeper layers show a higher concentration of H_2S than shallower layers. H_2S concentrations in the Black Sea's deeper layers are influenced by a variety of factors including water stratification, biological activity, and the input of organic matter from the surface. Generally in the anoxic (oxygen-depleted) layers of the Black Sea, hydrogen sulphide can be present, but its concentration typically ranges from very low values to a few micromoles per litre. H_2S concentrations in the Black Sea's deeper layers are influenced by a variety of factors, including water stratification, biological activity, and the input of organic matter from the surface. Eremeev, Konovalov and Romanov studied distribution of oxygen and hydrogen sulphide in Black Sea waters during winter-spring period using the method of spatial isopycnic analysis [87]. Spatial isopycnic analysis is useful to understand the structure and dynamics of stratified systems and distribution of various properties, such as temperature, salinity, or chemical composition, at constant density levels etc at different depths of the black sea. Soon after collecting samples, concentration hydrogen sulphide was determined by iodometric method. Variation of hydrogen sulphide distribution with relative density was also established. Their study indicates that, the ventilation process driven by winter-time conditions and the massive rotating ocean current, help to supply oxygen to specific layers of the ocean [87].

Thus, research activities are being pursued to ascertain the potential of the Black Sea as a source of hydrogen. It is possible to extract H_2S at a relatively high temperature and by lowering the pH of the water to 4. This process can be accomplished in two stages. In the first stage, hydrogen and polysulphides could be produced from seawater, and then they are produced through electrochemical production. Some researchers have also reported that the one-step thermal decomposition process is a convenient mode of hydrogen production from H_2S.

It is perhaps relevant to mention here that hydrogen production by the H_2S-splitting cycle is an environmentally friendly carbon-free process. Currently, no commercial technology exists for hydrogen production from H_2S.

Innovative technology must be initiated. Several routes to hydrogen production are already in place. Technology routes already established must be followed. SMR for hydrogen production and the Claus process for sulphur recovery are the appropriate routes in the initial phase. However, superadiabatic partial combustion is drawing a lot of attention and looks very promising in the future [88].

3.19 Hydrogen Production Using *Enterobacter cloacae*

Very little research has been conducted on the production of hydrogen using *Enterobacter cloacae. Enterobacter cloacae* are anaerobic, non-spore-forming bacteria that grow on organic substrates. They produce hydrogen only under anaerobic conditions. The effect of pH and temperature on hydrogen production by *Enterobacter aerogenes* (NCIMB 10102) utilising starch hydrolysate as a substrate has been studied [89]. In this study, an optimum pH range corresponding to batch runs of 6.1–6.6 was found to be optimum in studies carried out at different pH values. This study analysed fermentation and thermal inactivation. This study evaluated significant thermodynamic parameters such as enthalpy and entropy.

It is possible to generate hydrogen biologically through photosynthetic and fermentative routes. Between the two fermentative routes, it has been observed to be faster than the photosynthetic organism [90]. A study on continuous hydrogen production by immobilising *E. cloacae* IIT-BT 08 [91] using lignocellulosic agro-residue as a support matrix was carried out. This study was primarily conducted because lignocellulosic agro-residues are highly available in the environment. This research involved hydrogen production using different carbon sources and explored the possibility of using cheaper raw materials. Sucrose was observed to be the most suitable for hydrogen production. At a temperature of 36 °C and at a pH value of 6.0, a maximum hydrogen rate of 29.63 mmol per gram dry cell per hour could be achieved. In view of the change in the metabolic pattern of the organism, the pH values of fermentation broth under aerobic and anaerobic conditions were found to differ. The results of this study showed that SM-C had the most preferred hydrogen production rate. Apart from this, it had better feature in terms of cell retention, packing density, cell concentration, etc. [91]. The attainment of the substrate conversion efficiency of 28% shows the economic viability of the process. Yunli Ren et al. [92] investigated the hydrogen production performance of facultative anaerobes (*Enterobacter aerogenes*) from different monomeric sugars, such as arabinose, mannose, galactose and rhamnose. A comparative evaluation of the hydrolysis products of hemicellulose showed that, as far as hydrogen production is concerned, stereoisomers of mannose and galactose are relatively more effective than those of xylose and arabinose. An increase in mannose concentration resulted in an increase in hydrogen production as well as the maximum amount of hydrogen produced. It was also found from this study that rhamnose was not the appropriate resource for *E. aerogenes* to produce hydrogen. In fact, it had a much lower production efficiency of about 50% of mannose or galactose substrates.

Mohamed Hemida Abd-Alla and co-workers [93] investigated the potential of hydrogen production on a glucose medium with facultative anaerobic bacterial isolates, such as *Enterobacter aerogenes*, *Enterobacter cloacae* and *Cronobacter sakazakii*. The two most significant hydrogen-producing isolates, *E. aerogenes* (KY549389) and *E. cloacae* (KY524293), were studied to assess their potential of producing hydrogen using orange peel (OP) hydrolysate extracted with boiling water using facultative anaerobic *E. aerogenes* (KY549389) and *E. cloacae* (KY524293), respectively.

3.20 Hydrogen Production by Reforming Natural Gas and Bio-derived Liquids Using a Dense Ceramic Membrane

Although water dissociation is a well-established technology for hydrogen production, it is well known that water dissociation provides a very low concentration of hydrogen and oxygen. To improve the concentration level, a mixed-conducting (i.e. electron- and ion-conducting) membrane such as an oxygen transport membrane (OTM) is used to separate oxygen and hydrogen.

It is evident from the available literature that separation and purification play a very crucial role in hydrogen production through thermochemical processes. A close comparative assessment shows that membrane reactors can shift equilibrium. In contrast, the water–gas shift reaction converts carbon monoxide into hydrogen. Membranes have also proven useful for purification. There are two types of membranes used for hydrogen production and purification: (1) dense-phase metals and metal alloys and (2) porous ceramic membranes. Both are inorganic in character. As far as porous ceramic membranes are concerned, they are prepared by sol–gel or hydrothermal methods. They have certain very distinctive characteristics, such as high stability and durability in higher temperature ranges. They can withstand impurities and effectively operate in hydrothermal environments. At higher temperatures, microporous membranes have shown great promise in water–gas shift reactions.

Lu and co-researchers [94] discuss the broad characteristics of several membrane types. Inorganic membranes have the beneficial features of high flux and higher temperatures. Selective membranes exhibit different performances. Metallic Pd types, despite their wide range of selectivity, have some shortcomings. Metals are very expensive, and the selective film has low thermomechanical stability. Ceramic membranes, being microporous, are suitable for hydrogen separation. Modern technological developments are being pursued in both dense-phase metal and porous ceramic membranes. Apart from economic viability, a lot of attention is paid to the development of the optimum technical characteristics of membranes related to the zero-emission power technology.

By utilising mixed-conducting membranes to remove oxygen from water at relatively moderate temperatures (500–900 °C), the amount of hydrogen concentration can be significantly increased. Research work by Balachandran and co-researchers [95] confirms that 'hydrogen is produced by the dissociation of water, and not by permeation or leakage of hydrogen from the oxygen-permeated side of the membrane' [95]. The study also showed that as the membrane thickness decreases, the rate of hydrogen production increases. When thin membranes have a thickness of less than 0.5 mm, hydrogen production is limited by surface kinetics.

3.21 Plasma Reforming

It is perhaps relevant to mention here that the network of reforming reactions in plasma reforming is the same as that in conventional reforming. Plasma reforming is a promising technique used for hydrogen production from hydrocarbon feedstocks. It involves the use of plasma, which is a high-temperature ionised gas, to initiate and facilitate chemical reactions. Plasma reforming is typically carried out under non-equilibrium conditions, meaning that the temperature of the plasma is much higher than that of the surrounding gas.

However, in this case, the energy and free radicals that are used to carry out the reforming reactions are provided by plasma. A plasma is generated using a plasma generator, which could be achieved by subjecting a gas to high energy, such as through electric discharges, microwaves or lasers. The plasma reaches extremely high temperatures, often in the range of thousands of degrees Celsius. The high-energy plasma breaks down the hydrocarbon molecules into smaller fragments through various processes, including dissociation and ionisation. This results in the formation of reactive radicals and species. After the desired reaction occurred, the mixture was rapidly cooled (quenched) to prevent further reactions leading to unwanted side reactions. The resulting gas mixture is then processed to separate hydrogen from other by-products, typically through processes such as PSA or membrane separation.

The thermal plasma technology has been observed to be very useful in producing hydrogen or hydrogen-rich gases from a wide range of fuels, such as gasoline, diesel, oil, biomass, natural gas, jet fuel, etc. The unique capability of plasma devices or plasmatrons is that they can generate and accurately monitor very high temperatures (more than 20 000 °C). In such situations, the heat generated does not depend upon the reaction chemistry. Thus, it is possible to closely control operating conditions with varying compositions as well as over a relatively wide range of feedstocks. The plasma reforming technology has advantages over conventional hydrogen production technologies [90, 96–98]. There are many advantages of using the plasma technology. Even without a catalyst, high temperature and a high degree of dissociation can speed up chemical reactions. Plasma reformers are likely to obtain a much wider application due to some very distinctive positive features, such as high power density, higher conversion efficiencies, a fast response time (fraction of a second), etc. Moreover, as discussed earlier, this can be used with a broad range of fuels, including heavy hydrocarbons (crude) and high sulphur diesel. This is also a relatively inexpensive technology. This mode of hydrogen production is useful for both stationary and mobile applications. However, the only shortcoming of plasma reforming is its high pressure. High pressure, even if achieved easily, increases the possibility of erosion. Therefore, the electrode lifetime diminishes in plasma reforming.

Research and development continue on techniques such as thermal plasma gasification for enhanced hydrogen production [99] and glow-discharge plasma electrolysis [100].

A very exhaustive description of hydrogen production by plasma-driven solution electrolysis (PDSE) is provided by Bespalko and Mizeraczyk [101]. PDSE is also referred to as contact glow-discharge electrolysis (CGDE) or simply plasma electrolysis in the literature. The paper presents the relevance and discusses the subsequent effect of a series of parameters, such as 'the discharge polarity, the magnitude of the applied voltage, the type and concentration of the typical electrolytic solutions (K_2CO_3, Na_2CO_3, KOH, NaOH and H_2SO_4), the presence of organic additives (CH_3OH, C_2H_5OH and CH_3COOH), the temperature of the electrolytic solution, the active length and the immersion depth of the discharge electrode into the electrolytic solution', on the energy efficiency (%), energy yield [$g(H_2)$ kWh^{-1}] and hydrogen production rate [$g(H_2)$ h^{-1}] of PDSE [101]. A comparative analysis of hydrogen production energy yields with PDSE and those of AEL, SOEL, polymer electrolyte membrane electrolysis (PEMEL) and some of the plasma-assisted routes of hydrogen production is provided. The analysis shows that the PDSE of aqueous electrolytic solutions with alcohol (methanol) provides a 3.9 times greater energy yield than AEL and a 5.8% greater yield than natural gas reforming. In fact, this value is higher than many other processes. On the strength of the high energy efficiency of hydrogen production, the PDSE of alcohol–water mixtures seems to be a promising technology. The present route and the quantity of hydrogen production can still be improved with more stable materials, which will further enhance the lifetime of the discharge electrode. A more efficient mode of hydrogen separation can be accomplished by adopting a membrane or absorption separation technique [101].

3.22 Hydrogen from Nuclear Energy

During off-peak hours, the nuclear power plant produces more electricity than that is needed by consumers. It is the right time to generate hydrogen using excess electricity. Conventional sources, such as coal, water and biomass, can be conveniently used for hydrogen production by applying the heat generated by a nuclear reactor. At present, hydrogen is derived from nuclear energy through three different processes: (i) electrolysis, (ii) HTSE and (iii) thermochemical water-splitting cycles. In terms of the quality of hydrogen produced through these different routes, it has already been explained previously that hydrogen produced through coal reforming must be cleaned for chemical impurities. On the other hand, hydrogen produced by light water reactors (LWRs) does not need cleaning for chemical impurities. Thermochemical water splitting is divided into several phases of partial reactions, and each phase occurs at a relatively low temperature of 550–850 °C. In such reactions, water is split into hydrogen and oxygen by using heat. The efficiency of this process is between 40 and 60%. Iodine–sulphur (I–S), calcium–bromine (Ca–Br) and copper–chlorine (Cu–Cl) are common thermochemical processes.

Technologies such as the high-temperature thermochemical process, water electrolysis or HTSE can be feasible methods to use nuclear energy as the primary energy source for centralised hydrogen production. However, it is essential that optimum operating temperatures are maintained to improve the efficiency of thermochemical and electrochemical hydrogen production mechanisms with nuclear energy. Since high-temperature reactors, such as gas-cooled, molten-salt-cooled and liquid-metal-cooled reactors, can achieve high temperatures, these reactors are suitable for hydrogen production. Yildiz and Kazimi [102] conducted an exhaustive analysis of several such prospective candidates. The significant results of the analysis of these researchers show that HTSE coupled to an advanced gas reactor that is cooled by supercritical CO_2 (S-CO_2) and equipped with a supercritical CO_2 power conversion cycle is the most efficient system capable of providing greater energy efficiency within relatively much lower ranges of temperature compared to other systems.

Nuclear energy can play a significant role in fighting environmental pollution. As far as hydrogen production is concerned, nuclear energy can be used for thermochemical water decomposition. Nuclear energy, in combination with renewable energy, can significantly contribute to the problem of environmental degradation in difficult situations when wind does not blow or Sun does not shine or geothermal or hydropower energies are not available [103]. This chapter discusses several integrated systems of nuclear energy with various types of renewable energy from which hydrogen can be produced (Figures 3.30–3.35).

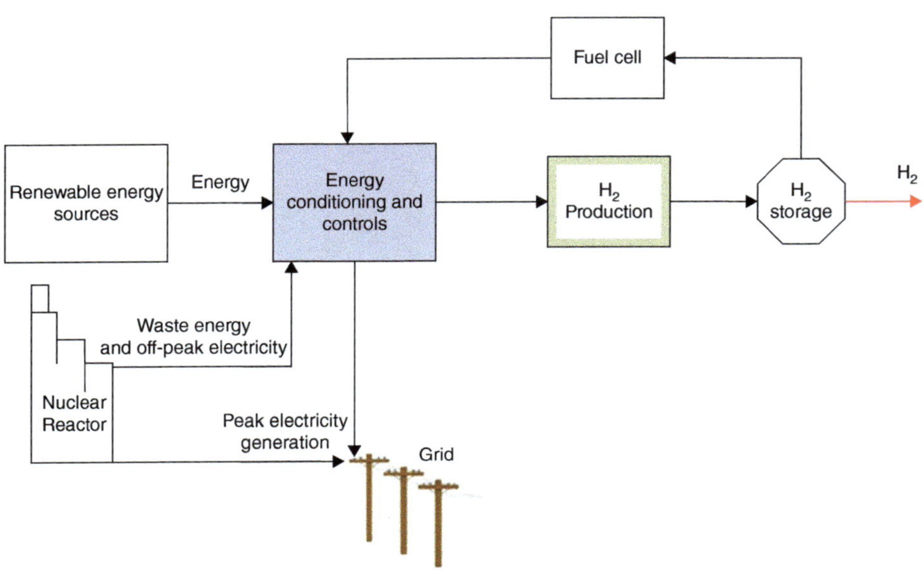

Figure 3.30 Nuclear and renewable energy coupled to a smart grid [103]/with permission of Elsevier.

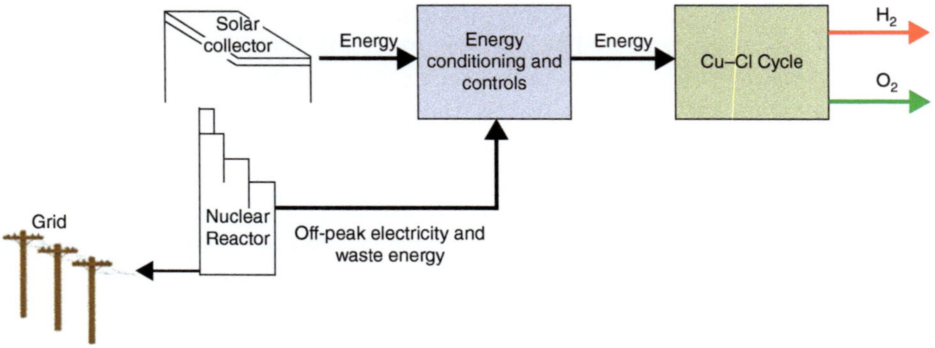

Figure 3.31 Nuclear-assisted hydrogen solar hydrogen generation [103]/with permission of Elsevier.

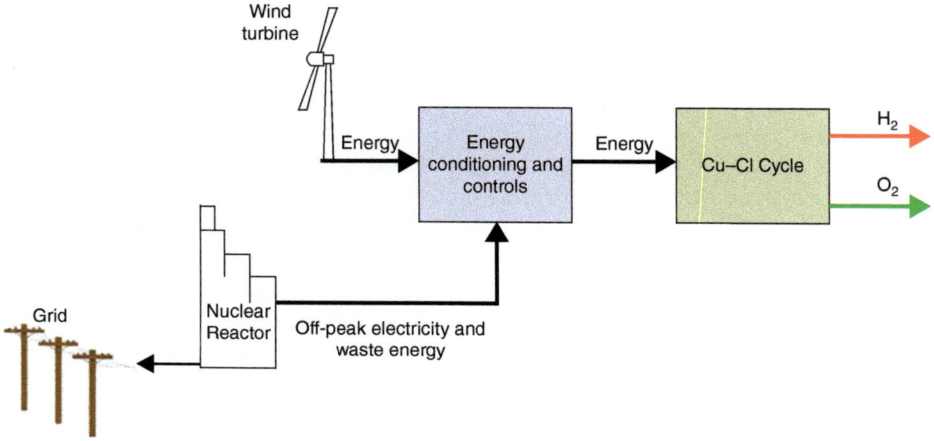

Figure 3.32 Nuclear-assisted wind–hydrogen generation.

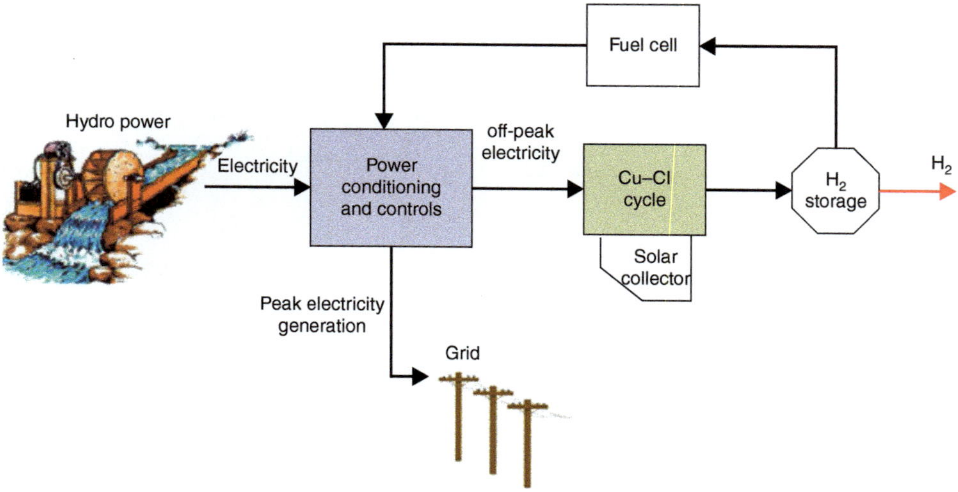

Figure 3.33 Integrated hydro-solar energy system for hydrogen production: Cu–Cl cycle [103]/with permission of Elsevier.

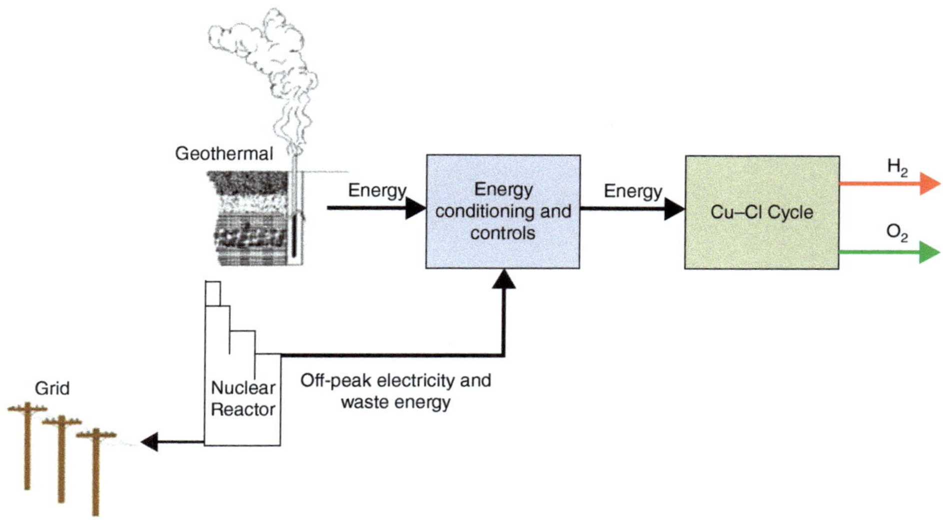

Figure 3.34 Nuclear-assisted geothermal-hydrogen production with the Cu–Cl cycle [103]/with permission of Elsevier.

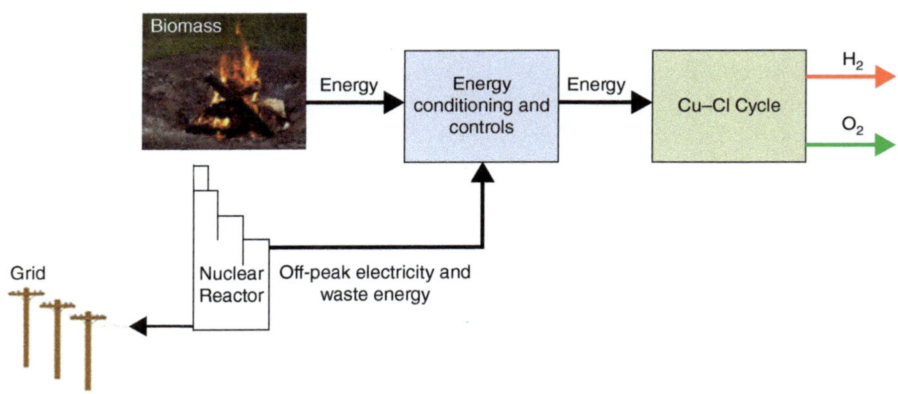

Figure 3.35 Nuclear-assisted biomass-hydrogen production with the Cu–Cl cycle [103], Mehmet et al., 2012/with permission from Elsevier.

3.23 Ammonia Dissociation

Ammonia (NH_3) plays a very critical role as a hydrogen (H_2) carrier for energy services in the context of the current thrust towards a zero-carbon future. The dissociation of ammonia to hydrogen (75%) and nitrogen (25%) was successfully carried out at a temperature of about 1000 °C with the aid of a catalyst. Nitrogen thus produced is in the mononuclear form (N) and not N_2. Nasharuddin et al. [104] conducted an exhaustive study on the techno-economic feasibility of NH_3 as an H_2 carrier for fuel cell vehicle (FCV) applications. A comparative study was conducted for hydrogen production for centralised hydrogen production as well as a distributed system, both from the points of view of technological feasibility and economic viability. It was observed that a centralised hydrogen system

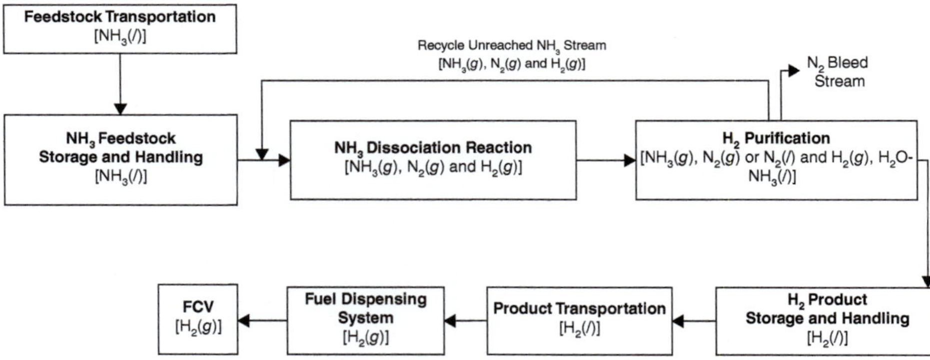

Figure 3.36 Centralised NH_3 dissociation for H_2 production [104]/with permission of Elsevier.

(Figure 3.36) is a more pragmatic system as distributed hydrogen production does not seem economical.

In their paper, Makhloufi et al. discussed [105] how the numerous challenges faced by hydrogen energy technology in terms of production and storage can be solved by storing renewable hydrogen 2 in the form of green ammonia (commonly referred to as power-to-ammonia) and using it. A technical and economic analysis carried out by the researchers shows that it is technically possible to produce pure hydrogen from ammonia on a large scale. It has been demonstrated that it is possible to obtain a thermal efficiency of 68.5% to manufacture 200 metric tonnes per day (MTPD) of pure hydrogen at 250 bar [105]. As far as the economic feasibility of the process is concerned, the levelised cost of hydrogen (LCOH) depends upon green ammonia costs. The characteristics and the process adopted to produce green ammonia from renewables ultimately determine the cost.

3.24 Hydrogen from Methane Hydrate

Studies are being conducted to obtain hydrogen from hydrates [106]. Methane hydrate mostly exists in sea sediments and geological structures below them. It is also known to exist in the deeply frozen layers under the Arctic region and at low temperatures and high pressures. Technology has recently been developed to produce methane on floating platforms in an endothermic process. It is possible to use current industrial processes on the platform to produce hydrogen and carbon dioxide. This will enable us to separate them from each other. A part of the methane could be diverted to generate heat energy for operating a power plant installed on the platform.

Thus, it would be possible to use the reforming process to produce hydrogen and carbon dioxide. Subsequently, effective carbon dioxide might be separated and sequestrated so that the hydrogen available is without carbon dioxide or other pollutants. Researchers have reported carbon dioxide sequestration [107–110]. Figure 3.37 shows the major components of the system platform [106].

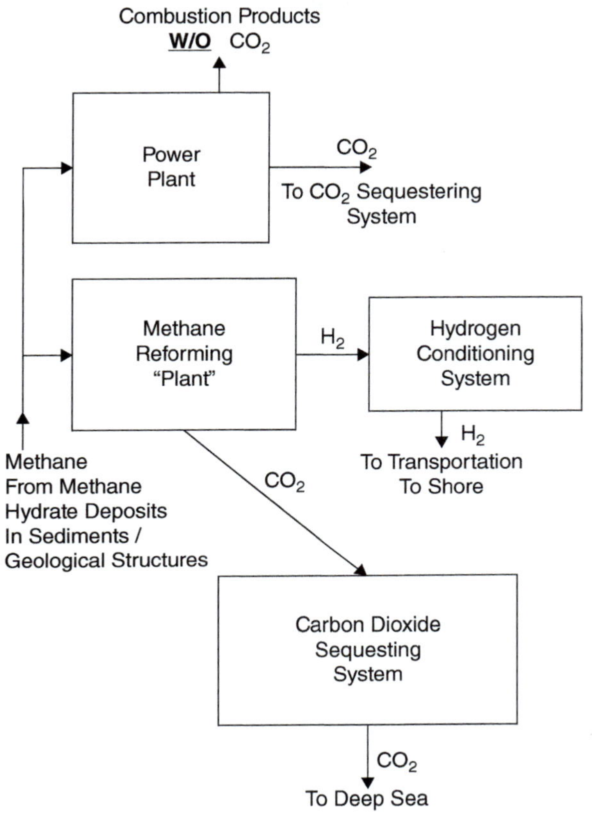

Figure 3.37 Major components of the parts of the system showing hydrogen from hydrates [106]/ with permission of Elsevier.

3.25 Improvements in Catalysts for Hydrogen Production

Apart from the well-established technological routes of hydrogen production, researchers worldwide are actively pursuing sustainable hydrogen production by various methods with an emphasis on improving the environment and reducing the GWP and effects of acid rain.

Some of the modest research experiments in the laboratory have really identified game-changing factors that need to be further studied and scaled up. Researchers at the University of Surrey investigated nano-carbons for hydrogen production. Their recent research results show edge-decorated nano-carbons as metal-free catalysts for the direct conversion of methane, leading to cost-effective methods of hydrogen production. Some traditional catalysts, used for hydrogen production, are reported to be poisoned by carbon. According to researchers, nitrogen- and phosphorus-doped nano-carbons are not affected by such problems. This process not only offers a superior sustainable alternative to traditional metal catalysts but also contributes to better climatic conditions by eliminating potent GHGs, such as methane, which has a much higher GWP than that of CO_2 [111].

PEM electrolysers have been proven to be suitable for the production of hydrogen from renewable intermittent sources such as solar energy and wind. A team of researchers from different institutions led by Argonne developed the National Laboratory of the US Department of Energy and successfully developed a technology that is very significant for PEM electrolysers in splitting water into hydrogen and oxygen at near-room temperature [112]. Scientists have developed game-changing catalysts for clean hydrogen from water.

Chunping Xu Emilia Paone, Daily Rodríguez-Padron et al. [113] discussed the role and significance of biomass-derived polyols in the context of hydrogen production. In the past, the potential application of the reductive catalytic conversion of polyols through hydrogenation, dehydrogenation, hydrogenolysis and aqueous phase reforming (APR) processes has proven to be very effective for the sustainable production of glycol, alkane and hydrogen. The paper discusses the relevance of processing biomass-derived C6 and C5 polyols, namely sorbitol and xylitol, for producing hydrogen. The role of the reliable catalytic system used for optimised reaction conditions has also been discussed. The production of renewable H_2 from biomass-derived polyols can be successfully accomplished by APR reactions that represent a sustainable, low-energy and demanding catalytic process that allows the formation of hydrogen in high yield with a low-carbon monoxide content. Catalytic hydrogenolysis, deoxygenation and APR are very well-known processes, and well-proven routes and efficient heterogeneous catalysts need to be developed to improve the production of hydrogen from biobased sorbitol and xylitol.

3.26 An Assessment of GWP and AP in Various Hydrogen Production Processes

At this point in time, when there is a dire need for an adequate amount of hydrogen, it is imperative to look at the environmental impact assessment of various hydrogen production processes. The negative effects of GHGs on human health and the overall ecological system are too well known to be described further.

Carbon dioxide emissions are the major source of GHGs. GWP and AP are the two crucial parameters used to assess the environmental effects of various production schemes. The GWP of a greenhouse gas is calculated by comparing the heat trapping capacity of the particular greenhouse gas to that of CO_2 over a specific time horizon (GWP (kg CO_2eq.). The acid deposition of acidifying pollutants on water, soil, materials, ecosystems and biological organisms leads to AP. AP is measured in (g SO_2 eq.) by assessing the acidity alteration using sulphur dioxide (SO_2) release on soil and into water.

The ecological footprint of various hydrogen production processes is shown in Figures 3.38, 3.39 and 3.40. Figure 3.38 shows a schematic diagram of non-renewables, whereas Figure 3.39 exhibits the AP and GWP of renewables. It is evident from Figure 3.38 that the AP and GWP of hydrogen production from natural gas and coal gasification are much higher compared to others. It may not be out of context to mention here that these are currently the more prevalent methods of hydrogen production.

The values of AP and GWP from renewable energy sources are shown in Figure 3.39. The normalisation values generated for this graph have been applied in the same way as used

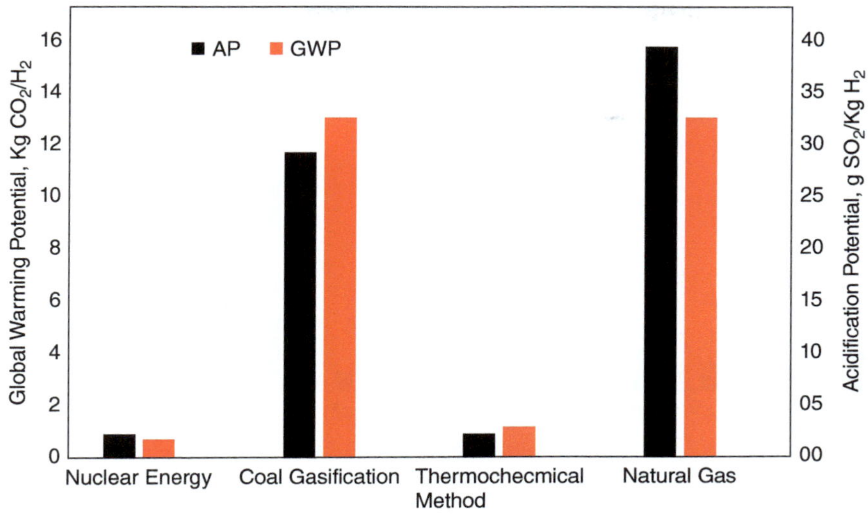

Figure 3.38 AP and GWP for hydrogen production processes from non-renewable energy sources per kg of H_2. Adapted from [86].

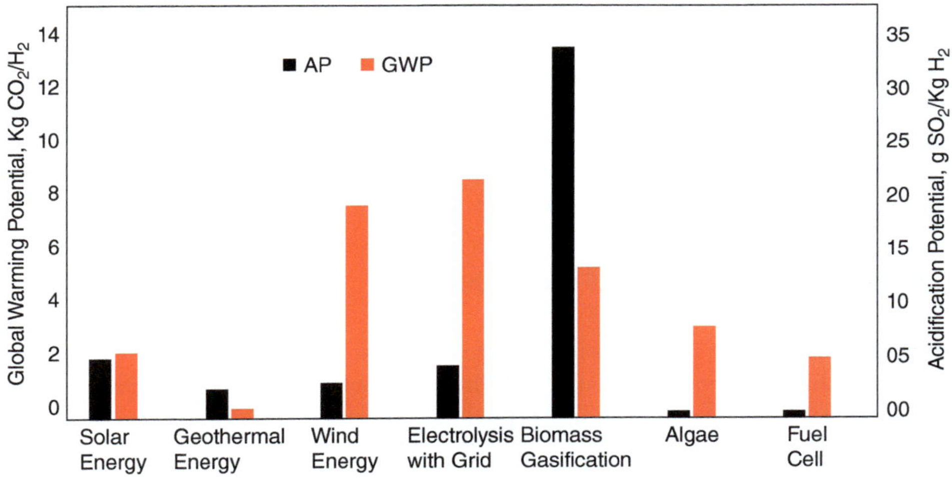

Figure 3.39 AP and GWP of different hydrogen production processes from renewable energy sources per kg of H_2. Adapted from [86].

for the renewable energy sources. As evident from the figure, GWP from wind and electrolysis is substantially higher compared to others. The GWP of biomass is also very high. As far as AP is concerned, biomass gasification demonstrates the highest value. Solar energy, algae and fuel cells show acceptable footprints, and geothermal energy seems to be the most environmentally friendly from the points of view of GWP and AP.

Figure 3.40 indicates a comparative depiction of the environmental impact assessment of all the processes of hydrogen production, including renewable and non-renewable processes.

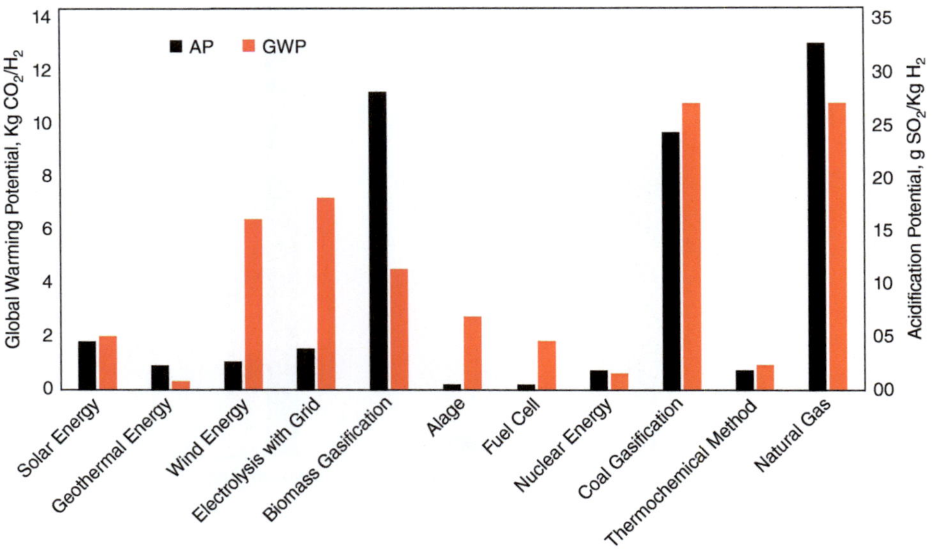

Figure 3.40 Overall comparability of GWP and AP of different hydrogen production schemes per kg of H_2. Adapted from [86].

This includes the methods widely accepted on a commercial scale as well as the technologies that need to be further investigated and optimised. Although impact is of paramount importance, practical implementation depends upon a series of factors, such as cost-effectiveness, efficiency, scalability, availability of resources, infrastructure compatibility, and the maturity level of technology for widespread application.

Concluding Remarks and Future Outlook

Hydrogen's significant role is clearly evident in our strides to net zero. It is extremely important that an adequate amount of hydrogen should be produced to cater to ever-increasing needs. At present, hydrogen is produced from natural gas (49%), crude oil (29%), coal (18%) and electrolysis (4%).

Demand for hydrogen is increasing at a rapid rate due to its ability to provide a solution to both energy and environment crises. Therefore, the production rate of green hydrogen requires to be increased. It is well known that hydrogen application has already reached maturity in certain sectors, such as ammonia and methanol production, and in some chemical and metallurgical industries. The possible applications of hydrogen are growing. In addition to ground transportation (IC engines and fuel cell electric vehicles), space applications, power sectors and domestic applications are also covered. The routes through which hydrogen can be produced from fossil fuels and hydrocarbon resources have been described. Processes such as alkaline, polymer electrolyte membrane, solid oxide and molten carbonate electrolysis as well as direct biophotolysis, indirect biophotolysis, photo-fermentation and dark fermentation have all been exhaustively tested. All the existing processes that produce hydrogen significantly contribute to the problem of fossil fuel starvation.

Some of the existing technologies need to be further improved. Further research on polymeric materials for membrane synthesis, such as polyether ketone, polyimides, and polyethylene, etc., needs to be carried out to assess the economic viability of hydrogen production by water splitting. The reaction thermodynamics of supercritical water gasification need to be further investigated to evaluate solutions of tar production and corrosion. Business models for improved efficiency and value chains for green hydrogen production need to be developed with international standards to find economically viable routes of sustainable green hydrogen production from renewable feedstocks. Efforts need to be devoted to find economically viable routes of green hydrogen production.

Abbreviations

DRI	Direct Reduced Iron
CNR	Catalytic Naphtha Reformer
CO	Carbon Monoxide
WGS	Water–Gas Shift
SMR	Steam Methane Reforming
PSA	Pressure Swing Adsorption
HTS	High-Temperature Shift
LTS	Low-Temperature Shift
PrOx	Preferential Oxidation
DFT	Density Functional Theory
EDTA	Ethylene Diamine Tetraacetic Acid
PEM	Proton Exchange Membrane/Polymer Electrolyte Membrane
POx	Partial Oxidation
CPOx	Catalytic Partial Oxidation
COS	Carbon Oxysulphide
ATR	Auto-Thermic Reaction
SR	Steam Reforming
SCG	Surface Coal Gasification
UCG	Underground Coal Gasification
GWP	Global Warming Potential
AP	Acidification Potential
CCS	Carbon Capture and Storage
GHG	Greenhouse Gas
MSW	Municipal Solid Waste

LFG	Landfill Gas
PGP	Plasma Gasification Process
ER	Equivalence Ratio
SBR	Steam-to-Biomass Ratio
SWGB	Supercritical Water Gasification of Biomass
PNS	Purple Non-Sulphur
AEL	Alkaline Electrolysis
SOEL	Solid Oxide Electrolysis
PP	Porous Propylene
KOH	Potassium Hydroxide
NaOH	Sodium Hydroxide
WRR	Water Reduction Reaction
PV	Photovoltaic
SCR	Selective Catalytic Reduction
DEA	Data Envelopment Analysis
OTEC	Ocean Thermal Energy Conversion
GHP	Geothermal Heat Pump
GHE	Geothermal Heat Exchanger
OTM	Oxygen Transport Membrane
PDSE	Plasma-Driven Solution Electrolysis
CGDE	Contact Glow-Discharge Electrolysis
LWR	Light Water Reactors
HTSE	High-Temperature Steam Electrolysis
LCOH	Levelised Cost of Hydrogen

References

1 Balat, M. (2009). Political, economic and environmental impacts of biomass-based hydrogen. *International Journal of Hydrogen Energy* 34 (9): 3589–3603.
2 Konieczny, A., Mondal, K., Wiltowski, T., and Dydo, P. (2008). Catalyst development for thermocatalytic decomposition of methane to hydrogen. *International Journal of Hydrogen Energy* 33 (1): 264–272.
3 Muradov, N.Z. and Veziroğlu, T.N. (2005). From hydrocarbon to hydrogen–carbon to hydrogen economy. *International Journal of Hydrogen Energy* 30 (3): 225–237.
4 World Energy Outlook. (2019). Report by International Energy Agency (IEA).
5 Energy Technology Perspective. (2020). Report by International Energy Agency.

6 Holladay, J.D., Hu, J., King, D.L., and Wang, Y. (2009). An overview of hydrogen production technologies. *Catalysis Today* 139 (4): 244–260.

7 Spath, P.L. and Mann, M.K. (2001). *Life Cycle Assessment of Hydrogen Production via Natural Gas Steam Reforming*. Golden, Colorado, NREL/TP-570-27637: National Renewable Energy Laboratory.

8 Advis, A.M., Pruder, B.B., Lim, C.J., and Grave, J.R. (1996). On the reported attempts to radically improve the performance of the steam-methane reforming reactor. *Canadian Journal of Chemical Engineering* 74 (2): 177–186.

9 Goula, M.A., Kontou, S.K., and Tsiakaras, P.E. (2004). Hydrogen production by ethanol steam reforming over a commercial Pd/γ-Al_2O_3 catalyst. *Applied Catalysis B: Environmental* 49 (2): 135–144.

10 Haryanto, A., Fernando, S., Murali, N., and Adhikari, S. (2005). Current status of hydrogen production techniques by steam reforming of ethanol: a review. *Energy & Fuels* 19 (5): 2098–2106.

11 Amphlett, J.C., Creber, K.A.M., Davis, J.M. et al. (1994). Hydrogen production by steam reforming of methanol for polymer electrolyte fuel cells. *International Journal of Hydrogen Energy* 19 (2): 131–137.

12 Agarwal, V., Patel, S., and Pant, K.K. (2005). H_2 production by steam reforming of methanol over Cu/ZnO/Al_2O_3 catalysts: transient deactivation kinetics modelling. *Applied Catalysis A: General* 279 (1–2): 155–164.

13 Almithn, A. and Alhulaybi, Z. (2022). A mechanistic study of methanol steam reforming on Ni2P catalyst. *Catalysts* 12: 1174.

14 Chen, G., Shen, Q., Zhang, X. et al. (2023). Methanol steam reforming over La1-xSrxCeO3-δ catalysts for hydrogen production: optimization of operating parameters. *Catalysts* 13: 248.

15 Mei, D., Qiu, X., Liu, H. et al. (2022). Progress on methanol reforming technologies for highly efficient hydrogen production and applications. *International Journal of Hydrogen Energy* 47 (84): 35757–35777.

16 Farrauto, R., Hwang, S., Shore, L. et al. (2003). New material needs for hydrocarbon fuel processing: generating hydrogen for the PEM fuel cell. *Annual Review of Materials Research* 33: 1–27.

17 Song, C. (2002). Fuel processing for low-temperature and high temperature fuel cells: challenges, and opportunities for sustainable development in the 21st century. *Catalysis Today* 77 (1–2): 17–49.

18 Krummenacher, J.J., West, K.N., and Schmidt, L.D. (2003). Catalytic partial oxidation of higher hydrocarbons at millisecond contact times: decane, hexadecane, and diesel fuel. *Journal of Catalysis* 215 (2): 332–343.

19 Lutz, A.E., Bradshaw, R.W., Bromberg, L., and Rabinovich, A. (2004). Thermodynamic analysis of hydrogen production by partial oxidation reforming. *International Journal of Hydrogen Energy* 29 (8): 809–816.

20 Nikolaidis, P. and Poullikkas, A. (2017). A comparative overview of hydrogen production processes. *Renewable and Sustainable Energy Reviews* 67: 597–611.

21 Wang, J.L. and Wan, W. (2008). Optimization of fermentative hydrogen production process by response surface methodology. *International Journal of Hydrogen Energy* 33 (23): 6976–6984.

22 Joensen, F. and Rostrup-Nielsen, J.R. (2002). Conversion of hydrocarbons and alcohols for fuel cells. *Journal of Power Sources* 105 (2): 195–201.

23 Ayabe, S., Omoto, H., Utaka, T. et al. (2003). Catalytic autothermal reforming of methane and propane over supported metal catalysts. *Applied Catalysis, A: General* 241 (1–2): 261–269.

24 Beer, J.M. (2000). Combustion technology developments in power generation in response to environmental challenges. *Progress in Energy and Combustion Science* 26: 301–327.

25 Verma, A. and Kumar, A. (2015). Life cycle assessment of hydrogen production from underground coal gasification. *Applied Energy* 147: 556–568.

26 Tezer, Ö. et al. (2022). Biomass gasification for sustainable energy production: a review. *International Journal of Hydrogen Energy* 47 (34): 15419–11543.

27 Koroneos, C., Dompros, A., and Roumbas, G. (2008). Hydrogen production via biomass gasification – a life cycle assessment approach. *Chemical Engineering and Processing* 47: 1261–1268.

28 Wang, J.L. and Yin, Y.N. (2019). Progress in microbiology for fermentative hydrogen production from organic wastes. *Critical Reviews in Environmental Science and Technology* 49 (10): 825–865.

29 Favas, J., Monteiro, E., and Rouboa, A. (2017). Hydrogen production using plasma gasification with steam injection. *International Journal of Hydrogen Energy* 42 (16): 10997–11100.

30 Lopez, G., Santamaria, L., Lemonidou, A. et al. (2022). Hydrogen generation from biomass by pyrolysis. *Nature Reviews Methods Primers* 2 (20).

31 Zhang, Y., Li, L., Xu, P. et al. (2019). Hydrogen production through biomass gasification in supercritical water: a review from exergy aspect. *International Journal of Hydrogen Energy* 44: 15727–15736.

32 Arregi, A., Amutio, M., Lopez, G. et al. (2018). Evaluation of thermochemical routes for hydrogen production from biomass: a review. *Energy Conversion and Management* 165: 696–719.

33 Reddy, S.N., Nanda, S., Dalai, A.K., and Kozinski, J.A. (2019). Supercritical water gasification of biomass for hydrogen production. *International Journal of Hydrogen Energy* 39: 6912–6926.

34 Lepage, T., Kammoun, M., Schmetz, Q., and Richel, A. (2021). Biomass-to-hydrogen: a review of main routes production, processes evaluation and techno-economical assessment. *Biomass and Bioenergy* 144: 105920.

35 Guo, L.J., Lu, Y.J., Zhang, X.M. et al. (2007). Hydrogen production by biomass gasification in supercritical water: a systematic experimental and analytical study. *Catalysis Today* 129 (3–4): 275–286.

36 Khan, M.A., Ngo, H.H., Guo, W. et al. (2018). Biohydrogen production from anaerobic digestion and its potential as renewable energy. *Renewable Energy* 129 (Part B): 754–768.

37 Zappi, A., Hernandez, R., and Holmes, W.E. (2021). A review of hydrogen production from anaerobic digestion. *International Journal of Environmental Science and Technology* 18: 4075–4090.

38 Hallenbeck, P.C. and Benemann, J.R. (2002). Biological hydrogen production; fundamentals and limiting processes. *International Journal of Hydrogen Energy* 27 (11–12): 1185–1193.

39 Wang, J.L. and Yin, Y.N. (2018). Fermentative hydrogen production using pre-treated microalgal biomass as feedstock. *Microbial Cell Factories* 17 (22).

40 Levin, D.B. and Chahine, R. (2010). Challenges for renewable hydrogen production from biomass. *International Journal of Hydrogen Energy* 35: 4962–4969.

41 Williams, C.R. and Bees, M. (2014). Mechanistic, modeling of sulphur deprived photosynthesis and hydrogen production in suspensions of *Chlamydomonas reinhardtii*. *Biotechnology and Bioengineering* 111: 320–335.

42 Melis, A., Zhang, L., Forestier, M. et al. (2000). Sustained photobiological hydrogen gas production upon reversible inactivation of oxygen evolution in the green alga Chlamydomonas reinhardtii. *Plant Physiology* 122: 127–136.

43 Yang, G. and Wang, J.L. (2018). Various additives for improving dark fermentative hydrogen production: a review. *Renewable and Sustainable Energy Reviews* 95: 130–146.

44 Basak, N. and Das, D. (2007). The prospect of purple non-sulphur (PNS) photosynthetic bacteria for hydrogen production: the present state of the art. *World Journal of Microbiology and Biotechnology* 23: 31–42.

45 Wang, J.L. and Wan, W. (2009). Factors influencing fermentative hydrogen production: a review. *International Journal of Hydrogen Energy* 34: 799–811.

46 Wang, J. and Yin, Y. (2018). Fermentative hydrogen production using various biomass-based materials as feedstock. *Renewable and Sustainable Energy Reviews* 92: 284–306.

47 Chen, Y., Yin, Y.N., and Wang, J.L. (2021). Recent advance in inhibition of dark fermentative hydrogen production. *International Journal of Hydrogen Energy* 46: 5053–5073.

48 Basak, B., Jeon, B.H., Kim, T.H. et al. (2020). Dark fermentative hydrogen production from pretreated lignocellulosic biomass: effects of inhibitory byproducts and recent trends in mitigation strategies. *Renewable and Sustainable Energy Reviews* 133: 110338.

49 Zhang, T., Jiang, D., Zhang, H. et al. (2020). Comparative study on bio-hydrogen production from corn stover: photo-fermentation, dark-fermentation and dark photo co-fermentation. *International Journal of Hydrogen Energy* 45: 3807–3814.

50 Asada, Y., Tokumoto, M., Aihara, Y. et al. (2006). Hydrogen production by co-cultures of Lactobacillus and a photosynthetic bacterium, *Rhodobacter sphaeroides* RV. *International Journal of Hydrogen Energy* 31: 1509–1513.

51 Pachapur, V., Sarma, S., Brar, S. et al. (2015). Co-culture strategies for increased biohydrogen production. *International Journal of Energy Research* 39 (11): 1479–1504.

52 Final report of the Roads2hycom project fuel cells and hydrogen in a sustainable energy economy edited by Nick Owen, Ricardo UK Ltd on behalf of the Roads2hycom Consortium, www.roads2hy.com. *Document Number: R2H8500PU.6*

53 Brauns, J. and Turek, T. (2020). Alkaline water electrolysis powered by renewable energy: a review. *Processes* 8: 248.

54 Li, C. and Baek, J.B. (2021). The promise of hydrogen production from alkaline anion exchange membrane electrolysers. *Nanomater Energy* 87: 106162.

55 Milani, D., Kiani, A., and McNaughton, R. (2020). Renewable-powered hydrogen economy from Australia's perspective. *International Journal of Hydrogen Energy* 45: 24125–24145.

56 Men Stano, L., Stano, M., and Durina, P. (2020). Separators for alkaline water electrolysis prepared by plasma-initiated grafting of acrylic acid on microporous polypropylene membranes. *International Journal of Hydrogen Energy* 45: 80–93.

57 Vermeiren, P., Adriansens, W., and Leysen, R. (1996). Zirfon: a new separator for Ni-H_2 batteries and alkaline fuel cells. *International Journal of Hydrogen Energy* 21: 679–684.

58 Lee, H.I., Mehdi, M., Kim, S.K. et al. (2020). Advanced Zirfon-type porous separator for a high-rate alkaline electrolyser operating in a dynamic mode. *Journal of Membrane Science* 616: 118541.

59 Vermeiren, P., Adriansens, W., Moreels, J.P., and Leysen, R. (1998). Evaluation of the Zirfon separator for use in alkaline water electrolysis and Ni-H_2 batteries. *International Journal of Hydrogen Energy* 23: 321–324.

60 Schalenbach, M., Lueke, W., and Stolten, D. (2016). Hydrogen diffusivity and electrolyte permeability of the Zirfon PERL separator for alkaline water electrolysis. *Journal of The Electrochemical Society* 163: F1480–F1488.

61 Lee, H., Dung, D., Kim, J. et al. (2019). The synthesis of a Zirfon-type porous separator with reduced gas crossover for alkaline electrolyzer. *International Journal of Energy Research* 44: 1875–1885.

62 Burnat, D., Schlupp, M., Wichser, A. et al. (2015). Composite membranes for alkaline electrolysis based on polysulfone and mineral fillers. *Journal of Power Sources* 291: 163–172.

63 Hosseini, S.E. and Wahid, M. (2019). Hydrogen from solar energy; a clean energy carrier from a sustainable source of energy. *International Journal of Energy Research* 44 (6): 4110–4131.

64 Saba, S.M., Muller, M., Robinius, M., and Stolten, D. (2018). The investment costs of electrolysis: a comparison of cost studies from the past 30 years. *International Journal of Hydrogen Energy* 43: 1209–1223.

65 Ursua, A., Gandia, L.M., and Sanchis, P. (2012). Hydrogen production from water electrolysis: current status and future trends. *Proceedings of the IEEE* 100: 410–426.

66 Stoll, T., Zafeiropoulos, G., and Tsampas, M. (2016). Solar fuel production in a novel polymeric electrolyte membrane photoelectrochemical (PEM-PEC) cell with a web of titania nanotube arrays as photoanode and gaseous reactants. *International Journal of Hydrogen Energy* 41: 17807–17817.

67 Giancola, S., Zaton, M., Reyes-Carmona, A. et al. (2019). Composite short side chain PFSA membranes for PEM water electrolysis. *Journal of Membrane Science* 570-571: 69–76.

68 Ampurdanes, J., Chourashiya, M., and Urakawa, A. (2019). Cobalt oxide-based materials as non-PGM catalyst for HER in PEM electrolysis and in situ XAS characterization of its functional state. *Catalysis Today* 336: 161–168.

69 Grigoriev, S.A., Pushkarev, A.S., Pushkareva, I.V. et al. (2017). Hydrogen production by proton exchange membrane water electrolysis using cobalt and iron hexachloroclathrochelates as efficient hydrogen-evolving electrocatalysts. *International Journal of Hydrogen Energy* 42: 27845–27850.

70 Sapountzi, F.M., Gracia, J.M., Weststrate, C.J. et al. (2017). Electrocatalysts for the generation of hydrogen, oxygen and synthesis gas. *Progress in Energy and Combustion Science* 58: 1–35.

71 Hnat, J., Plevova, M., Zitka, J. et al. (2017). Anion-selective materials with 1,4-diazabicyclo [2.2.2] functional groups for advanced alkaline water electrolysis. *Electrochimica Acta* 248: 547–555.

72 Ni, M., Leung, M.K.H., and Leung, D.Y.C. (2008). Technological development of hydrogen production by solid oxide electrolyzer cell (SOEC). *International Journal of Hydrogen Energy* 33 (9): 2337–2354.

73 Khan M.A., Tareq Al-Attas Soumyabrata Roy et al. (2021). Seawater electrolysis for hydrogen production: a solution looking for a problem? *Journal of Energy and Environment Sciences* 14 (9): 4831–4839.

74 Abuseada, M., Chuyu Wei, R., Spearrin, M., and Fisher, T.S. (2022). Solar–thermal production of graphitic carbon and hydrogen via methane decomposition. *Energy & Fuels* 36 (7): 3920–3392.

75 Bhosal, R.R. (2021). Solar hydrogen production via ZnO/Zn based thermochemical water splitting cycle: Effect of partial reduction of ZnO. *International Journal of Hydrogen Energy* 46 (6): 4739–4748.

76 Kamlungsua, K., Su, P.-C., and Chan, S.H. (2020). Hydrogen generation using solid oxide electrolysis cells. *Fuel Cells* 20 (6): 644–649.

77 Lim, C.K., Liu, Q., Zhou, J. et al. (2017). High-temperature electrolysis of synthetic seawater using solid oxide electrolyzer cells. *Journal of Power Sources* 342: 79–87.

78 Garcia-Valverde, R., Espinosa, N., and Urbina, A. (2011). Optimized method for photovoltaic-water electrolyser direct coupling. *International hydrogen of energy* 3 (6): 10574–10586.

79 Penconi, M. et al. (2015). Hydrogen production from water by photolysis, sonolysis and sonophotolysis with solid solutions of rare earth. *Gallium and Indium Oxides as Heterogeneous Catalysts, Sustainability* 7 (7).

80 Rezaei-Shouroki, M., Mostafaeipour, A., and Qolipour, M. (2017). Optimized method for photovoltaic-water electrolyser direct coupling. *International Journal of Hydrogen Energy* 42 (15): 9500–9510.

81 Kazim, A. (2005). Hydrogen production through an ocean thermal energy conversion system operating at an optimum temperature drop. *Applied Thermal Engineering* 25: 2236–2246.

82 Amin, M., Shah, H.H., Fareed, A.G. et al. (2022). Hydrogen production through renewable and non-renewable energy processes and their impact on climate change. *International Journal of Hydrogen Energy* 47 (77): 33112–33134.

83 Dincer, I. and Acar, C. (2015). A review on clean energy solutions for better sustainability. *International Journal of Energy Research* 39: 585–606.

84 Soltani, M., Farshad, M., Kashkooli, A. et al. (2019). A comprehensive study of geothermal heating and cooling systems. *Sustainable Cities and Society* 44: 793–818.

85 Soltani, M., Farzanehkhameneh, P., Kashkooli, F.M. et al. (2021). Optimization and energy assessment of geothermal heat exchangers for different circulating fluids. *Energy Conversion and Management* 228: 113733.

86 Tolga Balta, M., Dincer, I., and Hepbasli, A. (2010). Geothermal-based hydrogen production using thermochemical and hybrid cycles: a review and analysis. *International Journal of Energy Research* 34: 757–775.

87 Eremeev, V.N., Konovalov, S.K., and Romanov, A.S. (1998). The distribution of oxygen and hydrogen sulfide in Black Sea waters during winter-spring period. *Physical Oceanography* 9: 259–272.

88 Demirbas, A. (2009). Hydrogen Sulfide from the Black Sea for Hydrogen Production. *Energy Source Part A: Recovery, Utilization, and Environmental Effect* 20: 1866–1872.

89 Fabiano, B. and Perego, P. (2002). "Thermodynamic study and optimization of hydrogen production" by Enterobacter aerogenes. *International Journal of Hydrogen Energy* 27: 149–156.

90 Bromberg, L., Cohn, D.R., Rabinovich, A., and Alexeev, N. (1999). Plasma catalytic reforming of methane. *International Journal of Hydrogen Energy* 24 (12): 1131–1137.

91 Kumar, N. and Das, D. (2001). Continuous hydrogen production by immobilized *Enterobacter cloacae* IIT-BT 08 using lignocellulosic materials as solid matrices. *Enzyme and Microbial Technology* 29: 280–287.

92 Ren, Y., Wang, J., Liu, Z. et al. (2009). Hydrogen production from the monomeric sugars hydrolysed from hemicellulose by Enterobacter aerogenes. *Renewable Energy* 34.

93 Abd-Alla, M.H., Gabra, F.A., Danial, A.W., and Abdel-Wahab, A.M. (2018). Enhancement of biohydrogen production from sustainable orange peel wastes using Enterobacter species isolated from domestic wastewater. *International Journal of Energy Resource* 43 (1).

94 Lu, G.Q., Diniz da Costa, J.C., Dukec, M. et al. (2007). Inorganic membranes for hydrogen production and purification: a critical review and perspective. *Journal of Colloid and Interface Science* 314: 589–603.

95 Balachandran, U.B., Lee, T.H., and Dorris, S.E. (2007). Hydrogen production by water dissociation using mixed conducting dense ceramic membranes. *International Journal of Hydrogen Energy* 32 (2007): 451–456.

96 Bromberg, L., Cohn, D.R., and Rabinovich, A. (1997). Plasma reformer-fuel cell system for decentralized power applications. *International Journal of Hydrogen Energy* 22 (1): 83–94.

97 Hammer, T., Kappes, T., and Baldauf, M. (2004). Plasma catalytic hybrid processes: gas discharge initiation and plasma activation of catalytic processes. *Catalysis Today* 89 (1–2): 5–14.

98 Paulmier, T. and Fulcheri, L. (2005). Use of non-thermal plasma for hydrocarbon reforming. *Chemical Engineering Journal* 106 (1): 59–71.

99 Sikarwar, V.S., Peela, N.R., Vuppaladadiyam, A.K. et al. (2022). Thermal plasma gasification of organic waste stream coupled with CO2-sorption enhanced reforming employing different sorbents for enhanced hydrogen production. *Royal Society of Chemistry RSC Advance* 12: 6122–6132.

100 Yan, Z.C., Li, C., and Lin, W.H. (2009). Hydrogen generation by glow discharge plasma electrolysis of methanol solutions. *International Journal of Hydrogen Energy* 34: 48–55.

101 Bespalko, S. and Mizeraczyk, J. (2022). Overview of the hydrogen production by plasma-driven solution electrolysis. *Energies* 15 (20): 7508.

102 Yildiz, B. and Kazimi, M.S. (2006). Efficiency of hydrogen production systems using alternative nuclear energy technologies. *International Journal of Hydrogen Energy* 31: 77–92.

103 Orhan, M.F., Dincer, I., Rosen, M.A., and Kanoglu, M. (2012). Integrated hydrogen production options based on renewable and nuclear energy sources. *Renewable and Sustainable Energy Reviews* 16: 6059–6082.

104 Nasharuddin, R., Zhu, M., Zhang, Z., and Zhang, D. (2019). A techno-economic analysis of centralised and distributed processes of ammonia dissociation to hydrogen for fuel cell vehicle applications. *International Journal of Hydrogen Energy* 44 (28): 14445–14455.

105 Makhloufi, C. and Kezibri, N. (2021). Large-scale decomposition of green ammonia for pure hydrogen production. *International Journal of Hydrogen Energy* 46 (70): 34777–34787.

106 Rice, W. (2006). Hydrogen production from methane hydrate with sequestering of carbon dioxide. *International Journal of Hydrogen Energy* 31 (14): 1955–1963.

107 Brewer, P.G. et al. (1999). Direct experiments on the ocean disposal of fossil fuel CO_2. *Science* 284 (5415): 943–945.

108 North, W.J. et al. (1999). Hydrate formation from gaseous CO_2 and water. *Environmental Science & Technology* 33: 1448–1452.

109 Monastersky, R. (1999). Goodbye to a greenhouse gas. *Sci News* 155 (25): 342–394.

110 Teng, H. (1998). Could CO_2 hydrate deposited in the ocean always reach the seabed? *Energy Conversion and Management* 39 (10): 1045–1105.

111 Hydrogen. (2023). University of Surrey Researchers Make 'Game-Changing' Findings for Hydrogen Production. Publsihed on 09 February 2023, https://www.world-energy.org/article/29317.html

112 Hydrogen offers a clean and plentiful supply of energy, *The Week Magazine*, May 2023, https://www.theweek.in/news/sci-tech/2023/05/31/scientists-develop-game-changing-catalyst-for-clean-hydrogen-fro.html

113 Chunping, X., Paone, E., Rodríguez, D. et al. (2020). Reductive catalytic routes towards sustainable production of hydrogen fuels and chemicals from biomass derived polyols. *Renewable and Sustainable Energy Reviews* 127.

4

Hydrogen Storage, Transportation, Delivery and Distribution

If you can't fly then run, if you can't run then walk, if you can't walk then crawl, but whatever you do you have to keep moving forward.

– Martin Luther King, Jr.

CHAPTER MENU

Hydrogen Energy: Production, Safety, Storage and Applications, First Edition. L. M. Das.

4.1 Introduction

Fossil fuels are rapidly decreasing. Rapid transition to intermittent renewable energy is urgently needed. Industries have initiated efforts to reduce their dependence on conventional fuels [1–3] and switch over to alternative sources. In this context, hydrogen evolves as a very promising alternative, and thus the storage of hydrogen is very significant. Figure 4.1 shows a broad central hydrogen pathway system involving several stages starting with hydrogen production, storage, transport, delivery and finally utilisation. R&D activities are being pursued at each of the above stages to ensure its accelerated entry into the existing energy infrastructure. The choice of hydrogen storage mode for a specific application depends upon various factors, including storage capacity, weight, volume, safety, cost and operational requirements. Different technologies are being researched and developed to meet the diverse needs of hydrogen storage across various industries and applications.

Safe storage of hydrogen for several industrial applications, including the automotive and power sectors, is of paramount importance, keeping in mind the physico-chemical properties of hydrogen and its temperamental combustion characteristics.

4.2 Properties of Hydrogen Relevant to Storage

It is essential to have a closer look at some of the relevant properties of hydrogen in different phases (solid, liquid and gaseous) so that a more judicious approach can be adopted for the storage method depending upon the specific application. One kilogram of hydrogen gas occupies over 11 m^3 at room temperature and atmospheric pressure. Hydrogen has a normal boiling temperature of about 20 K and a critical temperature of approximately 33 K, above which a liquid state cannot be formed through the application of pressure. Since the

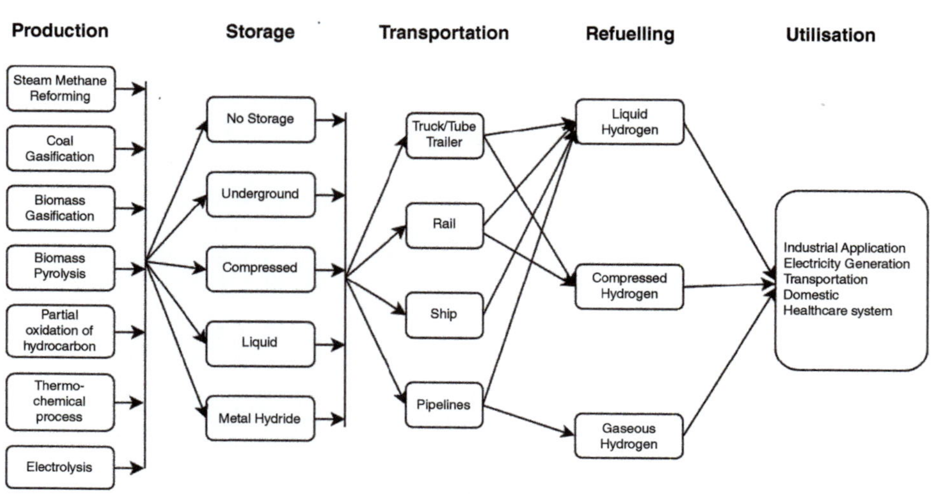

Figure 4.1 Hydrogen pathway system.

critical temperature is the maximum temperature at which a gaseous state can be liquefied under certain conditions, hydrogen behaves like an ideal gas under ambient pressure and temperature conditions. Hydrogen is the lightest of all elements, and since it behaves as an ideal gas close to ambient temperatures (~300 K) and pressure conditions, it has a very low normal density of 0.09 g l^{-1} (or 11 l g^{-1}) at 288 K and 0.1 MPa. Considering this normal density, with a low heating value of about 120 kJ g^{-1}, the typical energy density of hydrogen is 10 kJ l^{-1}. While gasoline has a specific energy of only about 42 kJ g^{-1}, its energy density is 32 000 kJ l^{-1}. This means that it has an energy density about 3200 times greater than that of normal hydrogen gas. Due to its low density, hydrogen gas is produced either in the form of liquid or compressed gas depending upon its usage. At cryogenic temperatures of about 20 K, it is usually produced near petroleum refineries. Alternative production methods involve compressed gas at higher pressure levels, approximately 15–25 MPa. The liquefaction of hydrogen consumes 30–40% of the lower heating value (LHV) of hydrogen, but the energy for hydrogen compression up to 35–70 MPa is in the range of 5–20% of LHV. An optimum hydrogen storage material must have (i) a high volumetric/gravimetric hydrogen storage capacity, (ii) fast absorption kinetics, (iii) a suitable room temperature and ambient pressure operation and (iv) lightweight materials.

Hydrogen storage options depend upon many factors such as whether storage is meant for stationary or mobile applications. Apart from weight and safety infrastructure, there are other practical considerations that should be taken into account. Transient operation and cost are extremely relevant factors, especially in the automotive industry, for which hydrogen is now being considered as a prospective fuel.

Hydrogen holds approximately three times more energy than gasoline. In comparison with gasoline, hydrogen requires 33 kWh kg^{-1}, while gasoline requires 12 kWh kg^{-1}. However, on a volume basis, the situation is reversed, and hydrogen requires higher volumes to store. It requires approximately 1 kWh l^{-1} for 700 bar hydrogen at 15 °C compared to 9 kWh l^{-1} for gasoline. Figures 4.2 and 4.3 show the mass-based and volume-based storage densities of hydrogen and other fuels, respectively [4].

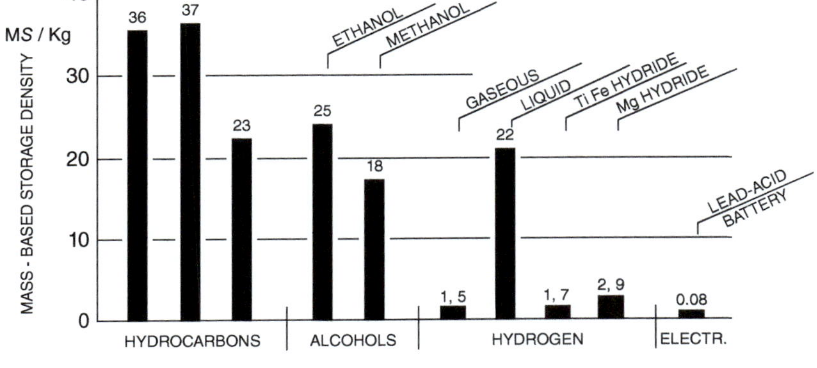

Figure 4.2 Mass-based storage density of hydrogen and other fuels.

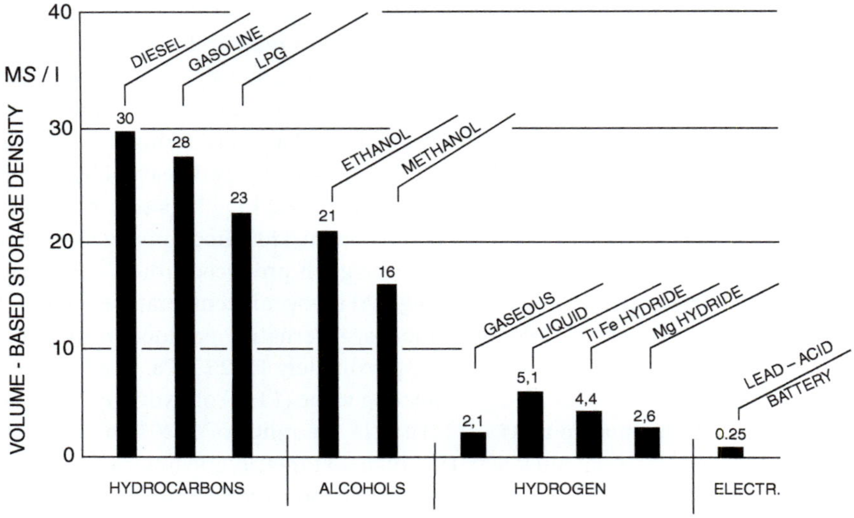

Figure 4.3 Volume-based storage density of different fuels.

4.3 Hydrogen Storage Criteria for Specific Application

With the rapidly increasing penetration of hydrogen in several emerging energy sectors, it has become essential that appropriate storage methods be developed for specific applications. This is because the demand for hydrogen broadly varies depending upon the application mode, such as mobile or stationary.

Hydrogen has high demand for certain applications. The hydrogen release response time should be much shorter in such cases. Additionally, a specific application might require a controlled rate of hydrogen storage. Thus, the capital, as well as operating costs related to the storage of hydrogen, depends upon the demand for a specific application [5, 6]. Broadly speaking, hydrogen applications could be either stationary or on the move as shown in Figure 4.4.

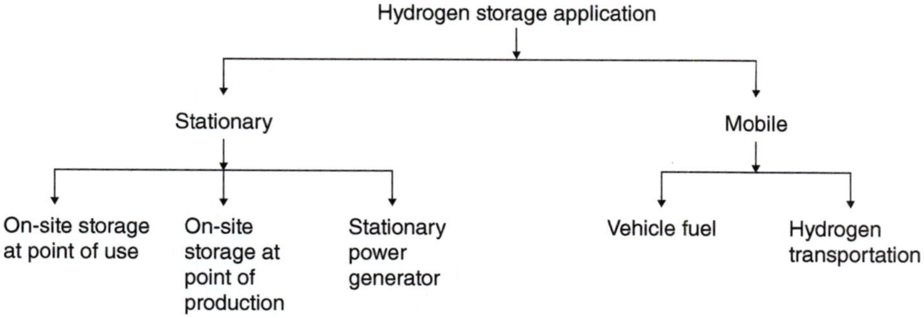

Figure 4.4 Hydrogen storage for stationary and mobile applications.

Depending on the application, the level of purity of stored hydrogen must be selected. A hydrogen-fuelled internal combustion engine can control some level of impurity, whereas a fuel cell requires a very high level of hydrogen purity to prevent catalyst poisoning. Thus, technological options for large-scale hydrogen storage provide a number of alternatives based on the quality and amount of hydrogen required for a specific application. The cost of storage is an important factor, particularly for large-scale storage facilities [7]. As reported in the literature, hydrogen in pure molecular form can be stored in a liquid or gaseous phase. The mode of storage of liquid hydrogen (LH) and gaseous hydrogen (GH), existing technological characteristics and strategies for industrial applications [8, 9] are important factors from the point of view of practical applications. The storage of hydrogen depends on the system where it has to be used. Large storage devices should be located at the production site or at the start or end of pipelines for stationary large storage systems. A small storage system can have facilities for the distribution of hydrogen to meet the demand for stationary applications. Mobile storage systems for transport and distribution include both large-capacity devices, such as a liquid hydrogen (LH) tanker-bulk carrier, and small systems depending upon the requirements. Several aspects of hydrogen storage systems are currently being researched and developed. Problems related to possible improvements in the conventional mode of storage, significance of the storage materials for stationary and mobile applications and, last but not least, detailed studies on storage for fuel cell have been discussed [10–16].

Large storage systems for stationary applications do not require rigid storage density and hydrogen release temperatures. Depending upon size and application, storage types may be differentiated. Hydrogen can be stored either as a pressurised gas or as a liquid or in the solid form as metal hydrides in physical combination with other materials. Compressed gas storage has the potential for large-scale transportation, while LH storage is used for air and space transportation. Metal hydrides contain metal atoms. In complex metal hydrides, it could be part of a complex ion that is bonded to a metal atom. On the other hand, chemical hydrides consist of non-metallic elements that are normally a combination of carbon, nitrogen, boron, oxygen and hydrogen. Metal hydrides are appropriate for small-scale transportation. The application of glass microsphere storage, chemical storage and cryo-absorbers is quite relevant in this context. Figure 4.5 [17] shows the energy density of hydrogen of other transportation fuels with respect to gasoline as the reference fuel. It provides appropriate guidelines for selecting storage space for a specific application.

Compressed gas storage is used for the storage of hydrogen in vehicles built on internal combustion engines. Metal hydrides have also been used for mobile applications in small engines. Recently, there has been a widespread interest in fuel cells. Different types of fuel cells are being manufactured for specific applications. Polymer electrolyte membrane (PEM) fuel cells are the most common fuel cells used for vehicular applications. It is important that these fuel cells are able to integrate heat into the storage system at an appropriate temperature interval. Whatever the mechanism, it is essential that the storage system must be able to provide the desired quantities of hydrogen under on-board conditions.

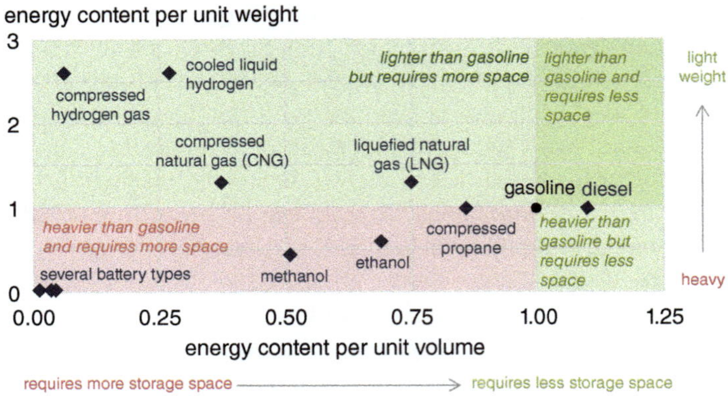

Figure 4.5 Energy density of hydrogen and other conventional and prospective fuels [17]/US Energy Information Administration/Public Domain.

4.4 Storage of Hydrogen as Compressed Gas

Storage and transport of hydrogen came into existence in 1880. During that time period, hydrogen transported in wrought-iron cylinders was used to inflate the war balloons. An important factor in a compressed hydrogen gas storage system is the storage components and the compressors. A compressor is necessary to raise the temperature of the hydrogen gas stored. The properties of desirable materials and operating costs are the major constraints for storing GH at 100 bar in the ground and 200 bar for underground storage [18]. Hydrogen is produced at relatively low pressures, but it is compressed for transport applications. The process of GH compression is carried out from an optimum thermodynamic point of view and taking the safety aspects into consideration. Andrzej Kwiatkowski et al. [19]discussed hydrogen compression stations depending upon specific applications and safety zones in the event of fire.

The performance of hydrogen storage options for on-board and off-board light-duty vehicles was analysed by Ahluwalia and his team of researchers [20]. After analysing several factors related to system efficiency, cost, environment and safety considerations for on-board and off-board conditions, it has been observed that compressed gas storage in high-pressure cylinders of Type III or Type IV with hydrogen gas at 350–700 bar will not be able to meet the 'near-term volumetric target' [20]. Compressed gas storage is commonly used in stationary applications, such as hydrogen refuelling stations, where a high storage capacity is required.

High-pressure vessels for storing hydrogen must be robustly designed. Hydrogen to be used for fuel-cell vehicles needs to have a pressure level of 700 bar, and therefore, the storage tanks for such applications use advanced vessel materials such as carbon fibres [21]. Compressed gas storage offers reliability, indefinite storage time, ease of use and affordability. Compressed hydrogen storage offers the least complex method of storing hydrogen. Hydrogen has one of the highest energy densities (120 MJ kg^{-1}) of all fuels. The energy content of hydrogen per volume is very low (0.0107 kJ m^{-3}). Under ambient conditions (300 K and 1 bar), 1 kg of hydrogen gas occupies 11 m^3. Consequently, hydrogen storage implies the reduction of an enormous volume of hydrogen gas. This is accomplished by

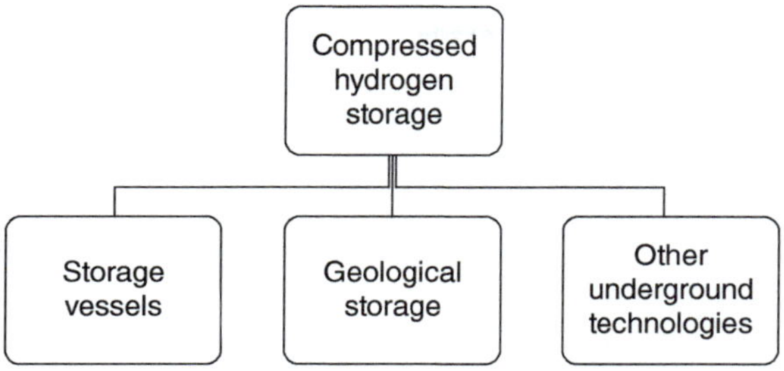

Figure 4.6 Storage modes of compressed hydrogen gas.

compressing the gas inside pressure vessels. Compressed GH can have three main types of storage: (i) storage vessels, (ii) geological storage and (iii) underground storage, as shown in Figure 4.6.

4.4.1 Types of Gas Cylinders

At present, there is not enough experience in the design and construction of large-scale systems in metallic vessels to store GH. In conventional practices, natural gas is stored in three types of metallic vessels [22]. Their design features and response characteristics under long-term operating conditions could provide some broad guidelines for hydrogen vessel design. It is crucial to select the right material considering the embrittlement phenomenon.

Cylinders used for the storage and transport of hydrogen can be classified into four types: Type I, Type II, Type III and Type IV, as shown in Table 4.1. Type I cylinders are made of steel or aluminium. These are the most common types of containers used by merchants for regular delivery and storage. The thickness of the cylinder walls and the maximum working pressure of the cylinder are determined on the basis of the tensile strength and the yield strength of the metal. Figure 4.7 shows the broad features of the design of a hydrogen cylinder.

Table 4.1 Types of storage cylinders.

		% Load taken
Designation	**Description**	**Metal vs composite**
Type I	A cylinder made wholly of steel or aluminium	100/0
Type II	A cylinder with a metal line of steel or aluminium and a hoop-wrapped composite overwrap	55/45
Type III	A cylinder with a thin metal liner of steel or aluminium and a fully wounded composite overwrap	20/80
Type IV	A cylinder with plastic liner and a fully wound composite overwrap	0/100

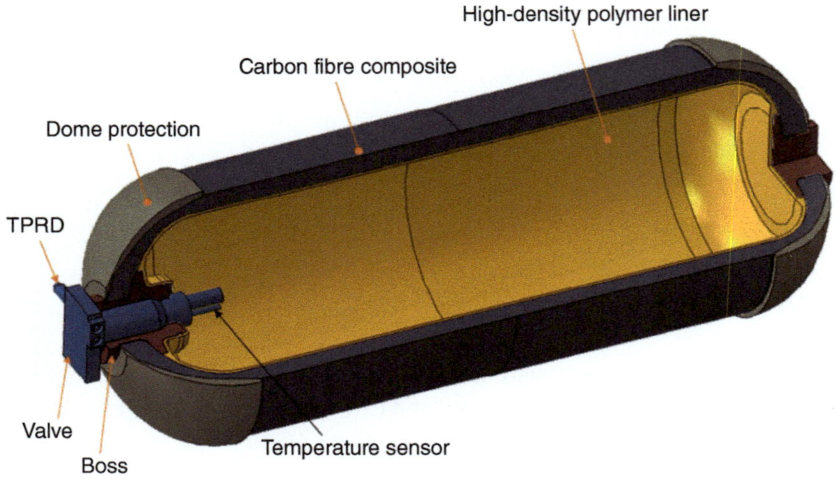

Figure 4.7 A typical hydrogen cylinder (*Source:* US Department of Energy/Public Domain).

Hydrogen tanks are equipped with thermal pressure relief devices (TPRDs) (Figure 4.8). These are designed to clear tanks and vessels when overheating. The design features are such that the glass ampoule is destroyed as the temperature reaches around 110 °C. This facilitates the immediate release of hydrogen from the tank through the safety vent. A check valve is installed to prevent the reverse flow. An automatic shut-off valve is installed to prevent the flow from the storage container to the engine or the fuel cell. Molkov et al. [23]assessed the response characteristics of TPRD mounted on a hydrogen gas cylinder. In the analysis, an approach of a non-adiabatic model of a tank–TPRD system under fire conditions was introduced from a safety point of view of the tank and TPRD.

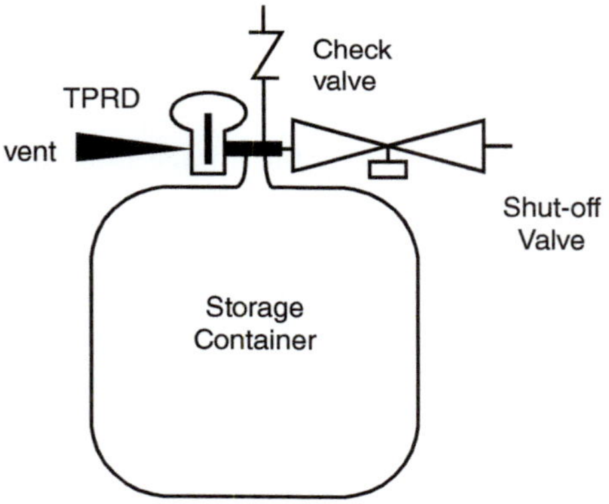

Figure 4.8 TPRD: thermal pressure relief device.

The hydrogen tank is a heavy storage container. A beneficial feature of TPRD is its rotatable nature and ability to be aligned according to the design of pipes. This becomes useful particularly because of the heavy weight of the tank and the position of the main valve. Type I tanks are the heaviest and can be built to a very large size. One of the major constraints for these Type I cylinders is that the mass of hydrogen stored to the mass of cylinder ratio is very low in these systems. This is because these cylinders can only hold 1–2% hydrogen storage compared to the cylinder mass.

4.5 Liquid Hydrogen Storage

In LH storage, when space between the inner vessel and outer container is evacuated, the whole apparatus is referred to as a dewar. A cryogenic dewar or LH tank is also a Type I hydrogen tank. Compared to GH, dewars are capable of storing higher density hydrogen. This helps in reducing the size of the vessel. Pressure relief valves, installed to reduce the pressure inside the tank and prevent any possible explosion, are sometimes responsible for causing hydrogen leakage, which is referred to as 'boil-off loss'. LH storage for on-board vehicular applications has a major drawback due to boil-off losses. Throughout the entire pathway, there are boil-off losses and transfer losses, as shown in Figure 4.9. Quite often, LH vaporises at extremely low temperatures and escapes the tank. As a result, the tank pressure increases due to the vapour in the tank. The process of pressure build-up continuously increases the weight of the tank. Unless the continuous rise in pressure is checked, this might even result in a tank explosion. LH storage is suitable for space exploration, where it is used as a propellant, and for some specialised transportation applications due to its high energy density.

4.5.1 Boil-off Losses

Boil-off typically occurs at a pressure of 1 MPa or less. Cryogenic LH tanks have been designed to prevent both convective and radiative heat transfer losses. The boil-off loss rate depends upon the quantity of LH stored in the storage tank. The surface area of the storage vessel is a crucial factor because the amount of heat leakage depends upon the surface area. Thus, LH storage tanks are constructed with double walls to maintain an extremely low

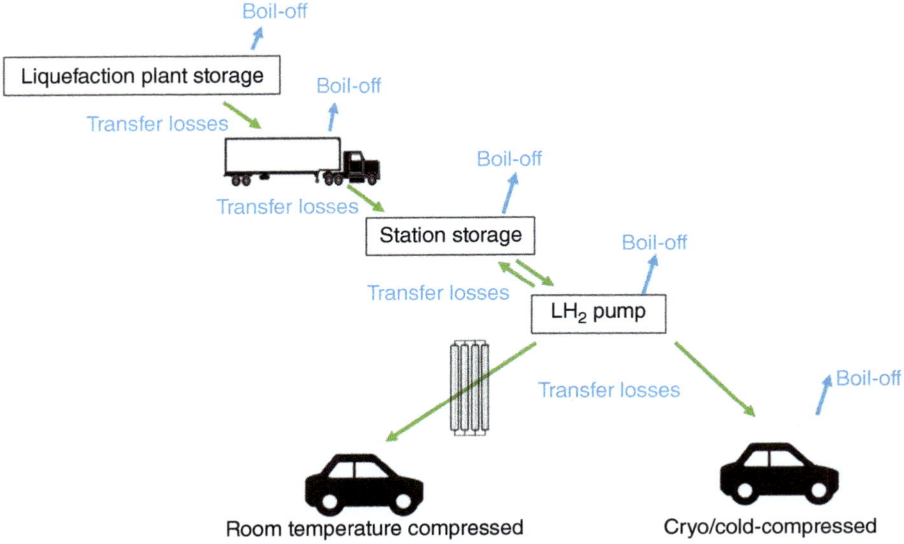

Figure 4.9 Transfer losses and boil-off losses over the entire LH_2 pathway (*Source:* US Department of Energy/Public Domain).

temperature with thermal insulation. Tanks are built up with multi-layer insulation (MLI). MLI consists of a thin metal layer that is also very effective in controlling both radiation and thermal irradiation between the layers. After a prolonged test period of about 13 months, cryogenic engineers at the National Aeronautics and Space Administration (NASA) implemented a method of integrated refrigeration and storage in which it is possible to continue to maintain LH in a zero boil-off state for an indefinite time period [24]. LH tanks have been installed in the trunk or rear of the vehicle and have been successfully tested. These tanks weigh about 50–60 kg and are usually made of aluminium alloy and fitted with vapour-cooled shields. Glass fibre or carbon fibre composites are used in the inner vessel to reduce the evaporation rate loss to about 1.5% per day. They are optimally instrumented with the relevant equipment, such as safety valves, tubes, heat exchangers and other subsystems installed at appropriate locations.

Considering the possible problems due to boil-off gas (BOG) in cryogenic hydrogen, the cryogenic temperature of hydrogen must be maintained to keep it in the liquid phase. When the temperature rises in the vessel, hydrogen boils and thereby changes liquid into the gaseous phase. Therefore, to minimise the BOG losses, the insulation level must be correctly calculated to ensure that the desired temperature is maintained. However, sometimes some amount of BOG occurs even in a well-insulated system. Therefore, as a precautionary measure, vents are always provided for the gases to escape. Under no circumstances, the pressure level in the vessel is permitted to exceed beyond the designed value. Type I cylinders are not suitable for automotive applications mainly because of weight and volume constraints. The Type II cylinder is a hoop-wrapped composite cylinder consisting of a metallic inner aluminium liner with an outer wrapping of a synthetic material because its physical strength improves and the cylinder can withstand higher pressure. Type III cylinders are fully wrapped composite cylinders with load-bearing metallic liners attached to them. Stored on board a vehicle as a gas bound with certain metals, as a liquid in cryogenic containers or as a highly compressed gas (690 bar) in ultra-high-pressure vessels, Type III cylinders derive most of their strength from the composite overwrap that is wound around the inner liner. This composite consists of high-strength fibres (usually carbon) that are wrapped around the cylinder in many layers and glued together by a resin such as epoxy.

Over the past several years, light Type IV composite cylinders have been developed, which can sustain higher levels of pressure up to 80 MPa. These cylinders offer low volumetric density that is almost half of that of LH at a normal boiling point. These typical features are of immense importance for the use of hydrogen in fuel-cell cars. The selection of a material with an appropriate tensile strength is a critical factor. Aluminium has a tensile strength of 50 MPa, while some steels have a tensile strength of 1100 MPa. The use of composite materials with lower densities offers the scope of increasing the value of tensile strength even further up to 2000 MPa. In the absence of any suitable alternative technology, in terms of gravimetric and volumetric storage densities, the use of composite gas cylinders seems to be very promising for transport applications. To ensure that fuel cells are the mainstay of power trains, appropriate infrastructure should be developed. Several transport systems using Type III (metal-line composite) and Type IV (all composite) have already been successfully demonstrated. However, the economic viability of these systems for large applications has yet to be established. Such compressed gas hydrogen storage tanks are expensive mainly because of the cost of carbon fibre. This is approximately 75% of the total vessel cost and about 50% of the cost of the entire storage system. Problems of mechanical failure due to high pressure in the storage tanks need to be closely monitored to improve the reliability and safety of the system.

4.5.2 Storage in High-pressure Gas Cylinders: Benefits and Challenges

As discussed, the storage of hydrogen as a compressed gas in metallic pressure vessels significantly contributes to the total weight of the system. This leads to low gravimetric efficiency. Hence, composite polymer pressure vessels in combination with increased working pressure up to 68.94 MPa are used to achieve high volumetric and gravimetric efficiency. Recently, the ever-increasing demand for hydrogen has led to the development and use of larger and more sophisticated equipment. Individual bundles of steel cylinders, large vessels/tubes mounted onto railway trucks and a variety of trailer trucks for road transportation, large cylinders mounted on semi-trailers and used standard gas cylinders have become common features of storage and transport.

High-pressure gaseous hydrogen (HPGH) is a mature technology with some very distinctive advantages. About 80% of the hydrogen refuelling stations worldwide have been using HPGH. Depending on the application area, HPGH storage vessels may be selected for stationary applications, vehicles or bulk transportation. Stationary storage systems are normally used in hydrogen refuelling stations where adequate large-scale hydrogen storage is required either for vehicular application or other applications such as operating decentralised power systems or for agricultural applications such as pumps. There are two types of hydrogen storage vessels: seamless hydrogen storage vessels and multifunctional layered hydrogen storage vessels. Seamless hydrogen storage vessels are made of high-strength quenched and tempered chrome–molybdenum alloy steels as per the prescribed specification. There is very little volume in these seamless storage vessels (about 0.411 m^3). Due to the size constraints of seamless thick-walled tubes, the volume of these vessels is limited. They have a higher maximum allowable working pressure of 65 MPa despite their lower capacity (volume). Therefore, a number of vessels need to be assembled when large volumes of hydrogen are required. It is possible to increase the range of maximum allowable working pressure to an even greater level. At the same time, increasing the operating pressure might give rise to embrittlement problems. Multifunctional layered stationary hydrogen storage vessels with valves and interconnected piping manifolds are used to meet high-volume requirements. The multi-vessel configuration consists of a flat ribbon wound cylinder, two double-layered hemispherical heads, two reinforcing rings and other necessary accessories.

A flat steel ribbon wound cylinder (Figure 4.10), as its name indicates, is a kind of vessel made from thin plates and steel ribbons. It is made of a thin inner shell, a layered shell and a protective shell. The most favourable design characteristic of the system is that there are no rigid restrictions on either the size of the cylinder length or the thickness of the shell. The failure of hydrogen cylinders often mainly occurs due to fatigue, which is accelerated by hydrogen embrittlement. Problems most often occur in bundles of cylinders that undergo a series of pressure cycles. Hydrogen embrittlement is an issue that should be analysed critically with materials in storage systems that come into contact with pure hydrogen.

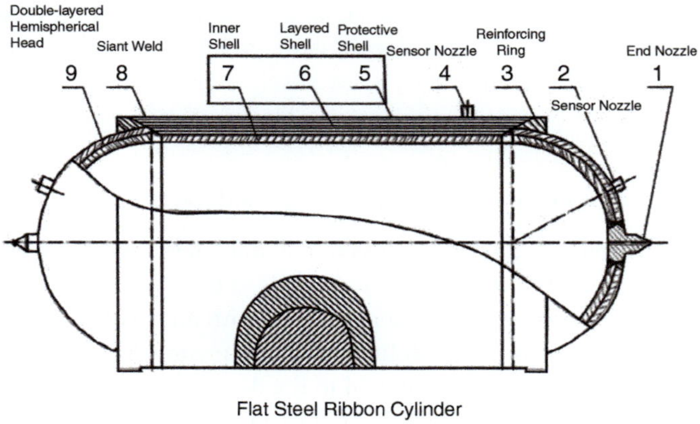

Figure 4.10 Features of a flat steel ribbon wound cylinder.

The desirable material is basically steel with limited tensile strength that does not have any harmful surface defects. The effect of embrittlement varies depending upon the design of a cylinder. Flat steel ribbon wound cylinders usually exhibit leakage under working pressure conditions but do not explode or burst. There is a provision for a protective layer shell to facilitate safety monitoring of the double-layered heads and a flat steel ribbon cylinder. There are three main parts of a flat steel ribbon cylinder: a thin inner shell, a layered shell and a protective shell. Double-layered heads are equipped with leakage monitoring facilities. In such designs, both layers have the same thickness. Under some circumstances, the outer head is capable of controlling operating pressure even if the inner layer leaks. Currently, HPGH hydrogen storage systems have been shipped in bulk via tube trailers from the source of production to end users. Normal pressures range from 16 to 20 MPa. However, the carrying capacity of these tube trailers is low. There has been a recent development of four full-winding storage vessels with a higher capacity. Each cylinder carries 150 kg of hydrogen, so the assembly of four vessels can carry a total of 600 kg of GH.

4.6 Underground Storage of Hydrogen

Underground storage of hydrogen is another technological option that has received a lot of attention. The major problems arise because hydrogen is chemically and biologically more reactive and thus needs to be handled carefully. It is perhaps worth recalling here that the geologic storage of natural gas and CO_2 has been adopted for decades and has attained technological maturity. Therefore, the decade-long experience gathered in storage and handling hydrogen-rich (natural) gases will be helpful in developing appropriate mitigation strategies to reduce the risk of biological and geochemical reactivity underground. It would be clear from the following discussions that underground hydrogen storage offers several advantages that make it an attractive option for storing large quantities of hydrogen. Underground storage facilities can accommodate large volumes of hydrogen, providing a reliable and scalable solution for storing surplus hydrogen generated during periods of low demand or from intermittent renewable energy sources. It minimises the land surface footprint compared to above-ground storage options, making it suitable for areas with limited available land or sensitive landscapes, such as depleted natural gas fields, salt caverns or aquifers. Such a storage system offers geological stability, reduces the risk of hydrogen leakage and enhances long-term storage reliability. Moreover, underground storage preserves the purity of stored hydrogen as it is isolated from the external environment and thus precludes chances of contamination. It can also play a vital role in balancing energy demand and supply. Excess hydrogen produced during periods of low demand or high renewable energy generation can be stored and later used to meet high demand or periods of low renewable energy availability. Underground hydrogen storage offers a practical and efficient means of addressing the challenges associated with hydrogen storage, making it a crucial component of the emerging hydrogen economy.

The method of underground coal gasification (UCG) to produce hydrogen is shown in Figure 4.11. In this gasification process, syngas (consisting of hydrogen, carbon monoxide and methane) is generated by igniting the coal. In this system, high pressure from deep UCG requires smaller and less costly plants. The composition of syngas is suitable for CO_2 capture. Combining UCG and carbon capture and storage (CCS) is a very attractive and less expensive approach to reducing carbon emissions. Underground storage or geological storage can broadly be classified into five categories: depleted natural gas and oil reservoirs, abandoned mines, aquifers, rock taverns and salt caverns.

The underground natural gas storage vessel in Skallen, Sweden, can provide broad guidelines for exploring the possibility of underground hydrogen storage. In this case, the major structural load is carried by the surrounding rock. The system is designed in such a way that a steel lining is provided all around, and the space between the inner steel lining and the surrounding rock is filled with concrete [25]. This existing natural gas storage facility is capable of withstanding a maximum pressure level of 200 bar. If such an appropriate system could be designed for hydrogen storage, it would be possible to store a large amount of hydrogen under similar conditions in depleted oil or gas fields and saline aquifers. Salt caverns, among them, are suitable due to several reasons. They are relatively less expensive. They can protect and insulate physical impact. In addition to providing secure storage systems, they are not prone to fires. Thus, the chances of hydrogen contamination are minimal. The specific amount of hydrogen that a salt cavern can store depends on several

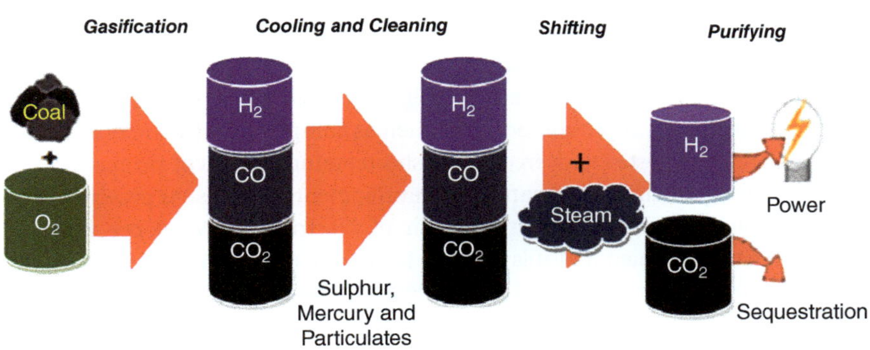

Figure 4.11 Underground coal gasification.

factors, such as the cavern's size, shape, depth and the desired operating pressure and temperature conditions.

Salt caverns permit a very wide volume range of hydrogen (9–6000 tonnes) to be stored under pressure. However, the inspection and maintenance of the underground storage system is cumbersome. Special care is needed to protect against corrosion of the external coating. Moreover, underground storage has to be location specific as it needs appropriate geological prerequisites. Several facilities for the underground storage of hydrogen are in operation or under development in different parts of the world. In-ground salt cavity storage systems in Teesside (United Kingdom) and Texas (United States) have been operational. These two countries are planning to further extend the underground storage facilities. It is important to note that the hydrogen industry is continuously evolving, and new projects are emerging. The H2Store project, located in an underground salt cavern in Bad Lauchstädt, Saxony-Anhalt of Germany, is one notable example of underground hydrogen. The H21 Project in Leeds of the United Kingdom aims at converting existing natural gas infrastructure to transport hydrogen and use an underground salt cavern for hydrogen storage. The Ulsan Hydrogen Energy Demonstration Project of South Korea includes an underground storage facility for hydrogen. Countries such as Japan, Australia and Canada and many other countries are exploring the use of underground salt caverns and other geological formations for hydrogen storage.

4.7 Liquid Hydrogen Storage

The main advantage of LH is its high density at low pressure. The density of hydrogen can be increased by liquefaction (condensation). LH has a specific advantage for applications where high density is required. Hydrogen has an extremely low boiling point of −253 °C at 1 bar. Moreover, it is not possible to cool down hydrogen by throttling processes (adiabatic and isenthalpic expansion) for temperatures above around 73 °C. Considering these factors, the thermodynamic characteristic storage of LH due to the process of liquefaction is very energy intensive, so liquefied hydrogen is expensive. The process of cooling and compressing results in the loss of about 30% of stored energy. LH has been used by NASA for a long time. Although there were initial failures, proper handling and effective utilisation have attained significant technical achievements. Hydrogen liquefaction was reported in 26 countries between 1898 and 2009. Designs of these plants of different capacities have evolved over the period of time. Efforts are being made to improve plant efficiency from 20–30 to 40–50%. Liquefaction plants are still designed on the pre-cooled Claude system. Improved designs of several major components, such as compressors, expanders and heat exchangers [26], are being tested to overcome the shortcomings. Plants of various sizes are being optimised through a range of research and development activities. At present, hydrogen liquefaction plants consume about 10 kWh of electricity (kWh_{el}) per kilogram. However, studies have indicated that it is possible to achieve 6 $kWh_{el}\ kg^{-1}$ if effective process improvements can be carried out [27–29]. Even if the demand for specific energy required for liquefaction can be substantially reduced, the liquefaction plant itself still significantly contributes to the major cost of the overall process of liquefaction even in the case of larger plants [27]. The volumetric ratio of LH to GH is 1:848. Therefore, it is

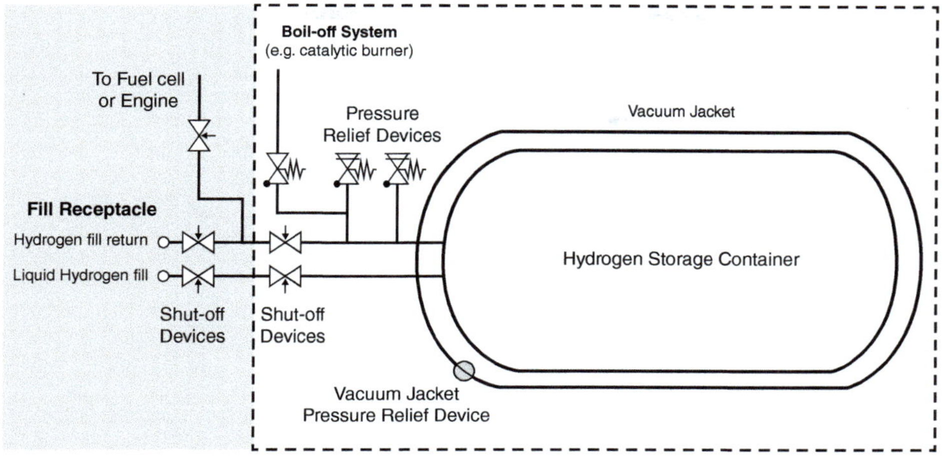

Figure 4.12 A liquefied hydrogen tank.

important that the storage tank be equipped with the appropriate closure devices, such as a pressure release device (PRD), along with an automatic shut-off valve and check valve installed at the appropriate locations. Figure 4.12 shows a typical liquefied hydrogen storage system (LHSS) along with the shut-off valves [30] and the arrangement of the boil-off system. The loss of stored hydrogen over a period of time is referred to as boiling off. The boil-off rate signifies the amount of stored hydrogen lost per day by this method.

Table 4.2 shows the density and the energy required by both GH and LH for on-board storage vehicles. Technology related to GH storage, even up to 700 bar, is developed even though in usual applications a much lower level of pressure is used for many other applications. LH storage units for on-board applications are much smaller in size and comparatively lighter than compressed gas storage systems at 700 bar. However, the technology has yet to be developed for compressed cryogenic LH to be used for such applications.

Figure 4.13 shows the pattern of variation in energy density for liquefied hydrogen and compressed gas hydrogen as a function of pressure [31]. As evident from this graph, LH is

Table 4.2 Density and the energy required by both gaseous and liquid hydrogen for on-board storage vehicles.

Phase	Temperature (K)	Pressure (bar)	Density (kg l^{-1})	Compressed/liquefied energy (MJ kg^{-1})
Compressed gas	300	350	0.0235	10.2
Higher pressure compressed gas	300	700	0.0387	18.5
Liquid	15–20	0.5–2	0.071	30–40
Compressed cryogenic liquid	25–50	300	0.08	Less than 1

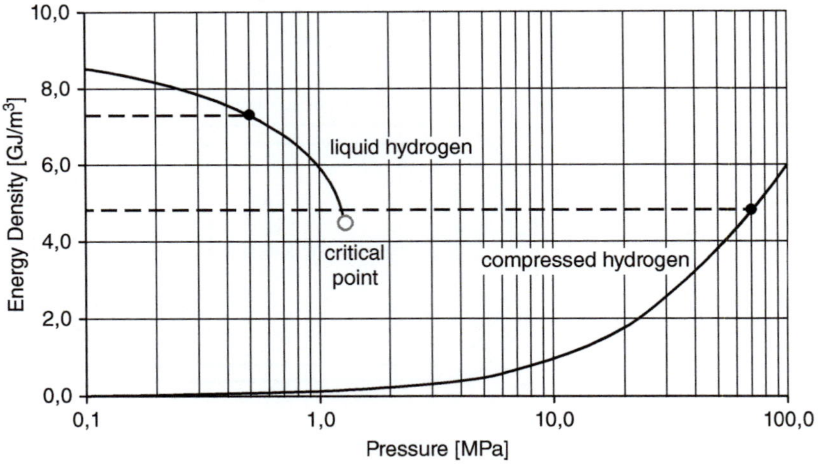

Figure 4.13 Variation of energy density of gaseous and liquid hydrogen [31]/Peschka, W./US Department of Energy/Public Domain.

preferred due to low storage pressure levels. As discussed earlier, boil-off is the major obstacle, and it is imperative to store the liquefied hydrogen safely so that evaporation is minimised. During the evaporation of LH, the energy spent on liquefying hydrogen is lost. Appropriate steps must be taken to vent the evaporated gas to avoid any pressure building up inside the storage vessel.

4.7.1 Design Features of Storage Vessels

The use of an appropriately designed storage vessel is crucial for the phenomenon of boil-off. A spherical storage vessel has the minimum surface-to-volume ratio. Hence, the boil-off can be used if LH is stored in spherical vessels. Moreover, proper insulation can be used to ensure minimum heat transfer through the walls of the storage tank. If the liquefaction plant and the LH storage system are installed in close proximity, then boil-off does not become a serious problem. In such cases, it is possible to use an ejector so that the cold BOG from the storage vessels can be injected back. Thus, the re-liquefaction process is less expensive. Boil-off losses can be drastically reduced for larger spherical vessels with robust effective insulation where the surface-to-volume ratio is low.

The cost of LH storage is not affected by the storage time because of the relatively lower cost of dewars. The boil-off factor becomes a point of consideration for the cost analysis only when the storage time is more than a week or two [32]. LH storage vessels are double walled with vacuum between the walls to minimise the heat transfer by conduction as well as convection. This vacuum sometimes creates difficulties and requires stronger container walls and more safety equipment. The inner surfaces of the vacuum spaces are sometimes silvered. In such cases, the vacuum space minimises conduction heating, and the silvering minimises radiant heating. The possible effect of heat transfer by radiation is also avoided by placing materials such as alumina-coated polyester sheets, i.e. aluminium and silica, in the space between the walls of the vessel. Alternating layers of aluminium foil and glass fibre have also been found to be effective in minimising radiation effects. Such useful technical information on thermal insulation along with storage, transport and distribution of LH has already been described in research publications [33–35].

From a safety perspective, the design of dewars is extremely significant. It is important that the system should be able to optimally manage the gas that is released as the liquid boils. In order to avoid any explosion, some dewars have an open top so that the released gas escapes. In some other simpler designs, there is a loose-fitting stopper that allows the released gas to escape. In some other advanced designs of dewars, there is an arrangement to trap the gas above the liquid. This arrangement maintains a relatively high-pressure level. This increases the boiling point of the liquid and thus allows it to be stored for a longer period. With the present level of technological advances, it is believed that large storage vessels with a capacity above 900 tonnes can be built at this time. According to a comparison study of large-scale hydrogen storage systems, LH tanks are more complex than pressurised gas tanks. However, LH storage is less expensive per weight of pressurised hydrogen gas stored.

4.8 Slush Hydrogen Storage

Slush hydrogen is a mixture of liquid and solid hydrogen at the triple point and thus is a cryogenic solid–liquid two-phase fluid. Here, solid hydrogen particles with diameters of several millimetres exist along with hydrogen. There is a distinctive positive feature of slush hydrogen that it possesses about 15% greater density compared to normal boiling LH. This can be attributed to the solid fraction in slush hydrogen. As indicated in Figure 4.14 of the Institute of Slush Hydrogen, Japan, this also increases the enthalpy (which is the refrigerant's heat capacity) by about 18% compared to LH at 20 K.

A very promising application of slush hydrogen emerges when large amounts of slush hydrogen are transported over long distances in combination with superconducting equipment. Such a situation facilitates both the transport and storage of hydrogen fuel and electrical power. Ohira's paper provides several aspects related to production, storage and transportation of slush hydrogen [36]. Slush hydrogen, due to its compact physical size, has a very high application potential as an ideal substitute to replace LH as rocket fuel. It needs smaller fuel tanks and can reduce the vehicle's weight. A given fuel load can be carried in tanks that are 13% smaller in volume. Broadly speaking, slush hydrogen, because it is a cryogenic solid–liquid two-phase fluid, has a wide range of applicability either as a solid or a liquid fluid. Due to the increase in density, it becomes possible to carry 15% greater loads.

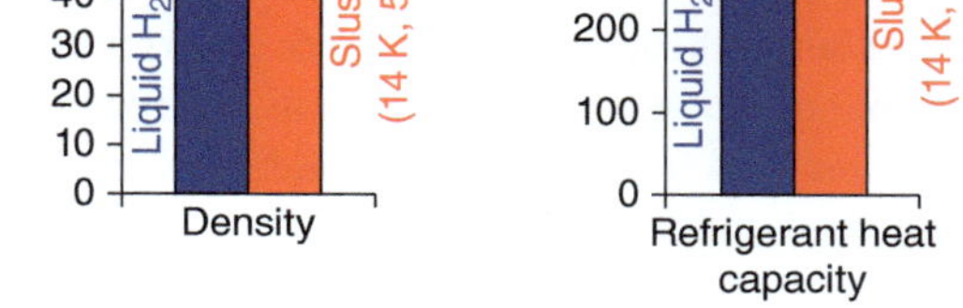

Figure 4.14 Density and refrigerant heat capacity of slush hydrogen (*Source:* https://slush-ish-english.com/technology8.html/last accessed 13 April, 2023).

The formation of slush hydrogen is initiated by repeating a freeze–thaw process in which the LH is brought closer to its boiling point. Subsequently, LH is made to boil off and vapourise by decreasing the pressure through a vacuum pump. In the process, latent heat is removed, and the temperature of LH is reduced. As the liquid is cooled and reaches its triple point, solid hydrogen is formed on the surface of the boiling liquid at the gas–liquid interface. Then, the vacuum pump is shut off, increasing the pressure, and the solid hydrogen that was formed on the surface starts melting. Repeated sequences of these series of reactions result in the formation of slush hydrogen. Some of the intrinsic constraints of LH storage are mainly due to its low density and temperature stratification, which do not occur in the case of slush hydrogen. The lower enthalpy of slush hydrogen reduces evaporation losses during storage and transport. This makes longer storage possible without venting. One of the main shortcomings of the freeze–thaw process is that it is a batch process. Moreover, since it operates at the triple point, it is quite likely that air is inadvertently drawn into the system at the absolute pressure level of about 0.07 bar. Sometimes this air could pose problems related to safety of hydrogen-operated systems.

4.9 Hydrides

As hydrogen energy technology emerges, storage has become one of the leading technical criteria to facilitate widespread hydrogen use. The selection of hydrogen storage materials is based on their ability to store a sufficient amount of hydrogen at an appropriate pressure in the operating range. Inherently safer than compressed GH or LH storage, metal hydrides are a very desirable option. A metal hydride is formed when certain metals absorb GH under high pressure and moderate temperature conditions. Hydrides are susceptible to contamination by impurities in the hydrogen gas, and 99.999% purity is necessary to ensure that hydrides have relatively long lives. Thus, hydrogen, which is chemically bonded to metal hydrides, is much stronger than the physical bonds involved in the adsorption of hydrogen. Therefore, more energy is required to release chemically bonded hydrogen. Considering the stronger bonding, it becomes possible to store hydrogen at a higher density even under ambient conditions [37]. Stored hydrogen is capable of absorbing and releasing large amounts of hydrogen gas several times without damaging the storage material or depleting it.

Figure 4.15 shows the mechanism of metal hydride formation, and Figure 4.16 shows how hydrogen is released during the de-hydrogenation process.

Metal hydrides absorb and release hydrogen gas when heated at low pressure and relatively high temperatures. Solid-state storage is either in the form of hydrides or absorption in nanoparticles. In both materials, hydrogen is absorbed under certain conditions and released under another set of conditions. The advantages of storing hydrogen in the form of metal hydrides include high-volume efficiencies, relative ease of recovery, indefinite storage capabilities without loss of hydrogen and a significant degree of safety.

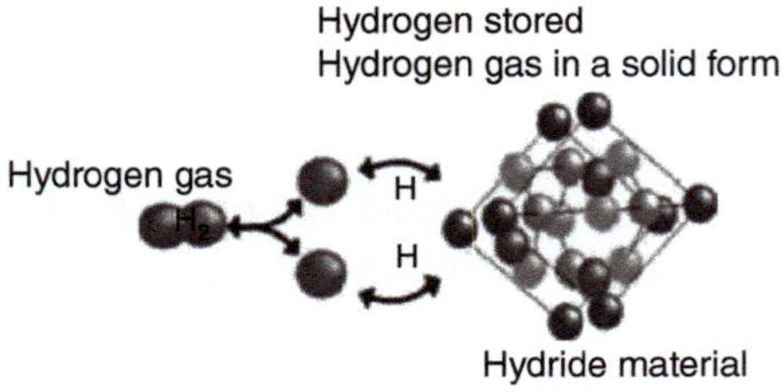

Figure 4.15 Metal hydride formation during hydrogen absorption.

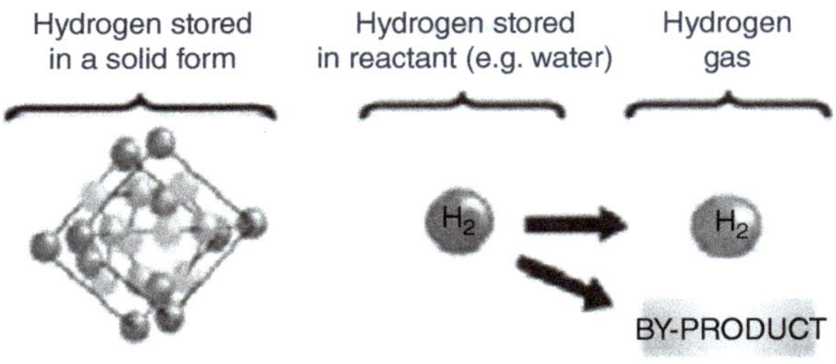

Figure 4.16 Hydrogen gas release during the dehydrogenation of metal hydride.

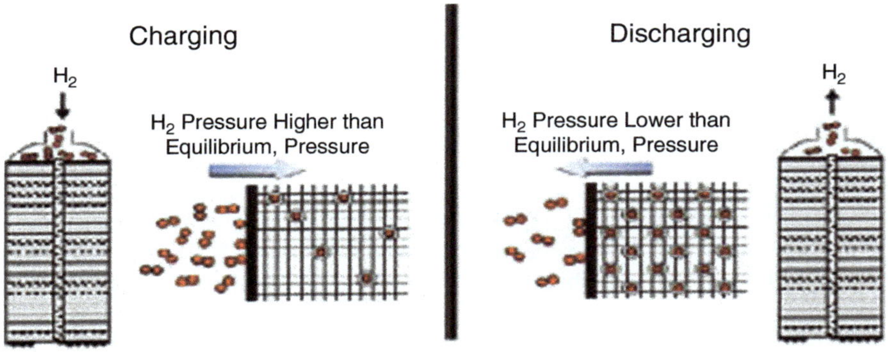

Figure 4.17 Charging and discharging.

Initially, hydrogen is pumped into a cylinder filled with small particles of metal alloy. In the following step, the alloy bonds to hydrogen and forms a metal hydride as a result of hydrogen absorption. During this process, heat is released. Appropriate steps must be taken to remove the material from the cylinder for the process to continue. Figure 4.17 shows the method of charging and discharging. A high level of hydrogen pressure, to the tune of 120 bar, is maintained during the charging process to form metal hydrides. Heat is added, and the tank pressure is allowed to decrease to about 35 bar to improve the purity of hydrogen in the tank. In this way, hydrogen is separated from hydrides and can be collected from the tank. As a result of its weight and volume restrictions, hydrogen storage is a problem for mobile applications in vehicles. In contrast, stationary storage systems have much wider flexibility in application as they can occupy a larger area and can be operated at higher temperature and pressure levels. The major challenge lies in the selection of the appropriate storage material.

Hydrides are currently being considered for use in vehicles. Vehicular hydride storage systems are usually long, thin, hollow cylinders that are tightly bundled and pressurised with hydrogen to about 34 bar. The most serious shortcoming of hydrides is their low mass and volumetric energy density which means that hydride vehicles are much heavier and less efficient than comparable gasoline. Figure 4.18 shows a comparative depiction of 4 kg of hydrogen stored in various modes. It is assumed that 4 kg of hydrogen is enough to cover a distance of 500 km.

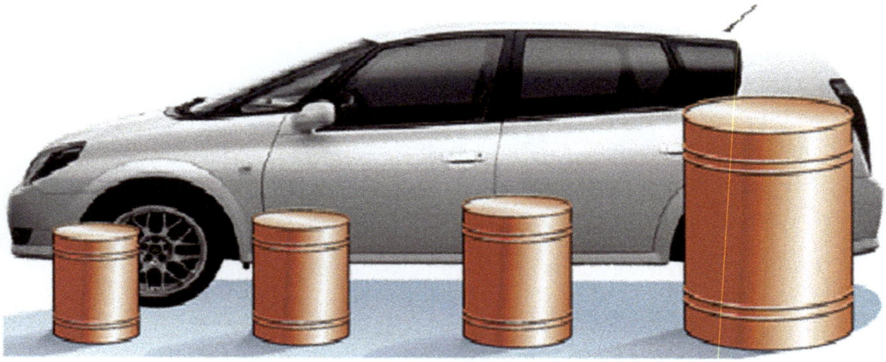

Figure 4.18 Volume and weight comparison of hydrogen storage in vehicles in different modes (*Source:* BirotiUMS).

Usually, the steel cylinder in which hydrogen is stored at 200 bar has a gravimetric density of approximately 1 wt% (1 wt% of stored hydrogen is equal to 186 Wh kg^{-1} of stored energy). Currently, high-grade carbon fibre composite cylinders have been developed, which can withstand a pressure level in the range of 700–1000 bar. These cylinders have a gravimetric hydrogen density of up to 10 wt%. On the other hand, the density of LH is close to 70.8 kg m^{-3} at −252.87 °C and 1.013 bar, so cryogenic liquid possesses a much higher gravimetric density than compressed gas. Therefore, it is essential that very robust insulation is provided in order to retain hydrogen in the liquid phase. As has been observed, despite very efficient insulation, boil-off rates are almost 1% per day for small tanks. Hydrogen is released from metal hydrides in two ways: thrombolysis and hydrolysis. As their names imply, thrombolysis is performed by heating, whereas hydrolysis is carried out by reactions with water. Both the approaches are different in their characteristics. The distinct features of thrombolysis are that it is a reversible endothermic reaction that occurs in solid phases at higher temperature levels. On the other hand, hydrolysis is an irreversible exothermic reaction that occurs in solutions at room temperature [38]. While many metallic elements can make binary compounds with hydrogen, most cannot store hydrogen due to thermodynamic limitations or insufficient storage capacity. Figure 4.19 shows the structure of a metal hydride.

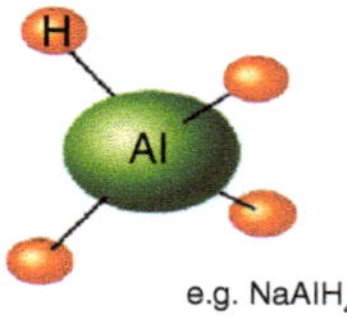

Figure 4.19 Structure of a typical metal hydride.

Only a few of them, such as magnesium hydride (MgH_2) and aluminium hydride (AlH_3), have been found to be suitable for the large-scale storage of hydrogen. Demirci, in his paper, discusses the level of technological achievement of producing hydrogen from certain boron–hydrogen (B(N) H) compounds [39]. One of the well-known compounds is ammonia borane (NH_3BH_3), which is a chemical compound consisting of boron, hydrogen and nitrogen. Ammonia borane has been extensively studied for use as a hydrogen storage material. Ammonia borane has a high hydrogen content of around 19.6% by weight.

Thus, it is an attractive candidate for hydrogen storage. However, there are challenges associated with developing practical hydrogen release systems based on boron hydrides, including the need for efficient catalysts and regeneration of boron hydrides after the release of hydrogen. Crivello [40] provides a detailed account of the development and application capabilities of magnesium-hydride-based materials. The availability of magnesium material at relatively low prices makes it an attractive hydrogen storage material. In addition, magnesium has a theoretical storage capacity of 7.6% (wt), which is very high. The role of magnesium-based materials in hydrogen storage has been discussed in several papers, highlighting the perspectives from several technological points of view [41–43]. Both hydrogenation and dehydrogenation reactions are largely affected by the gradual dissociation of molecular hydrogen on the magnesium surface. This is combined with the slow diffusion of hydrogen through the hydride phase. Thus, the reactions are very slow [41–43]. Considering the factors related to thermodynamic and kinetic constraints, it becomes necessary to maintain a temperature over 300 °C to ensure an effective dehydrogenation of pure MgH_2 [44]. Several research and development efforts have been devoted to overcome the constraints and improve the kinetics of dehydrogenation and hydrogenation reactions [39–45]. Pilot-scale storage systems using a variety of techniques have proven to be successful to some extent. One of the successful configurations has been pilot-scale storage systems based on particle size reduction by ball milling, alloying and subsequently adding metal additives. In order to increase the thermal conductivity of magnesium-based materials, several strategies have been implemented in order to make them a suitable material for the hydrogen storage process [46–48]. Long-term stability has been achieved, and reaction kinetics have been improved and made faster. Thermodynamic limitations related to heat transfer properties have been adequate. Approximately 6% (wt) hydrogen storage capacity has been achieved. MgH_2 pellets with volumetric hydrogen storage densities of 86 kg m^{-3} have been developed [45–47]. Despite all these remarkable improvements, the high temperature required for dehydrogenation and hydrogenation is the major stumbling block in the rapid and widespread use of magnesium hydride as a hydrogen storage material. Aluminium hydride (AlH_3) is another hydride that is considered a potential candidate for hydrogen storage. However, its properties are widely different from the hydrogen storage properties of MgH_2.

As a major limitation, aluminium hydride cannot easily be formed by a direct reaction between aluminium metal and hydrogen gas. For this reaction to be successful [49], it is necessary to optimise the mode of hydrogen regeneration from aluminium [50] so that aluminium hydride can be used as a hydrogen storage material. A wide range of studies have been conducted on aluminium hydride regeneration routes using thermochemical processes. Electrochemical regeneration routes have also been found to be successful to some extent [50]. Aluminium hydride as a hydrogen storage material has been critically reviewed [51]. High volumetric and gravimetric energy density, low decomposition enthalpy and rapid hydrogen release rates at low temperatures are some of the distinct characteristics of aluminium hydride. Furthermore, this hydride is kinetically restricted by particle size and surface coating. Hydrogen stability and decomposition could easily be adapted to small sensors, portable power systems, etc. The earlier system of ethereal reaction developed almost four decades ago still remains the preferred route of synthesis.

However, chemical and electrochemical regeneration technology, due to low energy consumption, is likely to have wider applications in the future. There are no major technical obstacles to their large implementation [52]. A family of complex hydrides essentially consists of group I or II metal cations, such as Li, Na, Mg and Ca, which are bonded to complex anions. These complex anions are bound to two or more hydrogen atoms, such as amides [NH_2], borohydrides (BH_4) and alanates (AlH_4). Some remarkable examples include $LiNH_2$, $LiBH_4$ and $NaAlH_4$ that have many potential practical applications such as for the storage of hydrogen in vehicles. For example, lithium borohydride ($LiBH_4$) has a total gravimetric capacity of about 18.5 wt% hydrogen and has proven to be useful for vehicular storage.

4.10 Hydrogen Storage in Zeolites

The storage of hydrogen in zeolites has shown that the storage capacity depends on different characteristics of zeolites, such as composition, specific surface area, storage temperature and pore architecture. Since gas molecules cannot penetrate zeolites at normal conditions of temperature and pressure, they are often forced into the channel at higher levels of pressures and temperatures. Therefore, gas molecules are usually entrapped in small sodalite cages as they have been observed to contain the highest concentration of sodalite. Active research has been carried out to study the synthesis of zeolite nanocrystals in a shorter crystallisation time, and efforts are being made to develop high-quality effective membranes at a lower cost. It has been reported that [53] Bikitaite zeolite exhibits excellent hydrogen storage. In both DDR and SAPO 34 zeolite membranes, H_2 separation performance from H_2–CO_2 and H_2–N_2 mixtures was quite satisfactory. However, this area of study needs to be further investigated before any large-scale commercial and industrial applications for some of these materials is contemplated.

4.11 Chemical Hydrides

Chemical compounds are also useful for hydrogen storage. In chemical hydrogen storage, hydrogen is stored in the form of chemical compounds that can release and capture hydrogen through chemical reactions. Chemical hydrides, such as metal hydrides, bond with hydrogen. The hydrogen content of these hydrides can be released through chemical pathways. The advantage of chemical hydrides vis-à-vis metal hydrides is that these are normally in a liquid state, and therefore, handling them during transport and storage is relatively easy. Hydrogenation and dehydrogenation also involve simple heat and mass transfers. Chemical hydrides are formed from chemicals such as methanol, formic acid and ammonia. Techniques adopted for this purpose include ammonia cracking, partial oxidation, methanol cracking, etc. Most of these chemicals are currently synthesised from natural gas. These chemicals already have a number of applications, not just for hydrogen storage. Widespread infrastructure exists to support the production, handling, transport and storage of these chemicals. However, an interesting feature of chemical hydrides in the present context is that chemical hydrides themselves can be alternatives to hydrogen in many applications rather than being used for hydrogen storage. Chemical hydrogen storage is still in the early stages of development, but it could find applications in areas where other storage methods are not feasible or economical. In the case of adsorbed hydrogen storage (physisorption), hydrogen molecules are absorbed onto the surface of high surface area materials, such as activated carbon or metal-organic frameworks (MOFs). This is suitable for portable and small-scale applications, where moderate storage capacities are required.

4.12 Nanomaterials for Hydrogen Storage

Following the recent discovery of carbon nanotubes (CNTs) in 1991, hydrogen storage technology in nanotubes has progressed very fast. The mechanism of hydrogen storage resembles that of hydrides. With the use of novel techniques such as nanocapillarity, several researchers have investigated hydrogen storage possibilities in carbon nanostructures and carbon nanotubes. Nanoscale materials provide many advantages and flexibility for improving a system's efficiency through their physical and chemical properties. Research efforts to store large quantities of hydrogen in several nanostructured systems, such as carbon nanotubes, boron nitride nanotubes, alanates, carbon nanocomposites, nanomagnesium-based hydrides, polymer nanocomposites, porous carbons, graphene, zeolites, MOFs, polymers and other materials, are in progress. Because hydrogen is a spherical small molecule, its dispersion is weak. Due to their high surface area, nanomaterials can be trapped in porous media and generate new active sites. Nanostructured materials have been found to be suitable for lightweight hydrogen storage, such as bulk absorption, intra- and inter-grain boundaries, and adsorption on the surface [54, 55]. Their microstructure allows them to increase the diffusion rate and decrease the diffusion length according to requirements. Thus, these materials have exceptional potential to influence the thermodynamics and kinetics of hydrogen dissociation and absorption. The storage of hydrogen in carbon nanotubes was first reported by Dillon [56]. In fact, since then, a lot of research has been carried out on several carbonaceous materials and their application potential for hydrogen storage. It has been reported that 'hydrogen can be physically adsorbed on activated carbon and it can be "packed" on the surface and inside the carbon structure more densely than if it has just been compressed' [57]. Researchers in the National Renewable Energy Laboratory, with their studies on carbon nanotubes, were able to achieve a hydrogen storage density of about 10% of the nanotube weight [58].

Carbonaceous materials are extremely suitable for hydrogen storage because of some of their characteristic features, such as high surface area, low mass density, adsorption capability and pore structure. Boron fullerene is being investigated as a potential candidate for hydrogen storage, and recently, borophene has drawn attention. Figure 4.20 shows several relevant properties of borophene and boron fullerene suitable for hydrogen storage.

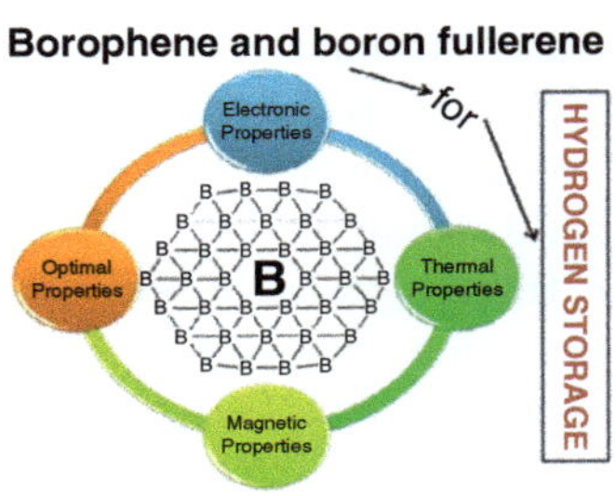

Figure 4.20 Borophene and boron fullerene for hydrogen storage [59]/John Wiley & Sons.

Researchers have examined doped borophene systems over pristine borophenes primarily for the 'enhanced properties' associated with metal atom decoration [59]. This work describes borophene- and boron-fullerene-based systems from the perspective of hydrogen storage mechanisms [59].

By using metal and fluorine with different levels of electronegativity, Im et al. [60] assessed hydrogen storage systems. The tests were carried out by introducing metal and fluorine into the 'activated carbon nanofibres by electrospinning, heat treatment and direct fluorination'. The KOH activation process was applied to develop the pore structure of the samples. Subsequently, nitrogen isotherms and micropore size distributions were used to analyse the results of the tests. In the metal–carbon–fluorine system, hydrogen storage was observed to increase by a large amount. The measured hydrogen content in this set of experiments was as high as 3.2 wt%. This happened because the electron of a hydrogen molecule was attracted to only the side of carbon nanofibres due to the electronegativity gap between metal and fluorine. Carbon nanotubes store hydrogen in the pores of microscopic tubes within the tube structures. Depending on their configuration and design, nanotubes can store hydrogen over a very wide range. According to the US Department of Energy, 6.5% by weight is a benchmark for the use of this storage technique in automobiles. Research and development activities are in progress, and a clear picture of its application potential has yet to emerge. It is not yet very clear as to under which temperature and pressure conditions, carbon nanostructures would be capable of storing the required amount of hydrogen for any specific applications, say for automobiles.

4.13 Hydrogen Storage in Hollow Microspheres

Hollow glass microsphere (HGM) is a porous material that stores hydrogen (Figure 4.21). The HGM has a diameter of 10–200 μm and a wall thickness of 0.5–2 μm. Hydrogen diffuses through the thin wall of the HGM at elevated temperatures and pressures. The gas is then trapped upon cooling to room temperature due to the poor thermal conductivity of a packed bed of HGMs. Slow release rates of hydrogen occur due to poor conduction of heat.

Heat can be used to increase the permeability of glass to hydrogen in a glass microsphere with a diameter of 50 m. Thus, it is possible to store 10 kg H_2 m^{-3} at a pressure of 25 MPa, and 20 kg H_2 m^{-3} can be stored at 62 MPa. Another desirable feature is that without increasing the permeability again, it is not feasible to release hydrogen. Several types of hydrogen storage technologies with their relative merits and disadvantages have been discussed including microsphere storage [61, 62]. It is certain that the future applications depend upon a number of factors for storage and processing and on the demand.

Hollow gas microsphere has the advantage of safely storing compressed hydrogen gases. It allows hydrogen to be stored at a much higher pressure. As compared to hydrides, it has a better weight density (hydrogen weight/total storage weight), while the energy required for this technique is almost one-eighth of the amount of energy needed for liquefaction. The geometry and overall design of the microsphere

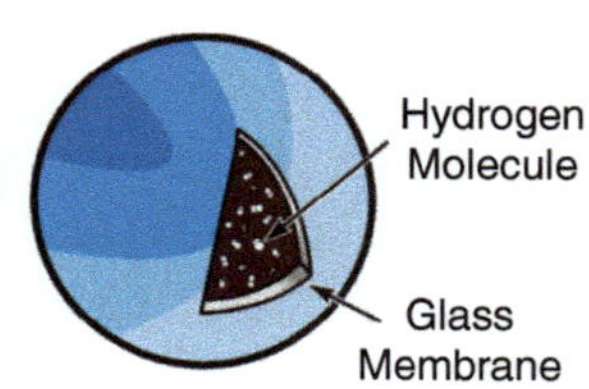

Figure 4.21 Hydrogen-filled hollow gas microsphere.

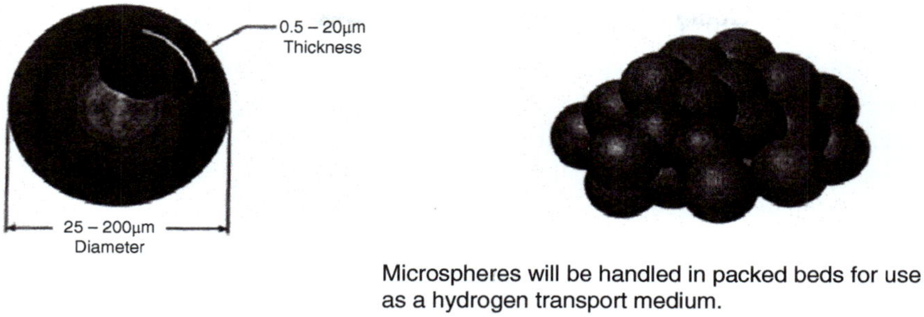

Figure 4.22 Glass microsphere for hydrogen storage.

for hydrogen storage affect the rate of hydrogen transfer. Relatively less expensive processes are required to develop monodisperse microspheres of a high-quality shell material. The microcapsule's material and dimensions are determined based on its strength and the weight of hydrogen to be stored inside. The permeability of the wall determines the time required to fill the storage cylinder. The shell material should be able to cater to short unloading times for on-demand delivery [62]. Hydrogen hollow microspheres are usually loaded and unloaded in batches. The existing method of coating should be improved so that the shell is strong enough to withstand a large number of loading and unloading cycles. The maximum storage pressure is a critical factor in the design of a cylinder. Figure 4.22 shows some typical hydrogen storage in glass microspheres.

The problems of safety, which are closely associated with single-storage high-pressure cylinders, are not applicable once these HGMs are filled with high-pressure hydrogen gas. However, one of the major constraints of the systems is that the hydrogen release rate is relatively much slower because of the poor thermal conductivity of the packed bed. The HGM system has the advantage of being able to continue refuelling as long as the glass membrane is intact. As compared to traditional heating methods, the time required to release hydrogen is substantially lower, mainly due to optical radiation of the appropriate wavelength. It is necessary that some materials be developed and strategies should be executed to unload hydrogen at a temperature below 100 °C. In such a situation, it would be possible to use the 'waste heat generated by a fuel cell or a simple heater powered by a battery during fuel-cell start-up' [62].

4.14 Hydrogen Transportation

Considering the potential application of hydrogen for use in several energy sectors, it is important to develop an optimum solution for an efficient and cost-effective transportation of hydrogen from its production site to the point of use. There is a direct correlation between hydrogen transportation and hydrogen storage systems. Broadly speaking, compact forms of hydrogen storage are relatively cheaper to transport than diffuse forms. At present, microscale hydrogen production systems are being developed primarily for local domestic applications that do not face any transportation problems. The transportation of hydrogen for large-scale production systems on a commercial scale must be closely studied and appropriate infrastructure developed.

4.14.1 Transport of Liquid and Gaseous Hydrogen

A stable and reliable infrastructure must be developed. This will enable the optimum delivery of either GH or LH from the production site to the point of use. This could be accomplished by either a dispenser or a stationary application site. There are several mechanisms, such as transfer by pipeline, cryogenic liquid tanker trucks on roads, gaseous tube trailers, railway or bars, depending upon the location of the end user.

Trucks that are used to transport LH are referred to as hydrogen tankers, and trucks that are used to transfer GH are called 'tube trailers'. The transfer of LH significantly differs from that of GH in some basic ways. Currently, LH is transported by super-insulated cryogenic tanker trucks of different sizes. The gross volume capacity of typical LH tankers, at present, is in the range of 28 400–49 200 l. On the other hand, GH, since it is produced at relatively low pressure (approximately 20–30 bar), needs to be compressed to about 180 bar after which they are put in a long cylinder on the trailer. Tube trailers are popular because they look like long tubes. Regulations and rules govern the normal pressure limit of tube trailers. The carrying capacity of tube trailers depends upon material and design features. Steel tube trailers, which are commonly employed, can carry about 350 kg of weight. Currently, some composite storage vessels have been developed that can carry about 560–900 kg of hydrogen per trailer [63]. A schematic diagram of a tube trailer is shown in Figure 4.23 [63]. They are part of the hydrogen highway and usually precede a local hydrogen station.

4.14.2 Hydrogen Transport Through Pipelines and Ships

Hydrogen pipeline transport is used to connect the point of hydrogen production or delivery of hydrogen with the point of demand. Figure 4.24 shows the hydrogen transfer from the production plant to the refuelling station. Pipeline transport costs are similar to those of compressed natural gas (CNG). However, most hydrogen is produced on demand sites,

Figure 4.23 Tube trailer carrying hydrogen [63], US Department of Energy https://www.hydrogen.energy.gov/h2a_delivery.html/last accessed 13 April, 2023/Public Domain.

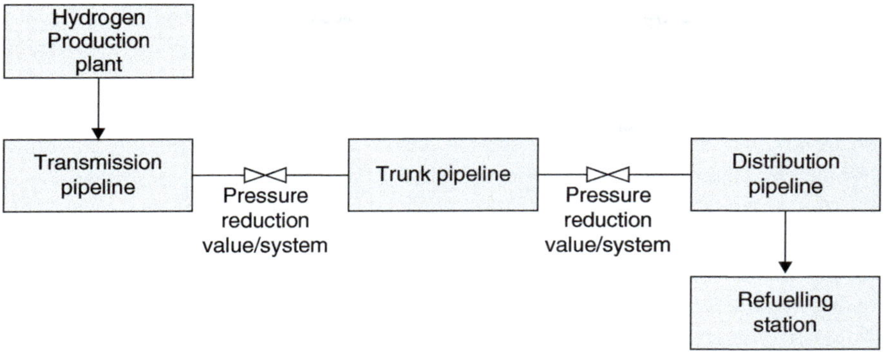

Figure 4.24 Hydrogen transfer from the production plant to the refuelling station.

80–160 km from an industrial production facility. Hydrogen can be transferred through pipelines either in a gaseous or a liquid form. It is well known that a widespread and stable network already exists in several parts of the world for the pipeline transfer of natural gas. In a similar way, it is possible to transport GH by pipelines. The pipelines installed for natural gas transport cannot be exchanged for hydrogen transport because of the fundamental difference in the specific properties of two gases. On the other hand, pipelines for hydrogen transport have to be designed on the basis of some specific properties of hydrogen. The density of hydrogen is much lower than that of natural gas. Therefore, the pipelines installed for transferring natural gas would transfer a much lesser amount of hydrogen under similar pressure conditions. Moreover, the energy density of hydrogen is almost one-third of that of natural gas on a volume basis. Therefore, to transfer an equivalent amount of energy as that of natural gas, the amount of hydrogen pumped should be increased to three times.

As far as material compatibility is concerned, pipelines, pumps, valves and all other accessories need to be compatible with hydrogen. Pipelines, in particular, must be resistant to embrittlement, and these pipelines need to prevent hydrogen permeation or leaks. The presence of any of these problems would result in a crack or rupture over time. Hydrogen pipelines are being successfully distributed using fibre-reinforced polymers (FRPs) in some parts of the world. FRP pipeline installation has also been found to be 20% cheaper than steel pipeline infrastructure. This is because FRP is available in much longer sections than steel, thereby avoiding a number of points to be welded for a long transmission system.

There are several hundreds of kilometres of hydrogen pipelines operating in several parts of the world. More than 2500 km of hydrogen pipelines are operating in the United States of America. These pipelines are laid mostly in the vicinity of large-scale hydrogen users such as chemical plants and petroleum refineries. Apart from the United States of America, some pipelines also operate in Canada and Europe. Some countries have established pipeline networks that can transport hydrogen-blended natural gas. A wide pipeline network could initially be used to transport natural hydrogen as a blended fuel. We will be able to use this information to design retrofitting seals, valves, gaskets and some fittings in order to prevent leaks from occurring. Based on the operating experience of these tests, suitable materials, leak detectors and other safety devices can be identified for the design of hydrogen-specific pipelines.

Rhine-Ruhr laid about 240 km of hydrogen pipes in 1938. As is well known, Rine-Ruhr is a highly industrialised, densely populated region of Western Germany that includes several major cities such as Bonn, Cologne, Dusseldorf, Dortmund and many others. Hydrogen steel pipelines are 25–30 cm in diameter and carry compressed hydrogen at a pressure of 210–220 bar. These are still in operation. Hydrogen pipelines were also laid out in Isberques in France in 1973, which were later extended to Zeebruqe in 1985. In 1997, pipeline connections were extended to Rotterdam. During the period 1997–2000, two separate hydrogen networks were developed near Corpus Christi, Texas, and between Freeport and Texas City. Another extension of the pipeline of about 240 km from Plaquemine to Chalmette was carried out in 2009. Thus, existing hydrogen gas pipelines operating in several parts of the world have shown that even these short expensive pipelines are worth in some specific locations.

Hydrogen ships are expected to be similar to the liquefied natural gas (LNG) ships that are widely used at present. All ships are expected to have an energy efficiency of 95%, though this would depend on the distance they travel. Transporting hydrogen by rail is similar to road transport. However, rail transport is viable over longer distances and in larger quantities. There are several advantages to transporting LH by air, such as it is lightweight and the delivery time is much shorter, and therefore, evaporation is not a serious problem. The transport of pressurised gas through air is not desirable due to safety and weight considerations. The transport of hydrogen by ship is a feasible option for situations where pipeline transfer is not possible. Because of its low energy density by volume, it is better that hydrogen gas is converted into a high-density liquid form and transported by ship.

4.14.3 Hydrogen Storage in Vehicles

Hydrogen is one of the most immediate and widespread applications. Transport, dispensing and delivery of hydrogen are extremely significant from a point of view of their utilisation. Figure 4.25 shows the transportation of GH and LH from the production facilities to the vehicle through trucks and pipelines.

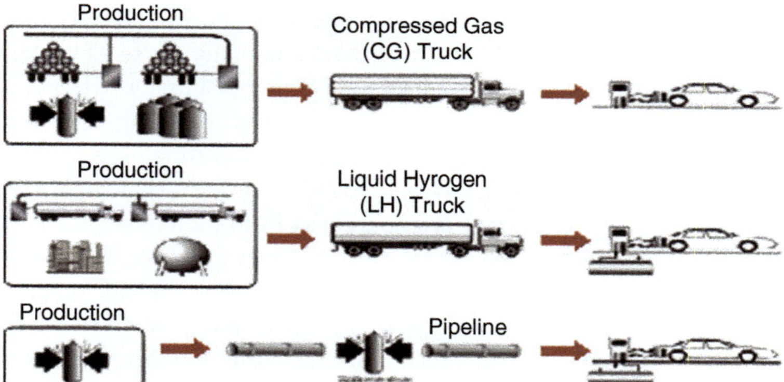

Figure 4.25 Gaseous hydrogen and liquid hydrogen transportation modes from production to vehicle [63]/US Department of Energy/Public Domain.

The below table gives a general idea of the factors that determine the phase of deployment of various modes of hydrogen transfer in liquid and gaseous forms.

	Capacity	Loss of energy	Regular fixed expenses	Variable expenses	Utility period
Tube trailer for gas	Low	Low	low	High	Near term
Truck trailer for liquid	Medium	Relatively high	Moderate	Moderate	Medium to long term
Pipelines	High	Low	High	Low	Medium to long term

Figure 4.26a indicates different stages of GH dispensing, and Figure 4.26b shows different stages of LH dispensing.

4.15 Transport of Gaseous Hydrogen

The transport of GH has been accomplished for different applications. Gaseous compressed hydrogen in pressure ranges of about 150–400 bar in long cylinders is usually stacked on a trailer, which is hauled by a truck. These tube trailers usually carry a number of large, high-strength steel tubes linked together through a common manifold. Tubes hold hydrogen from 63 to 460 kg. Steel tubes limit the carrying capacity of these tube trailers. Recently, composite storage vessels have been developed, which can carry 500–900 kg of hydrogen in one trailer. Several components constitute the storage system for fuelling a vehicle. Apart from the storage container and connectors, it is also essential to select piping and

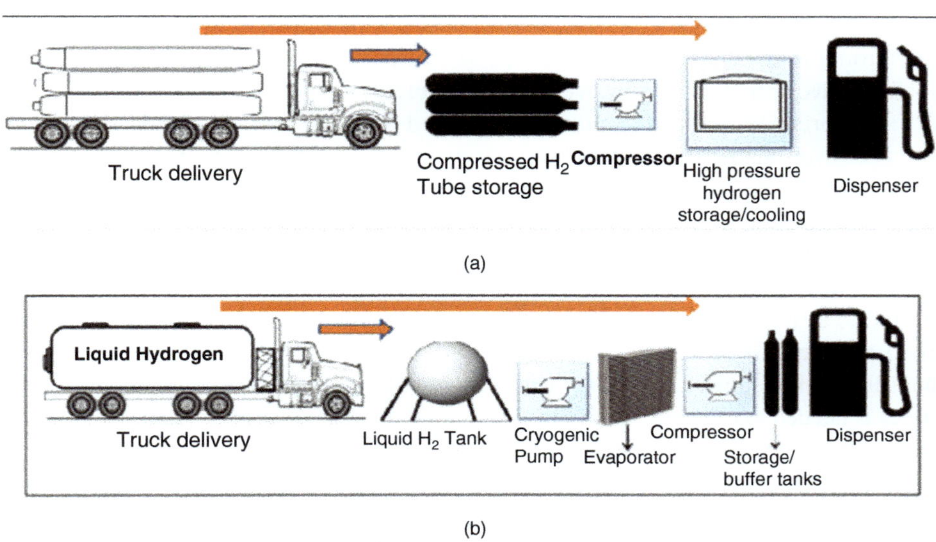

Figure 4.26 Hydrogen gas dispensing (a) and liquid hydrogen dispensing (b).

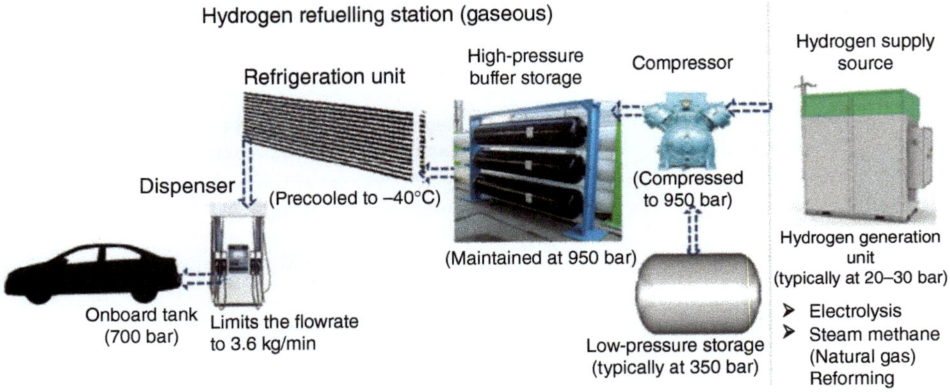

Figure 4.27 An integrated gaseous hydrogen refuelling station [64]/with permission of Elsevier and [64], Reddi et al., 2016/with permission from Elsevier.

vents that are technically robust. In order to control the volumetric flow rate in pipes, controlling devices such as regulators should be installed at the appropriate locations.

To prevent hydrogen accumulation in the event of leaks or for other reasons, adequate ventilation is essential. A typical hydrogen fuelling station with GH is shown in Figure 4.27. Fuelling small fleets with 350 bar hydrogen is extremely beneficial for composite mobile refuelling. One of the main advantages of these systems is that they can be implemented as soon as possible since they are cost-effective and at the same time reliable. Composite materials maintain vessel strength for high-pressure hydrogen storage while reducing the weight of the trailer.

4.16 Liquid Hydrogen

Apart from gas pipelines, hydrogen can also be transported in a pipeline as a liquid, but LH pipelines have been laid out only over a short distance. These pipelines have to be equipped with appropriate insulation to maintain the desired cryogenic temperature. Design criteria need to be evolved to avoid the possibility of two-phase (liquid/gas) flow formation. During exposure to air, sometimes the air liquefies and oxygen concentrates on the surface of the pipe. Such a situation increases the risk of fire. In the absence of pipelines, hydrogen is normally transported in the liquid form. Hydrogen is liquefied and cooled to cryogenic temperatures by the liquefaction process. Then, they are transported by trucks usually referred to as liquid tankers. LH is carried as a cryogenic liquid in double-walled, super-insulated vacuum-lined tanks. It has been observed that 70 kg of LH occupies a volume of about 1000 l and requires a container with a weight of 200–300 kg. A container of about 2–5 tonnes is needed to transport the same weight of hydrogen as gas or hydride. Thus, hydrogen transport in the liquid form is far more efficient than transfer as a high-pressure gas, particularly where higher volumes of hydrogen need to be transported. Usually, trucks can carry 360–4300 kg of LH, whereas railcars can carry about 2300–9100 kg of hydrogen.

Hydrogen production on-site plays a critical role in transportation and delivery. Sometimes, it is better to manufacture hydrogen at or near the location where it is used.

As a result, transport costs are minimised. The production of hydrogen at the site is always desirable in large industries requiring large amounts of hydrogen as chemicals. Ships are often used to transport large quantities of hydrogen over long distances.

Such hydrogen ships could be similar to existing LNG ships that already have a very established network. The efficiency of hydrogen transport on ships depends to a large extent on the distance it needs to be carried. Similar to ship transport, hydrogen can also be transported by rail or air. Energy efficiency is usually 95%. Hydrogen transport by rail is almost similar to the process of hydrogen transport by road. In contrast, transportation by rail is usually carried out when the amount of hydrogen is adequate, and it must be transported over a relatively long distance. From a safety perspective, LH has a definitive advantage. It is lighter than pressurised gas. In addition, evaporation is a problem of lesser concern with LH transport by air as the delivery time is much shorter.

4.17 Hydrogen Dispensing

There are definite steps that must be followed for hydrogen dispensing or fuelling. Any component used in the dispensing system should be approved by the corresponding administrative authority for safety purposes. Hydrogen is generated on-site or transported from the site of generation to the fuelling station during the first stage of the process. Furthermore, it varies depending on where hydrogen is stored and how it has to be fuelled (gas or liquid). Therefore, hydrogen stored at the dispensing site can be used in its original form. However, it has to be converted into the form the vehicle needs to be fuelled with. In the typical dispensing facility shown in Figure 4.28, hydrogen is in the liquid form and converted from liquid into gas for final distribution to vehicles [65].

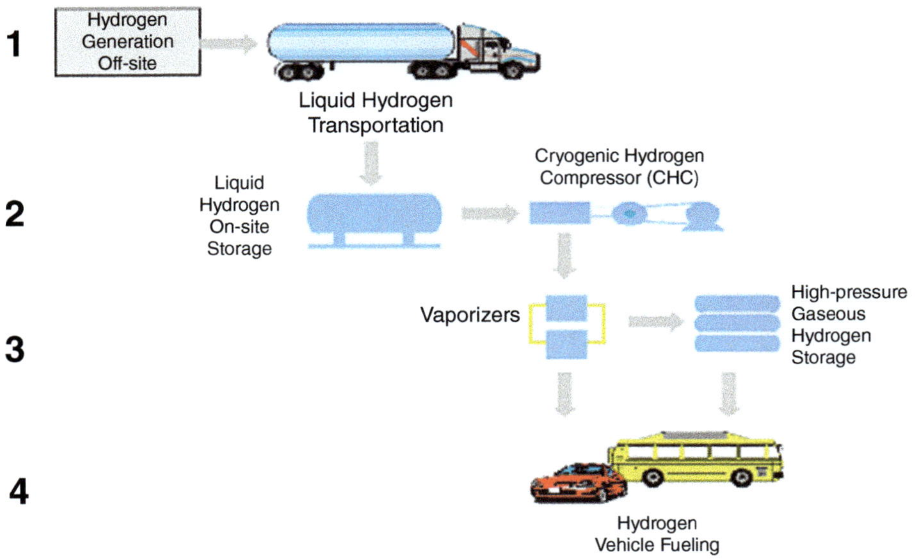

Figure 4.28 Hydrogen dispensing facility overview [65]/US Department of Energy/Public Domain.

4.18 Distribution and Delivery

The efficient transmission of hydrogen and its safe delivery form the backbone of the hydrogen economy. Hydrogen transport and distribution are very closely connected. As discussed, a hydrogen distribution system exists in the Rhine-Ruhr area, which includes 875 km of pipeline between a number of chemical and petrochemical plants. Hydrogen is produced by several methods, such as partial oxidation, electrolysis of water, catalytic cracking of methane, coal gasification, etc. Hydrogen is used to manufacture ammonia, fuel gas, methanol, etc. In laying out the distribution network, venting techniques are adopted to prevent explosions. Appropriate steps are taken to avoid problems due to leakage, and the unique property of hydrogen's diffusivity is very useful in preventing rupture. Pipe materials are ideally suited for hydrogen applications. It is very appropriate to design the distribution network and to centrally control the pipeline network in a facility.

As far as the vehicle is concerned, it is necessary that a delivery pathway starts at the location where hydrogen is produced and extends up to the vehicle's storage tank. It is possible to classify delivery in two stages. In the first stage, hydrogen could be transmitted from the production plant gate to the transfer point. Then, hydrogen can be carried from the transfer point to the refuelling station for dispensing into the vehicle's tank. There are several significant steps in the pathway that must be analysed separately, such as liquefiers, pipelines and refuelling stations. It mainly consists of GH delivery, LH delivery, and novel solid or liquid hydrogen carriers [66]. Figures 4.29–4.31 show the gaseous delivery pathway, liquid delivery pathway and carrier delivery pathway, respectively. Figure 4.32 depicts the gaseous fuel distribution.

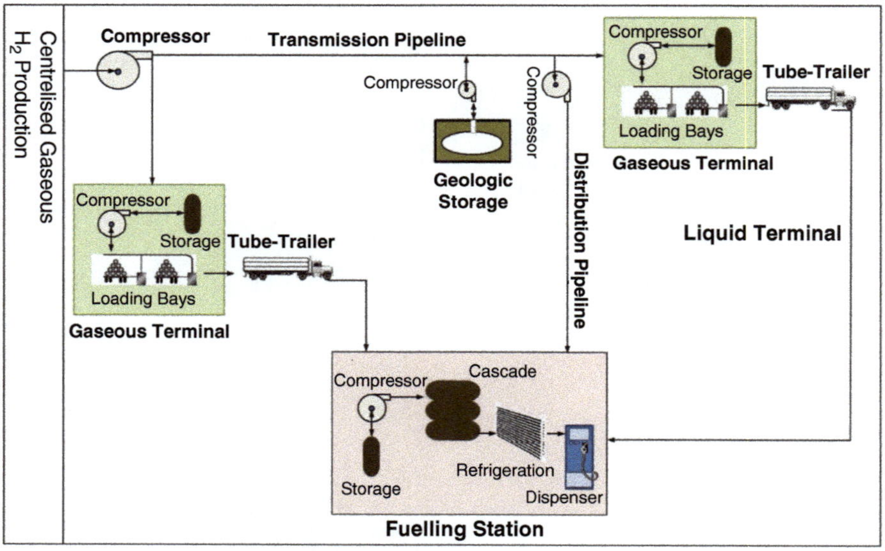

Figure 4.29 Gaseous delivery pathway [66]/US Department of Energy/Public Domain.

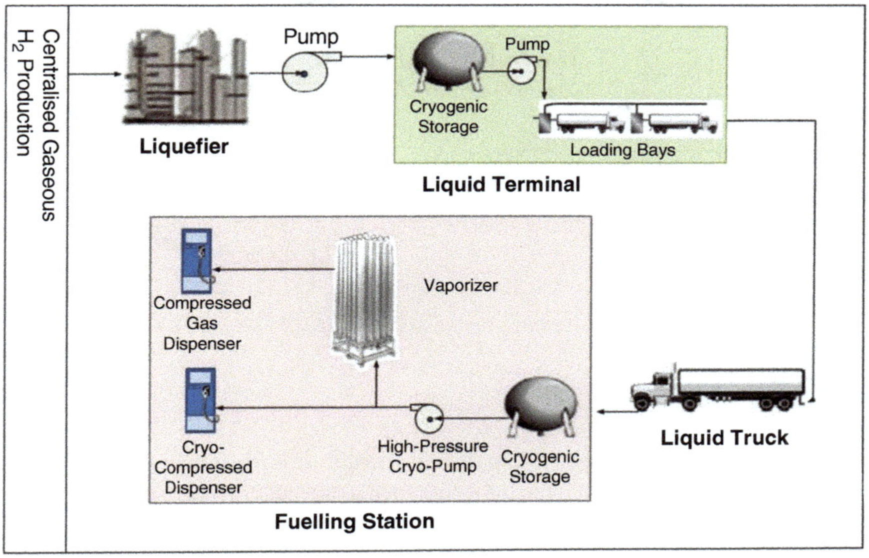

Figure 4.30 Liquid delivery pathways [66]/US Department of Energy/Public Domain.

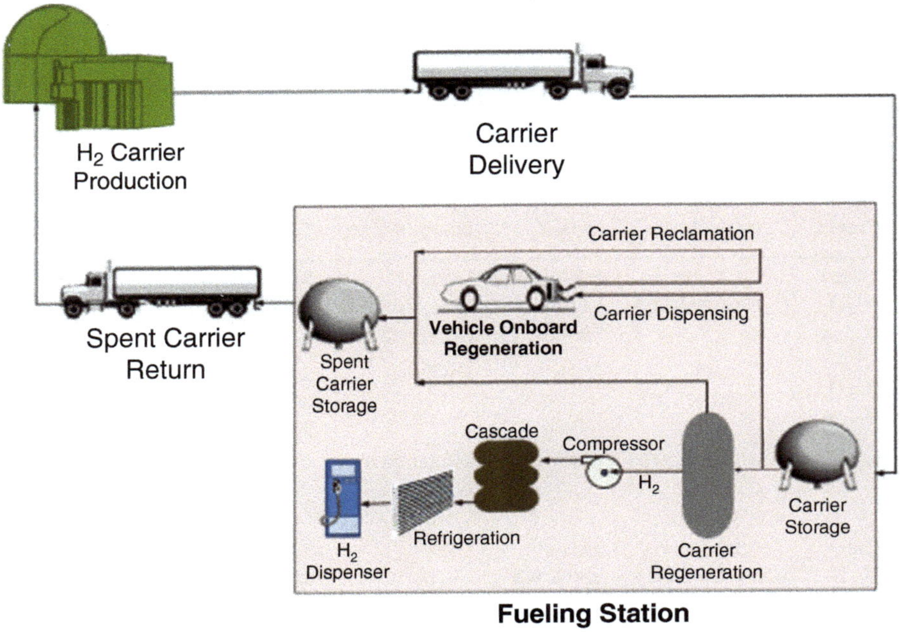

Figure 4.31 Carrier delivery pathway [66]/US Department of Energy/Public Domain.

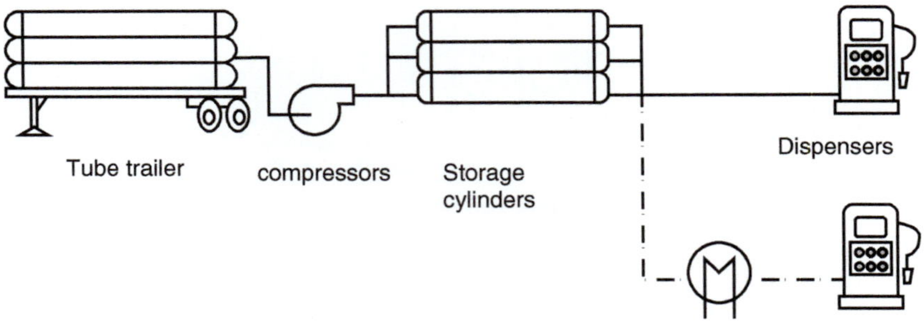

Figure 4.32 Tube trailer used for the gaseous fuel distribution [67]/MDPI/CC BY 4.0.

Concluding Remarks

Considering the rapid rate of hydrogen use in several energy and environment sectors, a range of research and development activities are being pursued in all areas of hydrogen energy technology, such as production, storage, transfer and safety at all stages. As far as storage is concerned, technology related to compressed gas storage and cryogenics is more mature, and as is well known, compressed hydrogen gas is being used in several industrial applications. However, they have their own limitations. Due to low density, compressed gas requires a large volume, whereas cryogenic gas is relatively inefficient. After some time, composite cylinders might be more useful for long-term applications once their reliability has been established.

The following table shows a comparative depiction of some of the major storage methods of hydrogen. The level of maturity has been indicated on the basis of how well established and how widely the method has been put to practical applications until recently.

Hydrogen storage mode	Storage method	Storage capacity	Energy density	Safety	Level of maturity
Compressed gas (CGH_2)	Compress hydrogen gas	High (up to 350–700 bar)	Low	Safe with proper handling	Well established
Liquid hydrogen (LH_2)	Liquefy hydrogen gas	High (cryogenic tanks)	High	Safe with cryogenic precautions	Well established
Metal hydride (MH)	Absorption in metal hydrides	Moderate	Moderate to high	Generally safe	Maturing
Chemical hybrids	Adsorption on chemical compounds	Low to moderate	High	Safety depends on the chemical used	Emerging
Adsorbed (physisorption)	Adsorption on porous materials	Low to moderate	Low to moderate	Safe with proper material selection	Emerging
Underground (salt caverns)	Store in geological formations	Very high	Low to moderate	Safe with proper geological assessment	Maturing

A clear picture of the transport and delivery of hydrogen has yet to emerge. In fact, it depends on the production site to the point of use. For small quantities of hydrogen transported over long distances, tube trailers are not an economically viable option. As a result, LH is preferred, but in that case, a substantial amount of energy, approximately 40%, is lost. Pipeline delivery seems to be very prospective and feasible, but adequate infrastructure must be built, which requires large amounts of energy. All these aspects of hydrogen energy technology are being analysed from the points of view of energy, ecology, environment and economics, and a universally acceptable solution is likely to emerge soon. As the hydrogen industry advances, further innovation, optimisation and integration with renewable energy sources will likely improve the efficiency, safety and cost effectiveness of these storage methods. Different storage techniques may find varying degrees of suitability depending on the specific applications, energy requirements and infrastructure availability. Advancements in storage methods continue to maximise the potential of hydrogen as a versatile and sustainable energy carrier.

Abbreviations

LHV	Low Heating Value
PEM	Polymer Electrolyte Membrane
TPRDs	Thermal Pressure Relief Devices
MLI	Multi-Layer Insulation
NASA	National Aeronautics and Space Administration
BOG	Boil-Off Gas
HPGH	High-Pressure Gaseous Hydrogen
UCG	Underground Coal Gasification
CCS	Carbon Capture And Storage
LHSS	Liquefied Hydrogen Storage System
DDR	Deca Dodecasil Rhombohedral
SAPO	Silicoaluminophosphate
KOH	Potassium Hydroxide
HGM	Hollow Glass Microsphere
CNG	Compressed Natural Gas
FRP	Fibre-Reinforced Polymer
LNG	Liquefied Natural Gas
GH	Gaseous Hydrogen
LH	Liquid Hydrogen

References

1 Ausfelder, F. and Bazzanella, A. (2016). Hydrogen in the chemical industry. In: *Hydrogen Science and Engineering: Materials, Processes, Systems and Technology* (ed. D. Stolten and B. Emonts), 19–40. Wiley.

2 Karakaya, E., Nuur, C., and Assbring, L. (2018). Potential transitions in the iron and steel industry in Sweden: towards a hydrogen based future? *Journal of Cleaner Production* 195: 651–663.

3 Otto, A., Robinius, M., Grube, T. et al. (2017). Power-to-steel: reducing CO_2 through the integration of renewable energy and hydrogen into the German steel industry. *Energies* 10 (4): 451.

4 Das, L.M. (1990). On-board hydrogen storage systems for automotive application. *International Journal of Hydrogen Energy* 21 (9): 789–800.

5 Luo, X. et al. (2015). Overview of current development in electrical energy storage technologies and the application potential in power system operation. *Applied Energy* 137: 511–536.

6 Taylor, J.B. et al. (1986). Technical and economic assessment of methods for the storage of large quantities of hydrogen. *International Journal of Hydrogen Energy* 11 (1): 5–22.

7 Carpetis, C. (1982). Estimation of storage costs for large hydrogen storage facilities. *International Journal of Hydrogen Energy* 7 (2): 191–203.

8 Stolten, D. (2016). Bulk storage vessels for compressed and liquid hydrogen. In: *Hydrogen Science and Engineering: Materials, Processes, Systems and Technology* (ed. D. Stolten and B. Emonts), 659–690. Wiley.

9 Barthelemy, H., Weber, M., and Barbier, F. (2017). Hydrogen storage: recent improvements and industrial perspectives. *International Journal of Hydrogen Energy* 42 (11): 7254–7262.

10 Durbin, D. and Malardier-Jugroot, C. (2013). Review of hydrogen storage techniques for on board vehicle applications. *International Journal of Hydrogen Energy* 38 (34): 14595–14617.

11 Paster, M. et al. (2011). Hydrogen storage technology options for fuel cell vehicles: well-to-wheel costs, energy efficiencies, and greenhouse gas emissions. *International Journal of Hydrogen Energy* 36 (22): 14534–14551.

12 Hwang, H.T. and Varma, A. (2014). Hydrogen storage for fuel cell vehicles. *Current Opinion in Chemical Engineering* 5: 42–48.

13 Eberle, U., Felderhoff, M., and Schuth, F. (2009). Chemical and physical solutions for hydrogen storage. *Angewandte Chemie (International Ed. in English)* 48 (36): 6608–6630.

14 Ren, J. et al. (2017). Current research trends and perspectives on materials-based hydrogen storage solutions: a critical review. *International Journal of Hydrogen Energy* 42 (1): 289–311.

15 Lai, Q. et al. (2015). Hydrogen storage materials for mobile and stationary applications: current state of the art. *ChemSusChem* 8 (17): 2789–2825.

16 Dalebrook, A.F. et al. (2013). Hydrogen storage: beyond conventional methods. *Chemical Communications* 49 (78): 8735–8751.

17 Today in Energy (2014, 3 January). US Energy Information Administration (EIA) (Source: US Energy Information Administration, based on the National Defense University).

18 Wolf, E. (2015). Large-scale hydrogen energy storage. In: *Electrochemical Energy Storage for Renewable Sources and Grid Balancing* (ed. J. Garche), 129–142. Amsterdam: Elsevier.

19 Witkowski, A. et al. (2017). Comprehensive analysis of hydrogen compression and pipeline transportation from thermodynamics and safety aspects. *Energy* 141: 2508–2518.

20 Ahluwalia, R.K., Hua, T.Q., and Peng, J.K. (2012). On-board and Off-board performance of hydrogen storage options for light-duty vehicles. *International Journal of Hydrogen Energy* 37 (3): 2891–2910.

21 O'Malley, K. et al. (2015). Applied hydrogen storage research and development: a perspective from the US Department of Energy. *Journal of Alloys and Compounds* 645: S419–S422. ISSN 0925-8388, https://doi.org/10.1016/j.jallcom.2014.12.090.

22 Tietze, V., Luhr, S., and Stolten, D. (2016). Bulk storage vessels for compressed and liquid hydrogen. In: *Hydrogen Science and Engineering: Materials, Processes, Systems and Technology*, 659–690. Wiley.

23 Molkov, V., Dadashzadeh, M., Kashkarov, S., and Makarov, D. (2021). Performance of hydrogen storage tank with TPRD in an engulfing fire. *International Journal of Hydrogen Energy* 46 (73): 36581–36597.

24 Notardonato, W.U. et al. (2017). Zero boil-off methods for large-scale liquid hydrogen tanks using integrated refrigeration and storage. *IOP Conference Series: Materials Science and Engineering* 278: 012012.

25 Tengborg P, Johansson J, Durup G. (2014). Storage of highly compressed gases in underground lined rock caverns more than 10 years of experience. *Proceedings of the World Tunnel Congress*. Foz do Iguaçu, Brazil.

26 Krasae-in, S., Stang, J.H., and Neksa, P. (2010). Development of large-scale hydrogen liquefaction processes from 1898 to 2009. *International Journal of Hydrogen Energy* 35 (10): 4524–4533.

27 Cardella, U., Decker, L., and Klein, H. (2017). Roadmap to economically viable hydrogen liquefaction. *International Journal of Hydrogen Energy* 42 (19): 13329.

28 Wilhelmsen, O. et al. (2018). Reducing the exergy destruction in the cryogenic heat exchangers of hydrogen liquefaction processes. *International Journal of Hydrogen Energy* 43 (10): 5033–5047.

29 Berstad, D.O., Stang, J.H., and Neksa, P. (2009). Comparison criteria for large-scale hydrogen liquefaction processes. *International Journal of Hydrogen Energy* 34 (3): 1560–1568.

30 United Nations Economic Commission for Europe (UNECE) (2013). Regulation on Hydrogen and fuel cell vehicles. *Global Technical Regulation no 13, ECE/TRANS/180/Add.13*

31 Peschka, W. (1997, August). State of Hydrogen Cryofuel Technology for Internal Combustion Engines, *"Hypothesis" Conference* (August 1997), Grimstad, Norway.

32 Amos, W.A. (1998). Costs of Storing and Transporting Hydrogen. *NREL/TP-570-25106, Prepared under Task No. HY914041 November 1998.*

33 Carpetis, C. (1988). Storage, transport and distribution of hydrogen. In: *Hydrogen as an Energy Carrier* (ed. C.-J. Winter and J. Nitsch), 249–289. Berlin, Heidelberg: Springer Berlin Heidelberg.

34 Godula-Jopek, A., Jehle, W., and Wellnitz, J. (2012). Storage of pure hydrogen in different states. In: *Hydrogen Storage Technologies*, 97–170. Wiley.

35 Peschka, W. (1992). Thermal insulation, storage and transportation of liquid hydrogen. In: *Liquid Hydrogen* (ed. W. Peschka), 71–103. Vienna: Springer Vienna.

36 Ohira, K. (2016). Slush hydrogen production, storage, and transportation. In: *Compendium of Hydrogen Energy*, Volume 2: Hydrogen Storage, Transportation and Infrastructure, 53–85. Woodhead Publishing Series in Energy.

37 Bellosta von Colbe, J. et al. (2019). Application of hydrides in hydrogen storage and compression: achievements, outlook and perspectives. *International Journal of Hydrogen Energy* 44 (15): 7780–7808.

38 Klebanoff, L.E. et al. (2014). Accelerating the understanding and development of hydrogen storage materials: a review of the five-year efforts of the three DOE hydrogen storage materials centers of excellence. *Metallurgical and Materials Transactions* 1 (2): 81–117.

39 Demirci, U.B. (2018). About the technological readiness of the H_2 generation by hydrolysis of B(N)H compounds. *Energy Technology* 6 (3): 470–486.

40 Crivello, J.C. et al. (2016). Review of magnesium hydride-based materials: development and optimisation. *Applied Physics A: Materials Science and Processing* 122 (2): 97.

41 Yartys, V.A. et al. (2019). Magnesium based materials for hydrogen based energy storage: past, present and future. *International Journal of Hydrogen Energy* 44 (15): 7809–7859.

42 Webb, C.J. (2015). A review of catalyst-enhanced magnesium hydride as a hydrogen storage material. *Journal of Physics and Chemistry of Solids* 84 (Supplement C): 96–106.

43 Aguey-Zinsou, K.F. and Ares-Fernandez, J.R. (2010). Hydrogen in magnesium: new perspectives toward functional stores. *Energy & Environmental Science* 3 (5): 526–543.

44 Sun, Y. et al. (2018). Tailoring magnesium based materials for hydrogen storage through synthesis: current state of the art. *Energy Storage Materials* 10: 168–198.

45 de Rango, P., Marty, P., and Fruchart, D. (2016). Hydrogen storage systems based on magnesium hydride: from laboratory tests to fuel cell integration. *Applied Physics A: Materials Science & Processing* 122 (2): 126.

46 Shao, H. et al. (2018). Progress and trends in magnesium-based materials for energy-storage research: A review. *Technology* 6 (3): 445–458.

47 Jehan, M. and Fruchart, D. (2013). McPhy-energy's proposal for solid state hydrogen storage materials and systems. *Journal of Alloys and Compounds* 580 (Supplement 1): S343–S348.

48 Nogita, K. et al. (2009). Engineering the Mg–Mg2Ni eutectic transformation to produce improved hydrogen storage alloys. *International Journal of Hydrogen Energy* 34 (18): 7686–7691.

49 Graetz, J. et al. (2011). Aluminum hydride as a hydrogen and energy storage material: past, present and future. *Journal of Alloys and Compounds* 509: S517–S528.

50 Fabian, T. et al. (2017). *Low-Cost Alpha Alane for Hydrogen Storage*. San Francisco, CA (United States): Ardica Technologies.

51 Graetz, J., Reillya, J.J., Yartys, V.A. et al. (2011). Aluminum hydride as a hydrogen and energy storage material: past, present and future. *Journal of Alloys and Compounds* 509S: S517–S528.

52 Graetz, J. and Hauback, B.C. (2013). Recent developments in aluminum based hydrides for hydrogen storage. *MRS Bulletin* 38 (6): 473–479.

53 Das, N. and Das, J.K. (2020). Zeolites: an emerging material for gas storage and separation applications. In: *Zeolites – New Challenges* (ed. K. Margeta and A. Farkaš). IntechOpen https://doi.org/10.5772/intechopen.91035.

54 Baburaj, E.G., Froes, F.H., Shutthanandan, V., and Thevuthasan, S. (2000). Low cost synthesis of nanocrystalline titanium aluminides. In: *Interfacial Chemistry and Engineering Annual Report* (ed. J.W. Grate). Oak Ridge, Tenn, USA: Pacific Northwest National Laboratory.

55 Schulz, R., Boily, S., Zaluski, L. et al. (1995). Nanocrystalline materials for hydrogen storage, innovation. *Metallic Materials* 529–535.

56 Dillon, A.C., Jones, K.M., Bekkedahl, T.A. et al. (1997). Storage of hydrogen in single-walled carbon nanotubes. *Nature* 386 (6623): 377–379.

57 Niemann, M.U., Srinivasan, S.S., Phani, A.R. et al. (2008). Nanomaterials for hydrogen storage applications: a review. *Journal of Nanomaterials* 2008: 950967. https://doi.org/10.1155/2008/950967.

58 Costa, P.M.F.J., Coleman, K.S., and Green, M.L.H. (2005). Influence of catalyst metal particles on the hydrogen sorption of single-walled carbon nanotube materials. *Nanotechnology* 16 (4): 512–517.

59 Joseph, J. et al. (2020). Borophene and Boron Fullerene Materials in Hydrogen Storage: Opportunities and Challenge. *ChemSusChem, Chemistry, Sustainability-Energy-Materials* https://doi.org/10.1002/cssc.202000782.

60 Im, J.S., Park, S.-J., and Lee, Y.-S. (2009). The metal–carbon–fluorine system for improving hydrogen storage by using metal and fluorine with different levels of electronegativity. *International Journal of Hydrogen Energy* 34 (3).

61 Yartys, V.A. and Lototsky, M.V. (2004). An overview of hydrogen storage methods. In: *Hydrogen Materials Science and Chemistry of Carbon Nanomaterials*, NATO Science Series II: Mathematics, Physics and Chemistry, vol. 172 (ed. T.N. Veziroglu, S.Y. Zaginaichenko, D.V. Schur, et al.). Dordrecht: Springer https://doi.org/10.1007/1-4020-2669-2_7.

62 M. Herr and J.A. Lercher. (2003). Hydrogen storage in microspheres. *Final Report, ET-TN-03-628*, 9 September 2003.

63 DOE H2. A Delivery Analysis: DOE Hydrogen Program https://www.hydrogen.energy.gov/h2a_delivery.html

64 Reddi, K., Mintz, M., Elgowainy, A., and Sutherland, E. (2016). Building a hydrogen infrastructure in the United States. *Compendium of Hydrogen Energy* 293–319.

65 Rivkin, C., Buttner, W., and Burgess, R. (2016). Guide to Permitting Hydrogen Motor Fuel Dispensing Facilities. *National Renewable Energy Laboratory Technical Report NREL/TP-5400-64042 March 2016, Contract No. DE-AC36-08GO28308.*

66 US DRIVE. (2017, July). Hydrogen Delivery Technical Team Roadmap: A document of the US DRIVE Partnership. United States Driving Research and Innovation for Vehicle Efficiency and Energy Sustainability.

67 Xiangmin, P., Hong, L., Wei, Z. et al. (2010). Design and operation of the hydrogen supply chain for fuel-cell vehicles in Expo Shanghai 2010. *World Electric Vehicle Journal* 4: 671–676. ISSN 2032-6653.

5

Safety, Sensing and Detection of Hydrogen

This is the greatest discovery of the scientific enterprise: You take hydrogen gas, and you leave it alone, and it turns into rosebuds, giraffes and humans.

– Brian Swimme

CHAPTER MENU

5.1 Introduction

Petroleum fuels have been used for more than a century and have passed through problems of different characteristics at different stages. Depending upon the nature and magnitude of the problem, solutions have evolved. Recent crises due to energy security

Hydrogen Energy: Production, Safety, Storage and Applications, First Edition. L. M. Das.

and environmental degradation have been of greater concern and need to be resolved. Considering its energy-efficient, low-polluting and renewable characteristics, hydrogen is emerging as a potential candidate. Thus, knowledge about the safety of hydrogen is very important [1]. Unknown and lesser-known factors always create doubts and uncertainty, which seems to be true in the case of hydrogen. Hydrogen as an energy carrier has been tried earlier as a choice of laboratory curiosity and research interest. During the process, it was observed that several accidents from minor magnitude to devastation occurred while handling hydrogen. There have been dozens of hydrogen airship disasters starting with LZ-4 (5 August 1908) and occurring intermittently until 6 May 1937 when the Hindenburg disaster occurred, putting a big question on the safe use of hydrogen. There have been several accidents involving gasoline and other hydrocarbon-based fuels. Over the period of time, along with the use of hydrogen in several industrial sectors, it has been understood that some of the combustion-related properties of hydrogen are widely different from conventional fuels and need to be handled in a different way. Let us discuss Hindenburg disaster and the characteristics of hydrogen bomb because a common man looks, on hearing these terms, with eyes wide open questioning the acceptance of hydrogen for public use. However, once we dig deeper and understand the versatile unique properties of hydrogen and know how to handle it properly, we realise that it is perhaps one of the most suitable and safe alternatives to conventional fuels.

5.2 Infamous Disasters Related to Hydrogen Safety

Currently, many countries worldwide have adopted several programmes leading to the early implementation of hydrogen energy mainly because of its vast resources and environmentally friendly features. Safety regulations and technologies are being developed [2, 3]. It is essential that the undeserved stigma attached to the safety of hydrogen must be erased to unfold its potential application in several areas.

Hydrogen, from a safety point of view, has received two distinct schools of thought. Hydrogen proponents look at it from its clean burning and rapid recycling characteristics, whereas the opponents amplify its explosive features.

Unfortunately, a common man's perception of hydrogen safety is still blanketed with the Hindenburg disaster. Hindenburg, named after the President of Germany, started its first commercial flight across the North Atlantic in March 1936 with 10 scheduled trips between Germany and the United States of America. Thus, it had finished several successful trips before the disaster.

As far as its dimensions and constructional features are concerned, it was 245 m long and about 41 m in diameter. It had four engines of 820 kW each and was designed to operate at a speed of 135 km h^{-1}. It was made with an aluminium framework and was fuelled with hydrogen instead of non-combustible helium, perhaps due to military restrictions or some other constraints. To prevent any explosion, the airship was coated with an envelope of specially designed cotton fabric impregnated with cellulose acetate butyrate resins containing iron oxide and iron powder. This special coating was specifically made to rule out any possible hazards caused by any possible spark from static electricity [4].

As far as the Hindenburg disaster is concerned, there have been several hypotheses by various experts over the infamous crash. These hypotheses range from deliberate sabotage to starting failure, lightening, fuel leakage, etc. One of the theories extended to explain the incident indicates that the atmospheric condition at the time of accident was due to explosion caused by electrostatic charge present in the thunderstorm. It is thought that the wet mooring lines, connected to the aluminium frame, when touching the ground, formed electrical discharge [4], and the airship was totally burned out as shown in Figure 5.1.

Therefore, perhaps improper selection of the material for the outer shell and incorrect design parameters were responsible for static electricity development. Even after more than eight decades, the cause of the accident has not been very clearly understood. No conclusive evidence has linked the tragic death of 62 people to combustion of hydrogen. Bain and Van Vorst [5], in a paper in later years, very clearly described the causes leading to the Hindenburg disaster and strongly believe that by reading the details given in their paper, 'Hopefully, readers will then – over time – help "spread the word" and reduce, if not eliminate, the negativity generated by the misinterpretation of the cause of the Hindenburg tragedy'.

Figure 5.1 Hindenburg disaster. *Source:* Hindenburg Wiki/CC BY SA.

Thus, it cannot be said if the mishap would not have occurred even with the use of non-combustible helium. The remaining 65% passengers survived in the midst of the flaming airship and reached Earth.

Addison Bain, one of the authors of the paper (paper sl. no. 5 referred in this article), who was associated with the Hydrogen Programs of Kennedy Space Center, National Aeronautics and Space Administration (NASA), USA, took up 'a long journey (physically as well as intellectually) in search of answers' to many conflicting theories being raised to find the cause of the accident. Figure 5.1 shows Hindenburg in a routine landing at the Lakehurst Naval Air Station in New Jersey, USA, in 1936 [5].

Figure 5.1 shows explosion in Hindenburg in its safe conditions, attempting to land. There have been several theories about hydrogen's role in this accident. The hydrogen flame should be colourless, but it seems that some witnesses have compared this flame to the 'flame of a firework display' [5]. Hydrogen, considering its diffusivity, is supposed to burn upwards. It is clear in the figure that something is burning downwards. These facts raise doubts whether hydrogen was really the culprit. Hindenburg was built in the early 1930s. Since this was before rocketry, before plastics and before thin metal cladding, a group of researchers realised that adequate attention might not have been paid to the problem of flammability of the substances used. Therefore, several material-oriented remedial steps were adopted quickly after the Hindenburg disaster. Calcium sulfamate, which is widely used as a *fire-proofing* agent in the textile industry, was added to the doping mixture. Doping, as is well known, is important for the aircraft fabric to make it taut and more resistant to airflow. After the Hindenburg incident, the existing doping compound was also changed from aluminium to powered bronze because bronze is far less combustible even if it is heavier than aluminium [5]. In spite of the successful application of hydrogen in several areas, the unsafe image of hydrogen emerges from the hydrogen bomb and the Hindenburg disaster, which dominates in the mind of the man in the street. The space shuttle challenger disaster was the first fatal accident in which an American spacecraft, in flight, was totally destroyed. On 28 January 1986, during its tenth launch, it exploded immediately after 73 s of lift off, and all seven crew members aboard were killed. The space shuttle challenger was built after a series of design and development activities for a period of seven years (1972–1979). The three engines were operated with liquid hydrogen as fuel, and liquid oxygen was used as an oxidiser. Catastrophic failure of two rubber O-ring seals in the right-hand solid rocket booster (SRB) was the cause of the disaster. The rubber seals could not withstand the severe cold weather when the temperature dipped much lower and rubber seals lost their elasticity. The reduced elasticity of the rubber seals was unable to seal the joints and resulted in the explosion of the hydrogen tank. It has been reported [6] that the ground-level temperature at Pad 39B of Kennedy Space Centres, Florida, was 2.20 °C, which was about 8.3 °C colder than all other previous launching programmes by the NASA. After this fatal incident, the space shuttle flights were grounded for nearly three years. Design improvements were incorporated into SRB. Several additional safety steps were implemented. There have been a few other accidents, such as vapour cloud explosion (VCE) at the polyethylene plant in Pasadena in 1989, which killed 23 people [7], and the rupture of the pressurised hydrogen tank at Hanau in Frankfurt in 1991. It is well known that the Fukushima Nuclear disaster was initially triggered by a magnitude of 9.0 earthquake and Tsunami on 11 March due to power loss in the Fukushima Daiichi plant. It was apprehended that hydrogen generated in the pressure vessel of the reactor was leaked through pipes and mixed with oxygen, resulting in an explosion. It would perhaps be appropriate to mention here that all fuels exhibit some degree of safety hazard if not properly handled. Most often, hydrogen-based accidents have occurred because of the liberal use of the fuel in many circumstances without considering its specific combustion characteristics that are widely different from conventional fuels.

There is absolutely no connection between the chemical reactions of hydrogen fuel and the thermonuclear explosive characteristics of hydrogen isotopes that are connected to hydrogen bombs. Figure 5.2 shows the schematic of a hydrogen bomb explosion. However, the explosive energy from this thermonuclear weapon should never be used for destructive purposes.

There is no room for doubt that, considering its widely flammable characteristics leading to undesirable combustion and sometimes even to explosion, hydrogen must be handled with appropriate care and safeguards. The evaluation of safety aspects of hydrogen needs to be assessed on the basis of technical data and comparisons with other conventional and prospective fuels. NASA has suggested relevant guidelines for several significant aspects of hydrogen technology, including system design and operation [7].

Figure 5.2 Schematic explosions from a hydrogen bomb. *Source:* The Official CTBTO Photostream/Flickr/CC BY 2.0.

Of course, modern technology-based high safety standards for energy supply, transmission and subsequent utilisation, for gasoline, LPG and natural gas, have been well established for each stage. The frequency of accidents with hydrogen has been 5–20 times higher than that with natural gas, mainly because hydrogen's distinctive properties, which are expected to be handled in a different way, have often been inadvertently ignored. The analysis of the causes of accidents at different stages has occurred due to several factors, such as poor ventilation, low purity, reaction with pollutants in air, failure of safety devices, etc. Mechanical failure resulting in embrittlement has been a major factor in some specific applications. However, some catastrophic accidents have taken place due to fire, explosion, failure in ignition system, etc. Human error resulting in accidents of course cannot be ignored because the operating staff have not adopted specific precautions essential for hydrogen considering its temperamental combustion characteristics.

This complex range of safety assessment activities involves safe and reliable stages of production, transmission, storage and appropriate infrastructure at the point of utilisation. Depending upon the quantity and nature of its utilisation, hydrogen can either be produced on-site in large quantities in production plants close to the filling stations or it could be transported to the filling stations through pipelines or by trucks. Thus, the risk can be reduced as the large amount of hydrogen would not be stored at the dispensing site. Hydrogen transported from the site could be either in the gaseous or liquid form. Markert et al. [8] have studied the future infrastructure and have given a broad and qualitative comparison on the safety aspects of centralised versus on-site local production of hydrogen.

It is well known that over the past several decades, hydrogen has been used in large quantities as an industrial chemical as well as for some medical applications. NASA's general recommendation on an approach for inherent safety system design and the approach given by Kletz [9] provide guidelines for the development of a relatively risk-free safe system. NASA, since its inception in 1958, has been using hydrogen for space exploration. It is being used as a rocket fuel and for several other purposes in space applications. NASA has a very safe record of handling hydrogen for space applications such as Centaur, Apollo and several space shuttle vehicles. Presently, with the effort for human missions to the moon and subsequently to Mars, the thrust is on innovative liquid hydrogen storage systems for space launch projects. Well-established infrastructure has been built to produce, store and use large amounts of hydrogen safely for many years. However, in all such applications the influence of some specific combustion-related properties of hydrogen has been very critically studied and implemented.

A closer look at the literature indicates that there have been several reasons and ways that have caused hydrogen-related accidents of different magnitudes. In the final technical report, Zalosh and Short [10]analysed a number of hydrogen-related accidents and observed that apparently minor problems such as undetected leaks and inadequate ventilation were the cause of accidents. Boiling water off-gas explosions and the malfunctioning of electrolysis have also been responsible. The study examined the source of ignition, production route and specific safety measures adopted at the location of the accident. Many investigations have been carried out since then to identify the specific causes leading to different degrees of hydrogen accidents. Federico, Nicola and Filippo [11], in a recent study, have observed that the loss of integrity phenomena is one of the reasons for hydrogen accidents. Researchers have discussed several aspects and the material selection criteria from a safety point of view.

It has been suggested that an effective collaboration among scientists and researchers from various research areas would be extremely useful. Inappropriate handling of hydrogen and human error has resulted in accidents [12]. Many accidents, including the recent The Challenger Space Shuttle Explosion of 1986 and the Fukushima Nuclear tragedy of 2011, very clearly show that hydrogen should be handled properly by considering its intrinsic properties. There have been many accidents, mainly due to embrittlement, material failure, corrosion, pressurised hydrogen tank rupture, release of hydrogen during maintenance work, etc. There have been several incidents affecting people due to hydrogen leakage, fire and explosion of hydrogen–air mixtures [13, 14]. With the emerging trend of the rapid application of hydrogen and fuel cell technologies, it is important that all stakeholders should have a clear understanding of the safety of the system and infrastructure.

5.3 Classification of Hazards

A closer look at the history shows that there have been several accidents in the past involving the use of hydrogen. These accidents of different magnitudes have not been restricted to industrial activities alone. These have occurred due to inadvertent human error, mechanical failure, corrosion, overpressurisation, permeation, damage to carbon fibre, resistance to fire and high temperature, rupture due to shock wave impact from explosions and the well-known phenomena such as enhanced embrittlement at low temperature, boiling liquid expanding vapour explosion (BLEVE), etc.

The characteristics of hazards can be physiological in nature, such as frostbite, respiratory ailment, asphyxiation, etc. This can also be due to phase changes, component failures and last but not least due to some of the typical distinctive combustion properties of hydrogen.

Physiological problems could be broadly due to asphyxiation, hypothermia, thermal burns and cryogenic burns. Sometimes damage is caused by overpressure. Chemical hazards could be broadly due to several properties of hydrogen, such as ignition energy auto-ignition temperature, flammability limit, quenching distance, thermal radiation and flame speed. There are a large number of other factors that are the outcome of some of these properties. Depending upon the specific system, the safety-related parameters vary. The safety requirement in a welding shop is different from the safety aspects to be considered for hydrogen use in vehicles. Therefore, it is important to discuss some of the relevant properties of hydrogen in detail. Considering its increased density and extremely low temperature, liquid hydrogen should be handled in a different way. The safe distance from the application location and the distance from the storage system should be maintained at each stage. All these properties provide relevant guidelines for the safe use of hydrogen.

5.4 Physiological Hazards

There have been incidents where some physiological problems have been experienced. Considering this discussion, these effects have been referred to here as 'hazards', indicating that these are the sources of danger. However, these sources of danger do not create any form of damage if appropriate precautions and preventive steps are taken. In such a case, these hazards may be termed as 'risks'. Without any sharp line of demarcation, the discussions here highlight the causes and possible effects of some such problems for which hydrogen is responsible.

5.4.1 Asphyxiation

It has been observed that hydrogen is released in a confined area and alters the composition of air by displacing oxygen. This situation leads to problems of asphyxiation. The effect of asphyxiation depends upon the percentage volume of oxygen present in air. Oxygen in the range of about 15–19% affects a normal person's ability to perform correctly. However, in some cases, it can create difficulties in the lungs and heart, and in some other cases, it has been observed to affect the circulatory system. The nature of the problem and the degree of severity depend upon the health of the individual. It has been reported in some literature that if the percentage of hydrogen is slightly lower and in the range of 12–15%, it can result in a fast pulse rate, deep respiration and poorer coordination. If oxygen slightly decreases in the range of 10–12%, it gives rise to giddiness and improper judgement. Frequent vomiting, nausea and unconsciousness leading to fainting and even mental failure have been observed in many cases if the oxygen level remains between 8 and 10%. When an individual is subjected to an oxygen concentration within the range of 6–8% physiological normalcy can be restored, enabling complete recovery within 4–5 minutes. In scenarios where exposure to this oxygen concentration persists and prompt intervention is initiated within six minutes, there is a 50% chance of recovery. Prolonged exposure under these conditions may lead to fatal outcomes, even death, after about eight minutes. An oxygen concentration as low as 4% is considered critically deficient, and its effects manifest in inducing coma in a mere 40 seconds. This can trigger convulsions and respiratory distress, with potentially lethal consequences if prolonged. Thus, the effects of asphyxiation within very small ranges have resulted in severe degrees of consequences.

5.4.2 Hypothermia

Hypothermia is caused if there is exposure to a large amount of liquefied hydrogen and the body temperature drops to an abnormally low level of 35 °C. Depending upon the severity of the case, hypothermia could be mild or moderate to extremely severe. The heart, nervous system and other organs fail to function if the body temperature decreases abnormally. This affects several parts of the human body and a person's mental and motor functions. Starting with physical tiredness and exhaustion, the range of symptoms could include shivering, faltering in speech, moving unsteadily, lack of coordination, confusion, loss of memory, etc. Quite often, it is difficult to identify hypothermia because in some cases it also leads to diabetes and heart problems that occur in human beings due to several other reasons. If the body temperature drops further below 32.2 °C, it becomes much more serious and even causes death.

5.4.3 Thermal and Cryogenic Burns

Radiant heat emitted by hydrogen fire is the main cause of thermal burns. Even long exposure to radiation intensity within a range of 1–6 kw m^{-2} practically does not cause any harm, and if the level of exposure is beyond 35%, it could be lethal even if the exposure time is about 10 s. LaChance et al. [15] studied and discussed the threshold dose of the ultraviolet and infrared radiation that can cause harm to a definite extent.

Radiation from hydrogen fire is significantly much lower than that from hydrocarbon fire primarily because of the absence of carbon and due to the presence of water vapour produced when hydrogen burns. Hydrogen fire lasts up to one-fifth to one-tenth of the time period of hydrocarbon fire. Hence, the damaging effects of hydrogen fire are much lower than those of hydrogen fire. Broadly speaking, absorbed radiant heat depends upon a number of factors, such as the size of the burning surface and burning rate and the humidity and wind speed under atmospheric conditions. Fire damage due to hydrogen fire is much lower because of the typical characteristics of hydrogen burning. Rapid mixing and a higher propagation velocity of hydrogen enhance the burning rate. A relatively higher buoyant velocity and the vapour generation rate also prevent excessive damage in hydrogen fire. The invisible hydrogen flame has been the cause of thermal burns in some cases. The flame becomes fatal, causing thermal burns when a person comes closer to the invisible flame. The degree of injury manifests itself in different forms. The initial mild phase of injury can result in the redness of the skin that was exposed to the flame and sometimes could have some swelling. A higher degree of injury could show some blistering of the skin in addition to pain and redness. Sometimes a much deeper degree of injury, affecting all layers of skin, damages the nerves and blood vessels. The features of all the stages of injury do not always remain in that stage. Often, the initial mild pain develops into the next stage, and in some cases, the injury gradually turns into severe conditions. The effect of thermal radiation intensity on human beings varies over a wide range depending upon the intensity of radiation and the time of exposure. Cryogenic burns of hydrogen often occur due to contact with extremely cold fluids. Usually, very robust constructions are made with adequate safety features installed at specific locations to reduce the possibility of human contact. Cryogenic hydrogen liquids are kept in well-designed, super-insulated double-walled containers. Similar to thermal burns, frost bites affect the human skin and cause various degrees of harm, ranging from simple itching to damaging the tissues, causing blisters and in some cases causing the loss of fingers and toes.

5.5 Properties Relevant to Hydrogen Safety

The physico-chemical properties of hydrogen are discussed in Chapter 2. The emphasis of discussions has been mainly on conventional fuels with the objective to assess whether hydrogen, in the range of these properties, can be considered as a substitute for conventional fuels. The discussions in this chapter highlights those properties that are closely related to the safety of hydrogen. All properties of hydrogen are not necessarily more severe than those of conventional fuels, but some of them are radically different. Properties such as flammability limit, minimum ignition energy, flame speed, auto-ignition temperature, flame visibility, thermal radiation from flames, burning velocity, detonability limit, etc. are very closely related to the safety of hydrogen. Therefore, it is necessary to carefully determine some of the properties (Table 5.1) of hydrogen that need to be properly handled to avoid any mishaps or accidents. Table 5.1 shows the properties of hydrogen vis-à-vis natural gas and gasoline.

Table 5.1 Properties of hydrogen, methane and gasoline.

Property	Hydrogen	Methane	Gasoline
Molecular weight	2.016	16.043	~107.0
Normal melting point (K)	14.1	90.68	213
Normal boiling point (K)	20.268	111.632	310–478
Critical temperature (K)	32.97–33.1	190	–
Critical pressure (MPa or atm)	1.8 or 12.8	4.6	–
Density of vapour at NBP (kg m^{-3})	1.338	73.4	–
Density of liquid at NBP (kg m^{-3})	70.78	423.8	745 (at STP)
Density of gas at NTP (g m^{-3})	82 (at 300 K) 83.764	717 651.19	5110 ~4400
Density of gas at STP (g m^{-3})	84 89.87	650 657 (at 298.2 K)	4400
Heat of fusion at 14.1 K (kJ g^{-1})	58	0.94 kJ mol^{-1}	–
Heat of vapourisation (kJ kg^{-1})	445.6 447	509.9	250–400
Heat of combustion (low) (kJ g^{-1})	119.93 119.7	50.02 46.72	44.5 44.79
Heat of combustion (high) (kJ g^{-1})	141.86 141.8 141.7	55.53 55.3 52.68	48 48.29 –
Flammability limits in NTP air (vol%)	4.0–75.0	5.3–15	1.0–7.6 1.2–6.0 1.4–7.6
Detonability limits in NTP air (vol%)	18.3–59.0 13.5–70	6.3–13.5 –	1.1–3.3 –
Detonability limits in NTP oxygen (vol%)	15–90	–	–
Stoichiometric composition in air (vol%)	29.53	9.48	1.76
Minimum ignition energy in air (mJ)	0.017 0.02 0.14	0.29 0.28 –	0.24 0.25 0.024
Auto-ignition temperature (K)	858	813	501–744 500–750
Adiabatic flame temperature in air (K)	2318 –	2148 2190	~2470 –
Thermal energy radiated from flame to surroundings (%)	17–25	23–33	30–42
Burning velocity in air (cm s^{-1})	265–325	37–45	37–43
Burning velocity in STP air (cm s^{-1})	346	45	176
Detonation velocity in NTP air (km s^{-1})	1.48–2.15	1.39–1.64	1.4–1.7

(*Continued*)

Table 5.1 (Continued)

Property	Hydrogen	Methane	Gasoline
Detonation velocity in STP air (km s^{-1})	1.48–2.15	1.4–1.64	1.4–1.7
Energy of stoichiometric mixture in NTP air (MJ m^{-3})	3.58	3.58	3.91
Velocity of sound of vapour (m s^{-1})	305	–	–
Velocity of sound of liquid (m s^{-1})	1273	–	–
Diffusion coefficient in NTP air (cm^2 s^{-1})	0.61	0.16	0.05
Diffusion coefficient in STP air (cm^2 s^{-1})	0.61	0.16	0.05
Buoyant velocity in NTP air (m s^{-1})	1.2–9	0.8–6	Non-buoyant
Limiting oxygen index (vol%)	5.0	12.1	11.6
MESG in NTP air (cm)	0.008	0.12	0.07
Quenching gap in NTP air (cm)	0.064	0.203	0.2
Detonation induction distance in NTP air	L/D-100	–	–

STP (standard temperature pressure): 273.15 K (0 °C), 101.3 kPa (1 atm); NTP (normal temperature pressure): 293.15 K (20 °C), 101.3 kPa; NBP (normal boiling point): boiling point at 101.3 kPa; MESG (maximum experimental safe gap).

Properties of hydrogen need to be considered from the point of view of a specific application as some of the properties strongly influence the safe performance of the overall system. Some of the properties need specific design of fuelling and associated engineering control. In certain applications, some specific materials become brittle when they come in contact with hydrogen. Properties such as density and diffusivity of the material are relevant with respect to leakage. The relevant physical and combustion-related properties are discussed here from a safety point of view.

5.5.1 Density, Buoyancy and Diffusivity

Hydrogen has a density of 0.0838 kg m^{-3} under NTP conditions. This is about 14 times lower than the density of air, which is about 1.205 kg m^{-3}. This property of low vapour density makes the hydrogen gas very buoyant. Figure 5.3 shows the density of hydrogen and other prospective fuels, such as methane and propane, relative to air.

The horizontal line parallel to the x-axis shows the reference specific gravity being unity, and 0.07 signifies the specific gravity of hydrogen with respect to air.

The specific gravity of air is 1.03×10^{-3}, and the specific gravity of hydrogen is 0.0696 and is thus approximately 7% of the density.

The low molecular weight of hydrogen results in high molecular diffusivity. It has about 3–8 times the diffusivity of air. When leakage occurs, hydrogen rapidly dissipates to the surroundings and does not accumulate near the ground. On the other hand, spilled gasoline remains in the ground, thereby increasing the threat of fire and explosion. Since the diffusion velocity varies with temperature according to $T^{3/2}$, the gases that are produced by the cryogenic liquid diffuse more slowly than those under normal temperature and pressure conditions.

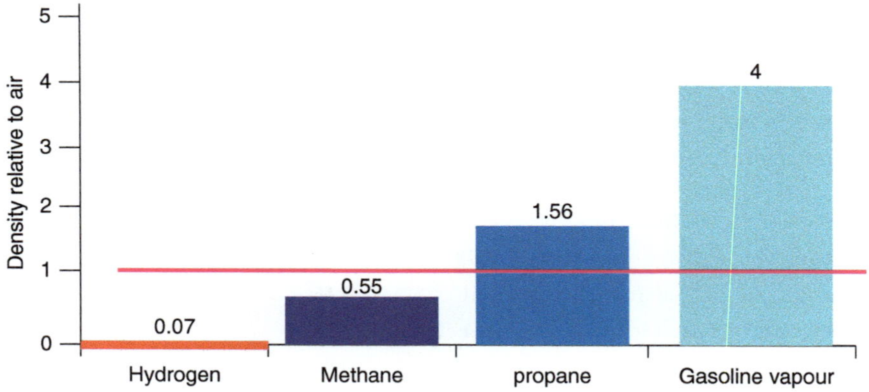

Figure 5.3 Density of hydrogen.

It is observed that natural gas is rapidly entering into several sectors of energy infrastructure and replacing the conventional fuels. Since hydrogen is also a gas, it is appropriate to have some comparative assessment of safety between the two. Considering the properties such as buoyancy, viscosity, density and diffusivity in air, it has been estimated that hydrogen leakage through holes or fuel line joints is about 1.26–2.8 times greater than the amount of leakage of natural gas in the same hole during the same time period. It has been reported that leaks from natural gas lines have the characteristics laminar. Under such circumstances, natural gas, which has about three times the energy density per volume of that of hydrogen, would release much more energy compared to hydrogen leakage under similar conditions. However, the leakage rate from high-pressure storage tanks is closely related to the sonic speed. Hydrogen has a sonic velocity of 2308 m s^{-1} compared to 449 m s^{-1} for natural gas. Thus, hydrogen, because of its higher sonic velocity, would leak much faster than natural gas in the initial phase.

Buoyant velocities under ambient conditions signify the velocity at which gaseous fuels rise in the air. There are several factors that need to be taken into consideration for the assessment of the buoyant force. Hydrogen has the highest buoyancy on Earth, and thus, buoyancy is one of the safest features of hydrogen. It is known that the disastrous explosions in Flixborough in 1974 [16] were caused by the large combustible cloud formed by heavy hydrocarbon fuels. In fact, in many real-life situations, hydrocarbons are responsible for fire and explosion hazards.

Buoyant forces depend upon the difference in the density of air and fuel. This causes the cold and dense gases, produced by the spillage of cryogenic liquid, to rise more slowly than gaseous hydrogen. Drag and friction forces oppose the buoyant forces, and the terminal velocities of buoyant gases are also affected by the shape and size of the gases. In addition, the turbulence of the atmosphere also affects the velocity of buoyant gases. When leakage occurs, the leaked hydrogen mixes with air and reaches the lower flammability limit (LFL; 4% by volume in air). This is very relevant for open or partially confined space. Even the inadvertent release of hydrogen from a storage tank or pipeline that ends up in deflagration liberates at best 1% of the total energy of the released hydrogen.

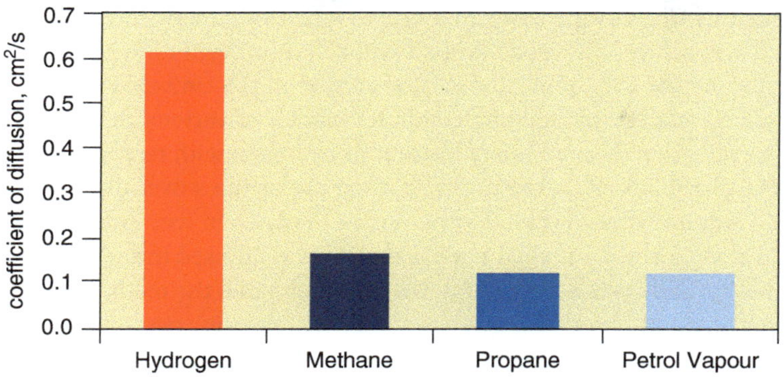

Figure 5.4 Diffusion coefficient.

Considering the small size of its molecules, the diffusivity of hydrogen is higher than that of other gases. Hydrogen's diffusivity is higher than that of helium and about three times the diffusivity of nitrogen under ambient conditions. However, it is expected that one who uses hydrogen does not liberally exchange these specific characteristics of hydrogen with those of petroleum fuels in any operating system. Figure 5.4 shows the diffusion coefficient of hydrogen with respect to methane and propane.

From a safety point of view, when hydrogen diffuses in a confined space, there could be two modes: filling box and feeding box scenarios. These scenarios describe two different ways in which hydrogen gas can disperse and accumulate within a confined space. Therefore, it is important to know how hydrogen behaves in a confined space, particularly regarding its dispersion and accumulation. Ventilation, leakage rates, confinement and potential ignition are important from a safety point of view. In a filling box scenario, hydrogen gas continuously leaks into a confined space, such as a closed room or container. As the gas leaks, it starts to fill the available volume within the space. If the leakage rate is relatively low and the ventilation or escape routes for the gas are limited, the hydrogen concentration may eventually reach the LFL of hydrogen in air, which is the minimum concentration at which gas can ignite. Thus, the filling box scenario is of concern if the concentration reaches the flammable range, and there is an ignition source to initiate combustion, which might lead to fire or explosion. The fading box scenario is different. It occurs when there is a sudden release of hydrogen gas into a confined space, followed by a gradual increase in concentration as the gas disperses and diffuses throughout the space. This scenario might occur, for example, if there is a sudden rupture or burst of a hydrogen container. Initially, the concentration of hydrogen is high near the source of the release, and then the gas gradually spreads and becomes more evenly distributed throughout the space. Similar to the filling box scenario, if the hydrogen concentration within the space eventually reaches the flammable range and an ignition source is present, combustion occurs. Depending upon the situation, it might lead to an explosion.

The buoyancy effects of gaseous hydrogen are neither similar nor applicable to the release of cryogenic hydrogen. Usually, heavy saturated hydrogen vapours remain closer to the ground. When the condensed water vapours of the atmosphere fall to the ground, both moisture and saturated hydrogen vapours form a visible mixture, thereby increasing the molecular mass of the mixture. This leads to a situation that prevents the cold, dense hydrogen gases, formed by the LH_2 spills, to rise. Researchers at the University of Miami performed some research analysis by building computer models of various geometries to assess a comparative evaluation of some safety features of hydrogen with respect to methane and propane. The possibility of convection was ruled out as the stoves did not have fans. The time period required to show cloud formation in the kitchen with each fuel was observed. A very thin cloud formed with hydrogen operation was diffused through the vent very soon, and practically, there was no trace of it. On the other hand, the methane-fuelled stove generated a small combustible cloud, whereas the stoves operating with propane had a large thick cloud. These studies closely agree with the literature discussing the safety of hydrogen.

It is an established well-known fact that hydrogen has a set of temperamental combustion characteristics that are radically different from those of conventional fuels. Properties such as buoyancy, flame visibility, low boiling point, vapour density, flame temperature and radiation from flame, flammability limit, lower ignition energy and higher flame speed need to be very judiciously handled; otherwise, they could be very damaging. However, it has been reported in the literature that these properties have often inadvertently resulted in explosive mixtures with air. Hydrogen is a non-toxic, colourless and odourless gas. In some applications, a minute amount of odorants is added to identify the presence of hydrogen gas to avoid any possibility of fire or explosion. For some applications using hydrogen, ethyl isobutyrate is used as the odorant. Odorants are non-toxic. However, the selection of a specific odorant for a particular application is important because some of the odorants contaminate some hydrides. Most of the safety-related problems arise due to the typical physiochemical characteristics of hydrogen. As discussed earlier, parameters such as buoyancy, diffusivity, stoichiometric concentration of hydrogen flammability limits, minimum ignition energy, auto-ignition temperature, quenching of flames, detonability, blow-off limits, etc. affect the performance, emission and combustion features.

5.5.2 Continuous Evaporation and High Vapour Density

Hydrogen, when stored in the liquid form in a vessel, continues to evaporate into the gaseous stage. During this stage, adequate precaution should be taken to prevent the possibility of air entering the storage vessel to eliminate the possible formation of any flammable mixture. It has been reported in tests at NASA that due to the high density of saturated vapour of the leaking hydrogen from the storage vessel, a hydrogen cloud forms and moves horizontally or downwards [7].

Apart from being odourless, hydrogen flame is also colourless and is nearly invisible in daylight although it is slightly visible in the dark and in subdued light. Sometimes, hydrogen flames become visible in the presence of impurities and moisture. However, most often, it is difficult to know if at all there exists a hydrogen fire. The location of hydrogen fires can sometimes be detected by 'hot ripples' and thermal radiation to the skin. Hydrogen flames are particularly weak compared to hydrocarbon flames; they are often referred to as almost invisible. Several studies have been carried out to examine the visibility of hydrogen flames. Schefer and Settersten [17] have carried out investigations and have compared the

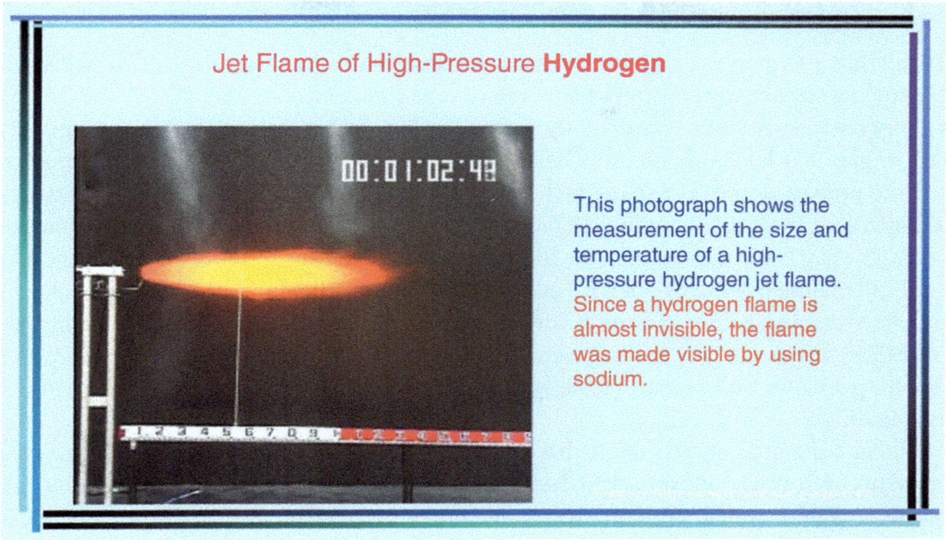

Figure 5.5 Colourant added to have visible hydrogen flame (http://www.unit.aist.go.jp/rces). *Source:* L.M. Das. (2016)/Elsevier [18].

diffusion and premixed flames and have observed them to be comparable in their fine scale structure. Sometimes, depending upon the situation, colourants are added to hydrogen flames, and these chemicals provide adequate luminosity to identify the presence of fire. Figure 5.5 shows the condition of a high-pressure jet flame in hydrogen combustion carried out in the Sandia National Laboratories. This photograph shows the measurement of the size and temperature of a high-pressure hydrogen jet flame. Since a hydrogen flame is almost invisible, the flame was made visible by using sodium.

5.5.3 Pressure Rise

When the liquid hydrogen, confined between two valves in a pipe, becomes warmer, it causes the pressure to rise, and sometimes the pressure goes up to a much higher level under different conditions. Considering these factors, design criteria have been laid out for liquid hydrogen storage systems, taking into account about 0.5% liquid leakage per day. If the liquefied hydrogen is assumed to be an ideal gas, then the pressure under 1 atm and 294 K is 85.8 MPa, whereas the pressure level increases to 172 MPa if the compressibility characteristics of hydrogen are taken into account. Liquid hydrogen, if spilled into the environment under normal temperature, flashes into the gaseous stage. However, situations of high temperature develop if the liquid hydrogen is warmed to ambient temperature in a confined space. Hydrogen gets solidified at a temperature of −259.16 °C. The solid thus formed results in plugging the pipes and orifices and choking the valves. Under certain circumstances, the volume of the condensing gases decreases. If the process of solidification continues for a long time, then a large amount of solids accumulates. During the maintenance of the system when heat is applied, the frozen material vaporises. This may give rise to an excessive rise in pressure and result in the formation of an explosive mixture.

5.5.4 Maximum Experimental Safe Gap (MESG)

Molkov has reported the results of exhaustive studies on the safety aspects of hydrogen [19, 20] and studied the effect of several relevant parameters, including MESG. The MESG is a very critical parameter, particularly in terms of flammable gases. This is the permissible narrow gap that has been specially designed, depending upon the hazardous location so that the propagation of the flame and spark are prevented. The clearance is so critically designed that if all the flame has to pass through the gap, it could pass through this gap safely. The gap is too narrow to allow the flame to ignite the surrounding mixture. In other words, the flame front does not have adequate energy to ignite the gas on the other side of the gap. Moreover, each explosive gas or vapour reacts differently under similar conditions. The MESG is measured by igniting a combustible air–fuel mixture inside an enclosure and obtaining a closer observation of the ignition of a similar type of fuel–air mixture in the surroundings.

Critical design parameters on the basis of the gas and its location have proven to be very useful in controlling explosions and flashback conditions. The narrowest MESG of 0.08 mm has been recommended to prevent the flame propagation, whereas MESG from propane is 0.92 mm and that for gasoline is 1.02.

5.5.5 Quenching Distance and Quenching Limit

Quenching distance is a very relevant parameter with respect to flame propagation. Apart from ignition energy, it depends upon the pressure, temperature, strength of the combustible mixture and electrode configuration. In the parallel-plate configuration, it is usually the distance between the two flat electrodes in which the flame is not able to propagate. Similarly, it refers to the diameter of the tube that prevents the ignition of the combustible mixture. Sometimes the term 'quenching diameter' is used for a circular tube.

A fast-burning gas such as hydrogen usually has a smaller quenching gap. Under normal temperature and pressure conditions, the quenching gap for hydrogen is 0.64 mm [21] in air. It has been reported in the literature that the lowest quenching distance for hydrogen could decrease to 0.076 mm. Ignition energy is a critical governing parameter for the quenching distance. A gap of about 1 cm becomes necessary for a high ignition energy of 10 mJ, whereas a low ignition energy of about 0.001 mJ needs a much smaller gap of the order of 0.01 cm. The quenching distance shows an increasing trend at lower pressure. The influence of mixture strength on quenching distance has not been adequately reported in the literature. However, the value of the quenching distance in the flammability range (between the upper flammability limit [UFL] and the LFL) remains almost the same for the given pressure condition. In the absence of relevant technical information on hydrogen–air mixtures, the available data related to the deflagration and detonation of acetylene can provide guidelines for an assessment of the effect of pressure as a function of tube diameter for hydrogen–air mixtures. The effects of temperature on LFL and UFL have been studied by Coward and Jones [22]. The quenching limit plays a very significant role in various configurations with leaked hydrogen. Turbines, internal combustion engines, burners, etc. have very different response characteristics Butler et al. [23] carried out research to identify the hydrogen leaks that can support combustion. The tests included a study on diffusion flames on round burners of diameters 0.006–3.18 mm with hydrogen, methane and propane. It was observed that mass flow rates at the quenching limits were independent of the diameter of the burner. This trend of results was very distinctly observed with burners of large diameters.

Very significant studies have been carried out on quenching and blow-off limits [24–28]. Measurements included limits of quenching and blow off for round burners with diameters of 0.006–3.18 mm. The measured mass flow rates at the quenching limits were found to be generally independent of burner diameter for relatively large burner diameters. In terms of the mass flow rate, hydrogen has the lowest quenching limit and the highest blow-off limit of the fuels considered, which means that there are high and low flow rates where hydrogen is able to support a flame while methane and propane are not able to. The quenching limits for hydrogen diffusion flames on round burners with thick walls were found to be higher than those for thin walls. The limits were also found to be independent of burner orientation; leaks with low flow rates are able to support flames independent of their orientation. The minimum mass flow rate of hydrogen that can support combustion from a leaking compression fitting was found to be 0.028 mg s^{-1}. This flow was independent of pressure (up to 131 bar) and about an order of magnitude lower than the corresponding methane and propane flow rates (Figure 5.6). The implications for fire safety are discussed with a smaller range of flow rates on the burners.

Excessive heat loss does not sustain combustion and causes quenching to occur. The other extreme conditions occur when the flow rate reaches the blow-off limit. In all cases, combustion conditions are fixed by quenching and blow-off limits. Studies have been carried out on the comparative assessment of hydrogen with methane and propane [24]. As reported in the literature, propane and methane do not cause any chemical embrittlement that is quite familiar to hydrogen. Hydrogen embrittlement is a function of pressure and exposure time [29]. A comparative assessment of volume and mass flow for the three fuels (hydrogen, methane and propane) has also been studied by Sunderland [28].

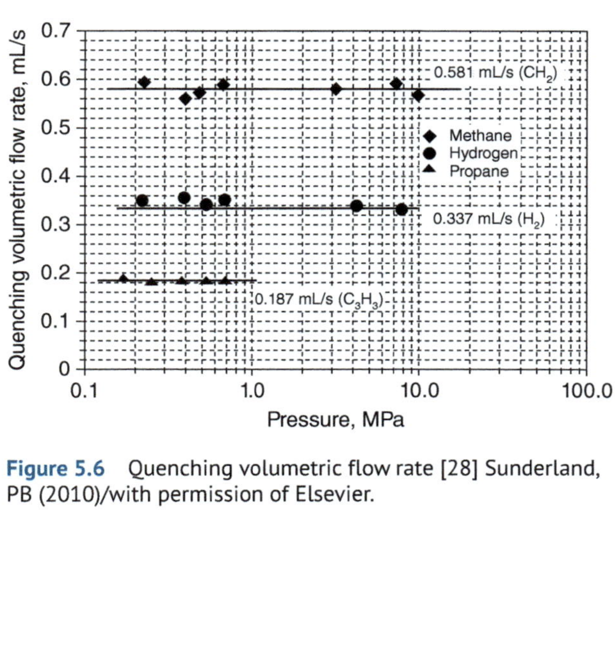

Figure 5.6 Quenching volumetric flow rate [28] Sunderland, PB (2010)/with permission of Elsevier.

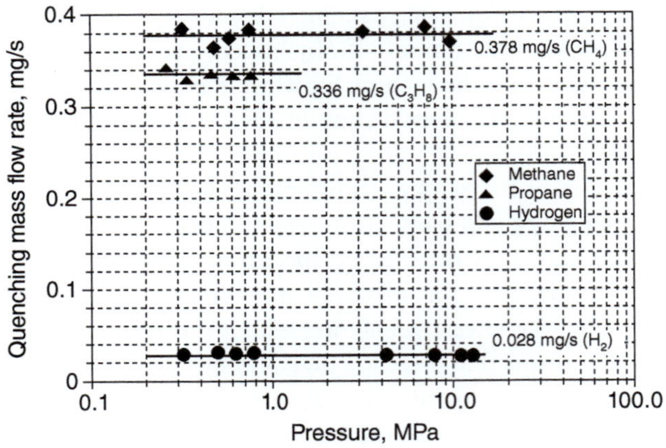

Figure 5.7 Quenching mass flow rate limit [23]/with permission of Elsevier.

A detailed discussion on several aspects of hydrogen combustion covers a much wider range of physical and thermodynamic parameters depending upon the system and is beyond the objective of the present discussion. Combustion phenomena involving safety-related problems and the methods to control them in other systems are discussed in other chapters of the book. Backfire and pre-ignition have proven to be of immense threat. An appropriate fuel induction technique has been observed to be quite effective in controlling the severity and frequency of backfire [30].

The quenching mass flow rate is depicted in Figure 5.7.

5.5.6 Ignition Energy

The *minimum spark energy for ignition* is defined as the lowest energy required for igniting the flammable concentration of fuel in air or oxygen. It gives a broader idea on the sensitivity of an explosive mixture to an electric spark. The minimum spark energy of hydrogen in air is 0.077 mJ at 1 atm, which is substantially lower than that of methane (0.29 mJ) or gasoline (0.24 mJ). However, the minimum spark energy for hydrogen also varies along with pressure and is 0.09 mJ at 0.05 atm and 0.56 mJ at 0.02 atm. Potential sources of ignition, such as open flames, heating systems, electrical installations and other associated equipment, should not be closer to the area or system where hydrogen is being handled. The ignition energy of hydrogen is so low that even weak ignition sources such as hot surfaces, open flames and sparks can cause ignition. While dealing with hydrogen, it is essential to eliminate the possibility of any unforeseen accidents. It is difficult to make a list of all possible sources of ignition. In some situations, or tests, it is unavoidable and ignition sources are essential to exist close to the hydrogen system installation. In such cases, adequate ventilation must be ensured. Specific precautions must be adopted while operating a hydrogen-fuelled engine so that accidental deflagration or detonation is controlled.

Figure 5.8 shows the energy required to initiate hydrogen combustion. As evident in the figure, this value is lower than that required for methane, propane and gasoline vapour. Electrical discharges between objects with different electrical potentials can create

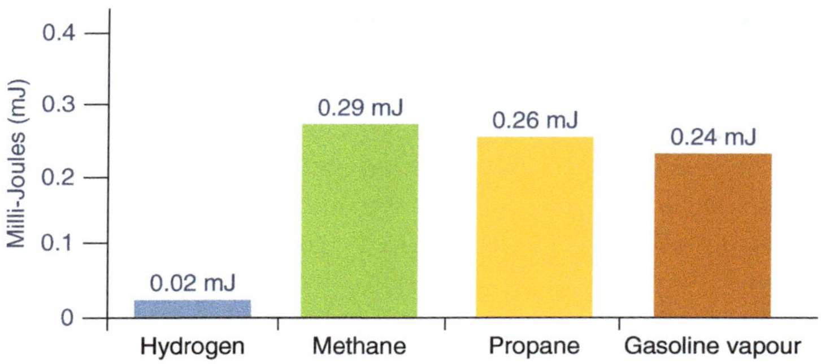

Figure 5.8 Minimum ignition energy of hydrogen and other fuels.

electrical sparks. There are several circumstances when static electricity is generated inadvertently by the common activities of people. Static electricity of higher voltage could be generated even while wearing synthetic clothing, walking on synthetic carpet or dry ground, combing hair and operating a conveyor belt. Quite often, it can be generated by the level of turbulence of the flow of hydrogen in ducts either in the gaseous or liquid form. There are occasions when static electricity charges have been induced by electrical storms. In our regular living pattern, there are many occasions when sparks develop due to friction between two hard objects, such as metal and stone. Two metals, two stones or one metal and a stone striking against each other can cause friction-generated sparks. In the course of some domestic activities or in workshops, sparks from hand tools are often low in energy, whereas drills and some mechanical tools generate high levels of sparks. These are some of the situations that are likely to occur in our usual working area, and under such situations, the presence or possible leakage of hydrogen needs to be avoided. For example, while working with a hydrogen engine in the laboratory, it is desirable that the system (engine) should preferably be purged with an inert gas such as nitrogen to ensure that a combustible mixture cannot develop. After purging, the system can be checked by gas chromatography or any other technique for residual oxygen before hydrogen operation. There are instances where inadequate purging has caused accidents in the laboratory with hydrogen engines. Usually, flame arrestors are installed upstream and downstream of possible ignition sources in piping systems containing flammable gases, such as hydrogen. Installation of flame arrestors at the appropriate location in the hydrogen engine test rig is very useful. Flame arrestors are high-pressure-rated components that are designed to have a large thermal mass and a high coefficient of heat transfer. They have been successful very often in preventing the propagation of flames. Under flame condition, the flame arrestor removes enough heat from the propagating flame front, which is often quite effective in quenching the flame. A review of the literature shows that the more common causes of possible ignition could be mechanical, electrical, chemical or thermal sources. Common thermal sources are mainly due to open flames, hot surfaces, exhaust from internal combustion engines, explosive charges, high velocity of jet heating, etc. Lightening, static electricity, electrical arc, electric short circuit, electric spark, etc. are the most potential sources of ignition. Sometimes mechanical impact, friction, fractures in metals and catalysts have been observed to be the source of ignition. An interplay between the pressure, temperature and concentration of hydrogen in a hydrogen–air mixture often determines the potential for ignition and the severity of the resulting reactions. Changes in pressure can influence gas density, collision rates and the likelihood of molecular interactions. According to collision theory, it is necessary that reactant molecules collide with sufficient energy and proper orientation so that the chemical reaction occurs. In situations of higher pressure, the frequency of collision increases, leading to a higher interaction of reactant molecules (hydrogen and oxygen) and initiating a combustion reaction. It is known that, for a chemical reaction to occur, the reactant molecules should possess adequate energy to go beyond the threshold of activation energy. At higher temperatures, gas molecules move faster and collide with higher frequency, thereby overcoming this barrier. As far as the concentration of hydrogen is concerned, hydrogen has a very wide flammability range of 4–75% by volume. It is important to note that hydrogen–air mixtures with very low concentrations of hydrogen (below 4%) or very high concentrations (above 75%) are less likely to ignite.

Hydrogen has a very wide range of flammability limit, and thus, the ignition energy of the hydrogen–air mixture varies over almost three orders of magnitude in the entire range. It has already been discussed that several types of ignition sources can cause many undesirable combustion phenomena. From a safety point of view, it is important to note that hydrogen's ignition energy is comparable to that of natural gas under conditions of LFL. However, on a volume-of-gas basis, hydrogen has a lower explosion potential than natural gas and propane, and it has the maximum explosion potential on a mass-of-fuel basis. Therefore, the hydrogen supply system consisting of piping connections, valves and seals should be designed to minimise the apparently vulnerable leakage points.

There are some distinctive characteristics associated with hydrogen ignition. Hydrogen–air mixtures can be ignited easily through open flames. Hydrogen–air mixtures or hydrogen–oxygen mixtures can initiate ignition at atmospheric pressure within temperature ranges from 500 to 584 °C. Sometimes, even pressures less than 1 atm can initiate ignition of relatively cooler objects if exposed to temperature of 590 °C (which is above the auto-ignition temperature of hydrogen) for a longer period of time. These specific combustion behaviours need to be considered very judiciously in engine combustion to avoid conditions leading to deflagration or detonation. It has been observed that both processes, such as spraying water in the exhaust system and addition of carbon dioxide to the water sprays, have sometimes proven effective in reducing the detonation pressures and avoiding any safety hazards due to these undesirable combustion phenomena. This, however, should not be taken as a reliable technique for avoiding detonation.

5.5.7 Thermal Energy and Radiation

The mechanical explosion of hydrogen tanks due to radiation or high temperature has been the cause of thermal hazards. There can be situations when the hydrogen tank becomes overheated due to fire in the neighbourhood, and the high temperature leads to the explosion of hydrogen tanks. Liquid hydrogen vessels have a much lower pressure of about 20 bar compared to the pressure of the hydrogen gas storage tank (400 bar). Therefore, the walls of the hydrogen liquid vessels are designed for much lower pressure compared to the pressure level for gas cylinders. When fire approaches a gas cylinder, the material loses its mechanical strength. The situation is different in the case of liquid vessels exposed to fire. In such a case, the portions of the vessel in which vapour exists will become heated much faster and the local temperature will increase in that area. Radiation from a hydrogen flame depends upon the moisture in the water vapour present in the atmosphere, which absorbs the thermal energy from the flame. In fact, depending upon the distance from the atmosphere, water sometimes reduces the intensity [31] as shown below:

$$I = I_0 e^{-0046wR}$$

where I is the intensity of radiation from the hydrogen flame (energy/time area), I_0 is the initial intensity (energy/time area), w is the weight of water vapour (%) and R is the distance in metres.

Hydrogen flames are almost invisible in daylight irradiated mostly in the infrared and ultraviolet regions. Sometimes they are visible only if there are some particles in the air. Any visibility of a hydrogen flame is caused by impurities such as moisture or particles in the air. However, hydrogen fires are readily visible in the dark, and large hydrogen fires are detectable in daylight by the 'heat ripples' and the thermal radiation felt on the skin. Thermal energy radiated from flames to surroundings ranges from 17 to 25%. As far as the radiation from flames to surroundings is concerned, it is well known that the heating value of the flame, whatever it might be, is ultimately dissipated by radiation. The maximum temperature that a hydrogen flame can have is of the order of 2130 °C. Hydrogen fires burn very rapidly and radiate very little heat and thus are relatively short lived. A person can be closer to an H_2 fire than a gasoline fire without being burned.

One cannot feel the heat unless it is almost inside the flame. Petroleum fuels produce carbon, soot and particulate matter during combustion, so the effect of radiant heat can reach a person who is farther off from the flame. Ethanol combustion produces CO_2, and thus, its flame emits almost twice the amount of heat that the same volume of gasoline flame emits. Persons and material are not affected by the low emissivity of the hydrogen flame. The Sandia National Laboratory in the United States of America has reported that the emissivity of hydrogen is <0.3, which is much lower than the emissivity of hydrocarbon flames, which is around 1. Since there is practically no radiation from hydrogen fire, there was no report of death in the Hindenburg accident due to hydrogen burns. Gasoline fire poses a threat due to smoke and fumes that can create health problems if inhaled. On the other hand, in the event of fire, hydrogen produces only water vapour. Hydrogen requires 18–59% oxygen to be explosive, whereas gasoline needs only 1–8% oxygen to be explosive.

5.5.8 Excessive Pressure and Blast Waves

Under some circumstances, a form of explosion takes place, leading to blast waves during hydrogen operation and handling, thus resulting in an abrupt rise in pressure. The character of the explosion depends upon the level of overpressure and the proximity of the human body close to the site of explosion. The magnitude of pressure varies over a very wide range starting from 12 or 13 kPa and going up to a level of about 1500 kPa. Accordingly, the direct damage to human body could be from ear drum rupture to severe lung haemorrhage. Apart from the direct effect, there could also be some indirect effect of the condition particularly when the human body is slightly farther away from the explosion site. Such situations usually occur when the debris of the explosion falls on human skin. There is also a possibility when the body of a human being is physically shifted because of the high impulse of the explosion. Damage still becomes severe if the shifted body strikes against a hard surface.

The situation could be much worse and can even lead to death if in any explosion the injury caused by overpressure is combined with the stringent thermal radiation. Such possible injuries due to overpressure and blast waves are not ruled out. The blast wave and overpressure resulting in some form of explosion also affect the structures, apart from human beings. This damage, which essentially depends upon the combined effect of blast wave pressure and thermal radiation, can be as minor as breaking a glass window to complete damage to the machinery or even to the total destruction of the building structure.

Let us examine the safety aspects in connection with hydrogen generation in nuclear power plants. In nuclear power plants, Zircaloy is used as the metal cladding for the reactors. Hydrogen is generated by the reaction of Zircaloy with water as given in the following equation:

$$\mathrm{Zr} + 2\mathrm{H_2O} \rightarrow \mathrm{ZrO_2} + 2\mathrm{H_2} + \Delta 5 \cdot H_{\mathrm{Zr}}$$

where ΔH_{Zr} is the heat of the reaction per mole of Zircaloy consumed. This exothermic reaction is often followed by deflagration or even conditions leading to detonation of hydrogen–air mixture. This condition could sometimes lead to major accidents. In an attempt to avoid accidents, the accumulated hydrogen is intentionally burned out, thereby avoiding an excessive rise in pressure. Once the pressure and temperatures are controlled, the situation becomes pretty safe. The application of an igniter is a very useful technique to monitor the severe thermal environment and conditions of high pressure [32].

There are some possibilities of safety hazards during liquid hydrogen storage. Liquid hydrogen storage vessels, with a capacity from 0.1 cubic metre to several thousand cubic metres, are being used for several applications. Depending upon the system configuration where they are used, specific safety steps are adopted. A large number of improved design parameters have already been used in liquid hydrogen tanks. The safe optimised well-insulated LH_2 tanks have been developed. These tanks are normally spherical in shape because of the sphere's minimum surface-to-volume ratio [33]. These spherical tanks are vacuum-jacketed and are constructed with double wall. Moreover, the spherical shape ensures a uniform distribution of stress and strain.

5.5.9 Burning Velocity

The burning velocity in air refers to the subsonic velocity with which the flame of a flammable fuel–air mixture propagates. The higher burning velocity of hydrogen is another typical characteristic that must be properly monitored. A higher burning velocity of hydrogen has both positive and potentially challenging features depending on the context in which hydrogen is being used. It enables rapid mixing with air, leading to efficient combustion. For internal combustion engines and gas turbines, it leads to higher energy conversion and reduced emission. In rocket engines, hydrogen's high burning velocity leads to higher thrust and higher propulsion efficiency. A higher burning velocity reduces the time period that reactants spend in the combustion zone. This can reduce the potential for the formation of harmful by-products or pollutants. However, there are some challenging situations where the burning velocity of hydrogen needs to be scrupulously handled to obtain benefits from points of view of both performance and emission and combustion. A higher burning velocity can sometimes pose challenges for hydrogen combustion in terms of flame stability and control. Rapid combustion can lead to flashback. Sometimes a higher burning velocity of hydrogen results in combustion instabilities, such as oscillations or pressure fluctuations within the combustion chamber. Sometimes, intense heat generated by hydrogen's fast combustion may create problems of material compatibility in combustion chambers, engines and other components exposed to this level of thermal stress. Thus, careful design and control of the combustion system is essential to derive benefits from hydrogen's property of high burning velocity.

The hydrogen–air mixture has a burning velocity ranging from 2.65 to 3.46 m s^{-1} depending upon the temperature, pressure and composition of the mixture. Thus, the burning velocity of hydrogen is almost eight times that of the methane and propane air mixture. It is quite difficult to arrest such a higher-burning flame. The problem is of greater relevance in closed systems where this flame cannot be confined and consequently might result in explosions. The possibility of detonation taking place in hydrogen–air mixtures is greater than the mixtures of other gaseous fuels. Damages due to detonation are more severe than that in ordinary explosions or deflagration. The flame temperature for 19.6% by volume of hydrogen in air is about 2318 K. Hytovick Chambers; Chin, Gamezo; Poludnenko and Ahmed [34] adopted advanced experimental technique such as direct measurements of pressure, flame, flow, and turbulent velocities through high-speed schlieren in hydrogen-air mixture to study the phenomenon of turbulence-induced deflagration-to-detonation transition. This study proposes some novel experimental results related to the hot spot initiation in the turbulent deflagration-to-detonation transition process. Problems such as deflagration or detonation have a more damaging effect in confined spaces or in congested atmospheres. The rapid rate of deflagration of hydrogen fire and diffusive characteristics often control the severity of the problem.

As evident in the graph (Figure 5.9), the maximum expansion ratio occurs at the stoichiometric mixture condition (29.5% by volume). The effect of the high molecular diffusivity of hydrogen is reflected in the figure. Because of this, the maximum burning velocity for the hydrogen–air mixture does not occur under stoichiometric conditions but rather in a rich mixture level at about 40.1% by volume. The maximum burning velocity of hydrocarbon–air flames occurs at a slightly richer mixture with an equivalence ratio of 1.1, whereas the maximum burning velocity for a hydrogen–air premixed flame occurs at an equivalence ratio of 1.6.

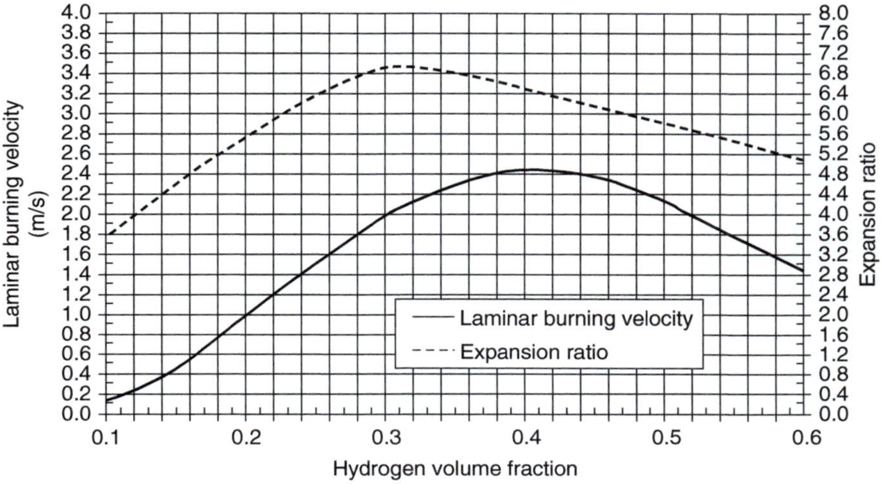

Figure 5.9 Effect of hydrogen concentration on the laminar burning velocity.

5.5.10 Flammability Range

The formation of flammable mixtures for hydrogen depends upon a number of factors. Figure 5.10 shows a comparative diagram of the flammability range of hydrogen along with other fuels. The wide flammability of hydrogen is an important property from a safety point of view. The hydrogen–air mixture has a much wider flammability limit (4–75%) by volume compared to natural gas (5.3–15%), propane (2.1–10%) and gasoline (1.5–8%) as shown in Figure 5.10.

Essentially, hydrogen is a temperamental fuel that provides flexibility and ample combustion benefits, and at the same time, demands meticulous attention to safety and combustion stability. To get the best out of its properties, they need to be very judiciously handled depending upon equipment and operating conditions. Arbitrary handling of some of the properties can lead to excessive damage.

A wider flammability limit makes it possible to be used in a variety of conditions, ranging from lean to rich mixtures. This flexibility is advantageous for designing combustion systems, such as engines and turbines, that can operate under varying conditions and load demands. Lean-burn combustion strategies in engines can achieve a higher thermal efficiency and lower nitrogen oxide (NOx) emissions, contributing to cleaner and more efficient power generation. Operating near the LFL or UFL requires the combustion system to be carefully designed to ensure stable and efficient operation. Depending upon the hydrogen-operating system, specific effective explosion mitigation strategies need to be implemented.

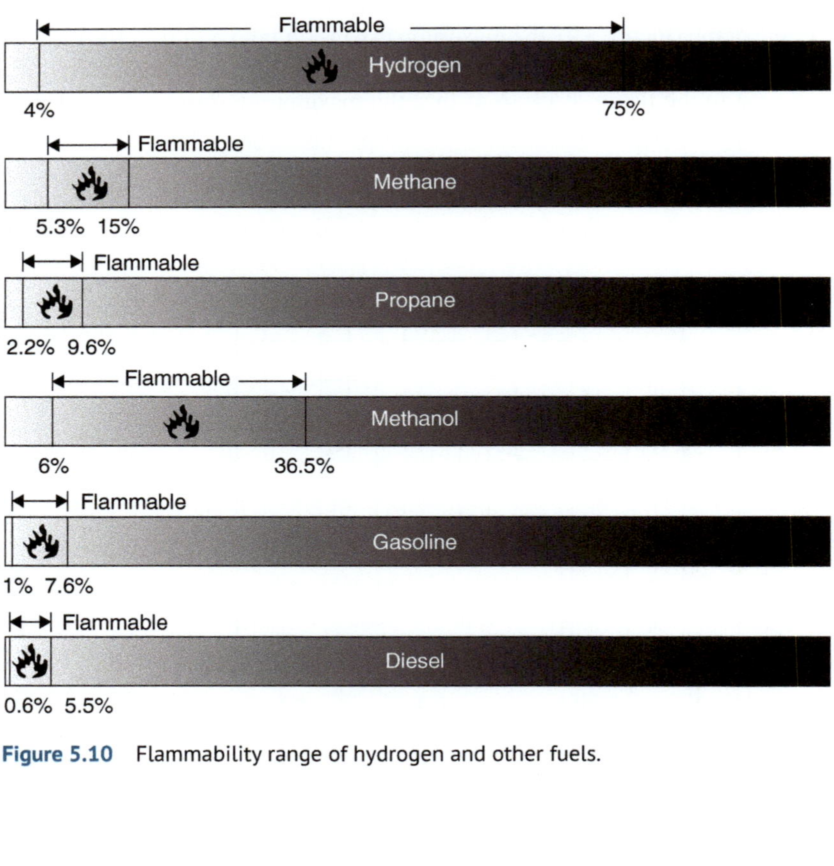

Figure 5.10 Flammability range of hydrogen and other fuels.

Flame arresters, pressure relief valves and other explosion-resistant devices need to be installed at the appropriate location to prevent the possibility of unintentional ignition within this wide range. The range of flammability is of lesser practical significance in the event of any leakage compared to LFL because it is the LFL that is at the lower end of the concentration responsible for causing ignition or supporting combustion. Under lower flammability conditions, hydrogen and natural gas have the comparable levels of ignition energy. Since hydrogen's flame velocity is almost seven times that of natural gas, it is quite likely that a progressive hydrogen flame would develop into deflagration or detonation. It is perhaps appropriate to discuss at this point the phenomena of explosion that has been so closely associated with hydrogen as far as its safety image is concerned.

5.6 Phenomena of Explosion

Unfortunately, the term 'explosion' has often been very liberally and interchangeably used without any technical identification of the cause and the character of the explosion. There are many occasions where hydrocarbons from large combustible clouds have resulted in disastrous explosions [16, 35]. The Buncefield Major Incident occurred on 11 December 2005. The group in the Investigation Board attempted to determine the cause of explosion from different aspects. The possibility of deflagration, detonation and a combination of both phenomena was analysed. The other possibilities leading to the acceleration of the flame was also analysed [35]. The Buncefield incident involved a massive explosion and fire at the Buncefield Oil Storage Depot in Hertfordshire, UK. There occurred a massive release of vaporised hydrocarbons, which formed a flammable vapour that got ignited, leading to a series of explosions and a large fire.

Broadly speaking, the entire range of explosions includes flash fire to deflagration, detonation, VCE in both confined and unconfined spaces, and BLEVE.

However, the cause of its occurrence and the character of explosion under a particular situation and in a specific environment must be clearly understood. The loose use of terms in the context of hydrogen has added an undeserved stigma to hydrogen with regard to its safety feature. In this context, it is essential to discuss the kind and character of the explosion phenomenon. Broadly speaking, an *explosion* is defined as a process that gives rise to a rapid rate of energy release and thereby raises the pressure. Explosion can occur in reactive systems in which premixed gases of fuel and oxidiser chemically react, resulting in a rapid energy release. Explosions due to excessively high pressure in gas vessels can be characterised as non-reactive explosions. There have been instances when pressure rises abnormally due to either deflagration or detonation. Sometimes, explosion has been referred to as a very narrow and restricted application, such as bursting the internal pressure of a container due to excessive pressure rise on account of deflagration. Although deflagration or detonation is transmitted through a wave, a gas explosion does not always need a medium. Under certain circumstances, the application of heat can create conditions of a volumetric explosion of the mixture in a vessel so that the mixture reaches the level of high temperature required for a fast reaction to occur [31]. As far as the occurrence of VCE is concerned, the velocity of the flame front depends upon the mixture strength: whether it is

Figure 5.11 The boiling liquid expanding vapour explosion (BLEVE) showing the rupture of a pressurised hydrogen tank due to fire contact or impact. *Source:* Initsys Ltd.

above the stoichiometric value. In some cases, the mixture strength is ultra-lean, the value of the equivalence ratio is very small. Under such circumstances, sometimes, the phenomenon of flash fire occurs, which is characterised by a short-duration fire with a small velocity and a small pressure level. As the terms suggest, if the explosion of the gaseous mixture takes place in the open or confined space, accordingly the phenomenon is termed 'unconfined vapour cloud explosion' (UVCE) or 'confined vapour cloud explosion' (CVCE), respectively. Under some circumstances, pressurised liquefied hydrogen is released into the atmosphere because the temperature goes beyond its boiling point, resulting in an explosion leading to the rupture of the storage vessel. This phenomenon is referred to as 'boiling liquid expanding vapour explosion'. Such a situation takes place when an external flame strikes the shell of a vessel. The vessel ruptures, and its flammable contents inside the vessel are released into the atmosphere and form the fireball (Figure 5.11). A close look at the cloud provides very interesting information. Ignition is initiated in the outer layer because of the presence of the air–fuel mixture within the flammable range, whereas only neat fuel exists in the inner core. The flame due to combustion radiates a large amount of heat, and because of the buoyancy of hot gases, the burning cloud moves upwards and gradually develops into the shape of a sphere.

5.7 Deflagration and Detonation

There is a basic difference between the characteristics of deflagration and detonation. Deflagration is a relatively milder form of explosion that occurs in the range of the subsonic flame speed. The mechanism of energy transfer from the reaction zone to the unreacted zone takes place by way of heat and mass transfer. Hydrogen remains buoyant even above the temperature of 22 K, and it remains so over its entire range of temperatures in the gaseous state. In the event of any inadvertent leakage of hydrogen, either by the failure of a storage tank or pipeline, a very small fraction of energy, within the range of 0.1–10%, is able to deflagrate. However, it is mostly less than even 1% of the total energy of the hydrogen released [36, 37].

In contrast, detonation is a high-order explosion phenomenon that occurs at the supersonic flame speed. The energy transfer here occurs from the reaction zone to the unreacted zone on a reactive shock wave. The detonation induction distance signifies the distance covered by deflagration inside a detonable fuel–air mixture and subsequently develops into detonation. The detonation induction distance, under normal temperature and pressure, is measured by spark or hot wire ignition at one end of the long cylindrical tube in which the mixture is kept. Appropriate instrumentation is performed on the entire length of the tube to identify the velocity of the flame front as it moves ahead inside the detonable mixture. The detonation induction distance is the distance between the igniters and the axial position in the tube where the flame front attains the detonation velocity. Detonation distance depends upon a number of factors such as pressure, temperature and the strength of the mixture. In addition, the geometry of the enclosure and the strength of the ignition source are also important. Figure 5.12 shows a comparative diagram of hydrogen's detonability range compared to conventional fuels.

Usually, the gaseous hydrogen–air mixture gives rise to deflagration that is accompanied by some explosions. As mentioned earlier, the deflagration of the gas–air mixture is relatively less severe in its explosive characteristics compared to detonation. Sometimes this deflagration develops into detonation in a confined enclosure, and the phenomenon is referred to as deflagration-to-detonation transition (DDT). A detailed mechanism of DDT has been observed to be a gradual process achieved through different stages, and it is a stepwise transition and not a continuous process [38]. With regard to hydrogen safety, DDT is of particular concern because of hydrogen's wider flammability range. In case a leak of hydrogen gas occurs and is ignited within a confined space, the flame accelerates and continues to propagate. This acceleration can lead to a pressure build-up and even a potential transition from deflagration (subsonic combustion) to detonation (supersonic combustion) under certain conditions. This transition can result in a much more powerful explosion.

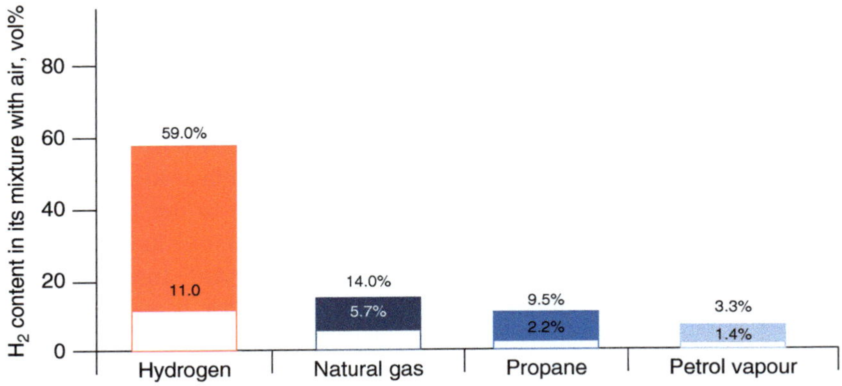

Figure 5.12 Detonability range.

DDT is a complex phenomenon by which a deflagration (which is a flame front moving at a subsonic speed) develops into a detonation, which is a supersonic shock wave accompanied by intense combustion. Several parameters, such as confinement, initial temperature and pressure, turbulence, mixture composition, ignition source and flame dynamics, are closely linked to this phenomenon. DDT is more likely to occur in confined spaces or channels where the combustion products are constrained, leading to increased pressure. This confinement enhances the interaction between the flame and the unburned mixture and thus leads to the transition to detonation. Turbulence in the flow of the hydrogen–air mixture can enhance mixing and increase the flame speed. Turbulence-generated vortices can compress the mixture, raising its pressure and temperature, which can facilitate the initiation of a detonation. Sometimes, elevated pressure leads to a faster flame speed, facilitating the mixture reaching a stage conducive for detonation. The concentration of hydrogen in a hydrogen–air mixture often determines the flammability and reactivity of the mixture. Stoichiometric mixtures with higher flame speeds often exhibit a greater tendency towards DDT. When the flame speed exceeds the critical value, it leads to shock wave formation ahead of the flame initiating detonation. Sometimes, obstacles or irregularities in the confinement create localised areas of higher temperature and pressure. This situation often creates conditions leading to shock wave formation and subsequently leading to DDT. It has been observed that the detonation takes place in a much narrower range of hydrogen–air mixtures, which falls within a broader limit of 18.3–59% of hydrogen. It was further observed that this range can be extended to a higher limit by a strong source of ignition. Similarly, even in the absence of a strong initiator, a deflagration can be developed to a stage of detonation by several other methods. Inducing turbulence in the flame front has been observed to be a very effective method. Studies related to DDT for hydrogen–air mixtures have been reported [39] in research work involving numerical studies. However, the condition leading to detonation depends on several factors, such as the air–fuel ratio, temperature and geometry of the confined space. Therefore, hydrogen detonation would not occur in an open atmosphere. Pre-ignition and backfire have been very frequently observed during hydrogen operation.

Fuyuan Yang et al. [40], on the basis of data between 1999 and 2019 from the Department of Energy and other sources, carried out a very informative analysis of the causes and consequences of several types of safety-related accidents at various institutes worldwide. The analysis shows that a maximum of about 38.3% of accidents occur in laboratories. About 10.6% of accidents occur in hydrogen fuel stations, whereas 9.0% accidents occur in hydrogen-related commercial facilities. A more detailed analysis shows that the accidents were caused by both human error and equipment failure. There have been incidents when hydrogen cylinders have been accidentally filled with air instead of additional hydrogen, thus charging the cylinder with an explosive mixture. In the 120 accident analyses caused by equipment, most of the cases involved piping rupture and fitting and valve breakdown, mainly because of the problem of embrittlement. In some cases, human error has been responsible for several incidents. There have been inadvertent design flaws in the choice and installation of safety equipment in the appropriate locations. Safety-related issues have also occurred due to incorrect assembly, improper movement and inappropriate replacements.

5.8 Safety at Different Stages: Production, Transmission, Storage and Application

Leakage, diffusion, ignition and explosion are the most important phenomena relevant to safety, which must be taken care during all stages of hydrogen production, transportation, storage and utilisation. With the rapid penetration of hydrogen into several sectors, research analysis and results have contributed to formulating safety regulations [41–44]. Safety studies have been conducted on several aspects of hydrogen energy technology, such as production [45–47], storage [48–50], refuelling stations [51, 52] and vehicular applications. Safety-related problems with regard to the leakage and environment have also been studied. The safety of hydrogen is dependent upon the phase in which it is used. Hydrogen in the gaseous form is the common mode, but liquid hydrogen is a more efficient option for transporting and storing hydrogen. It has been estimated that a unit of volume liquid hydrogen, when vaporised, expands to approximately 848 units of volume at STP. However, the materials used for liquid hydrogen service, such as pipings, vessels and other associated fittings, should be able to withstand low temperatures and temperature fluctuations and should also be resistant to embrittlement. Thermal expansion and contraction of equipment due to severe temperature fluctuation from ambient to LH_2 temperature is an important design constraint in some applications, such as transferring liquid from a delivery tanker to a storage vessel through pipes and other associated fittings.

The liquefaction of hydrogen is a very slow endothermic process that includes the removal of energy released by ortho–para conversion. In this case, the heat of conversion is about 1.5 times the heat of vapourisation. Sometimes a paramagnetic catalyst is used to accelerate the process of liquefaction. There have also been instances where the liquid hydrogen and cold boil-off gas, coming in contact with human skin, have caused severe burns. Splashed liquid hydrogen affects the delicate tissues of the eyes. Sometimes, the rapid evaporation of liquid hydrogen spills can develop into a flammable mixture resulting in subsequent fire even if the source of ignition is not present at the location of the spill. Sometimes rapid evaporation occurring in a liquid hydrogen spill results in a flammable mixture over a relatively longer stretch of distance. In such cases, even though ignition sources may not be present at the location of the spill or leak, fire could still occur if the flammable mixture moves forward to an ignition source. The significant fire damage parameters are usually thermal radiation, flame engulfment and smoke inhalation.

The safety-related problems that arise while dealing with hydrogen in the gaseous form also exist for hydrogen in the liquid state (LH_2). Problems with liquid hydrogen are due to its continuous evaporation, significant pressure rise and higher level of density of saturated vapour, and these characteristics need to be closely monitored. NASA's first effort to use liquid hydrogen in large quantities in the Centaur programme and the observations from the rocket's Saturn 1 block II vehicle employed specific design features for liquid storage, and the transfer system showed that a very small amount of hydrogen when combined with air can lead to an explosion because only a small amount of energy is required to initiate ignition. The other safety hazard associated with liquid hydrogen is that it causes cold burns if the cryogenic liquid continues to leak for a long duration of time. Another potential danger is a violent explosion of a boiling liquid expanding vapour (BLEV) in the case of a pressure relief valve failure. These are some of the safety-related issues that must be considered while handling liquid hydrogen.

5.8.1 Safety During Production

Dawood et al. [45] developed an innovative integrated model and proposed a square-shaped hydrogen square (HydS) model that interconnects the stages of the hydrogen life cycle.

The four corners represent production, storage, safety and utilisation (Figure 5.13). Thus, the safety aspects (third corner) have been discussed in the analysis. Subsystems, purification and compression have also been considered in the model. A major objective of the analysis in this paper has been to plan the production keeping the four corners (including safety) in mind. Chau et al. [46] conducted an exhaustive study on the production of hydrogen from a safety point of view. The study adopted a 'multi-criteria decision-making' (MCDM) integrated analytic hierarchy process and a 'life cycle index approach'. In this study, in the absence of adequate specific accident data and due to the varied differences in materials, safety was considered taking into account the availability of resources along with the process and the level of maturity of technology. Researchers adopted fuzzy logic 'to address subjectivity and uncertainty challenges in the data' [46]. Although electrolysis was the preferred route from an environmental point of view, it had a much higher level of

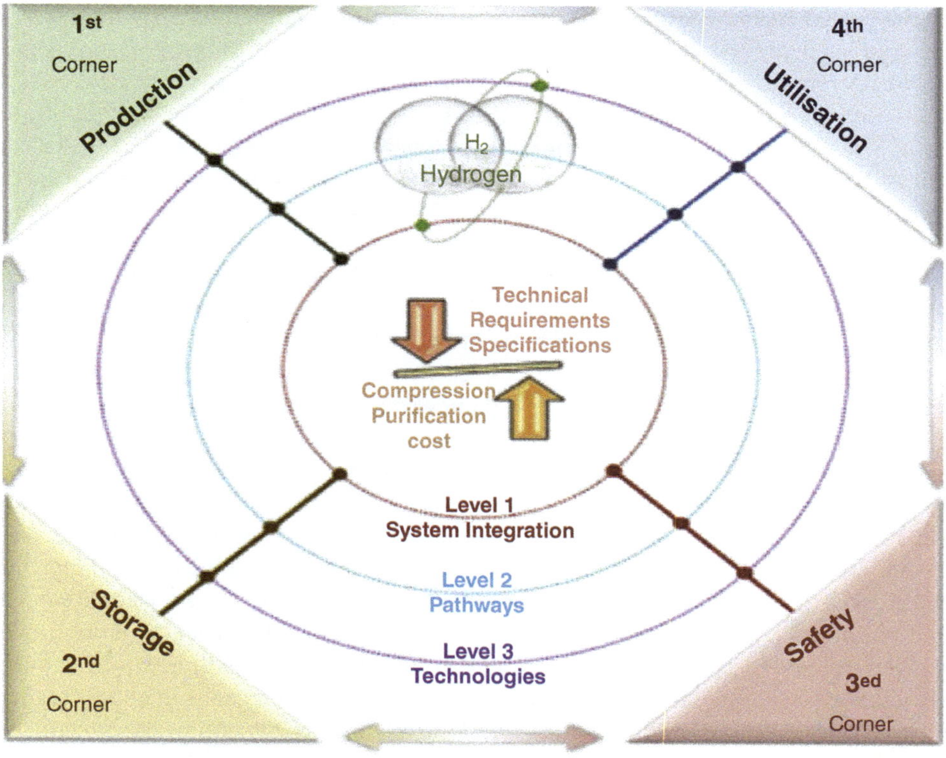

Figure 5.13 Hydrogen square (HydS): four-corner model [45] / With permission of Elsevier.

uncertainty compared to SMR. Both SMR and ATR are processes whose mechanisms have been studied. SMR is perhaps the most understood among all competing technologies. By using uncertainty analysis, it was also observed that SMR had a much more reliable and robust index [46]. Therefore, from an application point of view, steam methane reforming has been found to be the most desirable route not only from safety but also from the points of view of efficiency and infrastructure.

Dincer and the team of researchers [47] have discussed the problem of safety during hydrogen production. Proper handling under conditions of high temperature and pressure has been suggested. Researchers introduced the innovative 18S concept and discussed the role and relevance of each term starting with S ('source', 'system', 'service', 'scope', 'staff', 'scale-up', 'safety', 'scheme', 'sector', 'solution', 'stakeholder', 'standardisation', 'subsidy', 'stimulation', 'structure', 'strategy', 'support' and 'sustainability') for hydrogen. Their analysis, with an emphasis on safety, shows that renewables have the potential to satisfy the optimum conditions for hydrogen production. Geothermal and biomass sources have the highest sustainability performance, followed by hydro and solar sources. From a sustainability point of view, coal, nuclear and natural gas have been found to be in lower categories of ranking [46]. The use of leak detection devices and flame sensors, widespread pipelines for safe delivery, safety conditions in closed parking areas, trained personnel, community education, safety demonstration, etc. are some of the safety steps discussed in the paper. Since hydrogen application is gradually growing, specific safety measures need to be followed in building up widespread hydrogen delivery infrastructure.

5.8.2 Safety Criteria in Storage

Safety in storage is important for the use of hydrogen. Each stage of the system from production, storage and transmission should be optimised for high energy efficiency. Depending upon the demand of hydrogen for a specific application and specific location, bulk storage and transmission by pipelines becomes useful. At present, the major part of the hydrogen required in chemical industries and aerospace applications is stored in large quantities in the liquid form in insulated hydrogen LH_2 tanks, which are quite expensive. Moreover, loss due to evaporation decreases the efficiency of the process. The storage of high-purity liquid hydrogen has restricted its use in chemical industries and in aerospace applications. The leakage of hydrogen and the evaporation of liquid hydrogen are problems related to the storage of hydrogen. However, the problem is crucial and of great concern when storage is considered in vehicles. Moradi and Groth [48], in their discussion of risk and reliability analysis, have reported that problems related to leakage, contamination, resistance to fire, large-scale utilisation, etc. need urgent attention. Technology related to cryogenics and compressed storage has been used for a very long period and has matured. However, the poor energy efficiency of cryogenics and large volume for compressed storage are two major property-based problems. Material-based storage systems, which have been adopted only for a short period, need to be used for much longer period to establish their suitability for long-term use. Under ambient temperature and pressure conditions, 1 kg of compressed hydrogen occupies 11 m^3. Thus, an enormous amount of compression is needed to store hydrogen in compressed form. Although high-strength steel is a suitable choice from the point of view of strength, the diffusivity of hydrogen through steel is one of the major constraints for several applications, particularly for storage in vehicles. Since 5 kg of compressed hydrogen occupies a volume of 212 l, such an enormous quantity of compressed hydrogen cannot be accommodated in the existing tank of the vehicles. Since only 8.5% of the energy content of hydrogen is used for compression [49], it is difficult to design pressure vessels of a size that can be conveniently accommodated in the vehicle. Subsequently, the weight of the pressure vessel is another problem that needs to be considered. In fact, the net energy utilised for compressed hydrogen storage is incidentally much lower than that for liquid hydrogen and metal hydride [50]. The present trend of hydrogen storage system design is considered primarily from an economic perspective. Safety, which is extremely crucial in the context of hydrogen, does not receive the importance it deserves.

Son Tay Le et al. [53] carried out a quantitative risk assessment (QRA) to evaluate the influence of the relevant design parameters, such as mass flow rate, storage size, pressure, temperature of storage, etc. In this study, data were collected and possible hazards were identified. A series of analyses, such as frequency analysis, consequence analysis and risk analysis, were carried out to determine the risks associated with the storage system and to determine the possible ways to be adopted for the safety of the system. It was determined from this analysis that both storage capacity and pressure had a very strong influence on the safety of the system. It was observed that increases in storage pressure and storage capacity greatly affect the radiation, overpressure and other parameters. This situation sometimes also leads to leakage and even rupture. The results of this analysis clearly highlight the significance of implementing risk mitigation for hydrogen storage systems. Thus, QRA provides very important guidelines to design an optimum storage system from both economic and safety aspects [53].

5.8.3 Safety in Transmission

Safety is one of the topmost priorities of hydrogen transmission. At present, hydrogen is transported by pipelines and tube trailers for gaseous hydrogen and by trucks, rails and ships with cryogenic tanks for liquefied hydrogen. On several occasions, feedstock such as methanol, ethanol and biomass-derived liquids are transported to the point of use where they are reformed to hydrogen. Depending upon the properties of the feedstock, their mode of transport is selected, and these are still being developed. Economic viability is one of the major factors in working out the strategy for safe hydrogen transmission. Prior to transferring a large quantity of hydrogen, it is necessary to conduct a thorough risk assessment to identify potential hazards and develop a comprehensive transfer plan. If hydrogen is transferred in gas cylinders, it is desirable to inspect and ensure the integrity of transfer equipment, including cylinders, hoses, connections, valves and regulators. Similarly, for liquid hydrogen transfer, all transfer equipment, including transfer lines, valves and storage containers, should be designed for use with cryogenic liquids.

A pipeline network is safe and convenient when large quantities of hydrogen are transported over a long distance. This is mainly because the pipelines used in such systems are small in diameter and hydrogen is transferred at a relatively low level of constant pressure [51]. Such a situation of hydrogen transmission demands higher power and energy for the compressor. Moreover, while being used in fuel cell vehicles, the usual lubricant might become contaminated. Pipeline damage depends upon a number of factors such as hydrogen pressure, mechanical loading and the material used.

There are procedures to evaluate this pipeline damage tolerance by standard procedures normally adopted in the industries. Guidelines are already available to design barrier thin-film coatings that can minimise the permeation of hydrogen in the existing natural gas pipelines. To solve some of the problems described above, there is a need for an improved compressor with higher efficiency. As an alternative option, hydrogen transmission, in the absence of a compressor, is being considered using the high-pressure electrolysis technology.

5.9 Safe Handling, Storage and Use of Hydrogen in Vehicles

Considering its nearly zero-emission feature, hydrogen-fuelled vehicles are one of the most important areas of application and have received a lot of attention. It is therefore important that adequate attention must be given at every stage in the right sequence for the safety of hydrogen-operated vehicles. Appropriate and adequate safety measures need to be adopted during storage and transport of hydrogen in tanks and onboard storage, handling of fuelling stations and using vehicles. Problems related to safety are mostly related to fire and explosion. Any safety problem due to toxicity is ruled out because hydrogen itself is non-toxic and does not produce any toxic products upon combustion in vehicles. The largest amount of hydrogen is handled through tanks. The potential leakage points are cracks in the tank liner developed under conditions of higher stress. Vulnerable leakage points also exist at the connection from the tank to the fuel line. Sometimes leakage occurs due to the improper operation of the safety valve. The leakage from these sources is slow, and if identified early, the magnitude of hazard could be relatively low. Faulty operation of safety valves sometimes results in high hazards. The problem becomes worse if leakage is high, which most often occurs due to faulty operation of the pressure relief valve.

A hydrogen storage system in a vehicle should be designed to minimise potential leakage points. This is of greater relevance if compressed hydrogen gas is used to operate the vehicle. Any gas discharged from the system should escape from the vehicle through well-defined channels and as safely as possible, that is, without entering the passenger compartment or other sealed-off areas. On some occasions, massive leakage could be caused by the failure of the tank wall due to chemical reactions. Situations leading to the catastrophic rupture of the tank can take place due to manufacturing defects. Safety-related problems could arise due to incorrect handling of the tank and its subsystems. External fire associated with the accidental failure of pressure relief devices (PRDs) could prove disastrous. Depending upon the situation, appropriate design modification needs to be made to detect leaks and prevent them. Adequate ventilation and the design and installation of proper electrical system are very important from a safety point of view. Swain et al. provided a comparative depiction of the leakage of hydrogen along with other fuels, such as natural gas, liquefied petroleum gas and gasoline [52]. The study takes into account cases such as fuel tank leaks, fuel line leaks and in garages and refuelling stations. As far as onboard hydrogen tanks are concerned, the cylinders have to be designed to store the gas under pressure. Since the compressed gas tanks hold hydrogen, the age of the compressed gas tank with an expiration date is necessary for preventing possible problems due to corrosion and fatigue. The high-pressure regulator, tubing extending up to several metres and all the associated components that are likely to be subjected to the severe crash need to be inspected intermittently and maintained.

A simplified model was developed by Swain et al., which is useful in determining the health and safety hazards in the event of hydrogen leakage. The hydrogen risk assessment method (HRAM) can provide information and guidelines for venting in buildings [54].

There are four types of cylinders (types I, II, III and IV). Discussions about the cylinders are given in the chapter on storage. Usually, type III or type IV cylinders are used to store hydrogen gas. These are carbon fibre tanks with aluminium or plastic liners, and they can withstand pressure levels up to about 700 °C, and thus, the chances of their bursting in an accident are relatively low. The most severe case of an accident could be in a situation where the hydrogen is exposed to the fire coming out of another petroleum-fuelled vehicle after the crash. A well-insulated fibre composite tank is the best design to confront such accidents because it would restrict the rise in temperature and pressure. A situation could arise some time when the hydrogen is not vented out, and thus, the tank could burst. Therefore, a thermally activated PRD is usually provided.

The design of regulators is a critical consideration, and usually, the regulators are placed partially inside the tank so that the chances of them being turned off in a crash is relatively low. Sometimes if the gas from the cylinder is released rapidly and comes in contact with some ignition source, this could lead to a crash. In some situations, auto-ignition of the released hydrogen takes place due to static electricity. An effective measure to shut off the release of high-pressure hydrogen gas can be achieved by installing a solenoid valve inside the tank. As far as the safety aspects of vehicles are concerned, the mechanical energy stored in the compressed gas cylinder also plays a critical role. When leakage occurs, the impulse of the ruptured tank could produce a horizontal jet. Quite often, the impulse could initiate acceleration, particularly on a road where the effect of friction is low [55].

Very interesting and useful results have emerged from Miami researchers [13] on the safer aspects of hydrogen compared to gasoline with reference to vehicular applications, as shown in Figure 5.14. A comparative assessment of the degree of severity of hydrogen and gasoline fuel leaks leading to ignition was carried out.

Initially, a long, vertical jet flame was generated. The hydrogen car was not affected by the flame as the radiation from the hydrogen flame is very low. It was observed in these tests that after 60 s, the hydrogen flame became thinner, whereas the gasoline flame developed into a fire that continuously intensified. After 100 s, the hydrogen flame had vanished off, and the gasoline-fuelled car continued to burn until it was totally eaten up by the flame. The interior of the car remained undamaged and the temperature of inside back window was a mere −19.5 °C. The consequence of the fire from a hydrogen car would have been much more severe had it been inside a garage. BMW also carried out precautionary measures related to the safety of hydrogen vehicles. BMW has been pursuing hydrogen-fuelled vehicle development activities since 1979. In the process of developing the prototype vehicle, several precautionary measures were taken, and safety guidelines were identified [56]. Some of the effective steps suggest that vehicles should usually be away from any potential ignitable source. The presence of ignitable mixtures inside the vehicle or anywhere in an enclosed space should be avoided. Appropriate steps should be adopted to rule out any possibility of a critical amount of hydrogen leaking or escaping from the system. Since hydrogen does not have any lubricating effect, a suitable material for valve seat rings must be selected. Piston and ring assembly must be appropriately

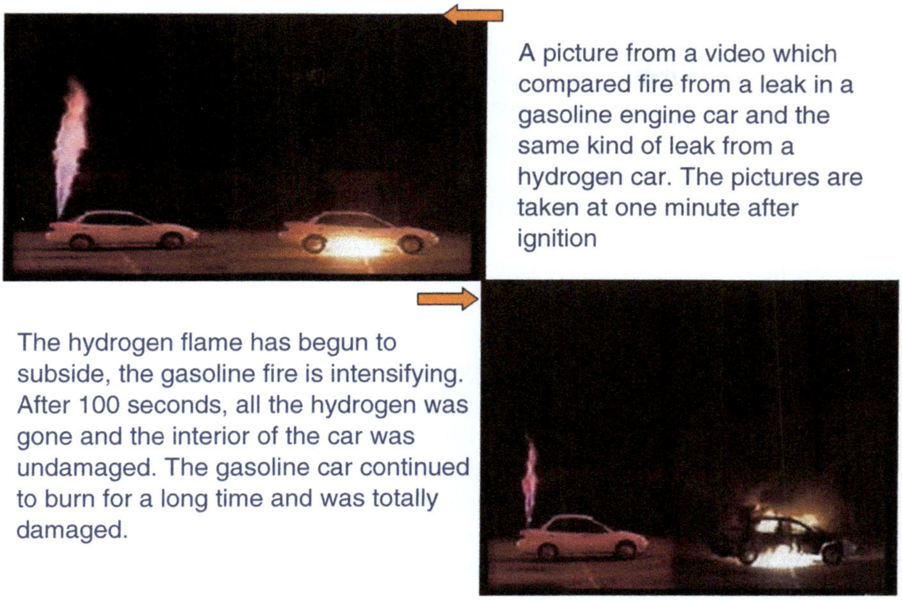

Figure 5.14 Effect of hydrogen fire and gasoline fire [13], Swain, 2001/US Department of Energy.

modified to suit hydrogen operation. Technological details of several hydrogen-fuelled vehicles based on internal combustion engines are described elsewhere [18, 57, 58]. Safety steps have been system specific and need to be broadly installed in the appropriate location of the vehicle.

Yongjun Li et al. [59] analysed the safety aspects in a mobile hydrogen-refuelling stations (MHRS) due to leakage. This aspect of hydrogen fuelling is very relevant in the present context when efforts are being initiated in commercialising the mobile hydrogen-refuelling station. The results of the simulation studies carried out by researchers have generated exact data that provide relevant guidelines for identifying the possible causes of hydrogen leakage accidents and how to prevent them.

The dispersion of hydrogen leakage and the minimum safe distance of explosion of the mobile hydrogen-refuelling station have been studied by simulation. The effects of several significant parameters, such as wind speed, wind direction, ventilation scenario of the truck container, horizontal diffusion distance, etc., were analysed. Experimental data are in close agreement with the simulation results. The results of this analysis have identified the specific conditions that enhance the risk of hydrogen leakage and diffusion. It is observed that the chances of leakage and diffusion are high under conditions when the side door and side widow are open, and there exists a low ambient wind speed with the direction of wind from the front to rear of the mobile refuelling station. The hydrogen dispersion height was the highest when the four exhaust fans were open, and hydrogen gas was well diluted in the truck container when all exhaust fans were open. Under no-wind conditions, the buoyancy effect was dominant when all hydrogen gas diffused vertically. The dispersion height was the highest in this condition. The minimum safe distance to avoid hydrogen explosion has been estimated to be 313.2 m. Useful data related to 'hydrogen volume in the open space and the hydrogen concentration in the four three-way corners of the ceiling in the truck container' [59] have been estimated to be 87.9 m^3 in air.

5.10 Hydrogen Leak Detection Technique and Sensors

In general, any liquid or gas is likely to leak, particularly during transportation, and hydrogen is no exception. In view of its typical combustion properties, such as low density (0.089 kg m^{-3}), boiling point (20.39 K), high diffusion coefficient (0.61 cm^2 s^{-1} in air) and buoyancy, it is essential that the presence of hydrogen and its concentration must be identified in the neighbourhood of its application. Highest percentages of hydrogen accidents of different levels of severity are caused by undetected leaks. Hydrogen can diffuse through joints that are normally referred to as 'airtight'. It is assumed that leaks can occur in any form of design of a component. Depending upon the system and the pressure and flow rate of hydrogen, sometimes expected mild leaks can be located by running inert gases such as helium through the system. Soapy water is also sometimes applied to identify vulnerable spots of mild leak. As far as possible, it is desirable that the number of joints and connections in a system should be reduced and welded joints should be used. These are certain basic steps meant to reduce the possibility of leakage. The choice of materials for welded joints sometimes becomes critical. Some metals, such as austenitic stainless steels and aluminium alloys, are more resistant to hydrogen embrittlement and leakage than others. The selected materials should have good hydrogen compatibility to minimise the risk of leakage. Moreover, hydrogen is a colourless, odourless, tasteless flammable gas, and its presence cannot be detected by human senses. Some odorants have been used for leak detection of hydrogen. However, these odorants do not serve the purpose when they have to travel inside a pipeline with a much lighter substance such as hydrogen because, being heavier, these odorants cannot disperse at the same dispersion rate. Moreover, these odorants are not recommended to be used in fuel cell as they contaminate the fuel cells. Instead of detecting hydrogen, these odorants create difficulties when used for odourising hydrogen gas [52]. Under such circumstances, it is essential that some mechanism should be used to detect the location and concentration level of hydrogen at a specific point. A team of researchers at NREL studied hydrogen contaminant detectors (HCDs) and developed a simplified test gas generation system as shown in Figure 5.15. In such a system, the gas concentration is generated by mass flow controllers. These mass flow controllers regulate the flows in the two gas lines. The pneumatic connection to either of the two HCDs is controlled by a three-way valve. In such a configuration, HCDs are installed in an enclosure. A dry nitrogen purge is supplied to HCD1 to remove background moisture and compressed air is supplied to HCD2 to actuate the pneumatic valve [60].

Nearly invisible characteristics of hydrogen flames have been a point of concern in several situations. Therefore, sensors are required that are sensitive and fast enough to indicate early detection of hydrogen leakage so that appropriate steps can be taken to prevent any possible conditions of explosion due to leaked hydrogen mixing with air. Hydrogen sensors are transducers that are used to detect the presence and concentration of hydrogen at a

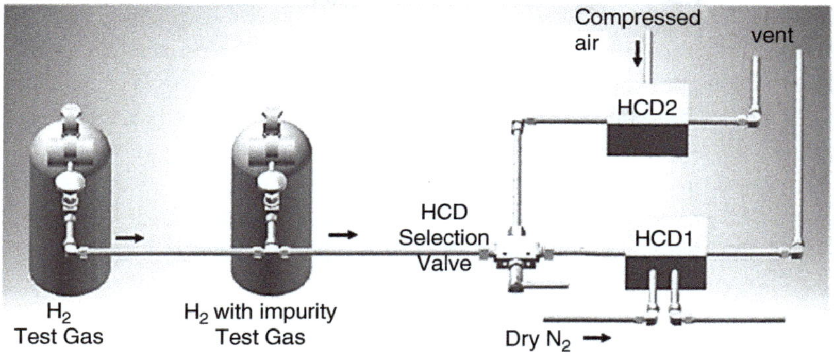

Figure 5.15 Test gas generation system contamination levels for HCD validation and calibration.

particular location. The changes in the electrical and mechanical properties as well as parameters such as temperature and refractive index are all transformed into electrical signals that identify the concentration. The design of sensor technology is based on several different principles, such as catalytic, thermal conductivity, electrochemical, work-function-based, mechanical, optical, acoustic, etc. A hydrogen sensor should have a good response time, and it should be sensitive with an accuracy within about 5% of full scale. Its operating range is normally between temperatures of −30 and 80 °C and 10–98% relative humidity. Last but not least, it must be resistant to hydrocarbon interference. BMW's prototype hydrogen cars (BMW Hydrogen 7) have an HSS-440 module that has been designed on field effect (FE) technology, specifically for use in fuel cell and hydrogen-operated vehicles. These are installed in the car cabin as well as under the hood in the trunk area so that the leakage of hydrogen is detected when it crosses the hydrogen's low flammability limit of 4% by volume. There are certain materials whose optical properties change when they come in contact with hydrogen. Optical hydrogen sensors are a type of hydrogen gas sensor that utilises light absorption or emission properties to detect the presence of hydrogen. They offer several advantages, such as high sensitivity, selectivity and the potential for remote sensing. One advantage of optical hydrogen sensors is that they are non-intrusive, potentially allowing for continuous monitoring without direct contact with the gas. They can also be designed to be immune to interference from other gases, making them suitable for applications even where gas mixtures are present. Fibre-optic sensors use optical fibres to transmit and receive light signals. They can be coated with materials that undergo changes in their optical properties when exposed to hydrogen. The interaction between the hydrogen and the sensing material causes a measurable change in light transmission or reflection, allowing the sensor to detect hydrogen concentrations. They offer high sensitivity and the ability to cover a wide range of hydrogen concentrations. They are used in applications where real-time monitoring and remote sensing are important.

It has been observed that the use of a fibre-optic sensor configuration provides a fast response and reliable results. The broad design feature of an optical sensor is an optical fibre, and a part of it is coated with palladium. When the palladium layer comes in contact with hydrogen, it expands. Thus, the effective optical pathlength increases. This change is usually detected by an interferometer. Silva et al. [61]discussed various types of optical fibre hydrogen sensors with an emphasis on palladium as an active element. A comparative evaluation of the sensing performance characteristics of the three types of sensors, such as interferometric, intensity and fibre-grating-based sensors, shows that intensity-based devices have better time responses. Interferometric-based sensors are potentially more sensitive, whereas fibre-grating-based sensors are robust structures and easily reproducible (Figure 5.16).

Raman-based sensors, using Raman spectroscopy, involve shining a laser on a sample and analysing the scattered light. The shift in the frequency of the scattered light provides information about the molecular vibrations of the sample. Certain materials exhibit changes in their Raman spectra when exposed to hydrogen, allowing for hydrogen detection. Photoluminescence sensors use the change in luminescent properties of a material in response to hydrogen exposure. For example, certain materials exhibit changes in their emission spectra when exposed to hydrogen, which can be measured and correlated with hydrogen concentration.

Chen et al. [62] reviewed seven types of hydrogen sensors based on metal hydrides. The strategies for improving the sensing material, sensor structure and sensing principles have been discussed. Hydrogen sensors based on metal-hydride-specific types of hydrogen gas sensors utilise the reversible absorption and desorption of hydrogen by certain metal hydride materials. These sensors work by measuring the changes in properties, such as electrical resistance or optical characteristics, of the metal hydride as it interacts with hydrogen. Metal hydride sensors offer advantages such as high selectivity, low cross-sensitivity to other gases and the ability to detect low concentrations of hydrogen.

The amount of heat transfer depends upon the thermal conductivity of the gas. Thermal conductivity sensors are designed to operate on the basic principle of the relatively high thermal conductivity of hydrogen (0.174 W mK^{-1} at 20 °C) compared to the thermal conductivity of air (0.026 W mK^{-1}). The function of the sensor takes into account the heat loss from a hot body to the surrounding gas. A thermal conductivity sensor consists of two resistor beads: a reference resistor and a detecting resistor; the reference resistor is sealed in a chamber with the reference gas, which is usually air. When both resistors are subjected to similar operating conditions and are exposed to the same gas, it shows zero reading.

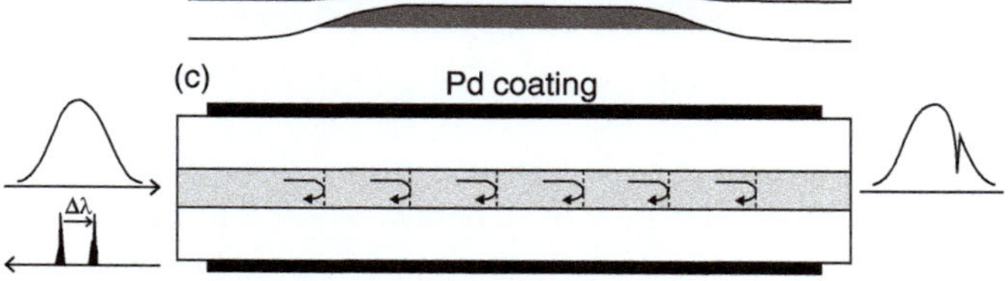

Figure 5.16 Optical-based hydrogen sensors: (a) interferometric-based sensors, (b) intensity-based sensors and (c) fibre-grating-based sensors.

However, when hydrogen is passed over the same resistor, some heat loss occurs depending upon the thermal conductivity of the reference gas. There also occurs some change in temperature and change in resistance. It was possible to measure the hydrogen contents within a range of 0–4% in small increments of 0.5% over a broad range of temperatures between −15 and 84 °C, taking into account the effect of humidity and moisture content.

Catalytic sensors are built on the principle that the reaction of a combustible gas with oxygen on the surface of the sensor releases heat. These sensors are designed to detect any combustible gases including hydrogen. The typical design shown here consists of two platinum coils, each embedded in a ceramic bead. Out of the two elements, the first is a detector element, and the second is a compensator element. The detector element has a catalytic material and is sensitive to the combustible gas (hydrogen). The other bead, the compensator element, does not have catalyst on its surface and thus is inert in the basic circuit as shown in Figure 5.17.

The micro-electromechanical system (MEMS) hydrogen sensor is a common type of sensor that functions quite well at room temperature. MEMS hydrogen sensors have gained popularity in recent years due to their small size, low power consumption and potential for integration into various devices and systems. MEMS technology allows for the miniaturisation of sensors, making them suitable for applications where size, weight and power constraints are critical. A particular type of MEMS-based hydrogen sensor is coated with a film consisting of nanostructured indium oxide (In_2O_3) and tin oxide (SnO_2). Studies on MEMS gas sensors for automotive applications show that air humidity influences the performance of the sensor. Transient and steady-state evaluation can be performed by micro-electromechanical system gas sensors [63]. While MEMS hydrogen sensors offer numerous advantages, they may also have limitations or challenges, such as sensitivity to environmental conditions, potential cross-sensitivity to other gases and the need for calibration.

There are also several other methods and techniques to detect and sense hydrogen under specific conditions. One of the distinctive properties of hydrogen is that when it is absorbed by a metal it occupies interstitial sites in the metal lattice. The lattice expands because of this, and the physical properties of the metal change. Based on this property, micro-fabricated palladium-coated cantilevers are developed where expansion of palladium takes place followed by hydrogen absorption. However, sometimes a problem could occur when the palladium film is dislocated due to expansion. Reversible and irreversible chemo-chromic hydrogen sensors have been quite effective in some cases. They contain a pigment paint that can easily detect the hydrogen leaks because the colour changes as the hydrogen is leaked.

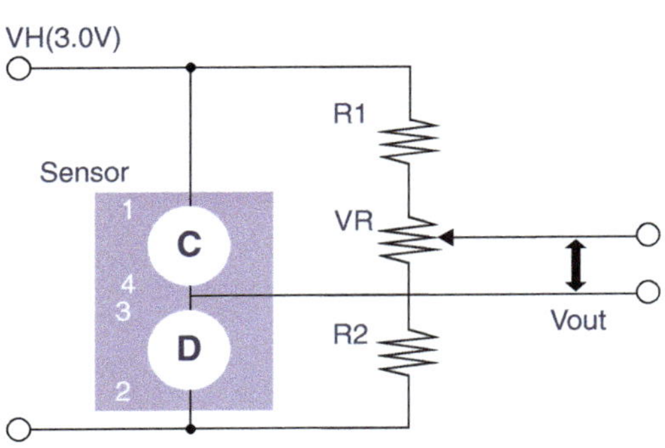

Figure 5.17 Basic circuit of a catalytic sensor: D, detector element; C, compensatory element.

Similarly, thin and thick sensors are also used in specific applications. Thin-film sensors are also used because of their specific properties. The formation of hydrides causes the palladium particles to expand and form new electrical connections. Such a situation gives rise to a large number of conduction pathways. Thick film sensors, on the other hand, contain a thick layer of semiconductor material such as SnO_2 and an upper layer of additive that is a noble metal (Pd, Pt, etc.) and metal oxides. This configuration accelerates the process of hydrogen oxidation, and thus, the response of the sensor becomes fast. A comparative evaluation shows that thick film sensors can have more stable signal drafting compared to thin-film sensors. However, they have a slower response characteristic because of the thick layer. Operation at a relatively high temperature is one of the major constraints of thick film sensors, and this condition does not suit many modern digital electronic systems. At present, there is a trend of rapid and effective integration of sensors into a vast number of electronic applications.

An acoustic sensor, also known as 'acoustic wave' sensors or 'surface acoustic wave' (SAW) sensors, can be used for detecting hydrogen gas by exploiting the changes in the acoustic wave properties caused by the interaction between the gas and the sensor surface. These sensors are also used for hydrogen detection by detecting the changes in the properties of acoustic waves due to an adsorbate. In an acoustic hydrogen sensor, SAWs are generated by a piezoelectric substrate with two transducers. These transducers are interdigital transducers (IDTs) with very well-defined functions. One transducer converts the applied electrical input signal into acoustic waves and the second transducer converts it back into an electrical output signal. Acoustic sensors can be highly sensitive to changes in the gas composition, allowing for the detection of low hydrogen concentrations. They provide real-time monitoring capabilities, allowing for a continuous and rapid detection of hydrogen. Devika et al. [64] developed and fabricated Pd nanoparticle-based SAW sensors. In their study, they observed that the sensing response mechanism was closely linked to the changes in the conductivity of the Pd film upon exposure to hydrogen. Rayleigh waves are surface waves that usually propagate in the direction of wave propagation along the surface of a solid. These waves are characteristically localised within one or two acoustic wavelengths of the surface. Any perturbation, such as change in the electrical properties, affects these waves. That is why Rayleigh waves are usually employed in hydrogen as sensors (and in others as sensing applications) because it is relatively easy to generate them using interdigital transducers (Figure 5.18).

Electrochemical sensors can detect low ppm of hydrogen due to electrochemical reactions occurring at a sensing electrode. This broadly consists of an array of electrodes surrounded by a conductive electrolyte. Electrochemical sensors operate at a constant applied voltage, and the sensor signal is controlled with a diffusion-limited capillary. As far as the design feature is concerned, the basic setup is composed of two electrodes: the sensing electrode

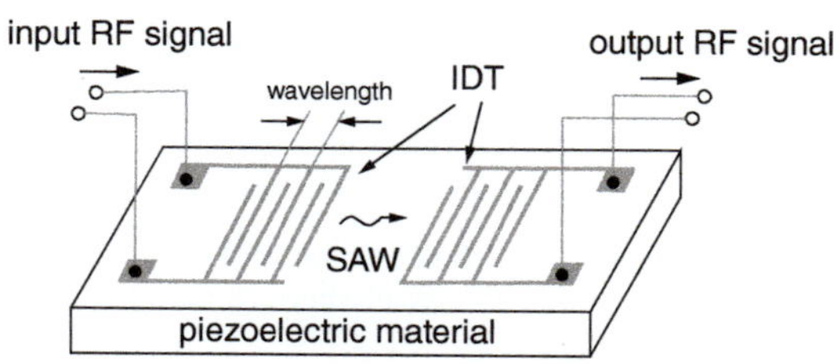

Figure 5.18 Interdigital transducer.

and the counter-electrode. The electrochemical cell is composed of a solid or liquid electrolyte to enable the transport of ions between electrodes. A permeable layer over the inlet to the sensing electrode is used to control the diffusion. Hydrogen diffused through this layer is oxidised at the sensing electrode, which changes the potential of the sensing electrode:

$$H_2 \rightarrow 2H^+ + 2e^-$$

Reduction of oxygen takes place in the counter-electrode:

$$\tfrac{1}{2}O_2 + 2H^+ + 2e^- \rightarrow H_2O$$

In the resistance-based semiconducting metal-oxide sensors, a metal-oxide film is applied onto an insulating material between two electrodes. When the reducing gases pass over metal oxides, they cause some changes in the electrical properties of the metal oxides. The usual operating temperature varies between 180 and 450 °C depending upon the particular metal oxide. Once the temperature of the film rises due to the heat, reaction with the reducing gas sets in. Water formed during the reaction is also removed. The resistance R depends upon the concentration of the reducing gas.

Sayago et al. [65] have shown that a novel type of resistive gas sensors based on single-walled carbon nanotubes (SWNTs) 'provides good response to hydrogen and an excellent selectivity with respect to the tested interfering gases such as octane, toluene and ammonia at room temperature'. It has also been observed that even though the temperature of operation of the sensor reduces the response to hydrogen, the process of detection is still faster, and the recovery times are also shorter.

The Schottky barrier is a potential energy barrier for electrons formed at a metal–semiconductor (MS) junction. When a metal is put in direct contact with a semiconductor, a so-called 'Schottky barrier' can be formed, leading to a rectifying behaviour of the electrical contact. A Schottky diode–based hydrogen gas sensor [66, 67] employs a palladium-alloy gate in which it is selectively absorbed, thereby lowering the Schottky energy barrier. It has been reported that it is possible to detect a concentration of 15 ppm of hydrogen in air in a Pd/InGaP metal–semiconductor Schottky diode. It has been observed that metallic La–Mg_2–Ni absorbs hydrogen near ambient conditions. It forms non-metallic hydride $LaMg_2NiH_7$.

There are some sensors whose operation is work-function-based. The construction of the semiconductor work-function-based gas sensors is such that they have three layers: metal, oxide and semiconductor (MOS). As far as specifically hydrogen detection is concerned, the metal chosen is a hydrogen-sensitive catalytic metal. This metal is deposited on an oxide layer. The oxide layer is deposited on a semiconductor substrate (Figure 5.19).

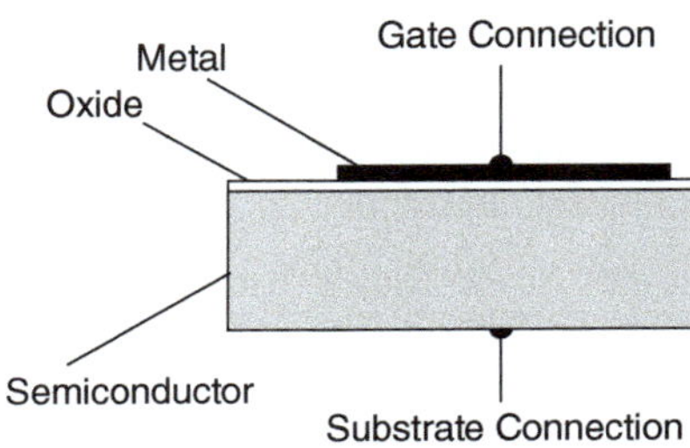

Figure 5.19 Work-function-based sensor.

In such structures, hydrogen, diffuses through the metal, is absorbed at the interface between the metal and oxide layer. Subsequently, a dipole layer is created by the hydrogen atoms polarised at the interface. As a result, the energy levels at the metal–insulator interface are shifted. The work function of the metal also changes. Thus, the device that can measure the change in metal work function is essentially a sensor that can also measure the hydrogen concentration.

Graphene is an excellent conductor of electricity and is known for its high surface area. Even small interactions with hydrogen molecules can lead to measurable changes in graphene's electrical properties. Graphene-based sensors can exhibit rapid response times, making them suitable for real-time monitoring of hydrogen concentrations. The fast response is particularly valuable in safety-critical applications. Huang and team of researchers have tested and found that graphene is an excellent material for 'multifunctional sensing' [68]. Tests with multifunctional graphene magnetic/hydrogen sensors show a 'high hydrogen response up to 32.5% when exposed to 1000 ppm hydrogen'. Testing during gas detection, to evaluate the hydrogen-sensing mechanism of Pd-decorated graphene, shows satisfactory performance. This work indicates graphene's ability not only to reduce the complexity of the manufacturing process but also to lessen the number of sensors in the system for generating the necessary technical information.

5.11 Hydrogen Embrittlement

Hydrogen embrittlement is the process by which hydrogen enters into the cracks of several metals, thereby making them brittle. The magnitude of the problems depends upon the amount of hydrogen accumulated over a period of time. The mechanism of embrittlement is not very clearly understood. The factors that are responsible for causing embrittlement are mainly pressure, temperature, concentration of hydrogen, exposure time etc. Embrittlement is of two types: internal hydrogen embrittlement (IHE) and external hydrogen embrittlement (EHE). IHE mainly occurs due to the residual hydrogen from various manufacturing processes, such as electroplating, pickling, etc. This is caused by the amount of hydrogen absorbed, which results in premature failure of certain metal EHE. As the name suggests, it occurs due to hydrogen from external sources such as in a hydrogen-rich environment with stress corrosion cracking. Hydrogen embrittlement occurs very frequently during pickling operations when the rate of corrosion of the base metal causes a very high rate of hydrogen evolution. The addition of suitable inhibitors has proven to be effective in reducing the base metal corrosion during pickling. The other well-known technique is based on using clean steel. Rimmed steel has many voids that cause embrittlement, and this can be reduced using clean steel that has fewer voids. Hydrogen embrittlement in steels has been observed to be almost a reversible process. Therefore, the mechanical properties of the treated material do not substantially vary if hydrogen is removed. This is a very advantageous feature and usually steel is baked to a relatively lower temperature of about 100–150 °C to remove hydrogen. High-strength steels are found to be susceptible to a high rate of embrittlement. Ni and Mo alloys do not have experience embrittlement. It is often recommended to use nickel-based alloys as they have a very low hydrogen diffusion rate, so embrittlement is not a problem of concern. Proper welding is a very relevant factor in connection with the problem of embrittlement, and it is always desirable to use low-hydrogen welding rods to reduce embrittlement. Water and water vapour, which happen to be the major source of hydrogen, need to be avoided and care must be taken to maintain dry conditions during welding.

Certain metals (such as martensitic and ferritic) are more susceptible to embitterment than highly alloyed austenitic steels. Metal degradation by hydrogen needs to be investigated by way of using suitable materials for hydrogen structures/sealing. Thus, the choice of appropriate material to use in specific locations in hydrogen-operating systems is very critical. Ferric steels and nickel-based steels have been found to be highly susceptible, whereas austenitic steels are moderately affected and low aluminium alloys and copper alloys are almost unaffected. Carbonaceous steel should be avoided as hydrogen can react with carbon in the steel to form CH_4 bubbles in the metal at high temperature. A detailed discussion on embrittlement is provided in Chapter 8.

Concluding Remarks

Generally, hydrogen is neither safer nor more unsafe than conventional fuels. It is only the improper handling of hydrogen that makes it unsafe because its physico-combustion properties are widely different from the fuels we commonly use. A common man should be aware of such properties as some of these properties could cause devastation. Handled scrupulously, we can get immense benefit out of it. Hydrogen flames are almost invisible and have a mass flow rate blown off that is much higher than that of other fuels such as methane and propane. The mishandling and liberal use of hydrogen flames has caused several accidents. Hydrogen's high energy content makes it a valuable fuel source for various applications, including transportation and power generation. The fast dispersion and lighter-than-air characteristics of hydrogen contribute to rapid dissipation and a reduced risk of ground-level accumulation. This is a very useful feature when leaks and accidental release occur. At the same time, meticulous safety measures are required to prevent accidental ignition, especially in confined spaces, transportation and storage facilities. Hydrogen's small molecular size can lead to material compatibility issues, including hydrogen embrittlement in metals. Special materials and coatings may be required for safe storage, transport and containment depending upon the specific application.

Judicious selection of an appropriate material is important because some materials, such as stainless steel, aluminium and silicon carbide fibres, show a much higher degree of fast corrosion when exposed to a hydrogen flame. The properties, such as buoyancy, diffusivity, thermal conductivity, etc., prove to be very safe assets of hydrogen if they are properly used in specific situations. A close life cycle analysis from cradle to grave for hydrogen production needs to be more elaborately studied. Green hydrogen has been receiving a lot of attention as this is the most appropriate form to move ahead to zero-emission goals. A large number of countries have strengthened their hydrogen-related activities, and several countries have initiated aggressive programmes on hydrogen energy. It would be extremely beneficial if all the activities separately carried out in different countries are integrated, leading to the early safe penetration of the technology.

Abbreviations

SRB	Solid Rocket Booster
NASA	National Aeronautics and Space Administration
LPG	Liquified Petroleum Gas
BLEV	Boiling Liquid Expanding Vapour
BLEVE	Boiling Liquid Expanding Vapour Explosion
NBP	Normal Boiling Point
NTP	Normal Temperature Pressure
STP	Standard Temperature Pressure
MESG	Maximum Experimental Safe Gap
UFL	Upper Flammability Limit
LFL	Lower Flammability Limit
VCE	Vapour Cloud Explosion
UVCE	Unconfined Vapour Cloud Explosion
CVCE	Confined Vapour Cloud Explosion
DDT	Deflagration-to-Detonation Transition
MCDM	Multi-Criteria Decision-Making
HRAM	Hydrogen Risk Assessment Method
PRD	Pressure Relief Device
HCD	Hydrogen Contaminant Detector
FE	Field Effect
HSS	High-Speed Steel
MEMS	Micro-Electromechanical Systems
SAW	Surface Acoustic Waves
IGT	Interdigital Transducers
MS	Metal–Semiconductor
MOS	Metal Oxide and Semiconductor
IHE	Internal Hydrogen Embrittlement
EHE	External Hydrogen Embrittlement

References

1 Federal Institute for Materials Research and Testing (2002). *Hydrogen Safety*, vol. 18, 21. Brussels: German Hydrogen Association.

2 Rigas, F. and Sklavounos, S. (2008). Hydrogen safety. In: *Hydrogen Fuel: Production, Transport and Storage* (ed. R. Gupta), 537–538. Boca Raton, FL: CRC Press/Taylor & Francis.

3 Rigas, F. and Sklavounos, S. (2005). Evaluation of hazards associated with hydrogen storage facilities. *International Journal of Hydrogen Energy* 30: 1501–1510.

4 Kirk-Othmer Encyclopaedia of Chemical Technology (1992). *Fundamentals and Use of Hydrogen as a Fuel*, 3e, vol. 4. New York: Wiley.

5 Bain, A., Van Vorst Wm. The Hindenburg tragedy revisited: the fatal flaw found. *International Journal of Hydrogen Energy* Volume 24, Issue 5, 1 May 1999, Pages 399–403.

6 Zuettel, A., Borgeschulte, A., and Schlapbach, L. (ed.) (2008). *Hydrogen as a Future Energy Carrier*, 14–20. Berlin, Germany: Wiley-VCH Verlag.

7 National Aeronautics and Space Administration. NSS 1740.16 (1997). *Safety Standard for Hydrogen and Hydrogen Systems: Guidelines for Hydrogen System Design, Materials Selection, Operations, Storage, and Transportation*, 20546. Washington, DC: Office of Safety and Mission Assurance.

8 Markert, F., Nielsen, S.K., Paulsen, J.L., and Andersen, V. (2007). Safety aspects of future infrastructure scenarios with hydrogen refuelling stations. *International Journal of Hydrogen Energy* 32: 2227–2234.

9 Kletz, T.A. (1978). What you don't have, can't leak. *Chemistry & Industry* 287–292.

10 Zalosh, R.G. and Short, T.P. Compilation and Analysis of Hydrogen Accidents Reports. *Final Technical Report Prepared for Division of Operational and Environmental Safety United States Department of Energy Washington D. C. 20545 T Under Contract No. Ee-77-C-02-4442.*

11 Ustolin, F., Nicola, P., and Berto, F. (2020). Loss of integrity of hydrogen technologies: a critical review. *International Journal of Hydrogen Energy* 45 (43): 23809–23840.

12 Federal Institute of Material and Testing (2002). *Hydrogen Safety*, 18–21. Brussels: German HYDROGEN ASSOCIATION.

13 Swain, M.R. *Proceedings of the 2001 DOE Hydrogen Energy Program Review, NREL/CP-570-30535, U S Department of Energy, Washington D.C.*

14 Das, L.M. (1991). Safety aspects of hydrogen-fuelled engine system development. *International Journal of Hydrogen Energy* 16 (9): 619–624.

15 LaChance, J., Tchouvelev, A., and Engebø, A. (2011). Development of uniform harm criteria for use in quantitative risk analysis of the hydrogen infrastructure. *International Journal of Hydrogen Energy* 36: 2381.

16 Health and Safety Executive (1975). *The Flixborough Disaster: Report of the Court of Inquiry*. HMSO. ISBN: 0113610750.

17 Schefer, R.W. and Settersten, T.B. (2009). Visible emissions of hydrogen flames. *Combustion and Flame* 156 (6): 1234–1241.

18 Das, L.M. Hydrogen-fuelled internal combustion engine. In: *Compendium of Hydrogen Energy* Chapter 7, *Hydrogen Energy Conversion*, vol. 3 (ed. F. Babir, A. Basile, and T. Nejat), 177–216. Elsevier Publishing.

19 Molkov, V. (2012). *Fundamentals of Hydrogen Safety Engineering, Part I.* bookboon.com & Ventus publishing ApS. ISBN: 978-87-403-0226-4 http://hdl.handle.net/123456789/475.

20 Molkov, V. (2012). *Fundamentals of Hydrogen Safety Engineering, Part II. bookboon.com.* Fundamentals of Hydrogen Safety Engineering II (weebly.com). ISBN: 978-87-403-0279-0.

21 ISO/TR 15916 (2004). *Basic Considerations for the Safety of Hydrogen Systems.* ISO Technical Committee 197 Hydrogen Technologies.

22 Coward, H.F. and Jones, G.W. (1952). Limits of flammability of gases and vapors. *Bulletin 503, Bureau of Mines,* p. 155.

23 Butler, M.S., Moran, C.W., Sunderland, P.B., and Axelbaum, R.L. (2009). Limits for hydrogen leaks that can support stable flames. *International Journal of Hydrogen Energy* 34: 5174–5182.

24 Michael, A. (2005, August). *Green Hydrogen Safety Issues Compared to Safety Issues with Methane and Propane*. Lawrence Berkeley National Laboratory, Berkeley CA 94720, LBNL-59002 SCMAG-854 MICE-PUB-GEN-131

25 Cheng, T.S., Chao, Y.-C., Wu, C.-Y. et al. (2005). Experimental and numerical investigation of microscale hydrogen diffusion flames. *Proceedings of Combustion Institute* 30: 2489–2497.

26 Kalghatgi, G.T. (1981). Blow-out stability of gaseous jet diffusion flames. Part I: in still air. *Combustion Science and Technology* 26 (5): 233–239.

27 Matta, L.M., Neumeier, Y., Lemon, B., and Zinn, B.T. (2002). Characteristics of microscale diffusion flames. *Proceedings of the Combustion Institute* 29: 933–938.

28 Sunderland, P.B. (2010). Hydrogen microflame hazards. *Proceedings of the 8th International Short Course and Advanced Research Workshop in the series "Progress in Hydrogen Safety", Hydrogen and Fuel Cell Early Market Applications* (11–15 October 2010). Belfast: University of Ulster.

29 Johnson, H.H. and Kumnick, A.J. (1975). Hydrogen and the integrity of structural alloys. In: *Hydrogen Energy*. New York NY: Plenum Press.

30 Das, L.M. (1990). Fuel induction technique for a hydrogen operated engine. *International Journal of Hydrogen Energy* 15 (11): 833–842.

31 ANSI. *Guide to Safety of Hydrogen and Hydrogen Systems*. American Institute of Aeronautics and Astronautics, American National Standard ANSI/AIAA G-095-2004, Chap. 2.

32 Lee, S.D., Suha, K.Y., and Jae, M. (2003). A framework for evaluating hydrogen control and management. *Reliability Engineering and System Safety* 82: 307.

33 Warren, P.J. (1997). *Hazardous Gases and Fumes*, 96–97. Oxford: Butterworth-Heinemann.

34 Hytovick, Chambers; Chin, Gamezo; Poludnenko and Ahmed, The evolution of fast turbulent deflagrations to detonations, *Physics of Fluids* Vol 35, Issue 4, April 2023.

35 Buncefield Investigation. (2010). The Buncefield Major Incident Investigation Board. Recommendations on land use planning and the control of societal risk around major hazard sites. Available from: https://www.icheme.org/media/13708/buncefield-miib-final-report-volume-2b.pdf

36 Biennial Report on Hydrogen Safety (BRHS). (2009). The European Network of Excellence: "Safety of Hydrogen as an Energy Carrier" (NoE HySafe). Available from: www.hysafe.org

37 Lind, C.D. (1975). What causes unconfined vapour cloud explosions? *Loss Prevention* 9: 101–110.

38 Lee, J. (1977). Initiation of gaseous detonation. *Annual Review of Physical Chemistry* 28: 75.

39 Eichert, H. (1987). Hydrogen–air deflagrations and detonations: numerical calculation of I-d tube combustion processes. *International Journal of Hydrogen Energy* 12: 171.

40 Yang, F., Wang, T., Deng, X. et al. (2021). Review on hydrogen safety issues: Incident statistics, hydrogen diffusion, and detonation process. *International Journal of Hydrogen Energy 46*: 31467–31488.

41 Dahoe, A.E. and Molkov, V.V. (2007). On the development of an international curriculum on hydrogen safety engineering and its implementation into educational programs. *International Journal of Hydrogen Energy* 32: 1113–1120.

42 Aprea, J.L. (2008). New standard on safety for hydrogen systems: keys for understanding and use. *International Journal of Hydrogen Energy* 33: 3526–3530.

43 Dorofeev, S.B. (2007). Evaluation of safety distances related to unconfined hydrogen explosions. *International Journal of Hydrogen Energy* 32: 2118–2124.

44 Marangon, A., Carcassi, M., Engebø, A., and Nilsen, S. (2007). Safety distances: definition and values. *International Journal of Hydrogen Energy* 32: 2192–2197.

45 Dawood, F., Anda, M., and Shafiullah, G.M. (2020). Hydrogen production for energy: an overview. *International Journal of Hydrogen Energy* 45 (7): 3847–3869.

46 Chau, K., Djire, A., and Khan, F. (2022). Review and analysis of the hydrogen production technologies from a safety perspective. *International Journal of Hydrogen Energy* 47 (29): 13990–11400.

47 Dincer, I. and Acar, C. (2017). Innovation in hydrogen production. *International Journal of Hydrogen Energy* 42 (22): 14843–14864.

48 Moradi, R. and Groth, K.M. (2019). Hydrogen storage and delivery: review of the state of the art technologies and risk and reliability analysis. *International Journal of Hydrogen Energy* 44 (23): 12254–12269.

49 Burke, A. and Gardnier, M. (2005). *Hydrogen Storage Options: Technologies and Comparisons for Light-Duty Vehicle Applications*. Institute of Transportation Studies University of California, Davis Research Report (2005). https://escholarship.org/uc/item/7425173j.

50 Sarkar, A. and Banerjee, R. (2004). Net energy analysis of hydrogen storage options. *International Journal of Hydrogen Energy* 30: 867–877.

51 Grasso, N., Pilo, F., Ciannelli, N. et al. (2009). Fire prevention technical rule for gaseous hydrogen transport in pipelines. *International Journal of Hydrogen Energy* 34 (10): 4675–4683.

52 Swain, M.R., Shriber, J., and Swain, M.N. (1998). Comparison of hydrogen, natural gas, liquefied petroleum gas, and gasoline leakage in a residential garage. *Energy & Fuels* 12: 83–89.

53 Tay Le, S., Ngoc Nguyen, T., Linforth, S., and Duc Ngo, T. (2023). Safety investigation of hydrogen energy storage systems using quantitative risk assessment. *International Journal of Hydrogen Energy 48*: 2861–2875.

54 Swain, M.R., Filoso, P., Grilliot, E.S., and Swain, M.N. (2003). Hydrogen leakage into simple geometric enclosures. *International Journal of Hydrogen Energy* 28 (2): 229–248.

55 Stephenson, R. (2003). Crash-induced fire safety issues with hydrogen-fuelled vehicles. Paper presented at the National Hydrogen Association's 18th Annual U S Hydrogen Conference, Washington D.C. (March 2003).

56 Dr Furst, S., Dub, M., Gruber, M., et al. (2004). Safety of Hydrogen-fueled Motor Vehicles with IC Engine. Hydrogen Tools.

57 Lapez et al. (2006). The design, development, validation and delivery of the Ford H2 ICE 450 Shuttle Bus Symposium on Hydrogen Internal Combustion Engine, pp. 20–33, BMW Hydrogen 8 Rescue Guidelines.

58 Natarajan, S., Abraham, M., Rajesh, M. et al. (2013). DelHy 3w-Hydrogen-fuelled HYalfa Three Wheelers, *SAE Technical Paper* 2013-01-0224.

59 Li, Y., Wang, Z., Shi, X., and Fan, R. (2023). Safety analysis of hydrogen leakage accident with a mobile hydrogen refueling station. *Process Safety and Environmental Protection 171*: 619–629.

60 Buttner, W., Koleva, M., and Post, M. (2020). *Hydrogen Contaminant Detector Kevin Hartmann*. National Renewable Energy Laboratory DOE Hydrogen and Fuel Cells Program 2020 Annual Merit Review and Peer Evaluation Meeting Project ID #h2

61 Silva, S.F., Coelho, L., Frazao, O. et al. (2012). A review of palladium-based fiber-optic sensors for molecular hydrogen detection. *IEEE Sensors Journal* 12 (1): 93–102. https://doi.org/10.1109/JSEN.2011.2138130.

62 Chen, K., Yuan, D., and Zhao, Y. (2021). Review of optical hydrogen sensors based on metal hydrides: recent developments and challenges. *Optics & Laser Technology* 137: 106808.

63 Berndt, D., Muggli, J., and Wittwer, F. (2020). MEMS-based thermal conductivity sensor for hydrogen gas detection in automotive applications. *Sensors and Actuators A: Physical* 305: 111670.

64 Sil, D., Hines, J., Udeoyo, U., and Borguet, E. (2015). Palladium nanoparticle-based surface acoustic wave hydrogen sensor. *ACS Applied Materials & Interfaces* 7: 5709–5714. https://doi.org/10.1021/am507531s.

65 Sayago, I., Terrado, E., Aleixandre, M. et al. (2007). Novel selective sensors based on carbon nanotube films for hydrogen detection. *Sensors and Actuators B* 122: 75–80.

66 Yu, J., Ippolito, S.J., Shafiei, M. et al. (2009). Reverse biased Pt/nanostructured MoO3/SiCMoO3/SiC Schottky diode based hydrogen gas sensors. *Applied Physics Letters* 94: 013504. https://doi.org/10.1063/1.3054164.

67 Yam, F.K. and Hassan, Z. (2007). Schottky diode based on porous GaN for hydrogen gas sensing application. *Applied Surface Science* 253 (24): 9525–9595.

68 Huang, L., Zhang, Z., Li, Z. et al. (2015). Multifunctional graphene sensors for magnetic and hydrogen detection. *ACS Applied Materials & Interfaces* 7 (18): 9581–9588. https://doi.org/10.1021/acsami.5b01070. Epub 2015 May 1. PMID: 25899539.

6

Application of Hydrogen Energy

The only thing ever achieved in life without effort is failure.

– St. Francis of Assisi

CHAPTER MENU

(*Continued*)

Hydrogen Energy: Production, Safety, Storage and Applications, First Edition. L. M. Das.

6.1 Introduction

Presently, our current energy infrastructure is dominated by fossil fuels, which cater to a major part of the global energy demand. The devastating effects of fossil fuels are being increasingly realised day by day. The harmful impacts of greenhouse gases (GHGs; carbon dioxide, nitrous oxide, methane, chlorofluorocarbons, etc.) causing global warming, ozone layer depletion and problems, such as thermal inversion, smog, acid rain, El Niño/La Niña and, last but not least, the unignorable damage caused by noxious pollutants, such as carbon monoxide (CO), unburnt hydrocarbons, oxides of nitrogen, particulate matter and sulphur dioxide, have necessitated the urgent need for a transition towards a carbon-neutral society. A strategic shift to the use of renewable energy is gaining momentum. Hydrogen plays a very significant role in enhancing this momentum towards a profound transformation to sustain the 1.5 °C climate target.

The technology of hydrogen production by electrolysis, the use of hydrogen in internal combustion engines (ICEs) for road transport, balloons and airships for air transport, etc., was performed before 200 years. Subsequently, hydrogen was also used for trips to the moon in the 1960s. Despite all these remarkable achievements, the idea remained confined to margins of discussion. It was intermittently used as an item of curiosity until the middle of the twentieth century when it again entered into widespread applications in oil refining. Gradually, its application started spreading to other industries and several other sectors. By now, hydrogen's undisputed image has been established as a no-carbon fuel in long-term strategies for use in industrial applications and transportation, electricity generation and buildings.

At present, hydrogen is used in petroleum refining to crack heavier oils to lighter ones for petroleum and petroleum products. It is used to lower the sulphur content. The other major industrial area of its use is in producing ammonia for fertilisers. Hydrogen is also being used in place of coal and natural gas in several industrial sectors, such as iron and steel industries, cement factories, food processing, forging, etc. Apart from industry, hydrogen is rapidly penetrating into several segments of the transportation sector. Hydrogen-operated vehicles (with both ICEs and fuel cells) have been developed and are operating. In power generation, it can be used in gas turbines. Linde plc (American–German Chemical Company), Air Products and Chemicals (USA), Reliance Industries Ltd. (India), Hydrogenises (Canada), Air Liquide International S.A. (France), Cummins Inc. (USA), Messer group GmbH (Germany), Parker Hannifin (USA), McPhee (France), Fuel Cell Energy (USA) and so many other leading companies are generating hydrogen and are also working out strategies for effective market penetration. Recently, several other small-scale and medium-scale industries have also initiated activities on several aspects of hydrogen energy engineering. An integrated hydrogen system comprises production, storage and transmission to the point of use. This chapter discusses the perspective and strategies of its application. With the acceleration at which the application of hydrogen energy is progressing, it is quite likely that by the time the book is published, there might have been several initiatives, the landscape might have evolved and many new projects might have been initiated in several parts of the world. Efforts have been made here to provide a few examples of existing systems and ongoing applications.

As mentioned earlier, globally, at present, hydrogen is used as a feedstock in emission-intensive industrial sectors such as oil refineries, steel production, petrochemicals, fertilisers, ammonia and methanol production. This is also used in oil and fat hydrogenation, electronics industries, glass and optical fibre manufacturing, nuclear reactors and electric power generation systems. The use of hydrogen in transport and several other sectors is rapidly growing worldwide. Considering the present thrust on a clean environment, it is important to determine its routes of production and areas of utilisation. Figure 6.1 indicates the sources of hydrogen production and its demands as a raw material in several industrial synthesis processes [1].

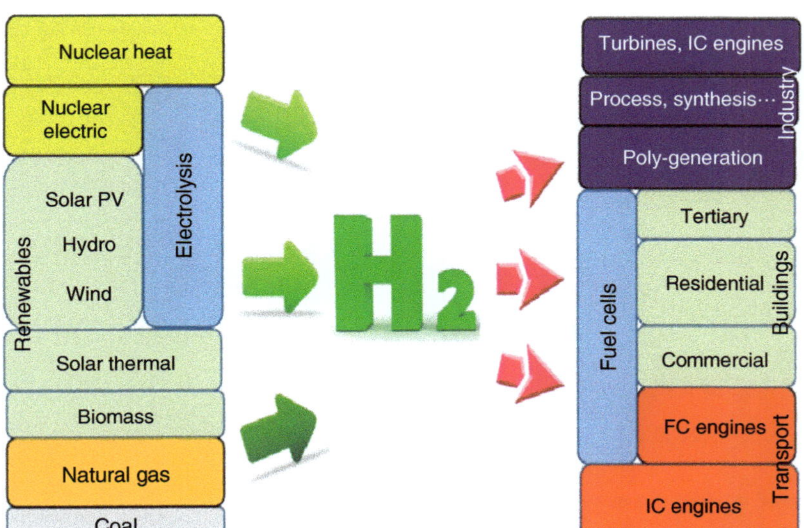

Figure 6.1 Production routes of hydrogen and its areas of demand [1]/KeAi.

Hydrogen covers a broad spectrum of applications ranging from the feedstock in chemical industries to ammonia production, petroleum processes, high-grade chemical products, reducing agents in metallurgy, coolants in large electrical generators, particularly in power plants, where the efficient removal of heat is crucial to maintain the integrity and performance of the generator, and protective gases in electronics, welding and cutting and food processing (hydrogenation of oils). Last but not least, its use as town gas and water gas dates back to 1780. Hydrogen has a variety of energetic and non-energetic uses for a long time. Approximately 500 billion cubic metres of hydrogen is used for fertilisers, hydrodesulphurisation of coal, petroleum fractions and fuels. In fact, it has a long history as an energy carrier rather than as a feedstock in chemical industries. Fertiliser industries mostly use a large amount of fossil fuels, primarily natural gas, in most of the countries worldwide. Figure 6.2 provides an idea of the global demand for hydrogen for different industrial applications [2]. In an effort to achieve total decarbonisation, the trend of increased hydrogen demand has started, and technological upgrades are being adopted to shift from coal and naphtha to natural gas where connectivity to natural gas pipelines exists. It may be relevant to recall that steam methane reforming (SMR) is currently a popular commercial method of hydrogen production.

The potential of hydrogen in transforming the world totally carbon free has been realised all around. About 45 countries have prepared hydrogen implementation strategies to cut carbon emissions and reach the net zero target. Global hydrogen demand is rising. Adequate amounts of green hydrogen need to be produced to cater to the needs of several sectors. To achieve this objective, several national and international programmes have been developed. Countries have prepared their time-bound goals for the emergence of green hydrogen technology from heavily locked traditional fossil fuel technology.

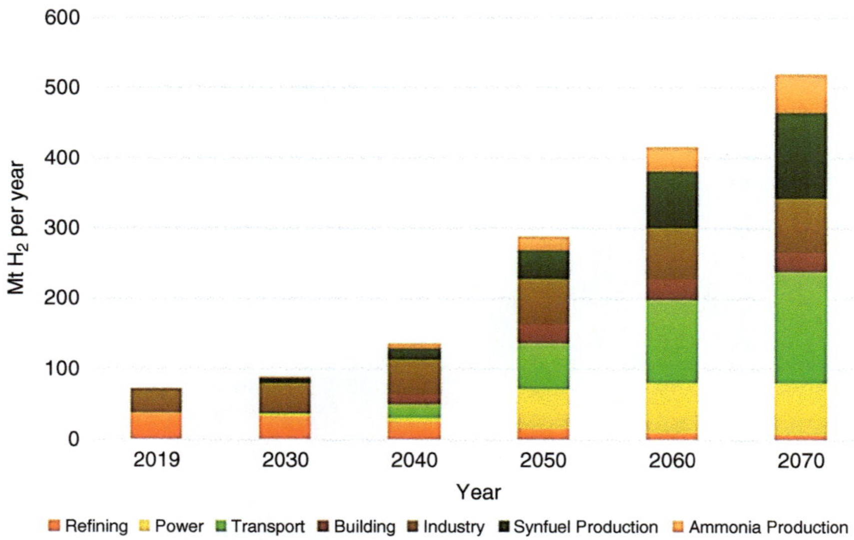

Figure 6.2 Global hydrogen demand by sectors.

6.2 Ammonia Production and Fertiliser Industry

Natural gas is used as a raw material and as a fuel for nitrogen fertiliser production. This is used to produce ammonia that forms the base for fertiliser production. At present, almost 80% of the ammonia produced per year is used for the production of fertilisers. Nitrogen is extremely essential as a source of nutrition for plants to grow, and ammonia has the highest nitrogen content among most of the fertilisers. Usually, plants take up two forms of nitrogen: NH_4^+ and nitrate (NO^{3-}). Thus, nitrogen is used as a plant nutrient in which nitrogen itself is directly applied to the soil. Sometimes, it becomes necessary that nitrogen be converted to a complex fertiliser with a couple of nutrients in addition to nitrogen. A number of common nitrogen-based fertilisers are now being produced. Hydrogen is used in the fertiliser industry in the form of ammonia, mainly to produce urea, diammonium phosphate (DAP) and some other complex fertilisers. Urea is one of the major nitrogenous fertilisers. Approximately 178 kg of hydrogen is required to produce one tonne of ammonia through the Haber–Bosch process. The balanced chemical equation for the Haber–Bosch process is as follows:

$$N_2 + 3H_2 \rightarrow 2NH_3$$

Among coal, natural gas, naphtha, etc., natural gas is the most hydrogen-rich feedstock suitable for hydrogen production. Recently, because of expanding natural gas connectivity, urea-based plants have switched over to natural gas. The annual global production of ammonia is about 180–200 Mt, and it requires approximately 30–40 Mt of hydrogen. Environmentally friendly character of ammonia depends on the hydrogen feedstock used for fertilisers. Usually, hydrogen is fed into the reactor along with nitrogen, and both hydrogen and nitrogen react together at high temperatures and pressures.

The present method of ammonia production is not a ‘green’ process. Usually, one tonne of ammonia needs about 8–10 MWh if SMR is used for hydrogen production. Hydrogen production through electrolysis requires about 15–20% additional energy. The amount of carbon dioxide (CO_2) emitted per tonne of synthesised ammonia using SMR can vary based on several factors, including the efficiency of the reforming process, the carbon content of the natural gas feedstock, and the specific technologies and practices employed. However, as a rough estimate, the production of one tonne of ammonia through SMR can result in the emission of approximately 1.8–2.2 tonnes of CO_2. Hydrogen production by coal gasification results in almost double the amount of CO_2 emitted by SMR. CO_2 can be totally eliminated if green hydrogen, with no carbon footprint, is produced by electrolysis using renewable energy. Depending upon the resource of renewable energy, steps for greening hydrogen could be adopted.

The conventional mode of hydrogen production results in the emission of a large quantity of CO_2, which, in turn, affects the quality of ammonia in the fertiliser industry. Substituting green hydrogen with hydrogen generated from existing fossil fuels is not an uphill task and is feasible without any major technical challenge. The thermal balance of SMR plants is not very much affected by electrolysis. Currently, there are some ammonia plants that use the by-product hydrogen that comes from chlor-alkali plants. Recently, many countries, such as Australia, the United States of America, India, Canada, Germany,

Egypt, Spain and others, have initiated or strengthened their green hydrogen production activities by using mega-projects. Several other countries have also announced their plans.

The Haber–Bosch process is the most common industrial route for ammonia synthesis. The process is designed to continuously run at relatively high temperatures and pressures. The process does not tolerate any fluctuation, so the required amount or renewable energy should be available to sustain this industrial process. In the current scenario, until an adequate amount of green hydrogen is available, a hybrid (running on both fossil and green hydrogen) process could be adopted. By hybridisation, green hydrogen derived by electrolysis from renewable energy could be introduced in conventional plants while using fossil fuels as the base feedstock. In this process, emission can be reduced depending upon the amount of green hydrogen used. This would be a transition phase until the solutions evolve to switch over to green hydrogen.

A rational cost analysis needs to be carried out to assess a country's demand for ammonia for the fertiliser industry and the optimised route of generating the required amount of green hydrogen to meet the need. To assess the amount of hydrogen to be used in the fertiliser sector, historical data on production and consumption should be reviewed, and the trend of fertiliser consumption in the future should be estimated. This is likely to provide background information for an economic analysis of fertiliser consumption and the amount of hydrogen required for that purpose. To calculate hydrogen demand from fertiliser production, the estimated ammonia and hydrogen requirements for different fertiliser products vary depending upon the fixed target.

Appropriate infrastructure needs to be there so that existing plants can be retrofitted to provide blue hydrogen as a step forward towards decarbonisation. Several small-scale projects have been initiated in different parts of the world using electrolysis in ammonia production to ensure a drastic reduction in pollutants in the fertiliser industries. The landmark Neom project on a green hydrogen plant is designed to produce 650 tonnes of hydrogen every day by the method of electrolysis. Since the present route of ammonia production by the Haber–Bosch process is highly carbon intensive, countries where high-quality solar energy is available can adopt the technology of hydrogen production from solar sources.

There are a number of problems to be resolved for the transition to low-carbon technology. One CO_2 molecule and two ammonia molecules are required to produce urea (CH_4N_2O). In the case of SMR using the natural gas route, it is possible that CO_2 recovered from the reformation process is recycled for urea production. Therefore, it is necessary to put an electrolyser in the existing natural gas plants and build additional facilities to utilise waste CO_2 for green ammonia production. Urea is an important nitrogenous fertiliser in several developing countries as it is manufactured from anhydrous ammonia with about 46% nitrogen. Almost 70% of emissions from an SMR process can be captured by urea. Under such circumstances, one of the beneficial features of conventional ammonia synthesis is that CO_2 emissions from the process can be captured and effectively used as feedstock for urea synthesis. In the future, plants could shift to using biomass for a CO_2 source to ensure low-carbon production. Over a period of time, the shift from urea would reduce the requirement for CO_2.

Storage and transportation are major factors for particular types of fertilisers. The transport and storage of urea has been relatively simpler to transport, store and handle because it does not require an optimised thermal environment. On the other hand, temperature and humidity should be very closely monitored for fertilisers, such as ammonium nitrate and diammonium phosphate, which is a very important and concentrated phosphate-based fertiliser. Needless to say that phosphorus is an extremely important nutrient along with nitrogen. It substantially contributes to the development of new plant tissues. It also helps monitor the synthesis of protein in crops. In several countries, there are already a large number of natural gas plants that can produce fertiliser products for many years. There could be a slow transition from natural gas plants to low-carbon fertiliser products. Emission reduction methods such as CCUS could be retrofitted, and technological options may be adopted for green hydrogen production. All these steps are to be carefully planned and implemented.

6.3 Production of Methanol

Dalena et al. [3] reviewed a wide range of processes of methanol production starting from the historical 'high-pressure method' called the 'BASF process' (which started at the beginning of the twentieth century and the low-pressure method of the ICI process that was introduced in the 1960s). Current methods of producing methanol from natural gas, biomass, char and CO_2 have been highlighted. The following reaction shows the CO_2-based methanol synthesis. The catalysts are usually Cu-based catalysts:

$$CO_2 + 3H_2 \rightarrow CH_3OH + H_2O$$

The overall conversion in the gas phase is an exothermic reaction ($\Delta H = -49.5$ kJ mol^{-1}). CO_2 has been identified as the most potent greenhouse gas. Carbon capture and storage (CCS) is an expensive process and is sometimes liable to leak. Therefore, one of the most desirable solutions is to convert the captured CO_2 into a value-added chemical such as methanol [4]. In the event of any process to produce methanol from green hydrogen, an external source of CO_2 will be necessary. A very effective approach would be to capture CO_2 from neighbouring industries and to recycle them. This is a temporary solution because carbon dioxide would continue to be emitted by industries. CO_2 captured from air or from biomass may provide a lasting solution.

Methanol synthesis from syngas is represented by:

$$CO + 2H_2 \rightarrow CH_3OH, \Delta H = -90.4\,\text{kJ}\,\text{mol}^{-1}$$

Cu-based catalysts, such as $Cu/ZnO/Al_2O_3$, were also originally used as catalysts for the synthesis of methanol from syngas, which consists of hydrogen and carbon monoxide [5]. This process is carried out at elevated pressures and temperatures, usually around temperatures of 200–300 °C and pressures of several tens of atmospheres. The exact pressure levels can vary based on the specific methanol synthesis technology and reactor design, but the typical operating pressure for methanol synthesis is in the range of 50–100 bar.

6.4 Hydrogen in Refineries

The demand for hydrogen in oil refineries is used for the process of desulphurisation. This demand, which was about 30–35% of the global hydrogen demand, is now increasing because of the stringent regulations on sulphur content in oil. Hydrotreating and hydrocracking are the refinery processes that consume most of the hydrogen (to the tune of 90%) in the refining sector to reduce the sulphur content in the finished products. Previously, naphtha was used to produce hydrogen in refineries, and subsequently, natural gas reformation was adopted as the common practice for producing hydrogen. The amount of hydrogen produced as a by-product of the refining process is extremely inadequate to cater to the needs of refineries. Several industries have initiated the introduction of green hydrogen in refineries. Several initiatives are exploring the introduction of green hydrogen in oil refining and fuel production processes. In 2022, Copse (Compañía Española de Petróleos, S.A.U.), a Spanish multinational oil and gas company, and the Port of Rotterdam have already signed a Memorandum of Understanding (MoU) to supply green hydrogen to Rotterdam and Algeciras, two major ports in Europe. Cepsa aims at green hydrogen production in Spain and Portugal, and Rotterdam plans to supply 4.6 million tonnes of green hydrogen to Northwest Europe by 2030. This collaborative project is likely to ensure fast decarbonisation of maritime transport in Europe. Preem, the largest fuel company in Sweden, has several national initiatives and projects related to green hydrogen fuel supply from renewables. Vattenfall and Preem are studying the feasibility of using offshore wind and hydrogen to switch over to fossil-free fuel by 2030. REFHYNE, the project funded by the European Commission's Fuel Cells and Hydrogen Joint Undertaking or FCH JU (now Clean Hydrogen Partnership), has developed a 10 MW polymer electrolyser membrane (PEM) at the Shell Energy and Chemicals Park in Germany. British petroleum in its plant in Lingen has started producing sustainable aviation fuel (SAF) from used cooking oil.

Hydrogen is used in the refining process for the removal of sulphur. The desulphurisation process is carried out in industries depending upon regulations permitting sulphur levels in different products. Currently, environmental regulations in most countries have put a limit on the permissible level of sulphur in diesel. Issues related to desulphurisation and the production of ultra-low sulphur diesel (ULSD) for transport applications have become necessary to control vehicular pollution. The production and consumption of hydrogen in refineries need to be upgraded by expanding the hydroprocessing facilities. In the hydrocracking process, the hydrocracker takes heavier fractions of gas oil and cracks heavy molecules into valuable products in the presence of hydrogen and a catalyst. Thus, the hydrocracker upgrades low-quality heavy gas oil to clean-burning gasoline, diesel or jet fuel. Figure 6.3 shows the hydrocracking process.

In fact, cracking crude hydrocarbons is the usual method of obtaining most of the hydrogen used for this process of conversion.

There is a constant demand for hydrogen in oil refineries because traditionally hydrogen used in hydrocracking processes is obtained from various sources, including natural gas, which is a fossil fuel. However, with the rise of green hydrogen production through electrolysis using renewable energy sources, hydrogen used in hydrocracking can become 'green', and in the current scenario, green hydrogen may not be available. Therefore, refineries need to

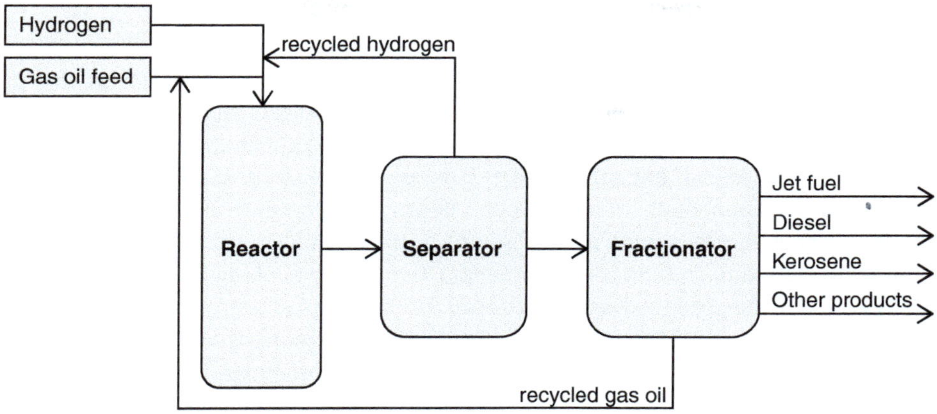

Figure 6.3 Hydrocracking process.

have adequate storage. Higher electrolyser capacities need to be installed to enhance the electrolyser capacity.

6.5 Hydrogen Use in Steel Industries

Although the technology of improving energy efficiency and at the same time reducing GHG emissions has been successfully implemented in several industrial sectors, ironmaking and steelmaking industries are still confronted with heavy challenges to achieve decarbonisation. The steelmaking process has been using coal for centuries to remove oxygen from iron ore. About 1.8 tonnes of CO_2 is emitted into the atmosphere during the production of 1 tonne of steel. The production of steel contributes to almost 8% of total GHG emissions. The use of green hydrogen can ensure total decarbonisation of the processes.

The technical feasibility of the reduction of iron ore requires a production chain different from the existing one with the provision of using green hydrogen. Usually, carbon (coke) is used as the reducing agent inside the blast furnace. It is possible to use hydrogen or hydrogen-containing compounds such as syngas or methane that can be injected into the blast furnace. This approach can be effective in reducing the emission of GHGs from blast furnaces to some extent. However, the total replacement of coke by other reducing agents has not been successful. There is a basic difference in the chemistry governing this reduction process. Reduction of hydrogen with iron ore is endothermic. On the other hand, the reaction is exothermic if the reduction process occurs with CO that is produced by the partial oxidation of coke. Therefore, if hydrogen is used as the reducing agent, there must be a provision for a continuous supply of heat from an external source. The direct reduction process (DRP) is another route of using hydrogen, where the removal of oxygen from iron ore is achieved without melting in the blast furnace. In such cases, hydrogen can be used as a reducing agent, and coke is completely avoided.

A team of Swedish companies joined hands through the concept called 'Hydrogen Breakthrough Ironmaking Technology' (HYBRIT), for which a pre-feasibility study of the concept was completed in the early 2018. The report describes the key problems to be sorted out and the strategies to be adopted for the steel production process by a fossil-free hydrogen (H-DR) process [6]. Hydrogen production by electrolysis, methods of storage and the test procedures along with the economic viability of the process should be analysed for producing good quality fossil-free steel. The overall analysis indicates that the H-DR process is not expensive and that a major part of the cost comes from the price of electricity. A novel and environmentally friendly process of producing fossil-free steel with hydrogen from low-carbon energy sources is in progress [7]. In this process, green hydrogen is used as the reducing agent, and the conventional practice of using coking coal in the blast furnace is avoided. The green hydrogen developed for this purpose is produced from renewable low-carbon sources. Thus, the possibility of CO_2 emissions is avoided by the basic design of the system, and, on the other hand, only water is produced.

6.6 Hydrogen in Agriculture, Healthcare, Food Industry and Several Other Sectors

Agriculture is vital for the socio-economic development of a country. The objective of modern agricultural systems has evolved in the midst of several problems related to natural disasters, food safety and environmental pollution. Wang et al. [8] described how 'hydrogen agronomy' can play a significant role in the growth of plants and protect several important agricultural organisms from abiotic and biotic stresses. Previously, the application of the biological effects of hydrogen was limited to only the 'mechanism of selective oxidants' [8]. Improvement in the nutritional value of agricultural products by the application of hydrogen in changing the microbial composition of the soil regulates the growth of plants. It is also possible that hydrogen-operated trucks, tractors, etc. could be used in farms. Ammonia fertilisers can also be developed for farms using hydrogen.

Russel et al. [9] discussed the potential and prospects of hydrogen being useful for human healthcare. Hydrogen, in its molecular form, has been in use for centuries for several medical applications in different forms and doses depending upon the situation. Although human beings and some animals and plants have used hydrogen to some extent, the potential of 'this electrochemically neutral and non-polar diatomic compound' [9] has yet to gain widespread entry into the healthcare system. Studies suggest that hydrogen gas inhalation could protect against the harmful effects of ionising radiation, potentially making it useful for radiation therapy and nuclear accidents. Hydrogen gas magnetic resonance imaging (MRI) techniques are being explored for their potential to provide enhanced contrast and imaging capabilities. A lot of research has to be carried out in this area to identify the primary target in a particular case on the basis of the character of illness and the innate history of the patient. The long-term effect should be critically analysed.

This information is necessary to obtain reliable evidence to ensure rapid implementation into medical infrastructure. Hydrogen fuel cells can be of great use for food preservation. Many reputed companies are using fuel cells to provide power and to heat and cool their food processing sites. Fuel cells are also used for refrigerated storage of food materials. Fuel cells have proven to be very useful during superstorms or in situations when grid power is intermittent or completely shut down. Although there is promising research, hydrogen-based therapies have not yet been established as mainstream medical treatments. Any medical use of hydrogen would need to go through rigorous scientific testing, clinical trials and regulatory approval processes before being widely adopted in clinical practices.

6.7 Hydrogen in the Welding, Cement and Paper Industries

Fydrychand and Łabanowski [10] carried out tests to estimate the diffusible hydrogen content in deposited metal by using two well-known methods: the glycerine method and the mercury method. A comparative evaluation of both processes indicates that the glycerine method is easier to perform in high-hydrogen processes, but the mercury method shows higher accuracy in the determination of low hydrogen content in deposited metal. Atomic hydrogen atoms are produced by the dissociation of dihydrogen with the help of an electric arc. This results in the release of a large amount of energy to the tune of 435.88 kJ mol^{-1} and a high temperature of about 4000 K, which is suitable for welding and cutting processes. Oxy-hydrogen flames offer the advantage of monitoring combustion during oxidation or reduction processes and can be useful in welding, brazing, cutting, etc. Cement industries contribute about 7–8% of plane-warming CO_2 emissions. The production of cement clinkers requires a high temperature of about 1450 °C, and this is where coal or natural gas is used as a source of heat to achieve these temperatures. The high-temperature process is essential for the chemical reactions that occur during clinker formation. In this process, hydrogen could be used instead of coal or natural gas. Pilot projects have been initiated to replace natural gas with green hydrogen during ongoing production.

6.8 Hydrogen for Electricity Generation

Looking closely at the properties of hydrogen, one can easily know that hydrogen can be combusted in ICEs, gas turbines, heaters, boilers, etc. Hydrogen can be used for electricity generation as well as for the purpose of co-generation where both heat and electricity can be generated. Hydrogen fuel cells are very efficient devices that produce electricity and heat. Enel's hydrogen project was initiated with two major objectives of achieving high efficiency and low NOx emissions. During the first phase, it was suggested to build up the gas turbine test rig and optimise the parameters to ensure smooth and safe operation. Since NOx is a pollutant of concern, the major objective of the next phase of activities was to evaluate and monitor NOx emissions [11]. The industrial-scale Fusina hydrogen power station in Italy has been successfully operating with 100% pure hydrogen. In the initial stage, the turbine technology was tested with methane. Subsequently, special pipelines were used for hydrogen, and the plant was switched over to total hydrogen operation. An integrated 16 MW gasification combined cycle (IGCC) consisting of a hydrogen-fuelled gas turbine and a heat-recovery steam generator (HRSG) was designed and installed at the coal-fired power plant at Fusina. Several power pants and pilot projects worldwide are exploring hydrogen-based electricity generation. The Emsland Power Plant in Germany uses hydrogen for its operation. In Japan, the Tomakomai Power Station has a hydrogen-based electricity generation project. It uses a mixture of hydrogen and natural gas to produce electricity. The Leeds H21 project aims at converting a natural gas combined-cycle power plant to run on 100% hydrogen. In the United States of America, West Virginia University's hybrid gas turbine research initiative involves modifying a gas turbine to run on a blend of natural gas and hydrogen. This gas turbine, located in the Irsching Power Station, Germany, has been converted to run on a mixture of hydrogen and natural gas. Although the Gigastack project in the United Kingdom does not have a specific power plant, studies are carried out to demonstrate the production and use of low-cost, zero-carbon hydrogen derived from offshore wind energy.

Co-generation has specific advantages due to flexibility and improved efficiency. At present, efforts are being made to design turbines for petrochemical plants that can be adapted to different fuels. A petrochemical plant in the Daesan complex in South Korea uses a gas containing 95% hydrogen as its primary fuel. The Magnum power plant in Vattenfall, the Netherlands, which was initially a coal-gasification plant, has now been developed into a gas-fired station with gas turbines. The Intermountain Power Agency (IPA), Utah, is expanding its activities related to the use of renewable energy resources. One main objective is to produce hydrogen for generating electricity. Hydrogen-based co-generation systems could be particularly useful in microgrid applications, remote areas or places with limited access to conventional energy sources. Co-generation systems can also be used to recover waste heat from hydrogen-related processes, such as hydrogen production through SMR. This recovered heat can then be used for other industrial processes, enhancing the overall energy efficiency.

Hydrogen can be utilised for generating electrical power with hydrogen fuel cells and hydrogen combustion engines. Both spark ignition and compression ignition engines can be modified to operate on hydrogen. Spark ignition engines can operate on total hydrogen, but compression ignition engines can be conveniently modified to run in hydrogen and diesel dual fuel mode. Electricity can also be generated by using hydrogen in gas turbines. In a hydrogen fuel cell, hydrogen combines with oxygen from the air to create electricity. Details about the role of hydrogen in ICEs, gas turbines and fuel cells are described later in this chapter.

Research initiatives are being explored to implement the concept of integrated hydrogen energy stations that combine hydrogen production through electrolysis with co-generation. These stations produce hydrogen through renewable energy-powered electrolysis and then use the generated hydrogen in fuel cells to produce both electricity and heat for various applications. Hydrogen fuel cells adopt the most appropriate technology for producing electricity by combining hydrogen and oxygen atoms. In a hydrogen fuel cell, hydrogen reacts with oxygen across an electrochemical cell to produce electricity, water and small amounts of heat. This principle is being adopted to generate power from hydrogen through fuel cells. Small fuel cells can power cell phones and laptop computers and can be used for some military applications. On the other hand, large fuel cells can supply electricity to power grids and to areas that are not connected to electric power grids. They can also be useful in supplying power to buildings in the case of an emergency. The Shinincheon Bitdream Fuel Cell Power Plant, Incheon, has a total capacity of 78.96 MW and an annual production of 700 GWh of electricity. In addition, the plant absorbs 24 000 tonnes of fine dust every year, thereby significantly contributing to the local problems of environmental pollution [12]. The present research activities have shown the way that efficient use of hydrogen can be achieved in solar power plants. The National Aeronautics and Space Administration (NASA) used hydrogen fuel cells to power electrical systems installed in spacecraft. The high-quality application potential of hydrogen–oxygen steam generators (HOSGs) has been studied for thermal power plants (TPPs) and nuclear power plants (NPPs). A feasibility analysis was carried out, which shows if the main steam-generating equipment does not function for some reason, the steam turbine can provide backup power. Researchers have carried out studies related to the design of the integration of HOSGs into the main power equipment of TPPs with thermal and nuclear power plants.

6.9 Hydrogen in ICEs

ICEs are broadly classified into two different families depending upon their ignition characteristics : spark ignition (SI) and compression ignition (CI) engines. Both types of engines are operated by petroleum-based fuels such as gasoline and diesel. Petroleum fuels are being depleted at an alarming rate, and ICEs emit noxious pollutants that degrade the environment. Many alternative fuels, both liquid and gaseous, such as ethanol, methanol, natural gas, hydrogen, dimethyl ether (DME), diethyl ether (DEE), propane, butane, liquefied petroleum gas (LPG), liquefied natural gas (LNG), biogas, producer gas, biomass-to-liquid (BTL), coal-to-liquid (CTL), gas-to-liquid (GTL) and electricity, are being closely studied for applications in different forms in several engines. Hydrogen is an ideal fuel for use in ICEs because it has the potential to provide solutions to both fuel crisis and environmental degradation due to the combustion of fossil fuels.

Conventional ICEs are fuelled by gasoline or diesel. ICEs have been dominating in the transport sector. In many of the developed and developing countries, the transport sector accounts for a major part of the total energy consumption. Figure 6.4 shows the projected energy demand in the transport sector in road, rail and aviation for passengers and freight.

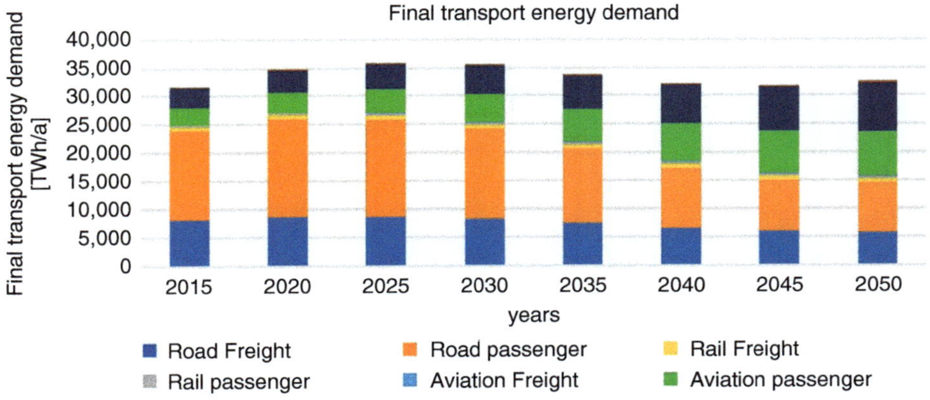

Figure 6.4 Final transport energy demand [13]/MDPI/CC BY 4.0.

A closer look at the energy consumption pattern in the transport sector shows that it takes up an increasing trend from year to year. Several studies have shown that the total energy consumption for all modes of transport is likely to increase substantially in coming years. It has been estimated that the total energy consumption in all modes of transport will increase by about 1.4% per year until 2040, and light-duty vehicles are expected to be a major consumer [13]. It is expected that the total combined energy demand for the two-wheelers, three-wheelers and buses would be taking a downwards trend, whereas the demand for oil consumption for air transport is likely to increase substantially during this period. Such a trend of increasing transport fuel consumption seriously affects the economic condition of several countries and brings a major drag on the balance of payments.

Many research and development activities and road test demonstrations with hydrogen-operated vehicles have been successfully carried out in different countries. Figure 6.5 [14] shows CO_2 emission levels across different transport sectors globally. It is quite likely that hydrogen would be deployed in road transport much earlier before it enters other modes of transportation such as rail transport, shipping and aviation. Thus, it would be appropriate to analyse the potential of hydrogen during this critical phase of transition towards a more sustainable and decarbonised transportation system.

The recent global trend in the use of hydrogen is rapidly shifting from ICEs to fuel cells. Infrastructure such as hydrogen fuelling stations has also been built by some organisations to run some of these fuel cell vehicles. The establishment of a number of refuelling stations and the cost of fuel cells are critical to facilitate the widespread use of these fuel cell vehicles. Since both ICEs and fuel cells are the two major systems for transportation, it is important to discuss the relative efficiencies of internal combustion engine, electric and hydrogen fuel cell drivetrains. Several light-duty hydrogen fuel cell vehicles have started operating on roads. Therefore, in addition to the system configuration, it is important to evaluate performance, emissions and long-term economic viability along with the total cost of ownership (TCO) of different drivetrains for different vehicle segments to obtain a correct picture of the long-term transport scenario.

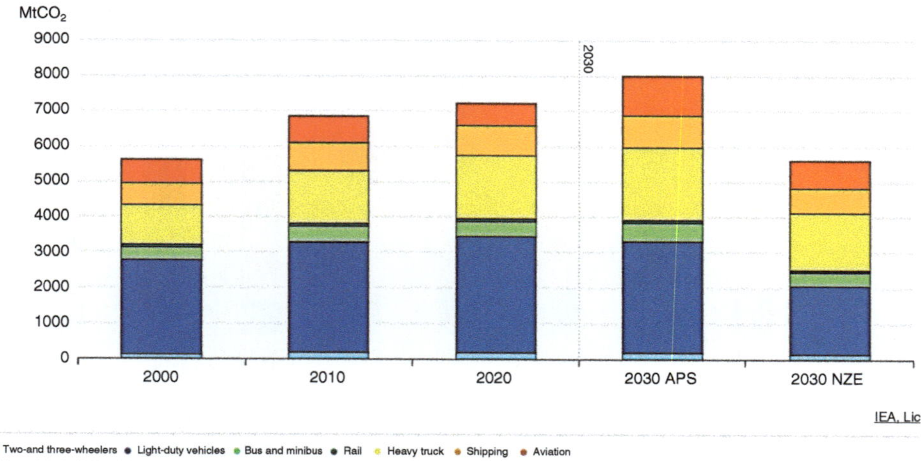

Figure 6.5 Global CO_2 emissions from transport by subsector, 2000–2030 [14]/IEA/CC BY 4.0.

6.10 ICEs

The use of hydrogen in ICEs is not a radically novel concept. Cecil [15] presented a paper entitled 'Hydrogen Gas As a Moving Power in Machinery' before the Cambridge Philosophical Society, where he said 'The engine in which hydrogen gas is employed to produce moving force may be inferior in some respects to many engines at present employed, yet it will not be wholly useless, if together with its own defects and it should be found to possess advantages also peculiar to itself' [16]. There was a large gap in this until the 1930s when as many as 1000 ICEs were reported to have been converted to operate on hydrogen and hydrogen gasoline blends [17]. Hydrogen engines were used during the world war when there was a dire need for petroleum fuel at a particular location and at a specific time and it was not available. Erren and Campbell [15] presented a paper at the Institute of Fuel in London mentioning the higher thermal efficiency of hydrogen engines and the severe problems of backfire. Erren successfully adopted direct cylinder injection and avoided backfire. Weil [18], who was the technical director of the 'Deutsche Erren Studiengesellschaft' in Germany until 1938, mentions about Erren's precious test results that the engines developed by Erren for hydrogen operation had 'no problems with detonation'. Perhaps the difference among detonation, knocking and backfire was not critically distinguished in hydrogen engines at that time. Weil mentions that Erren also tested hydrogen-operated diesel engines by way of high-pressure injection. Thus, it seems that Erren had almost initiated tests that are so specific to hydrogen engines and that researchers all over the world are still continuing their research and development activities. Unfortunately, detailed records of his investigations were totally destroyed by Allied bombing during the world war, and thus, researchers of later years could hardly obtain any benefit from his extremely valuable research results. All these efforts, spanning over a century, were sporadic in character and were restricted to laboratory curiosity. Direct cylinder injection of hydrogen intrinsically precludes the problems of backfire, which has literally been the 'Achilles heel' of hydrogen engines. Oehmichen [19] adopted direct cylinder injection early in the compression stroke and successfully ran a single-cylinder engine with hydrogen. The engine was operated at speeds ranging from 500 to 2000 rpm at compression ratios (CRs) from 5.4 to 15.4 and equivalence ratios from 0.29 to 0.67. Thus, he conducted the tests in the lean zone of the engine and was able to have a high thermal efficiency of 54% at an equivalence ratio of 0.4 and at a compression ratio of 15.4. He never encountered the problem of pre-ignition and backfire as his test was limited to lean operation. The chances of backfire increase with rich mixtures. It is beyond the scope of the present discussion to give a chronological account of the evolution of hydrogen engine technology.

Many researchers from academia, industries and research organisation worldwide have been working on the development of an optimum hydrogen engine that can give good performance and low emission characteristics without exhibiting any abnormal combustion phenomena. Relevant discussions on all these aspects can be found in several research papers published over the years [20–27]. Both SI and CI engines have been modified to operate on hydrogen fuel. R&D activities have been carried out on various aspects of engines, such as dual fuelling, hydrogen supplementation, operation strategies for NOx control, power improvement, fuel induction systems, optimised ignition timing and a range of equivalence ratios for smooth operation. It has been experimentally evaluated in the Indian Institute of Technology (IIT), Delhi, India, that it is possible to achieve a maximum thermal efficiency above 40% using timed manifold injection (TMI). The maximum brake thermal efficiency (BTE) of a hydrogen engine goes beyond a little above 40% using TMI. It has also been tested in the laboratory that the optimum engine operating condition adopting combined lean-burn and stoichiometric strategies could be achieved by operating the engine at variable lean mixtures at relatively lower load conditions. This operation provides good engine performance with high thermal efficiency and an extremely low concentration of exhaust emissions. However, beyond a certain level of engine power demand, the engine could be operated at a stoichiometric mixture in partial throttle mode. Researchers have identified the temperamental combustion characteristics of hydrogen for a long time. A lot of work related to the combustion of hydrogen in engines has been reported for a long time [28–30].

Many researchers have experimentally evaluated the operating conditions to avoid flashback (backfire) conditions. Since oxides of nitrogen are the only pollutant of concern in hydrogen engines, methods to control the NOx level have also been widely studied. Optimised hydrogen engines have been configured to prototype vehicles and road tested. Hydrogen-fuelled gensets have been designed and developed for use in decentralised energy applications and other applications. An alternative fuel has to be appropriately chosen for use in either SI or CI engines. These fuels cannot be arbitrarily and liberally exchanged without any modification either in engine operating conditions or basic engine hardware. Efforts to have an elaborate discussion on these aspects have not been made here. However, hydrogen is an ideal substitute for use in SI engines. Critical combustion properties, such as a wider flammability range, flame speed, auto-ignition temperature stoichiometric fuel/air ratio, quenching distance, minimum ignition energy, etc., have been observed to be best suited for use in SI engines.

6.10.1 Anomalies in Hydrogen Combustion Systems: Pre-ignition and Backfire

Some of the typical combustion-related properties of hydrogen are radically different from those of petroleum fuels. The major combustion-related problems that need to be critically analysed for smooth engine operation are identified by pre-ignition and backfire (flashback or backflash) and knocking. Pre-ignition takes place during the compression stroke with the intake valves already closed whereas backfiring occurs with the intake valves open. Quite often, pre-ignition is considered the predecessor for the occurrence of backfiring. In most operation strategies, the occurrence of backfiring is generally limited to external mixture formation concepts. Pre-ignition during an intake stroke takes place only when the charge remains beyond a particular level of temperature for a certain length of time, usually referred to as the induction time. The possibility of pre-ignition depends upon the size of the particle because a smaller particle cannot heat up a large volume within a small interval of time. The causes leading to combustion knock are different because knock occurs when pre-ignition occurs away from the fuel intake valve. In typical SI engines, knocking occurs when the end gas conditions (temperature, pressure and time) are such that the end gas spontaneously auto-ignites. As a result, a condition evolves when rapid release of the remaining energy generates very high amplitudes of pressure waves with typical pinging noise, which is referred to as engine knocking. In the case of a hydrogen engine, combustion knock badly impairs engine performance by lowering the power output and efficiency. The figure shows the value of intake pressure and the cylinder pressure at a time when pre-ignition has taken place. The speed of the engine (n) and the strength of the mixture at the time of pre-ignition are shown in Figure 6.6.

6.10.2 Phenomenon of Backfire: Causes and Control Techniques

The effects of backfire or flashbacks are much more devastating than those of pre-ignition. Broadly speaking, backfire is also characteristically a pre-ignition phenomenon where the combustion of fresh charge occurs during the suction stroke (in the combustion chamber [CC] or intake manifold) with the intake valve open. When pre-ignition occurs near the fuel intake valve in an engine with a premixed fuel induction system, the resultant flame can travel into the fuel induction system, resulting in flashback. Figure 6.7 shows a typical pressure–crank angle diagram in which the backfire zone has been clearly identified.

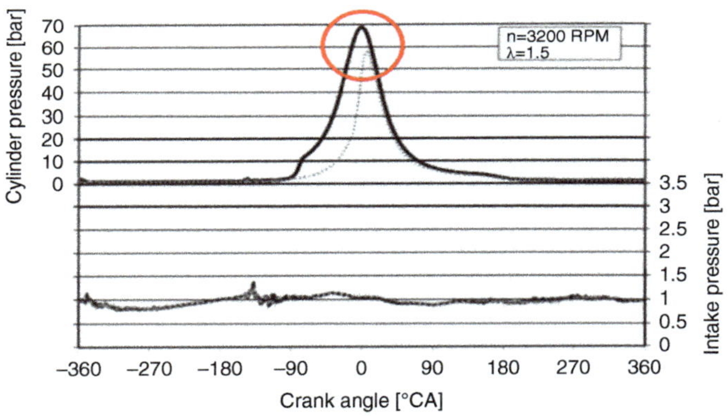

Figure 6.6 Pre-ignition in a hydrogen engine.

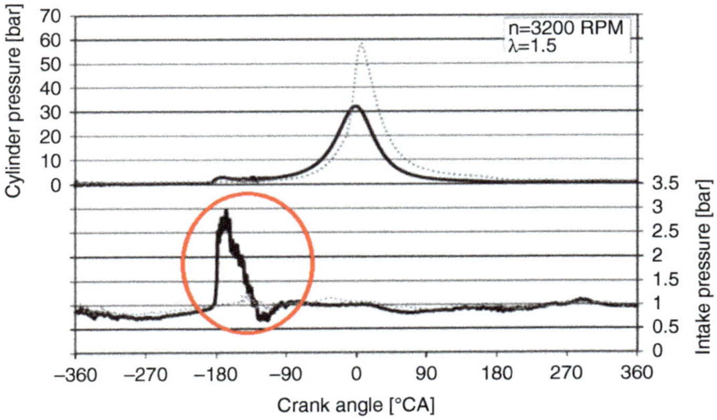

Figure 6.7 Backfire in hydrogen engines.

The severity of flashback varies over a very wide range. Quite often, it causes considerable damage to the carburettor and other engine components. Sometimes there is a loss of power and engine stalls. Extending the laboratory experience to the field, in a hydrogen-fuelled vehicle, flashback not only seriously impairs the drivability of the vehicle but also poses safety hazards if the vehicle stops in the midst of thick traffic. Considering these well-defined causes and subsequent effects, backfire conveys an entirely different picture and is usually discussed in the literature separately from pre-ignition.

The most erratic behaviour of backfire in the early years of hydrogen engine research was often experienced at an unscheduled point in the engine cycle during engine operation. Problems of backfire in the past have resulted from a simple misfire to devastation of an engine test rig. In some cases, the researchers discontinued their R&D activities, and in other cases, the sponsored project was withdrawn because the investigators were not able to find the cause of backfire and could never control it. However, over the period of time, the causes of backfire have been very clearly understood. Research and development activities have reached a stage to identify the root cause and eliminate this problem successfully to ensure smooth engine operation.

Backfire or flashback is a mistimed firing of the fuel, which sometimes results in firing out from the inlet side. It often occurs by certain hotspots or spots of high thermal energy of sufficient magnitude or intensity. Such hotspots could be deposits or hot particles in the cylinder, which are mostly the carbon residue from bunt lubricating oil leaking into the cylinder. Hotspots on the spark plug and cylinder walls, cervice volumes and hot valves have also been the cause of backfire. Crevice volume, the small gaps or spaces that exist between various components within an engine, such as the piston–cylinder interface, valve seats or other joints, can trap a mixture of air and fuel, creating a localised region that is rich in hydrogen fuel, which can lead to an unintended ignition source, causing backfire. In multi-cylinder engines, backfire often occurs when the crosstalk between spark plug wires fires two cylinders at the same time by induction. Communication of fresh charge with burning exhaust gases from another cylinder often causes backfire in multi-cylinder engines. Engine operation with very lean mixtures in the presence of still-burning gases from the previous cycle sometimes also initiates backfire. In certain other situations, pre-ignition occurs in the combustion chamber that advances and ultimately results in the backfire in the intake manifold. This is the most severe form of backfire. In some other situations, backfire also occurs due to very lean hydrogen–air mixtures during idling conditions. However, such a situation of backfire is very occasional and infrequent. In this kind of backfire, firing energy is not enough to stop the engine. Causes of backfire and effective methods to control them have been recorded in the literature [31–35].

The *backfire limit* (BFL) is defined as the equivalence ratio at which backfire occurs. One possible cause of backfire is the slow burning of hydrogen–air mixtures in the crevice volume. This slow burning would continue up to and during the intake process. During intake, the hot gases could flow out of the crevice volumes and ignite the fresh charge, thereby causing backfire. Coolant temperature also has some effect on the BFL equivalence ratio. At higher coolant temperatures, the BFL equivalence ratio decreases due to the lower heat-carrying capacity of the coolant. Sometimes it is possible to reduce the backfire by lowering the coolant temperature such that the cylinder temperature also decreases. However, this approach lowers engine efficiency and therefore needs to be adopted depending upon the situation. Several experimental approaches were adopted in the laboratory of the IIT (Delhi) to eliminate the problem of backfire. Visible improvement in the engine operation was observed by using cold spark plugs (spark plugs using very narrow gap settings and a non-protruding electrode configuration) and by reducing the spark plug gap setting and grinding/polishing the sharp edges and protrusions. Spark plug gaps could be as small as 0.25 mm. This can be increased to 0.5 mm in the event of difficulties due to a cold start at some weather conditions and at some locations. Spark plugs with platinum electrodes should be avoided as this can be a catalyst fir hydrogen oxidation. However, there could be some arrangement to use platinum in the exhaust so that unburned hydrogen could be oxidised.

Lean operation and exhaust gas recirculation (EGR) have also been found to be effective in controlling backfire to some extent. Since flashback occurs mainly at high equivalence ratios, usually above 0.8, it is desirable to operate the hydrogen engine preferably at equivalence ratios below their flashback value. The properties of a wider flammability limit and fast flame speed can be used to ensure lean operation of hydrogen engines because the excess air in the lean mode acts as an inert gas and helps decrease the combustion temperature. Lean operation, although it contributes to the improvement of efficiency, often results in lowering the power output.

There are some limitations in the application of the water injection mechanism although it has been tested as a method of controlling the backfire. In the laboratory in engines, it can be used to some degree of success, but for vehicles, an on-board storage method of water has to be evaluated along with the installation of a pump. Moreover, it is desirable that distilled water is used for the purpose. Sometimes, it has been observed that even the usual engine deposits lead to flashback if the water flow into the induction is not optimally monitored. EGR has been to some extent successful in lowering the frequency and intensity of backfire.

However, none of the above techniques are reliable enough to be adopted as a long-term solution for prolonged engine operation. These are good for research and laboratory studies but not to be adopted for regular use in prototype vehicles. The frequency and intensity of backfire can be drastically reduced and even can be eliminated by using an appropriately designed fuel induction technique. This mechanism of fuel induction to the engine has to be developed after taking into consideration the typical combustion properties of hydrogen, such as a wider flammability limit, minimum ignition energy, high flame speed, etc. Several engine configurations, using various modes of fuel induction, have been exhaustively tested in the laboratory of the IIT (Delhi) to identify the engine operating conditions that could successfully eliminate the problem of backfire [31]. It has been observed that flow timing and injection pressure are very crucial to smooth engine operation without flashback. Hydrogen engines can achieve higher thermal efficiency as they can run unthrottled, so pumping losses are avoided. However, in practice, some level of throttling becomes necessary at very low loads to maintain combustion stability and limit exhaust NOx emissions. Since backfire occurs if the air–fuel mixture remains in the fuel induction system before the intake valve closure (pre-IVC condition), backfire could be eliminated by designing a fuel induction system to avoid the possibility of formation of air and fuel mixture after the closure of the intake valve (post-IVC condition). Flow timing remains the most dominant condition for the design of an appropriate injection system. The range of tests carried out to evaluate the engine response characteristics to backfire is shown in Table 6.1 [31].

Table 6.1 Mixture formation and modes of fuel induction.

Modes of fuel induction	Classification	Hydrogen flow timing	Supply pressure
Continuous carburetion	Pre-IVC	Continuous flow	A little above atmospheric
Continuous manifold injection (CMI)	Pre-IVC	Continuous flow	Slightly greater than atmospheric
Timed manifold injection (TMI)	Pre-IVC	Mixture formation	1.4–5.5 atm
Direct cylinder injection	Post-IVC	Hydrogen flow commences after the intake valve closure and is completed before significant compression pressure rises	2.0–8.0 atm

The benefits of different configurations of several modes of fuel injection have been investigated by several researchers [31–40]. While conducting the tests, it is important to avoid dispersion of hydrogen that might leak in the engine that is not amassed at a particular location to form a potential hotspot. During tests, both carburetion and continuous manifold injection exhibit erratic symptoms of backfire. This is perhaps because of the presence of a combustible mixture in the intake system. Technically speaking, continuous carburetion is a form of pressurised carburation, and therefore, it is expected to cause backfire in the same way as carburation. On the other hand, in-cylinder injection precludes the occurrence of backfire because the air and fuel do not have a chance to mix at any such vulnerable point that can cause abnormal combustion. There are two major constraints that do not favour direct cylinder injection. It has been reported that during in-cylinder injection, which occurs at the end of the compressing stroke, the injector often fails in the severe thermal environment of the combustion chamber and starts leaking [31, 33]. Moreover, the mixture formation is not homogeneous because late injection offers much less time for mixing and preparing a homogeneous mixture. Low-pressure direct cylinder injection (LPDI) avoids backfire, but it shows incomplete combustion as it does not have the flexibility of flow controllability compared to TMI operation. The actuating mechanism of the injector and its fuel feeding capacity need to be optimally integrated for the TMI system. The valve timing of the engine, injection pressure, injection timing and injection duration play very crucial roles in ensuring an effective TMI system. It was observed that an appropriately designed TMI system embraces the combined capabilities of both the SI and CI engine configurations. It can adopt the quality governing a diesel engine, thereby providing higher thermal efficiency. At the same time, it has the potential to develop a higher specific power output of an SI engine. Laboratory tests have been carried out on three actuation mechanisms of injection systems: hydraulically operated, cam-actuated and electronically controlled solenoid-actuated injection systems. Apart from the actuation mechanism, an optimum injection strategy, in combination with variable valve timing for both intake and exhaust, is extremely important to achieve desirable engine characteristics with good performance and without any symptoms of undesirable combustion phenomena. There are certain well-defined features and limitations in terms of injection timing and

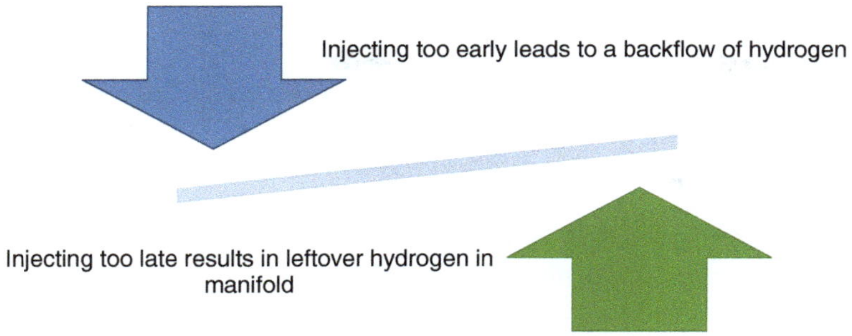

Figure 6.8 Optimum injection strategy for a hydrogen engine.

duration. Injection timing has to be limited to a fixed range of engine operation conditions with an earlier end of injection at lower engine speeds and richer hydrogen mixtures. Injection should not be started too early because, in that case, there is a possibility of backflow of hydrogen from the cylinder into the intake port at the end of the intake stroke. On the other hand, late injection often results in hydrogen remaining in the intake port because of insufficient time to reach the cylinder. Figure 6.8 shows the effect of injection either too early or too late and thus provides guidelines to adopt the optimum injection strategy.

6.11 Choice of Engine Configuration for Hydrogen Fuelling

As far as the use of hydrogen in ICEs is concerned, the basic and relevant discussions have already been made earlier in the beginning of this chapter. Diesel engines as such can only be operated in the dual fuel mode using hydrogen and diesel as fuels. This is because hydrogen has an auto-ignition temperature of 576 °C, and it is difficult to achieve CI of hydrogen. Researchers at the Cornell University have reported that [37] CI of hydrogen in diesel engines was not possible even at a compression ratio of 29.

The combustion of hydrogen in engines does not produce any noxious pollutants such as CO, CO_2, unburnt hydrocarbons, smoke, oxides of sulphur, particulate matter, lead or other toxic metals, ozone, benzene and formaldehyde, ketone, etc. The only pollutants that are formed are oxides of nitrogen (NOx) and water vapour. As shown later, oxides of nitrogen can be drastically reduced to much lower levels of concentrations by operating the engine/vehicle at a lower equivalence ratio.

The design of a hydrogen-specific engine has to take into consideration some of the critical combustion-related properties, such as a wider flammability range, flame speed, auto-ignition temperature, stoichiometric ratio, quenching distance, minimum ignition energy, etc. From a power output point of view, gasoline and diesel petroleum fuels are liquid and remain in the liquid form at room temperature. On the other hand, hydrogen remains in the form of a gas even at a much lower temperature of −253 °C. The heat of combustion of fuel per unit mass of hydrogen is much higher than that of gasoline. However, it has a much lower mass density. Therefore, the power output of a hydrogen engine is lower than that of a gasoline engine. However, the infrastructure built for supercharging may not be essential to compensate for power loss if an appropriate and optimum fuel injection system is used.

6.12 Performance of a Hydrogen-Operated SI Engine

The performance parameters of an engine are interpreted from parameters such as thermal efficiency, power developed and the rate of fuel consumption. Hydrogen-operated SI engines can operate at a higher compression ratio and at a much lower equivalence ratio than gasoline engines. The performance parameters are discussed here considering the experimental range of the compression ratio and the equivalence ratio.

Figure 6.9 compares the minimum ignition energy of hydrogen with methane. Since natural gas (with methane as the major combustible component) has already been established as a regular gaseous fuel for ICEs used in the transport sector of several developed and developing countries, some features of hydrogen (gas) fuel have been compared to other existing gaseous fuels, such as natural gas. It is glaringly evident from the graph that hydrogen needs a much lesser ignition energy compared to methane. This property makes it possible for hydrogen engines to run with a very lean mixture.

Hydrogen engines usually exhibit the problem of backfire closer to stoichiometric conditions. A wide range of flammability of hydrogen (Figure 6.10) is a distinctive property of the fuel, which brings out its ability to be used in lesser amounts and can also create favourable conditions for complete combustion because of its high flame speed.

Hydrogen can be combusted over a wide range in several systems as shown in Figure 6.11. In the present context, it is relevant to know the range of stable combustion in an engine system. The figure provides very useful guidelines to identify the window of equivalence ratio over which stable combustion can take place in an engine system.

The narrow window suitable for engine operation lies between an ultra-lean mode with an equivalence ratio of 0.26 and the stoichiometric range. However, tests in the laboratory have indicated that although it is possible to operate the engine at a very low equivalence ratio, misfire still occurs quite frequently. Stable engine operation can be achieved at an equivalence ratio of around 0.5 [36]. In fact, the problem of backfire has been the single major stumbling block in the development of hydrogen engines for decades. Moreover, all research studies have been carried out on modified engines that were initially designed to

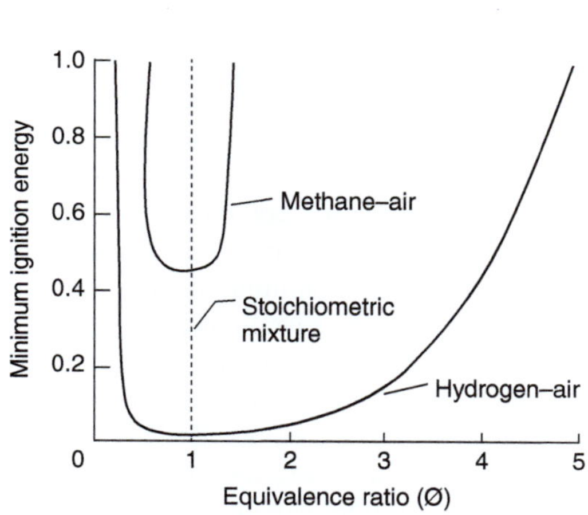

Figure 6.9 Minimum ignition energy of hydrogen and methane.

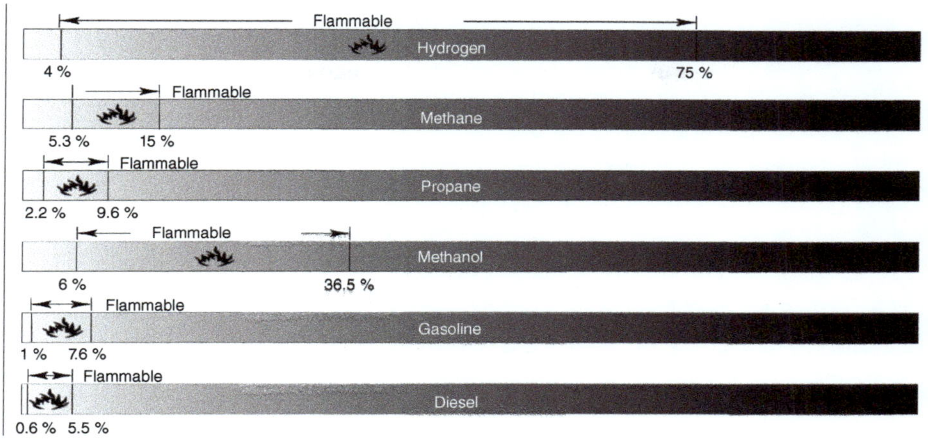

Figure 6.10 Flammability range of hydrogen compared to other fuels.

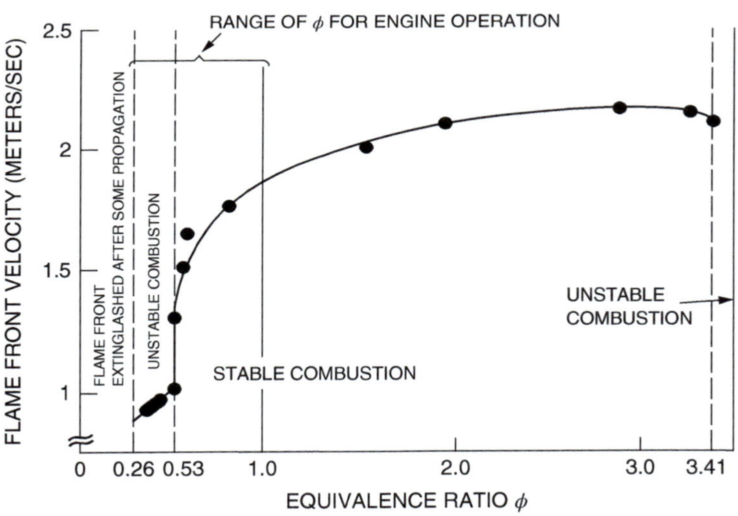

Figure 6.11 Range of equivalence ratio for stable combustion in engines.

operate on gasoline or diesel. During the initial tests on the SI engine, it was very clearly observed that an appropriate injection system is essential to feed the engine at the appropriate point in the engine cycle to totally eliminate backfire. The development of a suitable leak-proof, reliable injector with an appropriate actuating mechanism and adequate flow controllability is the most important factor to eliminate the problem of backfire. Tests with injectors having different actuation mechanisms (hydraulically operated; Figure 6.12: hydraulic-operated systems; Figure 6.13: cam-actuated systems; Figure 6.14: electronically controlled solenoid-actuated systems) have shown good performance and emission features, and the problem of backfire has been reported to be avoided during the entire range of operation.

Figure 6.12 Hydraulically operated injection system.

Figure 6.13 Cam-actuated injection system.

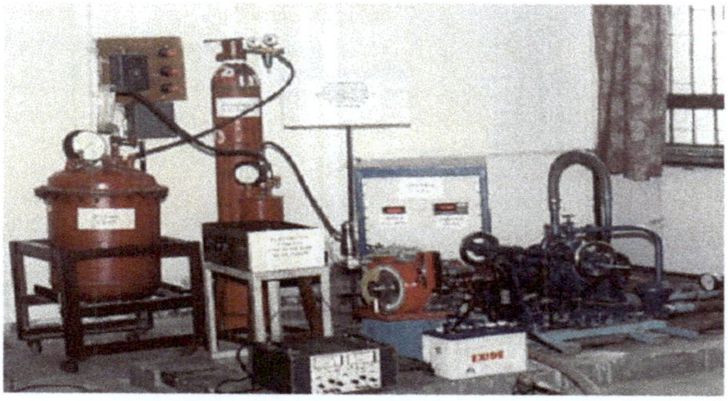

Figure 6.14 Electronically controlled solenoid-actuated injection system.

The solenoid-actuated injection system was successfully operated in a utility engine and hence can be adopted for regular hydrogen operation. Tests were conducted at different speeds, and hence, the engine was operated with different CRs. Figure 6.15 shows the test

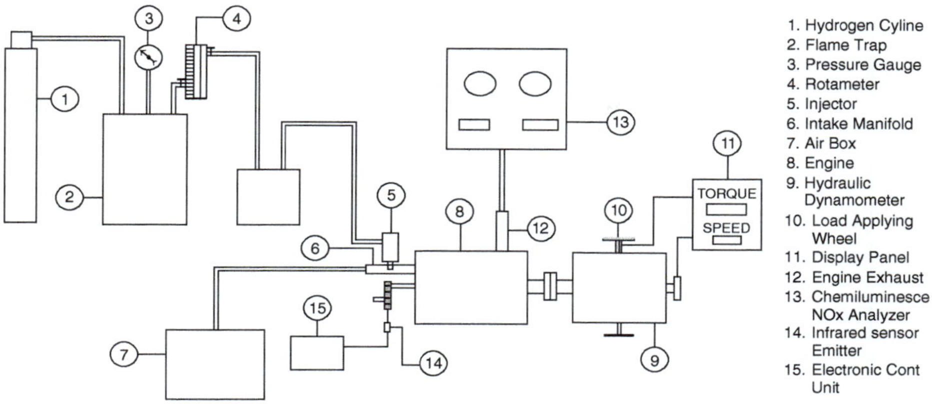

Figure 6.15 Hydrogen engine experimental setup in the laboratory.

rig in the laboratory in which the experimental data for the backfire-free operation of the engine were generated to evaluate the performance and combustion characteristics of the hydrogen engine under various operating conditions.

Figures 6.16 to 6.19 indicate some of the performance characteristics of the engine operating at different CRs and different equivalence ratios. The improvements in these characteristics are evident in the figures. These are only some of the sample graphs indicating the

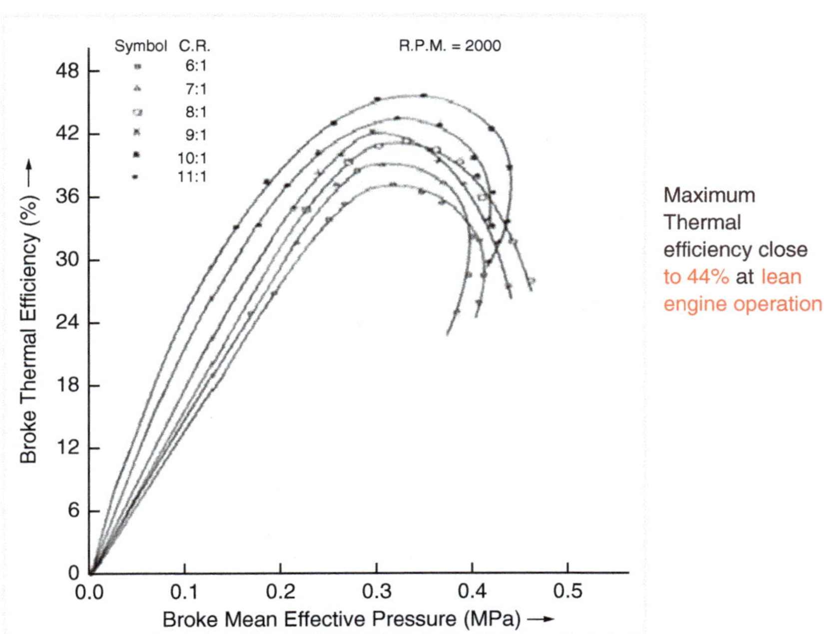

Figure 6.16 Brake thermal efficiency at various compression ratios.

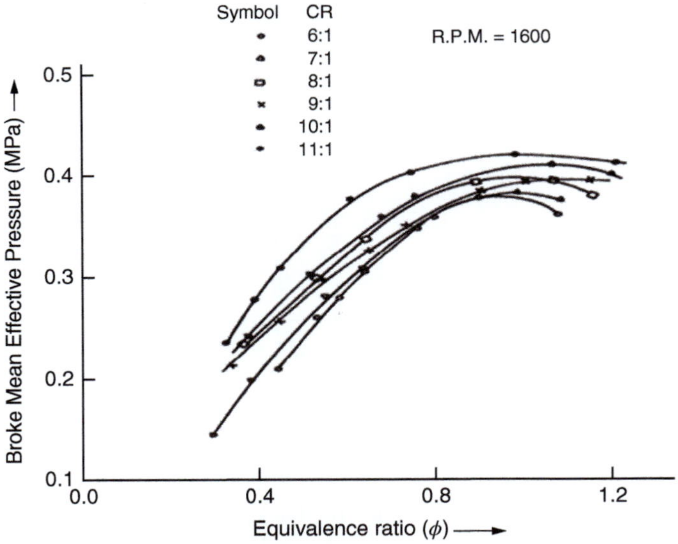

Figure 6.17 Variation of BMEP with equivalence ratio at different compression ratio.

trend of variation for a particular parameter achieved during the tests. It has been shown that a maximum BTE of close to 44% could be achieved at in lean engine operation (Figure 6.16).

These data were generated in a variable compression ratio research engine with the ability to alter some of the operating parameters (such as the compression ratio) under dynamic conditions when the engine was running. It has been reported in the literature that some old engines, which have been used for long on conventional fuels, often show the tendency of backfire. This is probably due to the deposits in the combustion chamber because of prior operation. Experience shows that new engines, developed and instrumented for hydrogen operation, can operate at an equivalence ratio even slightly above the stoichiometric ratio without backfire.

Under optimised operating conditions based on the ignition timing, compression ratio and equivalence ratio, it is always possible to have a thermal efficiency greater than 44% as indicated in the graph above. Apart from achieving a very high thermal efficiency, the hydrogen engine also shows a trend of lesser specific fuel consumption at different speed ranges.

The compression ratio and equivalence ratio have been found to be very significant parameters that also influence the brake mean effective pressure (BMEP). Variation of BMEP with equivalence ratio and BSFC with BMEP at various compression ratios are shown in Figure 6.17 and 6.18 respectively.

The compression ratio and equivalence ratio also influence the engine performance and emission characteristics. Figures 6.16–6.18 show the values of efficiency, power and fuel consumption of the engine operating at different CRs and at different equivalence ratios. The improvements in these characteristics are evident in the figures. Figures 6.19 and 6.20 are generated with some tests to broadly compare the power and emission of the same engine with gasoline and hydrogen operation. As shown in Figure 6.19, the power developed by

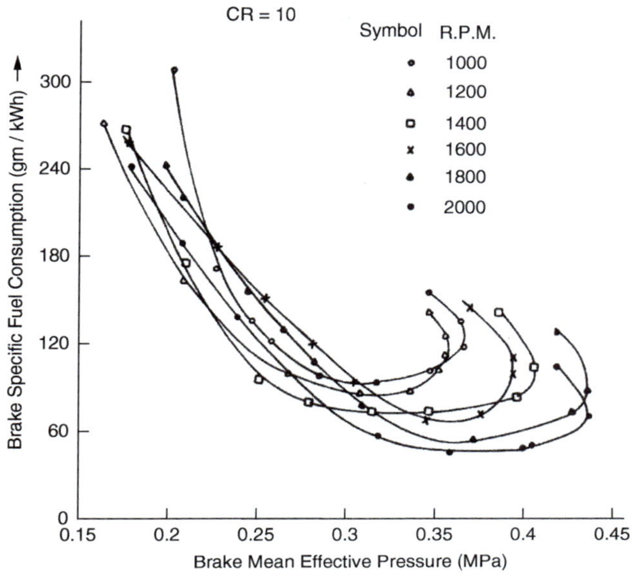

Figure 6.18 Brake mean effective pressure at various compression ratios.

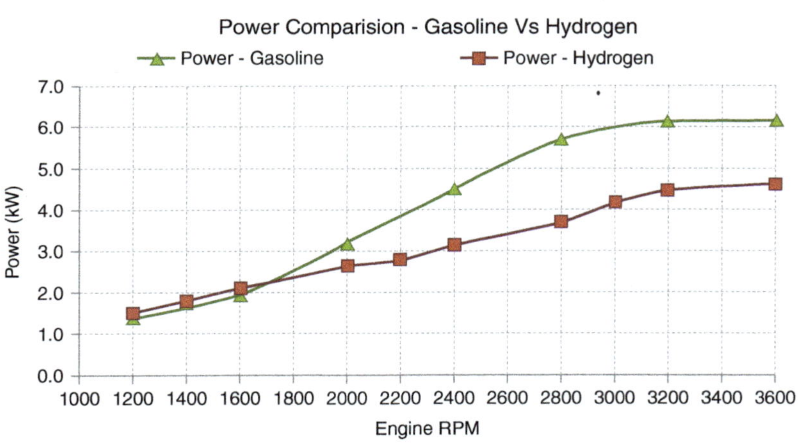

Figure 6.19 Power developed in an engine with hydrogen and gasoline.

hydrogen is lesser than that with gasoline operation. Low density of hydrogen reduces the power density significantly. The power output of a hydrogen engine can be improved by using fuel injection techniques or by adopting supercharging in combination with a variable equivalence ratio strategy. Although there is not supposed to be any hydrocarbon present during hydrogen combustion, it was observed in some tests that a very negligible amount of HC was present in the exhaust (Figure 6.20). This is suspected to be due to the lubricating oil. However, NOx emissions were negligible in hydrogen engines compared to gasoline engines (Figure 6.21).

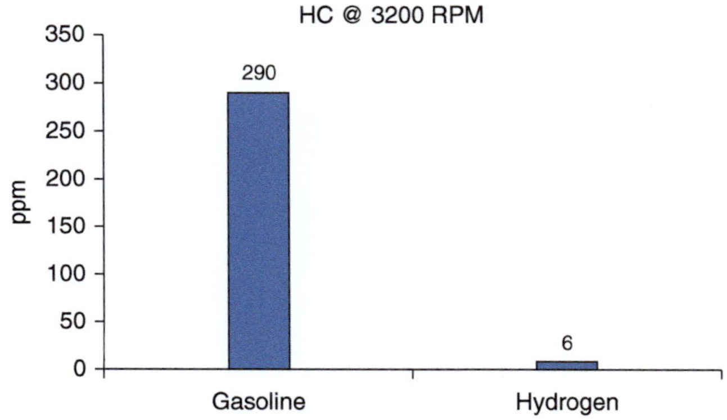

Figure 6.20 HC emission from an engine with hydrogen and gasoline operation.

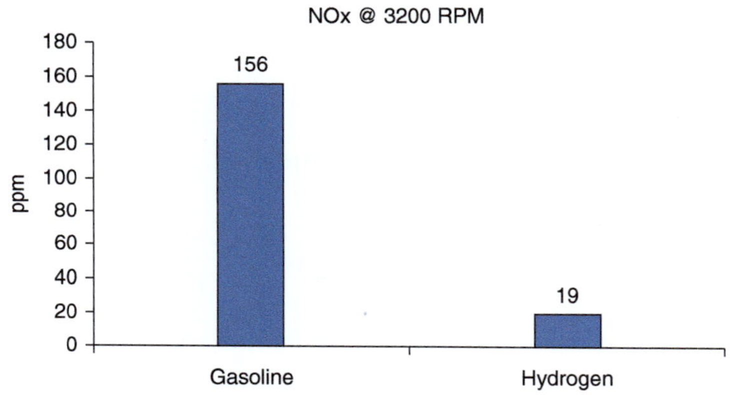

Figure 6.21 NOx emission from hydrogen engines.

6.13 Exhaust Emission Characteristics of Hydrogen Engine and NOx Control

Although there is not supposed to be any hydrocarbon present during hydrogen combustion, a negligible amount of HC sometimes comes out in the exhaust, which is suspected to be due to the lubricating oil. Thermal NOx formation is a strongly endothermic reaction with a significantly high activation energy. The formation is highly dependent on the temperature, with an increase in the temperature resulting in a significant increase in its formation. Several techniques, such as lean operation, installation of catalytic converters, exhaust recirculation, use of selective catalytic reduction (SCR) and advanced engine control techniques, have proven to be effective in bringing down the NOx level. The kinetics of NOx formation depend upon temperature, mixture strength and residence time. Lean operation of a hydrogen engine accompanied by slower chemical reactions weakens the kinetics of NOx formation [26, 41]. Figure 6.22 shows that NOx emission data generated in a TMI-operated hydrogen engine. It is evident that the engine is capable of operating at ultra-lean mode and there is hardly any NOx in the lower level of equivalence ratio. Thus, it is possible to achieve a 'close to a zero-emission' configuration of a hydrogen engine as shown in Figure 6.22.

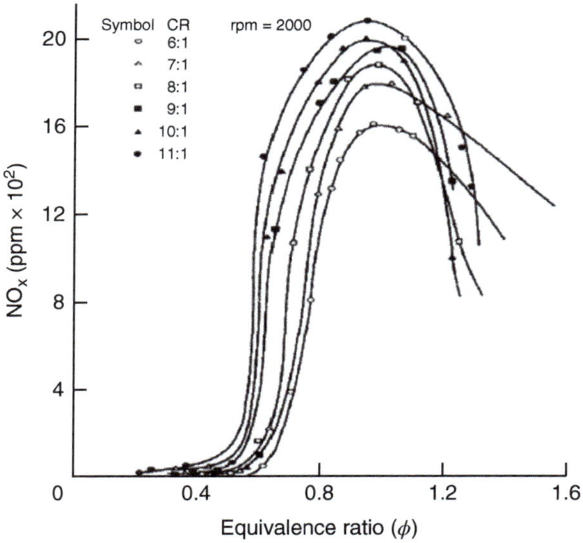

Figure 6.22 NOx emission with respect to equivalence ratio.

6.14 Exhaust Gas Recirculation

Exhaust gas recirculation (EGR) is a mature technology that has been used to reduce NOx emissions in ICEs operating with various fuels, such as diesel, gasoline and others [42–44]. During EGR, a part of the exhaust gas is recirculated back to the engine intake port. The exhaust gas of conventional fuels such as diesel and gasoline contains a significant amount of H_2O and CO_2, which have a high specific heat capacity. Thus, recirculating some of the exhaust gas back into the engine can reduce the peak in-cylinder temperature, thereby decreasing the thermal NOx formation. The same principle could be applicable in an H_2 ICE since its exhaust primarily contains H_2O, which has a high specific heat capacity. Thus, recirculating an H_2O-rich exhaust can be used to reduce the thermal NOx formation.

The effect of implementing EGR in an H_2-fuelled spark-ignited ICE is shown in Figure 6.23. It is observed that an increase in the level of EGR decreases the NOx emissions. The decrease in NOx emissions occurs upon implementing 14% EGR, which is attributed to the recirculation of H_2O-rich exhaust, which has a high specific heat capacity. However, a further increase in EGR does not result in a significant decrease in NOx emissions. It is well known that EGR can lead to increased heat transfer to the engine's coolant, which can affect engine efficiency. As EGR rates become too high, the increased heat loss can counteract the benefits of reduced peak temperatures for NOx reduction.

Additionally, very high levels of EGR can result in a decrease in the combustion quality, which can be inferred by the increase in the coefficient of variation (CoV) of the indicated mean effective pressure (IMEP). Thus, although high levels of EGR can be useful in reducing NOx emissions, the performance aspects need to be considered while selecting the EGR levels. The impact of EGR levels on the torque output is shown in Figure 6.24, wherein it is observed that though implementing EGR decreases the NOx emissions, it also decreases the torque output.

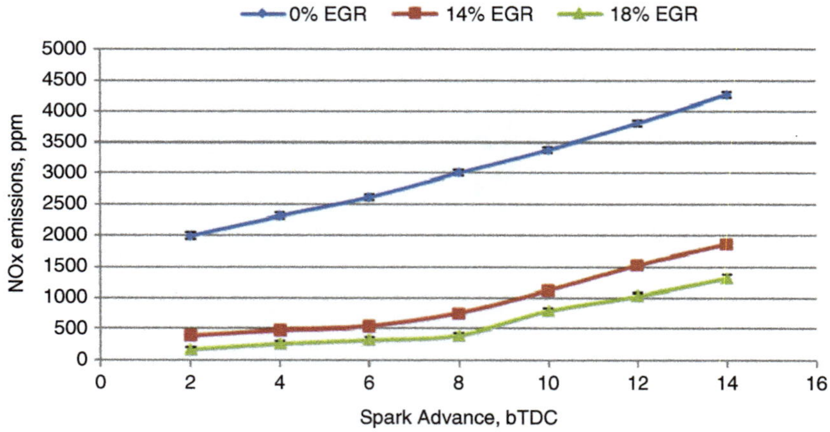

Figure 6.23 Effect of EGR on NOx emissions with varying spark timing (engine speed = 1200 rpm; equivalence ratio = 0.6). *Source:* Adapted from [45].

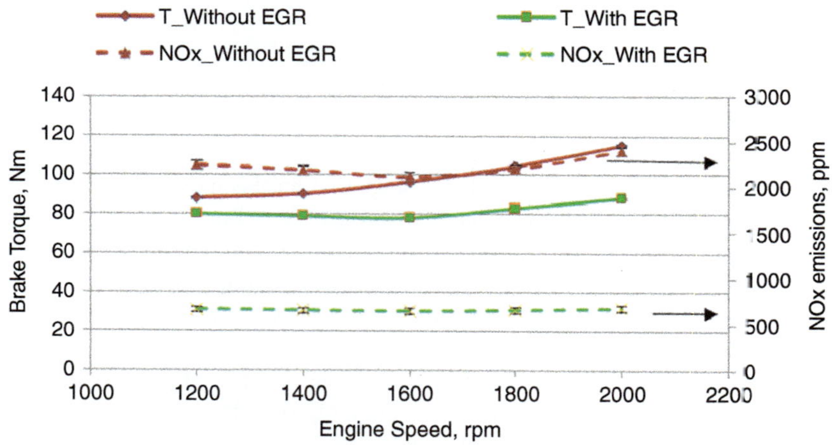

Figure 6.24 Variation of brake torque and NOx emissions with engine speed at 12% EGR (equivalence ratio = 0.6). *Source:* Adapted from [45].

6.15 Spark Timing

Methods such as retarding the spark timing (ST) in spark-ignited engines and retarding the fuel injection timing in CI engines have been used to reduce NOx emissions. The effect of changing the ST in an H_2-fuelled spark-ignited ICE is also shown in Figure 6.23. It is observed that advancing the ST increases the NOx emissions, and the observation is valid for EGR as well as non-EGR operation. Advancing the ST causes the combustion process to initiate earlier in the engine cycle. This leads to combustion occurring at a point when the piston is closer to the top dead centre (TDC), resulting in higher peak combustion temperatures. Higher temperatures promote the formation of thermal NOx, which occurs when nitrogen and oxygen in air combine at elevated temperatures. The other reason may be due to hydrogen's high flame speed, meaning that it burns rapidly. Advancing the ST can cause the hydrogen–air mixture to ignite and propagate faster, potentially leading to a faster and hotter combustion process. This can contribute to increased peak temperatures and, consequently, higher NOx emissions.

Thus, retarding the ST can be used as an effective method to reduce the thermal NOx emissions. These methods and others that work on the principle of reducing the in-cylinder temperatures could be implemented in H_2 engines as well.

6.16 Catalytic Methods

6.16.1 Use of Unburnt H_2

In a gasoline vehicle, the concentration of exhaust components such as CO, unburnt hydrocarbons and NOx is reduced by using a three-way catalytic converter. In addition to the oxidation of CO and hydrocarbons, the reduction of NOx takes place over the catalyst [46, 47]. Species such as H_2, CO and hydrocarbons function as reductants to reduce the concentration of NOx. A similar method can work for an H_2 engine as well since unburnt H_2 may be present in the exhaust. Thus, if the exhaust containing NOx and unburnt H_2 is passed over a catalyst, the following reaction (reaction (6.1)) involving H_2 and NOx as reactants can be useful in reducing the concentration of NOx. Moreover, it can reduce the safety issues involved in the leakage of H_2 into the atmosphere:

$$H_2 + NO/NO_2 \rightarrow H_2O + N_2 \tag{6.1}$$

The effect of placing an NOx reduction catalyst downstream of the engine is shown in Figure 6.25. The NOx emissions decrease upon implementing EGR. If an NOx reduction

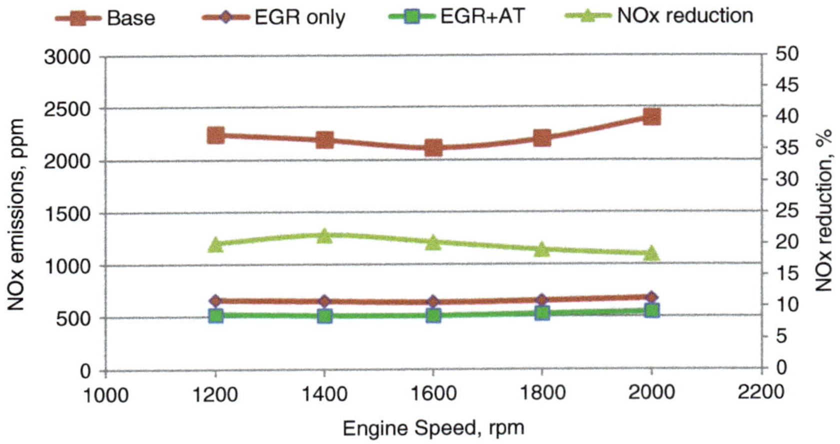

Figure 6.25 Variation of NOx emissions with engine speed under the influence of EGR and catalytic reduction without external reductant supply (equivalence ratio = 0.6). *Source:* Adapted from [45].

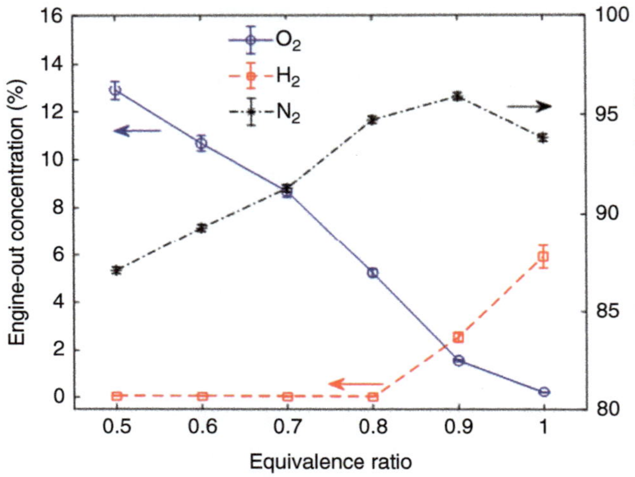

Figure 6.26 Variation of exhaust gas composition (unburnt H_2, O_2 and N_2) with equivalence ratio [48]/with permission of Elsevier.

catalyst is used along with the implementation of EGR, the NOx concentrations decrease further. The NOx reduction (in %) obtained in the catalytic converter is also shown in Figure 6.25. The decrease in NOx concentration can be explained by the reduction of NOx by unburnt H_2. This also serves as indirect proof of the presence of unburnt H_2 in the exhaust. This is because in the absence of H_2 or any other reductant, the only possible method for NOx reduction is its dissociation to N_2 and O_2. However, commercial catalysts are not active for NOx decomposition. Thus, the reduction in NOx concentration by using a catalyst shows the presence of an NOx reductant in the exhaust. The presence of H_2 in the exhaust has also been shown by measurement of the exhaust composition using gas chromatography (Figure 6.26).

6.16.2 Dosing of External H_2

The use of urea-SCR technology is widespread for the reduction of NOx emissions. It involves dosing the required quantity of urea solution in the exhaust pipe [49]. The hydrolysis of urea results in the generation of NH_3, which functions as the reductant for NOx. Since it is shown that unburnt H_2 can be used for NOx reduction (Figure 6.25), the dosing of additional H_2 as a reductant can increase the NOx conversion. In contrast to the dosing of urea solution for diesel vehicles that require a separate urea tank and refilling, the H_2 for NOx reduction in an H_2-fuelled vehicle can be dosed using the already existing fuel tank. The effect of dosing H_2 as an external reductant for NOx is shown in Figure 6.27. Due to the high amount of H_2 available in this case, the NOx reduction in the catalytic converter (plotted on the secondary y-axis) is higher than the case when H_2 is not injected externally (Figure 6.25).

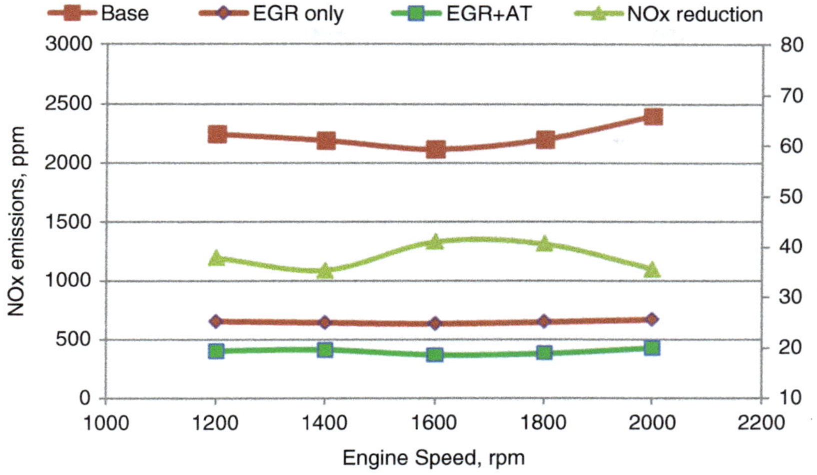

Figure 6.27 Variation in NOx emissions using EGR and aftertreatment (AT) catalyst with an external supply of hydrogen (equivalence ratio = 0.6). *Source:* Adapted from [45].

6.17 Operation at a High Equivalence Ratio

The NOx formation within an H_2-ICE goes through a maximum with an equivalence ratio, and the maximum is reported to be for an equivalence ratio in the range of 0.75–0.85 [50, 51]. Thus, operating at low equivalence ratios (0.5–0.6) or high equivalence ratios (>~0.8–0.9) can decrease the formation of NOx. At very low equivalence ratios, although the NOx formation is low, the power output is also low and hence may not be desirable from an operational perspective. Operation at high equivalence ratios can be used to achieve the required torque output levels, while also resulting in NOx concentrations lower than the peak values achieved at equivalence ratios in the range of 0.7–0.8. Moreover, a high degree of NOx conversion can be achieved at close to stoichiometric conditions in a catalyst. This is because NOx reduction over a typical commercial catalyst is favoured under stoichiometric conditions and richer than stoichiometric conditions but is not favoured under lean conditions. This can be seen from the low NOx conversions observed at an equivalence ratio of 0.6 despite the dosage of external H_2 as a reductant (Figure 6.27, secondary *y*-axis). The low NOx conversion under lean conditions is due to the competition between O_2 and NO for H_2. Under lean conditions, H_2 could react with the excess O_2 at the expense of NO (reaction (6.2)). Moreover, the reduced form of the catalyst is active for NOx reduction whereas the active metal may be in an oxidised form under lean conditions. In summary, NOx reduction using a catalyst is not favoured under lean conditions, and high equivalence ratios can result in the required reduction in NOx emissions. However, operation at high equivalence ratios can result in instability of the engine functioning and issues such as engine misfire, which can be prevented by using substantial levels of EGR:

$$H_2 + \tfrac{1}{2}O_2 \rightarrow H_2O \tag{6.2}$$

The effect of the equivalence ratio on the NOx emissions is shown in Figure 6.28. The maximum engine-out NOx emissions with equivalence ratio and the low amount of NOx formation at high equivalence ratios are observed. In the presence of externally dosed H_2, the tailpipe NOx emissions are also plotted. The catalytic converter is effective in reducing NOx emissions by using unburnt and externally dosed H_2, and results in the highest NOx conversion at the highest equivalence ratio of 1 (Figure 6.29).

Overall, the low amount of NOx formation at the equivalence ratio of 1 combined with the highest NOx conversion efficiency under these conditions results in very low tailpipe

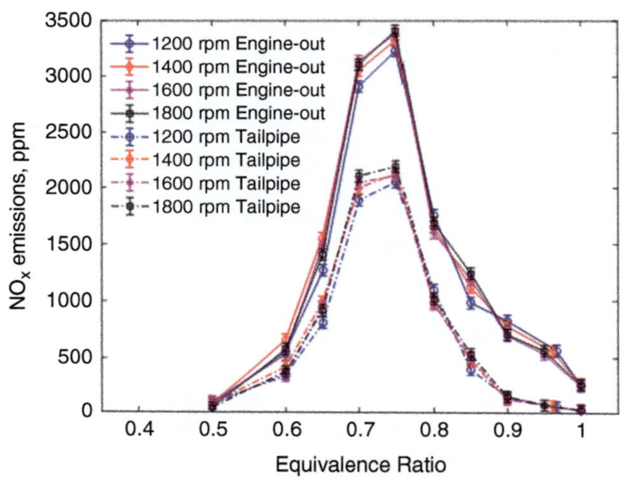

Figure 6.28 Variation of engine-out and tailpipe NOx emissions with equivalence ratio [48]/with permission of Elsevier.

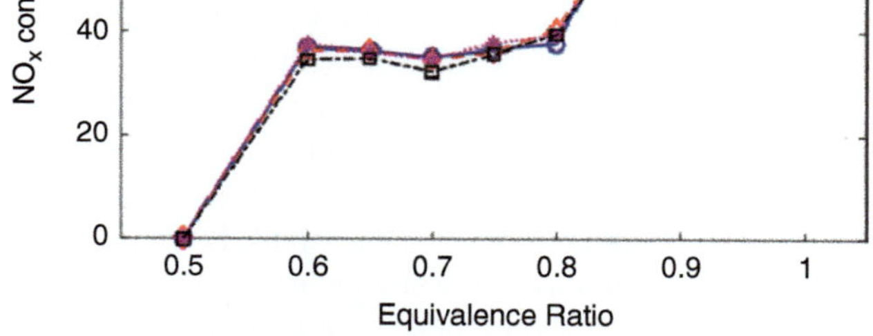

Figure 6.29 Variation in NOx conversion efficiency of the aftertreatment device with equivalence ratio [48]/with permission of Elsevier.

NOx emissions. Additionally, operation at high equivalence ratios results in a high torque output. Thus, to obtain a high torque output and low NOx emissions, it is suggested to operate under near-stoichiometric conditions with high EGR levels, combined with the usage of a downstream NOx reduction catalyst with external dosing of H_2. The comparison of the torque output and NOx emissions obtained under the optimised stoichiometric conditions and those obtained under the lean conditions is shown in Figures 6.30 and 6.31, respectively. The clear advantage of obtaining a high torque output and low NOx emissions under the suggested conditions is reported.

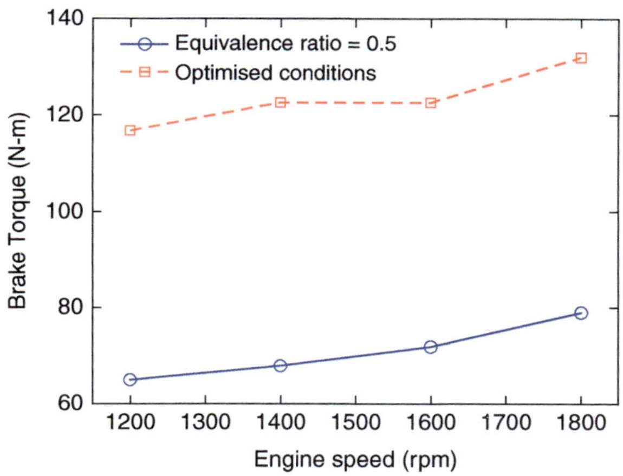

Figure 6.30 Comparison of brake torque generated under optimised and lean conditions at optimised spark timing [48]/with permission of Elsevier.

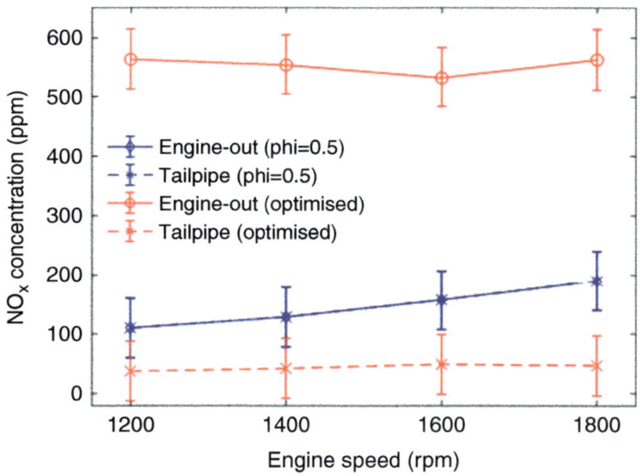

Figure 6.31 Comparison of NOx emissions under lean and optimised conditions [48]/with permission of Elsevier.

6.18 Development of Hydrogen Engine (Both SI and CI Engine) Gensets

Hydrogen-fuelled engine generators do not need any substantial modification in existing hardware of the engine. The NOx emission level can be reduced by controlling the operating conditions. Both SI and CI engine gensets were developed at the IIT (Delhi) [14]. As shown in Figures 6.32 and 6.33, two hydrogen-operated engine gensets (SI and CI engines), developed in the laboratory, are mounted on trolleys. Thus, they could be shifted to any location where there is a need for electric power. These systems can also be used for agricultural applications, such as for irrigation, or can be used to operate a pump set, a lawn mower, etc., for small-scale domestic applications. It is important to note that the SI engine genset is fully operated on hydrogen, whereas the diesel engine genset has to be operated

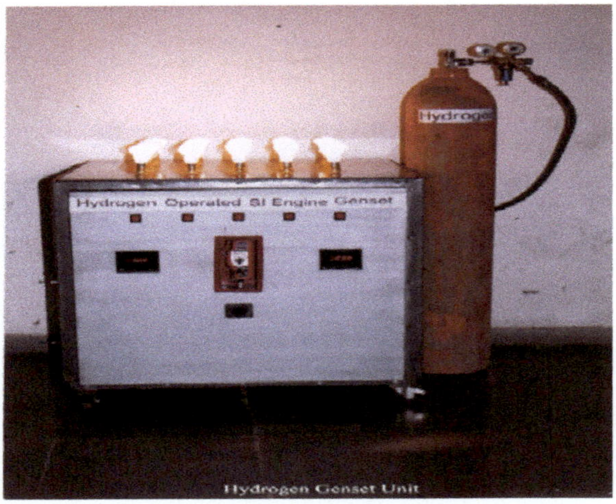

Figure 6.32 Hydrogen-operated spark ignition engine genset.

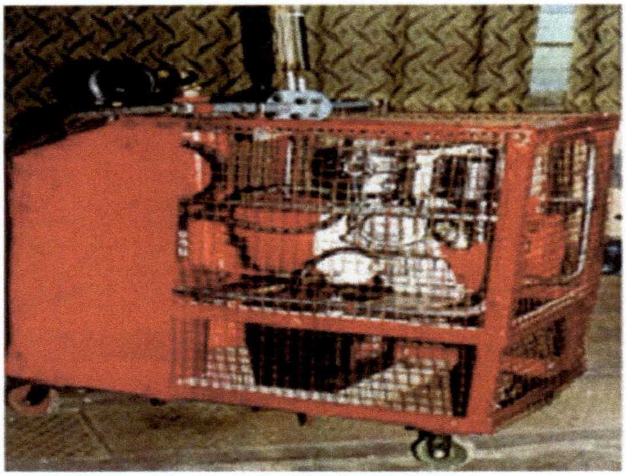

Figure 6.33 Hydrogen-fuelled CI engine genset.

on dual fuelling. It has been discussed earlier that diesel engines cannot achieve CI by hydrogen operation alone mainly because of the high auto-ignition temperature of hydrogen.

6.19 Combustion in Hydrogen-fuelled SI Engines

A lot of literature is available on hydrogen combustion in several systems primarily because the combustion properties of hydrogen are very different, and thus, the application of hydrogen has to be judiciously used depending upon the particular system. It has already been discussed that hydrogen combustion is possible over a wide range, but stable combustion conditions need to be determined depending upon the particular system. Steady and unsteady combustion depends upon the application range of the system. Steady combustion of hydrogen manifests itself distinctly in systems ranging from simple-flame burners (for heating) to catalytic heaters, steam and gas turbines, and rocket engines or jet engines. Unsteady combustion can be predominantly applied to ICEs.

It has been widely reported in the literature that the hydrogen–oxygen reaction mechanism has a very pronounced influence on hydrogen combustion [52, 53] in several systems, including rockets and some other space applications. As far as engine studies are concerned, it has been possible to obtain conditions of smooth combustion provided that the operating conditions are judiciously chosen and implemented such that backfire and combustion noise are eliminated. A typical pressure–crank-angle diagram generated in the TMI-operated hydrogen engine is shown in Figure 6.34. As shown in this figure, the combustion has been smooth without any symptoms of abnormal combustion.

Several diagnostic techniques related to laser ignition for engine combustion have been reported, which provide extremely significant information about the typical combustion behaviour and pattern in hydrogen engines.

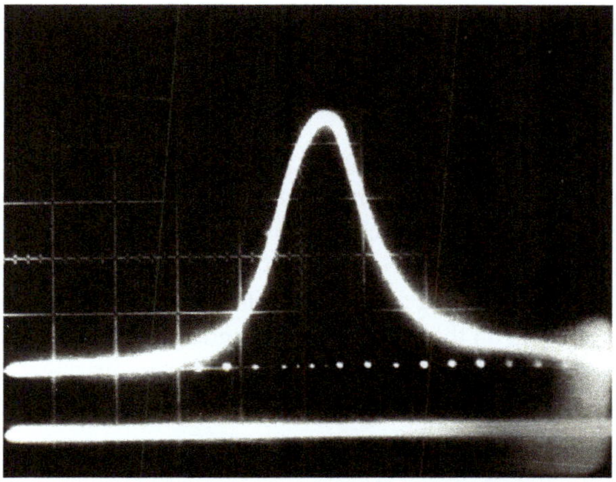

Figure 6.34 Pressure–crank-angle diagram of a hydrogen engine obtained from an oscilloscope.

6.20 Significant Contribution of Laser Ignition to Engine Combustion

Developing the state-of-the-art ignition systems is necessary to improve the performance of stationary, natural gas-fuelled and lean-burn engines [54]. There is a need to increase the engine power density and lean operation to improve efficiency and reduce exhaust emissions. The use of hydrogen in ICEs can potentially address the CO_2 emission reduction at the tailpipe [55]. The current state of hydrogen technology in fuel cells (FCs) is expensive. It requires pure hydrogen and a high-capacity compressor to supply compressed air or batteries for electricity storage. Non-purified hydrogen produced at low cost with advanced combustion and control strategies can efficiently operate engines such as FC powertrains [3]. Hydrogen fuel ICEs generate a specific amount of NOx (high combustion temperature) and traces of hydrocarbons (burning of lubricating oil). The exhaust emissions can be controlled using a suitable aftertreatment system (SCR catalysts and particulate filters), lean mixtures and specific lubricating oil [56].

For better and continuous operation of the spark plug, a lean combustible mixture at high initial pressures needs to be rationally handled. The supply of high voltage by conventional spark plug technology reduces the spark plug lifetime. Using a focused plasma beam by a pulsed laser ignition system provides ease with the combustion of a lean fuel–air mixture, reducing NOx emissions [57].

6.20.1 Laser Ignition

Laser ignition can be divided into spark creation (by plasma formation and shock wave generation) and ignition based on energy balance (develops flame kernel and propagation) inside the combustible mixture [57]. Laser ignition provides advantages such as flexible positioning of plasma, precise regulation of ignition timing and energy, multipoint lean mixture ignition and avoidance of NOx formation [58]. Moreover, the absence of quenching effects and precise ignition timing are some of the distinctive features achievable in laser ignition. The laser ignition system has a significantly higher lifetime (10 000 h) due to no direct physical contact with high temperature–pressure conditions and hence the absence of erosion. The possible multipoint ignition (two or more plasmas) can shorten the total combustion duration with laser systems [59].

In laser ignition, a powerful laser beam (irradiance 1012 W cm^{-2}) interacts with a combustible mixture and generates a plasma of high temperature (106 K) and pressure (100 MPa) [59]. The plasma is generated by an electron cascade, i.e. absorption of photons from a laser beam, ionisation of other gas molecules on impact and breakdown of gas in the focal region [60]. The impurities (e.g. dust, aerosol or soot particles) in the gas mixture provide the initial seed electrons (depending on the effective wavelength) by absorbing the laser radiation and starting the avalanche process [59]. Figure 6.35 shows an overview of the processes in laser-induced ignition (nanosecond) for a combustion duration of hundreds of milliseconds.

The deposition of laser energy (a few nanoseconds) leads to a shock wave generation, ignition delay (5–100 ms) and combustion duration (100–2000 ms). It depends on the gas mixture ratio, initial pressure, temperature, energy, size and plasma position in the combustion

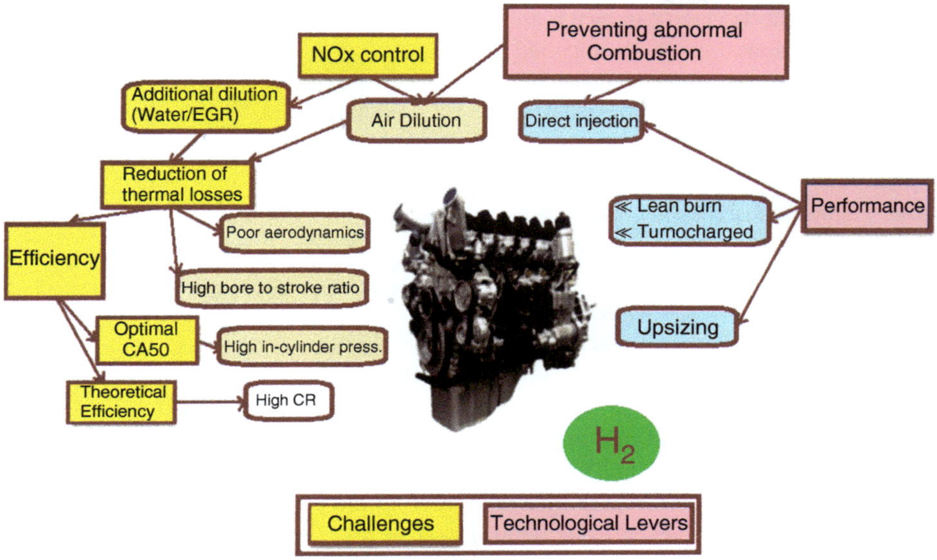

Figure 6.35 Technological assembly concerning key challenges [56], Onorati et al. (2022)/Sage Publications, Inc.

chamber [59]. The minimum pulse energy for ignition (MPE) increases with lean mixtures, and laser energy higher than MPE shortens the ignition delay, reducing the total combustion duration [58]. Weinrotte et al. carried out a comparative evaluation of laser-induced ignition and spark plug ignition. Tests were conducted under engine-like conditions at a peak pressure of 25 MPa. To demonstrate the better ignition features of laser ignition over conventional spark plug ignition, statistically relevant results were generated in a series of tests with over 1800 ignitions [61]. Figure 6.36 shows the timescales of processes involved in laser-induced ignition [61]. Avinash Kumar Agarwal, Akhilendra Pratap Singh and Anuj Pal [62]

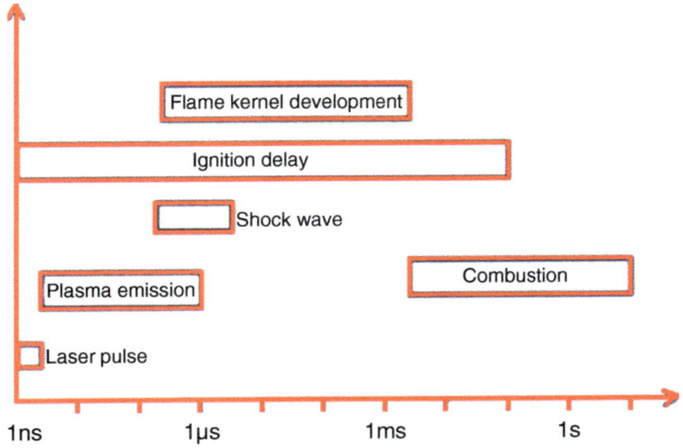

Figure 6.36 Timescales of processes involved in laser-induced ignition (box length indicates durations of processes). Source: Adapted from [61].

investigated the effect of laser parameters on particulate emissions from a laser-ignited hydrogen engine. The study used several significant parameters, such as laser pulse energy (E), ST and different CRs, to evaluate the particulate emissions from the hydrogen engine. Figure 6.37 shows the laser spark plug used, and Figure 6.38 shows the schematic diagram of the experimental setup. This study shows that the particulate number concentration increased with increasing load. Adopting a higher compression ratio and advancing the ST also led to a higher particulate number concentration [62].

Figure 6.37 Laser spark plug [62], Agarwal et al. (2017)/with permission from Elsevier.

The laser ignition of a lean mixture increases the flame velocity in the lean gas mixture compared to conventionally ignited [63] mixture. The early flame kernel growth of the hydrogen–air mixture dominates the gas motion produced by laser ignition [58]. The present challenges with a practical laser ignition system are the high cost of a laser, optical hardware and the transport of an aligned laser beam into the CC [64].

6.20.2 Hydrogen-fuelled Laser-ignited Engine

Hydrogen-fuelled engines require a small ignition energy to initiate combustion. At the same time, they suffer from the problems of backfire and hotspot (overheated valve or a

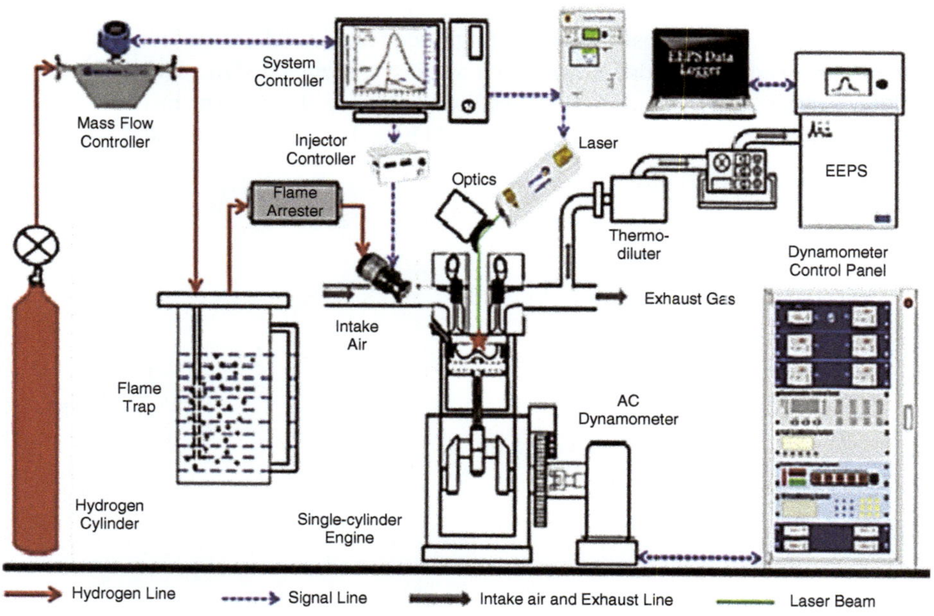

Figure 6.38 Schematic of a hydrogen-fuelled laser-ignited single-cylinder engine [62]/with permission of Elsevier.

spark plug, glowing CC deposits) ignition [56]. The use of laser ignition improves the pressure rise and engine combustion efficiency and achieves minimum NOx emissions. Hydrogen has a high calorific value (w/w) and auto-ignition temperature and generates more mass energy than conventional fuels. Hydrogen has low density, knocking tendency, ignition energy, high diffusivity, flame speed and wide flammability [62].

Dale et al. [65] demonstrated the first laser ignition-fired ICE in 1978. The high surface area to volume ratio of the spark plug electrode leads to a heated surface and becomes the site of pre-ignition. Thus, an electrode-less laser ignition system is suitable for hydrogen–air mixture combustion initiation [56]. The peak chamber pressure decreases with the leaning of the air–fuel ratio, while multipoint laser ignition shows fast combustion [64]. The fuel properties of hydrogen bring in important feasible aspects for engine operation. The wide flammability range enables hydrogen engines to operate efficiently in highly lean mixtures, achieving low combustion temperatures, NOx formation, improved fuel economy and high thermal efficiency [56].

Additionally, a stoichiometric hydrogen–air mixture requires low ignition energy and shows a higher flame speed and diffusivity than other fuels. The ignition of lean fuel–air mixtures with high intake air density using lasers can meet the demand for increased thermal efficiency and stringent emission norms. The use of laser ignition shows good feasibility due to the low lean limit of ignition with high ignition pressure and results in increased efficiency and low NOx emissions compared to spark plug ignition [64]. A short pulsating laser beam is focused on the focal point using a converging lens and generates high-intensity plasma at the focal point. For laser ignition of an ICE, short laser pulses of a few nanoseconds are required (Q-switched lasers).

Hydrogen port fuel injection (PFI) has drawbacks of pre-ignition, knocking, backfiring, low volumetric efficiency and compression loss. It limits the engine's operating power and BTE [3]. Using a turbocharging system or liquid hydrogen injection inside the CC improves the power density, engine performance and ignition timing control [66]. Fuel injection, ignition, and mixture formation are critical factors in preventing abnormal combustion. In 2009, BMW demonstrated 42% efficiency with a hydrogen-fuelled DISI engine operating at 300 bar injection pressure [67].

Similarly, the US Department of Energy showed a DISI engine operating at 2000 rpm and 13.5 bar BMEP with 45% thermal efficiency [68]. Mixing-controlled combustion in a hydrogen-fuelled CI engine requires 600 bar injection pressure, and the commercial availability of a fuel injection system is the critical limiting factor in practical application. This suggests redesigning combustors due to low knocking, high reactivity and diffusivity, i.e. significant flame speed, wide flammability limits, high auto-ignition and instabilities, with hydrogen combustion. Pre-ignition and backfiring result in abnormal combustion, and prevention of these phenomena is the first challenge. It affects the engine's efficiency and durability, while direct hydrogen injection, supercharging and advanced combustion techniques provide solutions to such problems [56].

A naturally aspirated diesel engine (CR12) is modified into port-injected hydrogen and a laser-ignited prototype engine. The flame arrestor ('flame trap'/water trap) absorbs heat from the flames (backfire event) backwards to the hydrogen cylinder. The laser ignition system consists of a laser spark plug fitted on the engine cylinder (converges laser beam into CC) and a laser stand for holding the laser source (protect from engine vibrations).

The laser spark plug comprises a lens holder and a window holder (a sapphire window passes the laser beam). The lens holder houses a converging lens and focuses the laser beam inside the CC. The laser stand holds a beam reflector and a housing comprising diverging and collimating lenses. The laser beam reflects through the beam reflector such that the laser beam is aligned with the axis of the laser spark plug and then converges at a focal point inside the CC to form a plasma [64].

The combustion analysis and exhaust emissions (CO, CO_2, NO and THC) of hydrogen-fuelled laser ignition engines are explained below.

6.20.2.1 Combustion and Performance Characteristics

Figure 6.39 shows the variation in in-cylinder pressure (P) and heat release rate (HRR), with crank angle position for varying BMEP and laser pulse energy (E). For analysis of combustion parameters, an average data set of 200 consecutive engine cycles at every test condition is taken. The test results show an increase in the maximum pressure P_{max} and HRR with increasing BMEP for both laser pulse energies, while at low BMEP, both pulse energies show the same P_{max}. At a high laser pulse energy, the HRR is relatively high. Figure 6.40 shows the variation in BTE and λ with BMEP for laser ignition of a hydrogen–air mixture at different laser pulse energies. As the BMEP increases, the hydrogen supplied to the CC in each engine cycle increases. The P_{max} and HRR increase, and combustion starts earlier, close to TDC, producing the desired power. A laser beam propagates into the interiors of plasma and absorbs energy up to a certain threshold. At high laser pulse energy, combustion starts earlier due to the increased energy density of the plasma. Therefore, at a constant BMEP, a high laser pulse energy increases P_{max} and HRR. For high laser pulse energy, the combustion duration decreases with increasing BMEP and remains stable (fuel continued to burn).

For different laser pulse energies, engine performance is evaluated by the BTE and air–fuel ratio (λ). For high laser energy, the start of hydrogen combustion advances, i.e. near TDC, and a large fuel–air mixture density results in superior combustion and improved combustion efficiency. The excess air availability in the engine CC is determined by λ, and its value decreases at high loads. The improved BTE results in a slightly high λ requirement to generate the same power at high laser energy. A lower hydrogen quantity is required per unit power output for increased BTE.

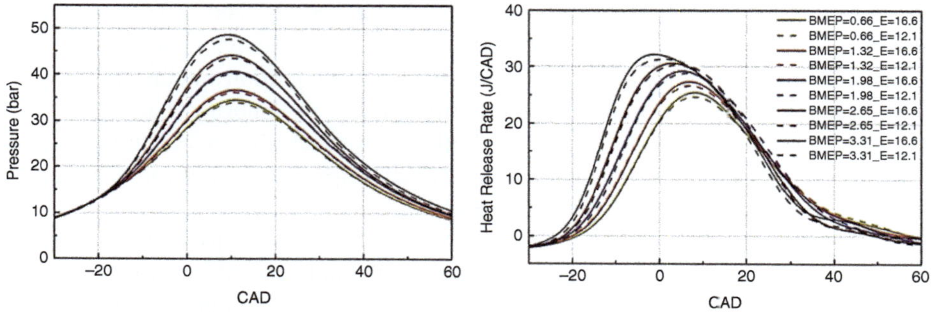

Figure 6.39 Variations in P and HRR with crank angle position for laser ignition of hydrogen–air mixtures with varying laser pulse energies.

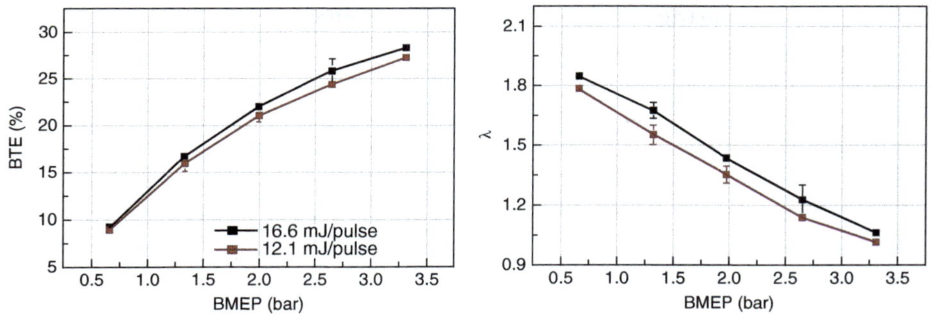

Figure 6.40 Variations in BTE and λ with BMEP for laser ignition of hydrogen–air mixtures with varying laser pulse energies.

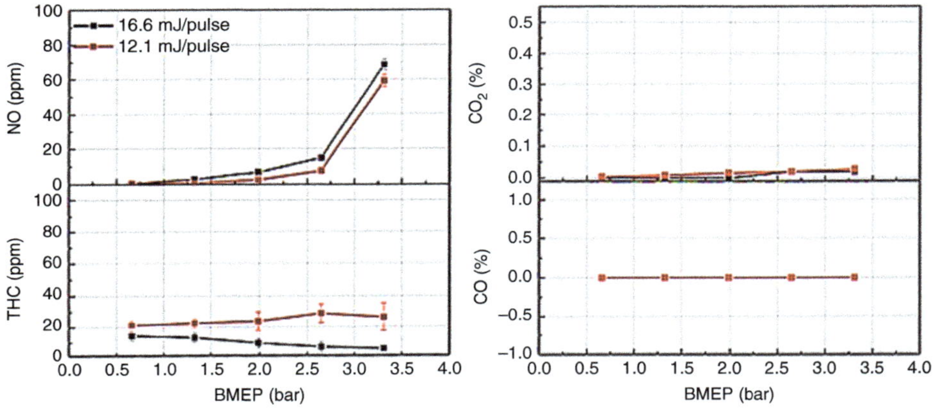

Figure 6.41 Variations in (a) mass emissions (g kwh^{-1}) and (b) raw emissions (ppm/%) with load for laser ignition of hydrogen–air mixtures with varying laser pulse energies.

6.20.2.2 Emission Characteristics

The exhaust emission (NO, THC, CO and CO_2) analysis of a laser-ignited hydrogen engine with two different energies is presented in Figure 6.41. For high laser energy, the NO emissions are relatively high due to high peak in-cylinder temperatures. For both laser pulse energies, CO emissions are almost zero (absence of fuel-bound carbon), while THC and CO_2 emissions are minor (combustion of splashed lubricating oil). The overall carbon-based emissions are negligible compared to gasoline-fuelled SI engines.

6.21 Hydrogen Use in CI Engines

Since the properties of hydrogen are better for use in SI engine configurations, most of the research activities in several parts of the world have been carried out on SI engines.

It has also been discussed earlier with respect to its auto-ignition temperature (576 °C). It is not suitable to achieve CI of hydrogen engines. However, a CI engine can be made to run on hydrogen if some triggering devices, such as glow plug, are installed to initiate ignition. Ikegami and his team of researchers, during tests on hydrogen-fuelled CI engines, observed that [69, 70] even inadvertent leakage from the injector could initiate ignition in diesel engines. However, several researchers, including Ikegami, have carried out exhaustive tests on CI engines using pilot injection and have reported their experience with hydrogen use in diesel engines [69–76].

The interest in hydrogen-operated diesel engines stems from the fact that many developing countries use diesel for agricultural activities (irrigation, pumping, etc.), public transport systems (buses) and decentralised energy applications (genets). Switching over hydrogen from diesel for such applications is quite promising from an energy–environment point of view. Installation of low plug or creating a hotspot in the combustion chamber does not provide a long-term and reliable solution for hydrogen use in CI engines. Immediate use of hydrogen in diesel engines could be achieved by dual fuelling where both diesel and hydrogen could be used together. Diesel engine operation in a dual fuel mode is achieved by way of hydrogen–air mixture induction during suction and injecting of a pilot diesel charge in the conventional manner. This configuration has the flexibility of switching back to total diesel operation if required. Tests in the laboratory show that any effort to increase the percentage of hydrogen substitution under full load conditions beyond a certain limit and for a longer period often results in a rapid rate of pressure rise and subsequently leads to knocking. To eliminate knocking conditions and ensure stable engine operation with a higher level of hydrogen, the use of diluents is effective [73–77]. It has been reported that knocking could be totally eliminated by the use of some inert gas as diluents, such as helium, argon and nitrogen, along with water vapour (Figure 6.42). The amount of hydrogen was monitored along with the small quantity of diluents to ensure good performance and low emission characteristics of the engine operating under stable conditions without showing a knocking tendency or any other combustion problems. Optimum engine operating conditions without knocking or any other combustion problems have been experimentally evaluated by altering the quantity of diluent, flow of hydrogen, etc. Since smoke and NOx are the major pollutants from the exhaust of the CI engine, these data were recorded.

An optimum configuration of a small horsepower portable hydrogen–diesel portable genset needs about 38% energy to improve the performance and emission features. No erratic abnormal combustion, such as knocking, was encountered during the tests. An overall comparison of the long-term applicability potential of helium, nitrogen and water vapour shows that 2640 ppm of water vapour ensured smooth engine operation with good performance. NOx and smoke emission were drastically reduced with improved emission features. The figures indicate the engine operating conditions with diluents and water vapour such that knocking conditions never occurred. Figure 6.43 shows the performance characteristics of thermal efficiency, exhaust emission of NOx and smoke from the diesel engine. The influence of the three diluents (helium, nitrogen and water vapour) on the optimum performance of the engine is shown in Figure 6.44. The thermal efficiency level achieved separately using the three diluents and neat hydrogen operation is shown in Figure 6.45.

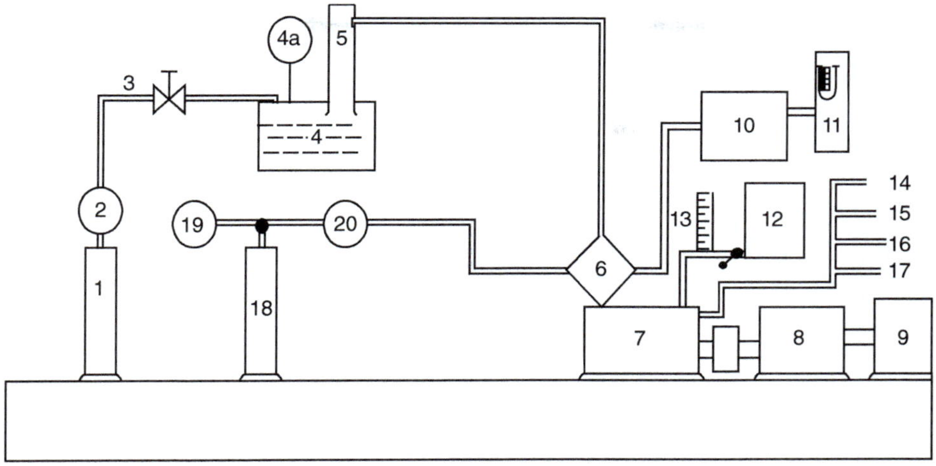

1. Hydrogen cylinder
2. Pressure gauge
3. Non-return valve
4. Flame trap

4a. Pressure gauge

5. Hydrogen flow meter
6. Intake rtrofit

7 and 8. C.I. engine/genset system

9. Electric lamp load system
10. Air surge tank
11. Water manometer
12. Diesel fuel tank
13. Diesel fuel flow meter
14. Exhaust outlet for NO_X measurement
15. Exhaust outlet for smoke measurement
16. Exhaust outlet for temperature monitoring
17. Exhaust outlet for other provisions
18. Diluent cylinder/storage unit
19. Pressure gauge
20. Diluent flow metering device

Figure 6.42 Test rig with diluents for hydrogen–diesel engine operation.

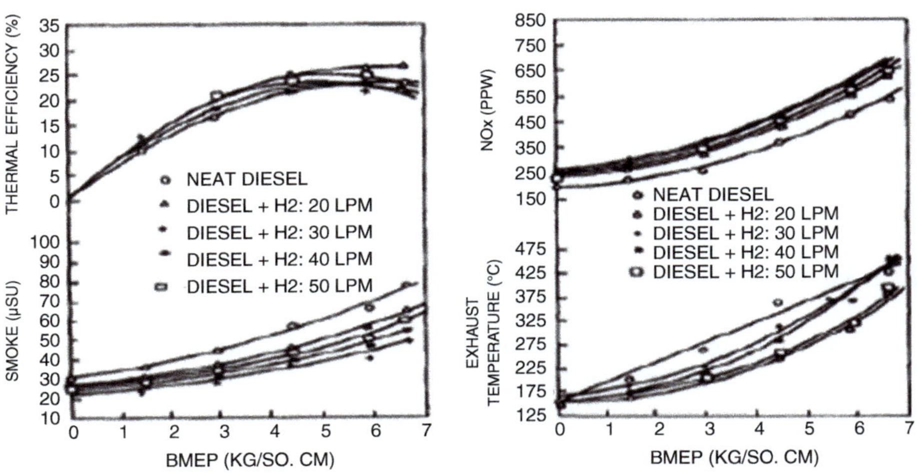

Figure 6.43 Hydrogen-supplemented diesel engine operating at 1500 rpm.

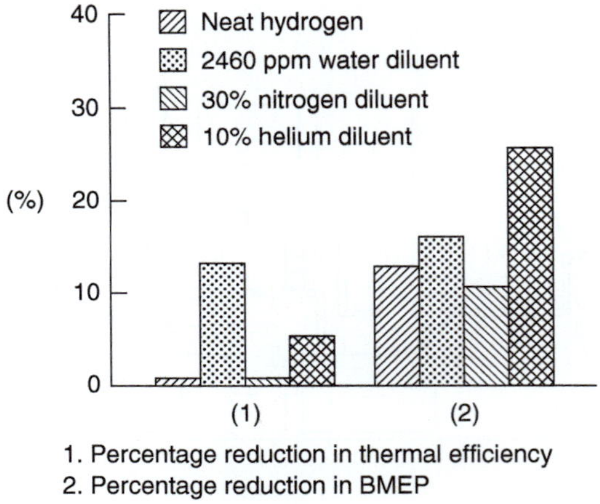

Figure 6.44 Effect of diluents on optimum performance parameters.

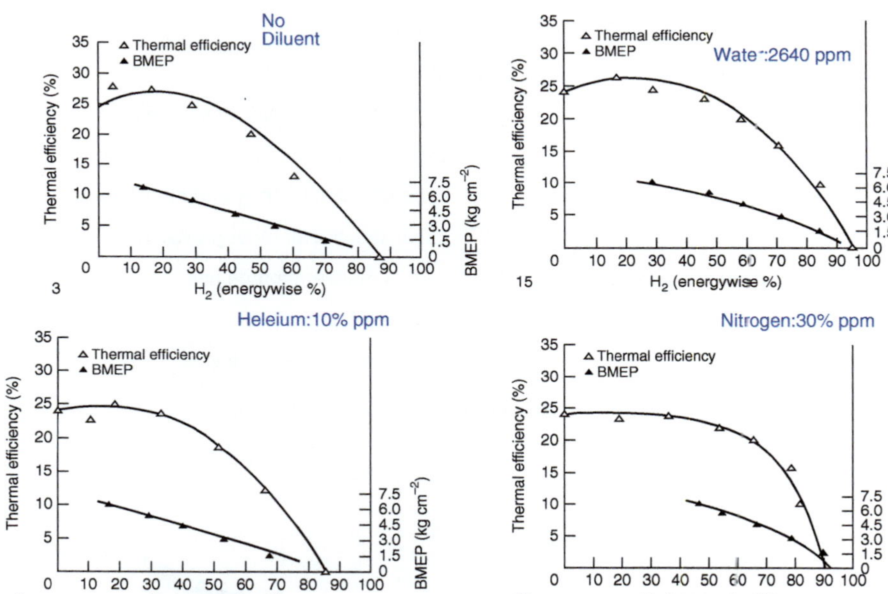

Figure 6.45 Thermal efficiency and power with and without diluents.

6.22 Use of Hydrogen in the Rotary (Wankel Engine)

Apart from conventional reciprocating engines, Wankel engines have also been investigated for operation with hydrogen. Rotary engines have some inherent design features so that the problem of backfire does not occur. In this case, the combustion chamber has a relatively higher surface-to-volume ratio. Therefore, it is colder at the TDC compared to conventional reciprocating ones. After the opening of the inlet valve, the blowby is sucked away from the following cycle and does not have any possibility of entering the engine. Therefore, there is no possibility of backfire. Thus, the engine can run on a much wider range of equivalence ratios without backfire. The researchers of Mazda have reported a power output of about 75%, which is much higher than that obtained in reciprocating engines [78]. Rotary engines have a unique design that involves moving parts sealing against the housing, which can lead to increased wear and maintenance requirements. The design can also lead to challenges in sealing and oil consumption. The development of rotary engines has been relatively limited. As the world transitions toward hydrogen and the infrastructure for refuelling is expanding, rotary engines have not been a focus of this transition.

6.23 Use of Hydrogen in ICEs with Natural Gas

Considering clean-burning characteristics of hydrogen, the use of hydrogen in combination with several other fuels in ICEs is being investigated in several parts of the world. The most prominent and common among these is natural gas.

Natural gas is used as transportation fuel, and millions of vehicles are being operated worldwide. Robust infrastructure has already been built, sources have been identified, production has increased and fuelling stations have been established. Since the common man has already accepted compressed natural gas (CNG) as a transport fuel, vehicles have been developed and are running on roads by adding a small fraction of hydrogen to the existing CNG-operated vehicles. This transition phase can continue until hydrogen infrastructure is developed and gradually the CNG-fuelled vehicle can be modified for total hydrogen operation. Depending upon the percentage of hydrogen (by volume) added to natural gas, two terms, 'Hythane' and 'hydrogen-blended compressed natural gas' (HCNG), are referred to in the literature. Both Hythane and HCNG aim to harness the benefits of hydrogen's clean combustion properties while leveraging the existing infrastructure and technologies associated with natural gas. The choice between these two fuels may depend on factors such as the availability of hydrogen, the desired emissions reduction, engine compatibility and overall fuel-supply considerations. Discussions related to both Hythane and HCNG are provided later.

Methane, the major combustible constituent of CNG, has a lower flame speed than several other hydrocarbons, whereas the flame speed of hydrogen is much higher. Therefore, the flame speed of the hydrogen and CNG blend (HCNG) will be higher than that of CNG alone. Another important feature of the blend is that the hydrogen to carbon ratio of HCNG is higher than that of CNG alone, and this situation helps in reducing the concentrations of the major carbon-based pollutants, such as hydrocarbons, carbon monoxide and carbon dioxide, in the exhaust. The engine can thus be operated in a lean mode, and the thermal efficiency improves. Since hydrogen has a low minimum ignition energy and a very wide range of flammability limits, the flammability characteristics of the blend are expected to be dominated by the relevant properties of hydrogen. Last but not least, CNG comes predominantly from fossil fuel resources, whereas hydrogen is renewable in character. Hythane ('Hythane': a registered trademark of Hydrogen Consultants Inc., USA) is a blend of hydrogen and CNG, where 20% by volume of hydrogen is blended into compressed natural gas.

Table 6.2 Properties of hydrogen, CNG and Hythane.

Properties	Hydrogen	Hythane	CNG
Flame speed (cm s^{-1})	237	120	42
Diffusion coefficient (cm^2 s^{-1})	0.61	0.25	0.16
Higher calorific value (MJ kg^{-1})	142	92	55
Stoichiometric A/F (ratio)	34.3:1	22.8:1	17.2:1
Flammability limit (vol%)	4–75	5–35	5.3–15
Minimum ignition energy (mJ)	0.02	0.21	0.28
Auto-ignition temperature (K)	858	825	813

Therefore, adding hydrogen would increase the lifespan of fossil-fuel-based natural gas. Table 6.2 shows the values of the relevant properties of Hythane compared to neat hydrogen and neat CNG. It is evident from these data that Hythane has a higher flame speed, wider flammability range and higher calorific value than CNG. Hydrogen and natural gas are miscible in all proportions and hence depend upon the availability, and hydrogen in small proportions can be added to CNG for running ICEs. The available literature in this area has not reported any problem due to embrittlement in vehicular applications. Hydrogen blended with CNG (referred to as 'hydrogen-added natural gas' [HCNG]) is being studied by several researchers using different quantities of hydrogen added to CNG-operated engines and vehicles [78–85]. HCNG-fuelled automobiles using various proportions of hydrogen blended with CNG have been successfully demonstrated in Vancouver (Canada), Washington DC, Pennsylvania, Phoenix, Los Vegas (USA), New Delhi (India) and several other places. There are several vehicles operating on roads with blends of hydrogen and natural gas (HCNG).

Several road tests on Hythane-fuelled vehicles in the Denver Hythane Project (1992), the Montreal Hythane Bus project (1995) and SunLine's second Hythane project have shown good road performance and substantial reduction (45–50%) in NOx emissions. The blending of different proportions of hydrogen to CNG for an engine has been tested in IIT Delhi [78, 82, 85]. Smooth engine operation with good performance and low exhaust emissions has been achieved in the laboratory with 30% by volume of hydrogen added to CNG. It has been observed that increased concentrations of hydrogen to CNG engine increases power output and at the same time decreases the cyclic variability. Adding hydrogen to natural gas allows leaner operations with less spark advance. Adopting the lean-burn technique can decrease burning velocity and increase misfire probability. Lean operation makes it possible for the engine to run in higher ranges of CRs without giving rise to knock. Heat losses would also be lower. Thus, it is very important to identify the critical percentage of hydrogen that needs to be added to CNG to obtain the best results from the system. The addition of hydrogen to the CNG requires increasingly less advance in ST. Most of the research activities carried out with different percentages of hydrogen blended with CNG have reported similar observations. With increasing concentrations of hydrogen to CNG, the levels of HC, CO, and CO_2 in the exhaust show a steady decreasing trend for the same equivalence ratio. However, this rate of decrease does not occur proportionally with the addition of hydrogen to methane. To ensure stoichiometric operation of a naturally aspirated HCNG engine, it is necessary that recalibration of ST should be carried out along with adjustments to the air/fuel ratio. Lynch [86] carried out research on hydrogen engines and subsequently investigated 'Hythane'. Figure 6.46 shows a Hythane-operating system in which the process of blending and ultimate dispensing of Hythane to the vehicle are represented. Tests have shown that existing designs of gasoline engines can be operated on Hythane without any substantial modification in the engine hardware except for the installation of the injector.

Studies related to NOx emissions show the same trend of results as discussed earlier for hydrogen combustion. The lean-burn strategy has been useful in lowering the NOx level, but at the same time, it causes a significant amount of power loss. Compared to CNG operation, the Hythane engine shows a substantial reduction in the major pollutants. NOx emissions decrease by 40–50%, whereas CO emissions decrease by 40–50%. The non-methane hydrocarbon (NMHC) level decreases by about 25–30%, and CO_2 decreases by about 8–10%.

As part of a sponsored project funded by the Government of India, several automobile industries and oil companies in India carried out extensive tests with different percentages of blends of hydrogen added to CNG to fuel their vehicles. Their performance, emission, combustion, endurance and tribological investigations showed that the existing CNG engines can be operated with 18% hydrogen by volume. Keeping this in mind, extensive tests were carried out at IIT Delhi to have a comparative assessment of the effect of CNG and 18% HCNG on a retrofitted gasoline genset engine. Tests were carried out to evaluate the performance, exhaust emissions and combustion characteristics of the engine [81]. Another major objective was to determine the deposits and the condition of lubricants in the combustion chamber after continuous prolonged operation. The results of long-duration tests were carried out for 60 h [85] in the laboratory with experimental test rig shown in Figure 6.47. As evident from the graphs, brake-specific fuel consumption (BSFC)

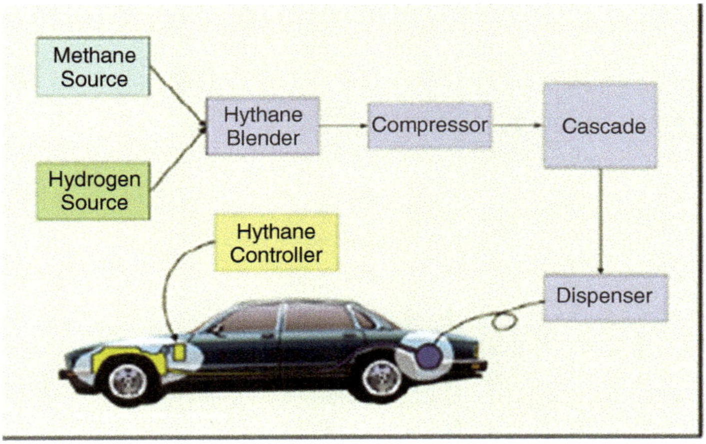

Figure 6.46 Schematic diagram of a Hythane-operating system.

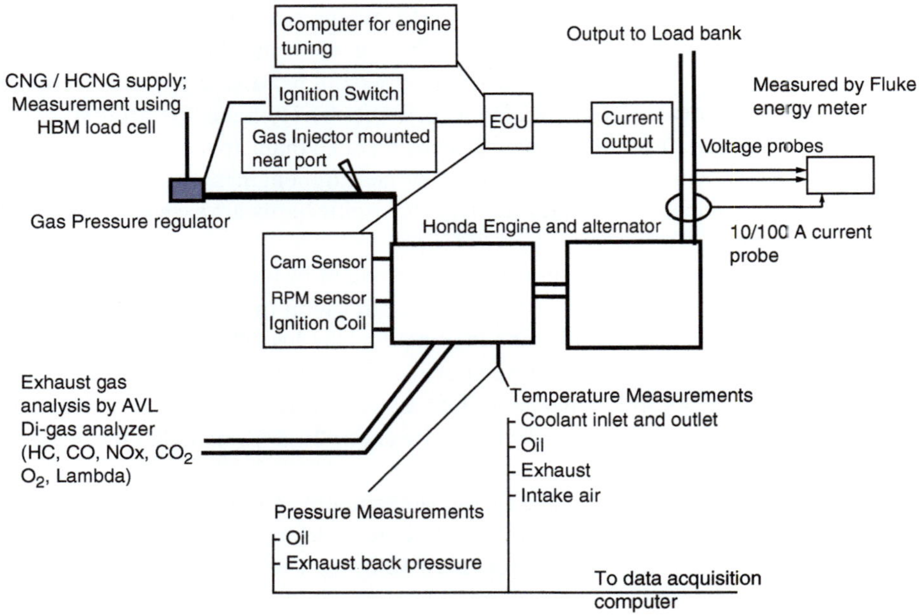

Figure 6.47 HCNG engine test bed.

seemed lower (Figure 6.48) compared to BSFC for CNG operation. Both CO and HC were lesser in concentration, while NOx emission increased with HCNG operation (Figure 6.49).

Spectrochemical analysis (ASTM D 6595) was carried out by using Multi-element Oil Analyser II Plus equipment to estimate the wear of metals Fe, Cu, Al, Cr and Pb. There was a distinct variation in Cu and Fe due to the effect of wear, whereas other metals were not practically affected. This is perhaps due to the decrease in viscosity, which subsequently forms a low oil film thickness between the piston rings and liner [85].

Deposits on the piston and cylinder head showed higher deposits on the HCNG piston because of the degradation of lubricant at higher temperature levels. There was a lacquer

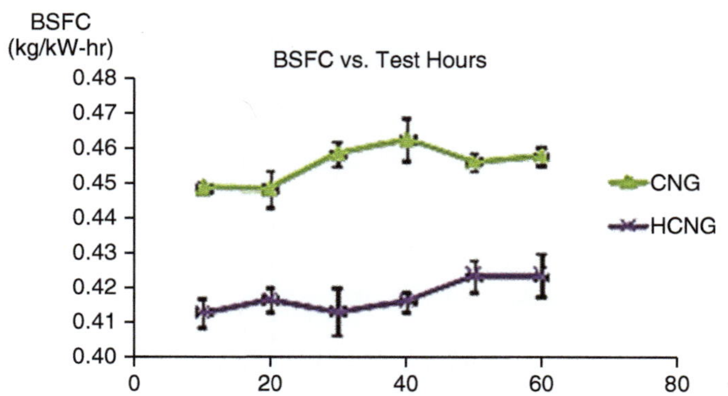

Figure 6.48 Evaluation of the fuel consumption rate.

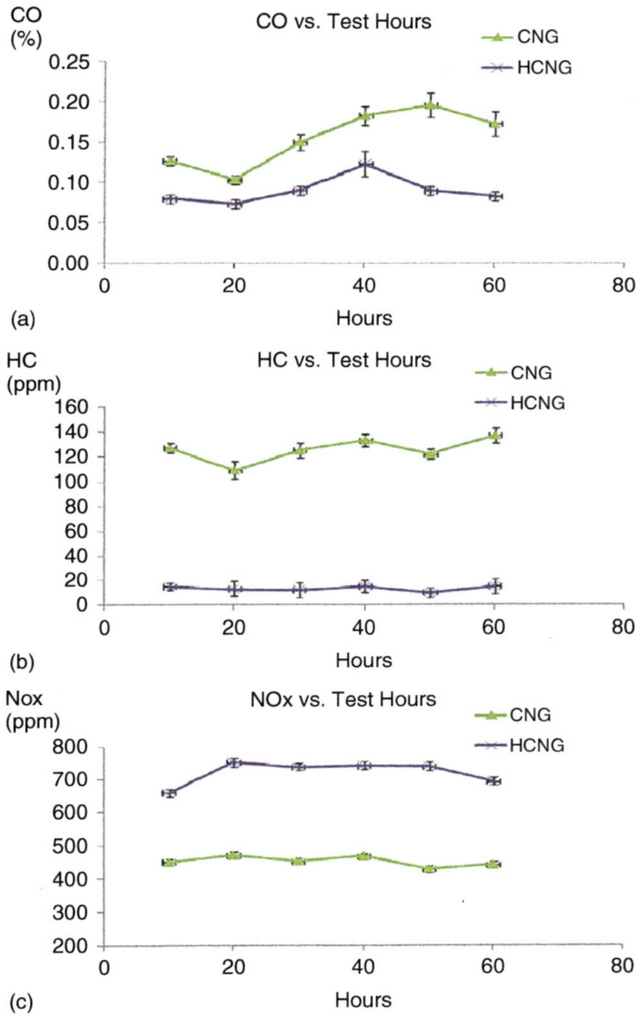

Figure 6.49 HC, CO and NOx emissions from the HCNG-fuelled engine.

on the land of the HCNG piston, which clearly shows that the lubricant oil film had been degraded in the severe thermal environment of high combustion gases, which was also responsible for increasing the thermal stress. Thus, the thermal stability of the lubricant was destroyed.

Deposits on the cylinder bore and spark plugs were also inspected after the tests. The HCNG engine had a greyish-reddish colour marked on its spark plug and cylinder bore. Figures 6.50 and 6.51 show the deposits on the cylinder head and spark plug, respectively, after the prolonged operation of the engine with HCNG. Scanning electron microscope tests showed that the spark plug for CNG operation had only 1% iron content, whereas the spark plug for HCNG operation had exceedingly high levels of iron of 12.9 and 17.1% at two different locations. Such a trend is due to the high temperature that might have caused

Figure 6.50 Cylinder head deposits [85], Reji, 2012/Indian Institute of Technology, Delhi.

Figure 6.51 Deposits on the spark plugs [85], Reji, 2012/Indian Institute of Technology, Delhi.

corrosion and wear [85]. The effect of using CNG and HCNG for engine operation separately is shown in Figure 6.52.

The success of ongoing global research in different countries and wide-scale use of vehicles using hydrogen and CNG blend very strongly advocate in favour of HCNG as a transition fuel until the infrastructure to shift to total hydrogen has been built up. Performance, emission and combustion characteristics have been reported by several researchers. However, long-duration engine tests and associated tribological studies need to be exhaustively studied to develop dedicated lubricants for HCNG-specific engines.

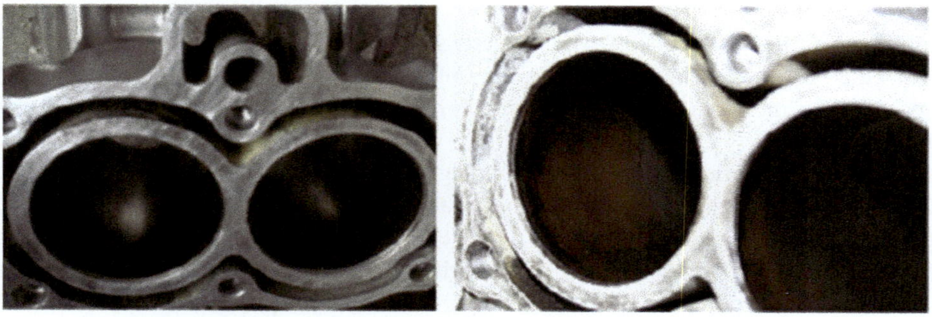

Figure 6.52 Effect on cylinder bore by CNG and HCNG fuelling [85], Reji, 2012/Indian Institute of Technology, Delhi.

6.24 Hydrogen in Combination with Other Fuels for ICEs

A host of other fuels, such as ethanol, LPG, biodiesel, DME, and biogas (whose major component is methane), are being studied in combination with hydrogen for use in ICEs. Both the SI and CI engine configurations are being appropriately modified to operate on the desired blend of the particular type of fuel (either liquid or gaseous) considering the performance, emission and consumption features.

6.24.1 Hydrogen with Ethanol

Since ethanol has already been studied for a long time and engines designed for ethanol operation have already been developed, it is worth trying to ensure engine operation by hydrogen–ethanol blends. Mohammed and the team [87] carried out some tests with a gasoline–ethanol blend and studied the engine performance. Tests were carried out with ethanol mixed with gasoline in different proportions (10% ethanol + 90% gasoline; 20% ethanol + 80% gasoline; 30% ethanol +70% gasoline; 40% ethanol + 60% gasoline) using an ultrasonic bath to ensure perfect mixing. In their studies with 20% ethanol added to 80% gasoline, it was observed that the engine's performance and emission deteriorated. However, the response characteristics of the engine show improved performance and exhaust emission characteristics. The thermal efficiency improves by hydrogen addition to the gasoline–ethanol blend. The concentrations of exhaust emissions of HC, CO and CO_2 are reduced. The combustion conditions are smooth and stable. The range of combustibility is wider than that of gasoline or gasoline–ethanol blends, and thus, the combustion efficiency improves.

The results showed that power, BSFC and thermal efficiency are improved with an increase in the ethanol concentration. On the other hand, ethanol was found to have a negative effect on volumetric efficiency. Furthermore, adding ethanol reduces harmful exhaust gases. It is found that more ethanol is accompanied by less exhaust gases. Finally, the research octane number (RON) and the motor octane number (MON) are found to show significant improvement with ethanol blends. Although the lower heating value was found to be higher for pure gasoline, all other parameters are enhanced by adding ethanol to the engine fuel. Studies on a single-cylinder engine indicated that performance characteristics such as thermal efficiency and BSFC were improved, but the volumetric efficiency decreased. BTE improved from a minimum value of 20.21% (gasoline engine) to a maximum value of 38.88% (engine using blended fuel E40). Tests with the fuel blends show that both RON and MON were higher for ethanol blends. Ethanol improved the emission features, and exhaust gas concentrations were reduced. It was observed that the engine speed had a very significant role, and with increasing engine speed, the BSFC showed a decreasing trend. It was also evident in this study that the BSFC decreases by 17.21% compared to gasoline fuel at an engine speed of 2500 rpm. This could be mainly due to the increased power required to overcome friction power. As far as the exhaust emissions at different ranges of blend are concerned, the maximum drop in CO emissions was found in E30 26.33%, whereas maximum CO_2 emissions (25%) were observed in E40. HC emissions decreased to 25% in E40, whereas NOx emissions decreased by 20.91% in E40.

Parthsarathy et al. [88] reported their results on the effect of hydrogen on the performance and exhaust emission characteristics of an ethanol–biodiesel blend in a diesel engine. In their initial tests, biodiesel named tamanu methyl ester (TME) was used in an unmodified diesel engine. After a series of tests, it was observed that ethanol can be blended with biodiesel up to 30% in the existing design of diesel engines. However, in all these configurations, problems of low thermal efficiency and high smoke emissions were encountered, which could be successfully overcome by introducing hydrogen in a port-injected mode into the engine. The use of hydrogen resulted in higher levels of NOx emissions. By adding 4 lpm of hydrogen to the TME–ethanol blend, the BTE showed a remarkable improvement. There was no substantial difference in the engine operation with either TME80E20 (TME 80% and ethanol 20%) or TME90E10 (TME 90% and ethanol 10%). Studies with different blends of TME and ethanol showed very interesting trends in the results. The specific energy consumption of the engine obtained for hydrogen with TME90E10 was lower than those obtained for neat diesel, TME80E20 and TME70E30. The lowest percent of CO exhaust emission was obtained for engine operation with hydrogen and TME90E10. The HC emission of hydrogen with TME80E20 was close to that of TME70E30 and diesel, while that of TME90E10 was slightly higher. The smoke level of biodiesel blends along with the use of hydrogen and ethanol was always lower than that of neat diesel operation. Engine operations with TME80E20 brought about the lowest NOx level. Thus, these experimental studies have shown different blend conditions among diesel, biodiesel and hydrogen depending upon the situations and requirements.

Some studies were carried out by Al-Baghdadi [89] to evaluate the performance and emission behaviour of a four-stroke SI engine using 2, 4, 6, 8, 10 and 12% by mass of hydrogen in hydrogen–ethanol blends. Specific fuel consumption, power and emission of CO and NOx were experimentally evaluated under different operating ranges of equivalence ratio (up to 1.2) and compression ratio (in the ranges of 7–12). There was a distinct improvement in power, and the fuel consumption was reduced when hydrogen was added to the ethanol–air mixture. Figure 6.53 shows the engine test rig with hydrogen and ethanol supplied to the engine.

A wider range of combustibility of ethanol was obtained, and the combustion efficiency was improved. The addition of hydrogen helps improve the efficiency of ethanol because its high heat of vapourisation decreases the peak temperature inside the engine cylinder.

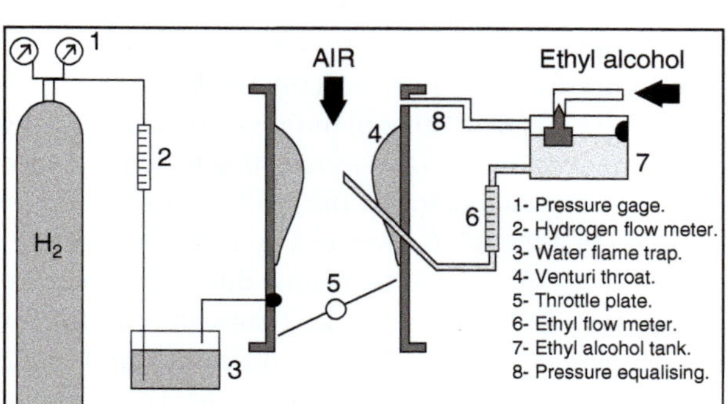

Figure 6.53 Hydrogen and ethanol supplied to the engine.

As a result, NOx emissions show a decreasing trend. Within the operating ranges of the engine, the addition of hydrogen showed an increase in NOx. Interestingly, NOx emission was still lower in situations where ethanol fuel was at a CR of 12 and hydrogen addition was at 0–3.5% by mass compared to gasoline operation at a compression ratio of 7. In the entire range of various operations, with all percentages of hydrogen added to the ethanol engine, the power developed was always greater compared to neat gasoline operation.

Yamine and co-researchers [90] carried out simulation studies on hydrogen and ethanol addition as supplementary fuel to gasoline engines. The study aimed to reduce fuel consumption by reducing the gasoline in the mixture. The analysis was carried out at an equivalence ratio of 0.8. One part of their analysis was carried out by changing the ratio between gasoline and the supplementary fuels in the fuel mixture so that the desired equivalence ratio was maintained. Performance and exhaust emission data were evaluated. The second set of simulation studies on availability analysis was performed with the test data collected on the Ricardo E6/T (variable compression ratio engine) engine, where the test data were generated operating the engine at 2750 rpm. The test data were verified to be in agreement with the model. Both CO and NOx were observed to be much lower under lean operating conditions. Such a trend very closely agrees with the reported literature. A comparative assessment between hydrogen and gasoline showed that the HRR was substantially improved with hydrogen operation. Due to the fast burning speed of hydrogen, it was observed that the heat released was often close to the TDC. On the other hand, the addition of ethanol often results in improved power and efficiency but at the cost of increased heat loss.

6.24.2 Hydrogen and DME

Hydrogen's high octane rating and fast combustion properties can lead to improved engine efficiency and power output. DME also has a high cetane number, which can improve the ignition quality and efficiency of the fuel. The addition of hydrogen and DME can help prevent knocking or pre-ignition in the engine, allowing for higher CRs and more advanced ignition timing, which can further enhance efficiency. However, while the use of hydrogen and DME in CI engines holds promise for reducing emissions and improving efficiency, addressing these challenges requires significant research and development, which has been relatively limited until now.

Several researchers have investigated the possibility of using DME and hydrogen in a CI engine in the optimised mode of fuelling the engine [91]. The researchers investigated the mechanism of premixed-charge compression ignition (PCCI) combustion of dimethyl ether (DME) by adopting an injection strategy in which the DME main injection was replaced by hydrogen. Such modifications were incorporated keeping in mind the objective of lowering the CO_2 emission level. Tests were conducted with the injected amount of hydrogen to obtain a definite picture of the effect of hydrogen. Many interesting observations were made during the tests with DME PCCI combustion and those with hydrogen–DME PCCI combustion.

The sequence of engine operation with the DME PCCI configuration consisted, firstly, of direct injection of DME into the engine cylinder at −120 crank angle degrees (°CA) after the top dead centre (ATDC). This simulated the homogeneous charge conditions first, and subsequently, DME was injected with variable injection timing. It was observed that the engine performance and emission characteristics were affected by injection timing in the second stage. Conditions of retarded injection in the second stage resulted in increased NOx emission. On the other hand, in the hydrogen–DME PCCI mode of operation, hydrogen was injected at the intake port with fixed injection timing. In these tests, combustion was also varied from −120 °CA to TDC, as in the earlier DME PCCI configuration.

In the entire range of engine operation, the total supplied heating value was maintained at 400 J. Very useful information could be collected from these tests with hydrogen–DME PCCI operation. The exhaust emissions showed a decreasing trend as the amount of hydrogen increased during the tests. Conditions of optimised engine operation could be evolved by decreasing the amount of DME, monitoring DME injection timing and the quantity of hydrogen addition. Least emission and higher power could be achieved at a DME injection timing of −30 °CA ATDC.

6.24.3 Hydrogen with Propane and LPG

Chaudhuri and co-workers [92] have reported some test results with a diffusion flame of a mixture of hydrogen and propane. This was a fairly exhaustive study on the flame appearance and inflame profiles of temperature and composition. The emission features of CO, NO and NOx were also determined. The experiments were useful in generating data on the comparative flame characteristics of neat hydrogen with 20 and 30% by volume of propane mixed with hydrogen. It was observed that the increased concentration of propane in the hydrogen–propane mixture resulted in the luminosity of the flame as well as the increased visible length of the diffusion flame. An increase in the propane fraction caused an increase in the CO concentration as well as the soot emission level. The concentrations of NO and NOx were decreased. The heat release due to radiation, which showed an increasing trend in the beginning, was observed to level off beyond 30% by volume.

Sevrouk and co-workers [93] carried out numerical and experimental analyses of propane–hydrogen mixture ignition in air. They have carried out studies mainly from the point of view to assess the fact that hydrogen addition to several hydrocarbons results in increasing the efficiency and at the same time brings down the pollutants. Previously, we discussed these technical features with respect to hydrogen combustion along with gasoline, CNG, ethanol, etc. This paper, using numerical analysis, discusses the ignition delay time of hydrogen–propane mixtures, and efforts have been made to identify the mechanism that causes the acceleration of this chain reaction of hydrogen-added propane. Their study involved ignition of a stoichiometric C_3H_8–H_2–air mixture with different amounts of H_2 in fuel at temperatures ranging from 920 to 1900 K and pressures varying from 5.2 to 11.2 bar. Only at compositions of a higher mole fraction of hydrogen in fuel, a significant reduction occurs in ignition delay. Interestingly, the ignition delay time decreases to about 50% compared with that of pure propane. A closer numerical analysis clearly showed that the branching reaction $H + O_2 = OH + O$ is very crucial to the acceleration of the ignition process by the addition of hydrogen to propane. Investigators have observed that there is an intimate reaction between the kinetics of the oxidation of C_3H_8 and H_2.

Karagöz and Köten [94] reported their research results on a CI engine using hydrogen and LPG. The study had put a lot of emphasis on the exhaust emission aspects of the engine system. Full load and constant speed (1200 rpm) tests were carried out on a diesel engine (with a torque of 70 N m) fuelled with neat LPG and a blend of LPG and hydrogen (in varying % as 155, 30 and 45%). The test setup is shown in Figure 6.54.

As shown in the figure, hydrogen and LPG are supplied separately from the high-pressure tank, and were very closely monitored so that the hydrogen flow was fixed at 20% by volume of the total gas mixture fuel consisting of hydrogen and LPG. As shown in the test setup in the diagram, a line regulator, shut-off valve, check valve, relief valve and flowmeters were installed at the appropriate locations to check any problems such as backfire and at the same time to ensure that the desired quantity of fuel goes into the system.

It has been observed that (Figure 6.55) BTE shows a decreasing trend by 0.8, 2.1 and 2.9% in efficiency with the addition of 15, 30 and 45% of gas fuel, respectively. The reason for this trend could be that the basic engine test rig, which is a CI engine, was not appropriately designed/modified to effectively actuate LPG injection. However, the typical combustion

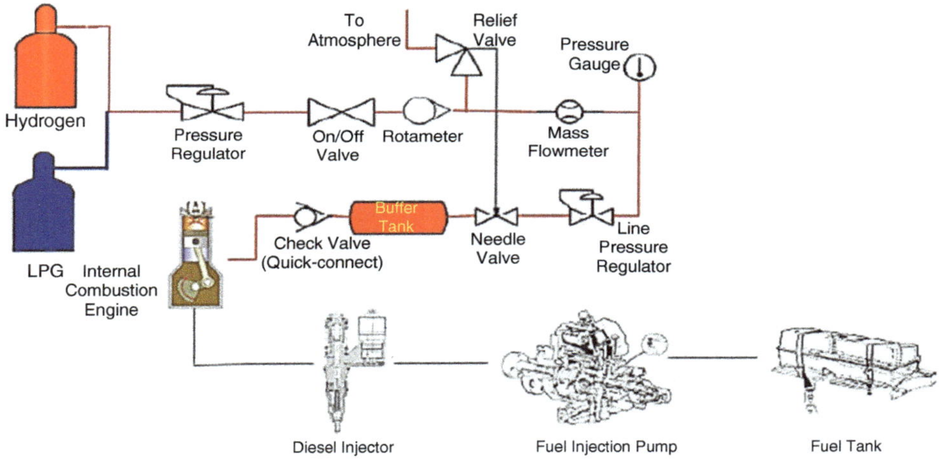

Figure 6.54 Test rig with LPG and hydrogen along with diesel [94]/Yildiz Technical University Press.

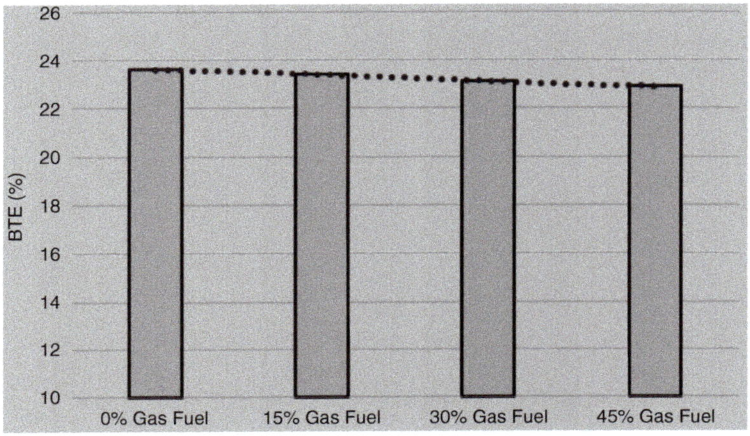

Figure 6.55 Brake thermal efficiency with various fuel blends [94]/Yildiz Technical University Press.

features of hydrogen, such as high diffusion and high flame speed, partially compensated for the further decline in BTE. As far as the exhaust emissions are concerned, CO emissions increased by 4, 12 and 16%, respectively, with 15, 30 and 45% of LPG + hydrogen addition compared to neat diesel fuel. Hydrocarbon concentrations were increased by 8.3, 12.5 and 25%, respectively, with 15, 30 and 45% of gas fuel (LPG + hydrogen addition) compared to neat diesel. On the other hand, the NOx emission level showed a gradually decreasing trend along with the increase in the percentage of hydrogen and LPG. NOx emissions were reduced by 1.8, 2.8 and 3.9% with 15, 40 and 75% LPG + hydrogen addition, respectively. Figure 6.56 shows the NOx emission level with several blends of fuel during engine operation.

Smoke emissions were also reduced by 13.6, 20.3 and 28.8% with 15, 40 and 75% LPG + hydrogen addition, respectively, compared with neat diesel fuel.

As shown in Figure 6.57, the filter smoke number (FSN) takes up a downwards trend as the composition of gas fuel (blend of hydrogen and CNG) increases.

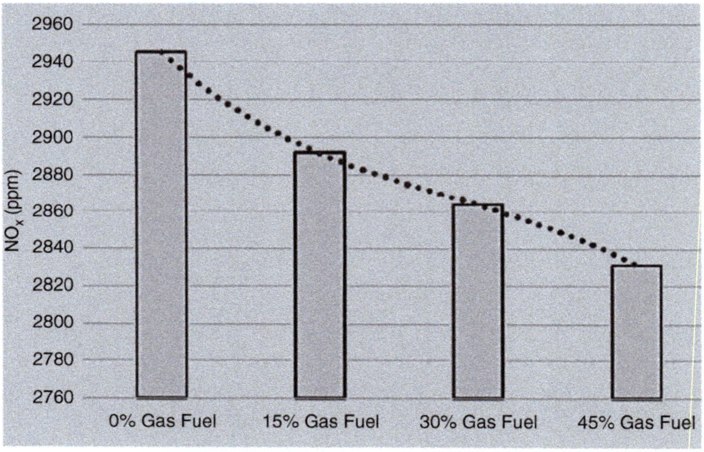

Figure 6.56 NOx emission with various ranges of blended fuel operation [94]/Yildiz Technical University Press.

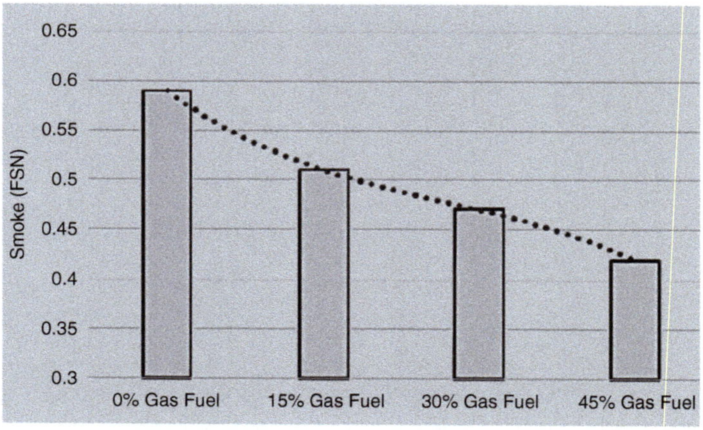

Figure 6.57 Smoke emission from the engine blended with LPG and hydrogen [94]/Yildiz Technical University Press.

6.24.4 Hydrogen Addition to Biogas–Biodiesel Engine

The literature shows great prospects for biodiesel–biogas-based dual fuel engines due to their renewable nature. However, adequate research has not been reported in the literature to overcome accompanying challenges such as reduced efficiency and increased emissions. Tests were carried out in IIT Delhi [95], where hydrogen enrichment in biogas was investigated as a possible technique to improve the limitations associated with biodiesel–biogas dual fuel engines. A simpler simulation methodology was presented to optimise hydrogen injection parameters for improved performance and safe operation of the engine. Another significant aspect of this study is based on the application of exergy analysis and irreversibility quantification of various dual fuel engine processes.

A single-cylinder diesel engine was modified to operate in dual fuel mode using *Jatropha curcas* biodiesel as the pilot fuel and biogas as the main fuel. An electronic control unit is developed in-house to study 5–20% hydrogen enrichment in biogas using the TMI technique. A three-dimensional computational fluid dynamics-based simulation methodology is adopted for optimal selection of TMI parameters to ensure efficient and safe operation of the engine. Subsequently, the optimised injection conditions are used for the experimental evaluations of the performance and emission characteristics of the engine at low and high engine loads. Engine performance is analysed based on energy and exergy analyses. Hydrocarbon (HC), carbon monoxide (CO), oxides of nitrogen (NOx) and smoke emissions are analysed for emission characterisation [95, 96].

The simulation results show that the injection angle and injection pressure play a critical role in mixture formation. An increase in the injection angle directs the flow towards the combustion chamber and increases the turbulent kinetic energy (TKE), which results in better mixing. It is found that late injection timing is an effective approach to avoid backfire. However, extremely late injection timing results in poor mixture formation. Therefore, an optimum injection timing should be selected with the TMI of hydrogen. In this work, a combination of an injection angle of 60° and an injection pressure of 150 kPa offers good mixture formation. Experimental results show that at 20% hydrogen enrichment, the exergy efficiencies of the dual fuel engine are increased from 8.4 to 10.1% at a low load and 23.3 to 25.5% at a high load.

However, maximum reductions in HC and CO emissions of 35.6 and 50.0%, respectively, are calculated at low loads. Therefore, a small amount of hydrogen enrichment can significantly improve the performance and emission features of biodiesel–biogas dual fuel engines. Researchers of this experimental investigation have suggested the extension of this work where the potential application of engine exhaust could be used for biogas reforming and on-site hydrogen production.

Biogas and biodiesel are both considered renewable fuels because they are derived from organic materials. When used in combination with hydrogen, they can further reduce greenhouse gas emissions compared to conventional fossil fuels. Hydrogen has a high flame speed and wide flammability range, while biogas and biodiesel contribute to the fuel's energy density and combustion stability. This combination can enhance the overall combustion efficiency and power output. Similarly, hydrogen's high octane rating and biogas or biodiesel's cetane number can improve knock resistance, allowing for higher CRs and more advanced ignition timing. This can lead to increased engine efficiency and power output. Biodiesel has lubricating properties that can help reduce the engine wear and extend the lifespan of the engine components. Thus, there are several possible benefits from such systems whose technological feasibility needs to be further studied.

6.25 Homogeneous Charge Compression Ignition Engine (HCCI)

HCCI is a concept in which a lean homogeneous mixture of air and fuel is compressed and ignited. The most fascinating characteristic of the system is that the compression of the mixture leads to auto-ignition in multiple locations, which gives rise to very fast combustion compared to conventional SI and CI engines. Such features make it possible to adopt better combustion technology that is controlled by chemical kinetics. This mode of combustion maintains a relatively lower peak temperature, which, in turn, controls the NOx emissions. The engine performance and emission characteristics with the HCCI configuration are not achievable with the present designs of SI or CI engines. This is precisely why, in recent years, the HCCI engine has emerged as one of the most prospective candidates for widespread use in internal combustion engine applications. It can also be operated with a series of alternative fuels. Several R&D activities are in progress in conventional (such as gasoline and diesel) and alternative fuels such as hydrogen and biofuels. The intrinsic features of the HCCI configuration show good volumetric efficiency. Moreover, the emission levels of NOx and particulate matter (PM) are extremely low, even while operating with conventional fuels. Alias and co-authors in their review paper [97] mention that engine operation with hydrogen and biodiesel blend can adopt several injection strategies and as such can use several techniques to improve the efficiency and at the same time bring down the emissions.

Parthasarathy et al. [98] reported their work on the first phase of experiments when tests were carried out in a diesel engine using diesel *Euglena sanguinea* (ES) and their blends (ES20D80, ES40D60, ES60D40 and ES80D20). The highest BTE that was attained with ES20D80 operation was still lower than neat diesel operation by 4.1%. However, the engine configuration indicated that the concentration of pollutants in exhaust needed to be lowered down. With an effort to improve the emissions, the researchers carried out the next phase of tests in the HCCI mode with ES20D80 having different ranges of graphite oxide (GO) nanoparticle concentrations. Additionally, hydrogen at a rate of 3 lpm was introduced at the intake port. Thus, the engine test had acquired the desired optimum configuration of hydrogen-enriched gas and GO-added ES20D80, and the engine was operated in the HCCI mode. It was observed that the NOx level was lowered by 75.24%, and the concentration of smoke was reduced by 53.07%.

Rao et al. [99] discussed the positive features and prospects of hydrogen as an ideal fuel for internal combustion and reported the distinctive features of the HCCI mode of operation with hydrogen to reduce NOx and PM. However, it is suggested that uncontrolled auto-ignition, high peak pressures and poor high load need to be critically analysed as challenges for the HCCI mode of operation. This review paper discusses the several techniques adopted for developing the appropriate test rig for HCCI operation. The HCCI configuration has to be designed to avoid knocking conditions. Emissions of unburned HC and CO can be reduced with a catalytic converter. The combustion pattern in the HCCI mode can be numerically analysed, and operating parameters can be optimised to develop the optimum HCCI engine.

Apart from experimental work, it has also been reported that analysis through multi-dimensional modelling can provide several guidelines to assess the expected performance of the systems. Swirl, along with other operating parameters, has been suggested to take up the challenges encountered during the HCCI mode of operation. As far as the use of non-petroleum fuels is concerned, hydrogen, produced from biomass, is the best choice. In addition, hydrogen–DME and biofuels can also be considered for application in the HCCI mode in engines. The authors have mentioned that negative valve overlap is a better route than external EGR for controlling NOx emission. Apart from bringing down the NOx concentration level due to low combustion temperature, this approach can also ensure improved combustion by slowing down the HRR and decreasing the rate of pressure rise.

In their experimental studies, Ibrahim and Ramesh [100, 101] achieved HCCI operation of a neat hydrogen engine at a compression ratio of 16:1 and intake charge temperatures in the range of 130–180 °C. However, it is very difficult to control the rapid rate of combustion at a high equivalence ratio that often gives rise to abnormal combustion symptoms. On the other hand, at low equivalence ratios, it is difficult to ensure smooth combustion without misfiring.

However, there are some beneficial features if hydrogen is added to a diesel-fuelled HCCI engine. Smooth engine operation has been possible with a diesel engine having a common rail injection system where hydrogen is being inducted and diesel is injected into the combustion chamber. Monitoring the combustion in these systems has resulted in higher thermal efficiency (Figure 6.58). NOx emissions were also observed to be lower compared to diesel engine while running in the HCCI mode. As expected, HC emissions are highly reduced in this hydrogen HCCI engine compared to the HCCI diesel.

Jeon, Bae and Johnson [102] carried out studies on hydrogen–DME injection in the HCCI configuration. In a single-cylinder direct injection diesel engine, the effect of a second DME injection was investigated for HCCI combustion. In these tests, both hydrogen and DME were used as the main fuels. The combustion of the HCCI configuration was closely monitored by varying the amount as well as altering the DME second injection timing. During the tests, it was observed that the optimum injection timing variation was from 100° crank angle before top dead centre (BTDC) to about 5° crank angle ATDC, and the corresponding quantity of DME injected was between 5 and 9 mg. Stable combustion conditions were achieved, and the injection timing did not have any effect, particularly on tests that were conducted using DME as the main fuel. However, when hydrogen was used as the main fuel, the ignition delay of the system was accordingly controlled by varying the injection timing of the DME. It was also necessary to monitor the amount of DME injected. This was primarily because of the lower auto-ignition temperature and higher certain number of DME with respect to hydrogen.

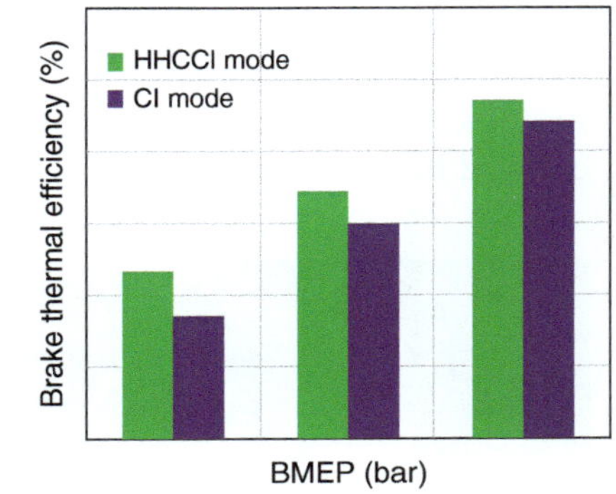

Figure 6.58 Comparison of diesel (CI) and hydrogen HCCI modes: brake thermal efficiency. Source: Adapted from [100].

6.26 Hydrogen-fuelled Vehicles (ICE Based)

As discussed earlier, several researchers, after their successful tests in the laboratory, have made efforts to develop prototype vehicles on hydrogen. The pioneering efforts of Erren have already been described earlier. Furuhama and his team had been consistently working on hydrogen engine considering the successful application of the engine technology on actual vehicles. Several engine configurations were tested in the laboratory [23, 103, 104], and as early as 1975, a hydrogen car was developed, and its smooth operation was demonstrated. The UCLA car [105] tested by a group of dedicated enthusiastic students definitely inspired several other researchers to put their efforts into the field application of hydrogen vehicles. Of interest here that one of the students of the UCLA team pursued hydrogen engine research and contributed significantly in later years. Lynch, who encountered the teething problem of backfire during his tests, discussed the causes and control techniques in his publications. The parallel induction technique has been suggested by him as an effective technique to tame backfire problems in hydrogen engines [106]. To date, several groups from academia and industries have already developed the operation of hydrogen-operated vehicles.

BMW and Ford have developed dedicated hydrogen vehicles built on ICEs [107–110]. Technical problems related to backfire and the rapid rate of pressure rise in hydrogen-fuelled engines are thoroughly understood, and several techniques have been adopted to eliminate backfire and obtain the intrinsic benefits of hydrogen engines in terms of high thermal efficiency, low fuel consumption and almost-zero emission characteristics.

BMW has developed hydrogen cars based on ICEs. These are the traditional BMW 7 series of vehicles that have bene operating on hydrogen (Figure 6.59).

Research and development activities in the IIT (Delhi) were initiated as early as 1980. Several engine configurations (both SI and CI) were developed and appropriately instrumented to evaluate the conditions leading to optimum performance and low emission characteristics along with tribological studies and long-term endurance tests. After carrying out exhaustive laboratory tests for decades, efforts were made to transfer the optimised technical features to a prototype vehicle. As part of a sponsored project funded by the United Nations Industrial Development Organization (UNIDO), an industry–academia collaborative project was taken up with the IIT (Delhi), Mahindra and Mahindra and Air Products (USA) as partners. The objective of this project was to develop a hydrogen-operated three-wheeled (which is a common transportation vehicle in many developing

Figure 6.59 BMW cars fuelled with hydrogen. (a) *Source:* Claus Ableiter/Wikimedia Commons/CC BY-SA 4.0. (b) *Source:* Adrian Padeanu/BMWBLOG.

countries) and demonstrate the utilisation potential of hydrogen-operated vehicles on roads. Tests on a typical three-wheeler vehicular engine were carried out in the laboratory of IIT, Delhi, under widely varying operating conditions. Adequate preventive and curative steps and appropriate safety features were adopted to avoid the occurrence of any undesirable combustion symptoms, such as backfire, higher maximum engine pressure and rapid rate of pressure rise. These optimised conditions for high efficiency, low emission and smooth vehicle operation were implemented on the vehicle. Subsequently Mahindra and Mahindra carried out field tests on actual vehicles (both passenger and cargo types) that already adopted the optimised operating conditions developed during laboratory tests. As part of the project, a hydrogen refuelling station was also set up by air products to refuel the field tests of the vehicles without any disruption. Thus, vehicles developed, modified and instrumented on the optimised conditions were successfully launched during Auto Expo 2012 in Delhi. After the phase of the project sponsored by UNIDO was over, the Ministry of New and Renewable Energy (MNRE) of the Government of India sponsored the continuation of field tests. Successful field operation of these vehicles for a long period without any problems has established the feasibility of using hydrogen as a vehicular fuel and have advantages from the points of view of performance and emission. Figure 6.60 shows the hydrogen-dispensing facilities, whereas Figure 6.61 shows the arrangement of a bank of cylinders installed in the fuelling station. Figure 6.62 shows that the NOx emissions from these vehicles are negligible compared to gasoline operation.

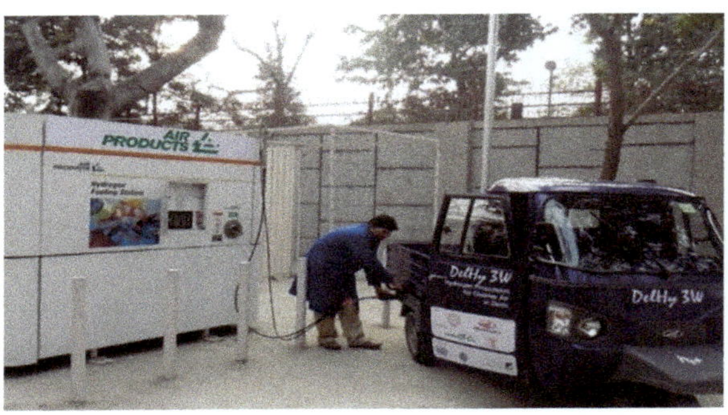

Figure 6.60 Hydrogen-fuel-dispensing station.

Figure 6.61 Bank of hydrogen gas cylinders.

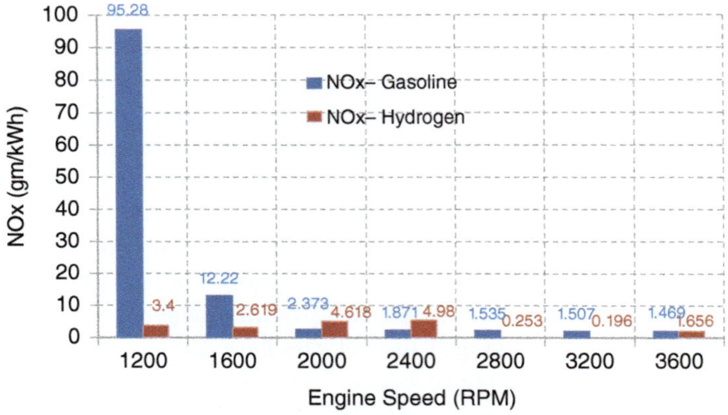

Figure 6.62 NOx emissions from gasoline- and hydrogen-operated vehicles.

Figures 6.63 and 6.64 show two types of vehicles (for passengers and for cargo) developed for operation on hydrogen.

After successful field tests and demonstrations of hydrogen-operated three-wheelers, there was a spate of enthusiasm to develop the technology so that vehicles with multi-cylinder

Figure 6.63 Hydrogen-powered three-wheeler for passengers.

Figure 6.64 Hydrogen-operated three-wheelers for cargo.

engines (used for public transportation systems) could be operated on hydrogen. MNRE (Government of India) sponsored this multi-year mission mode project assigned to IIT Delhi and Mahindra and Mahindra. In a similar sequence, a large number of tests were carried out in the laboratory of IIT Delhi to optimise the vehicular engine operating conditions to obtain good performance and low exhaust emissions. Of course, the major thrust on the development activities was concentrated on avoiding any symptoms of abnormal combustion. A specially designed TMI system was developed for the vehicular engine. Since NOx is the only pollutant, several steps, such as adopting EGR, retarded ST and installation of catalytic converters, were taken to ensure smooth engine operation and at the same time control the NOx emission level. The appropriately instrumented engine was exhaustively tested in the laboratory. Subsequently, trial runs were conducted by Mahindra in their premises with the vehicles that had in it the optimised laboratory conditions very meticulously installed. The hydrogen-specific bus engine has been subjected to extensive field tests and thus establishes the feasibility of hydrogen as a regular transportation fuel. Figure 6.65 shows the minibus with specific technical features for hydrogen operation.

6.27 Hydrogen in Fuel Cells

The most promising technological options of using hydrogen as a clean and efficient tool for power generation and mobile application through environmentally benign operation can be achieved by fuel cells. Fuel cells convert the fuel (hydrogen or hydrogen-based fuels) directly into energy. This is an electric cell that is continuously fed with fuel, and therefore, the output of this cell can be available for an indefinite period. This electrochemical device consists of an electrolyte, an ion-containing solution and liquid or solid in contact with two electrodes. The oxidation of fuel takes place in the anode, and the reduction of oxygen occurs at the cathode. This makes it possible to convert the fuel directly to electrical energy by chemical reaction. In fact, the process carried out at the fuel cell is the inverse of electrolysis as denoted by the following reaction:

$$2H_2 + O_2 \rightarrow 2H_2O + \text{energy}$$

In this process, the chemical energy of hydrogen (fuel) and oxygen (oxidant) produces water and heat (energy). Figure 6.66 shows a typical acid electrolytic fuel cell.

Figure 6.65 Hydrogen-fuelled mini bus.

Fuel cells have received the attention of scientists, engineers and researchers since they were invented by Grover in 1839. However, the work was rather sporadic and not at all coordinated. A renewed spate of interest surged in the 1950s and 1960s after Bacon's successful alkaline fuel cell (AFC) systems. Exploratory space programmes, such as the Gemini and Apollo programmes, infused a lot of enthusiasm. R&D activities on almost all modern fuel cells were initiated during this period.

Intensive fuel cell research and development activities started again from Bacon's pioneering engineering and success of his AFC systems. Highly publicised and successful fuel cell applications in space explorations, notably US Gemini and Apollo programmes, led to high hope and overoptimism in terrestrial and commercial fuel cell power systems. Many development programmes were initiated with public and private funding, and almost all modern fuel cells and beyond were under investigation and development, including AFCs, phosphoric fuel cells, molten carbonate fuel cells (MCFCs), solid oxide fuel cells (SOFCs), proton exchange membrane fuel cells, direct hydrocarbon/alcohol fuel cells and nitrogen-compound-based fuel cells (e.g. hydrazine–ammonia fuel cells). AFCs have mobile electrolytes for terrestrial and military (submarine) applications. Phosphoric acid fuel cells (PAFCs), MCFCs and SOFCs were chosen for commercial applications. Unfortunately, due to some reason or other, development programmes suffered a severe setback, and there was no visible development work even if the technology had survived.

Interests in fuel cells developed in the 1980s with a more realistic set of objectives when the depletion of natural resources caused an energy crisis, environmental degradation due to the combustion of fossil fuels and public concern for nuclear power safety became grave problems. This brought in the need for sustainable development with renewable resources. The goal is to improve the performance (i.e. energy efficiency, power density, reliability, lifetime, etc.) and reduce the capital cost for second-wave fuel cells. Notable characteristics of third-wave fuel cells are that the electrolytes are immobilised (or at least semi-immobile) for all applications, with increasing interest and focus on polymer electrolyte membrane fuel cells (PEMFCs), MCFCs, SOFCs and direct methanol fuel cells (DMFCs).

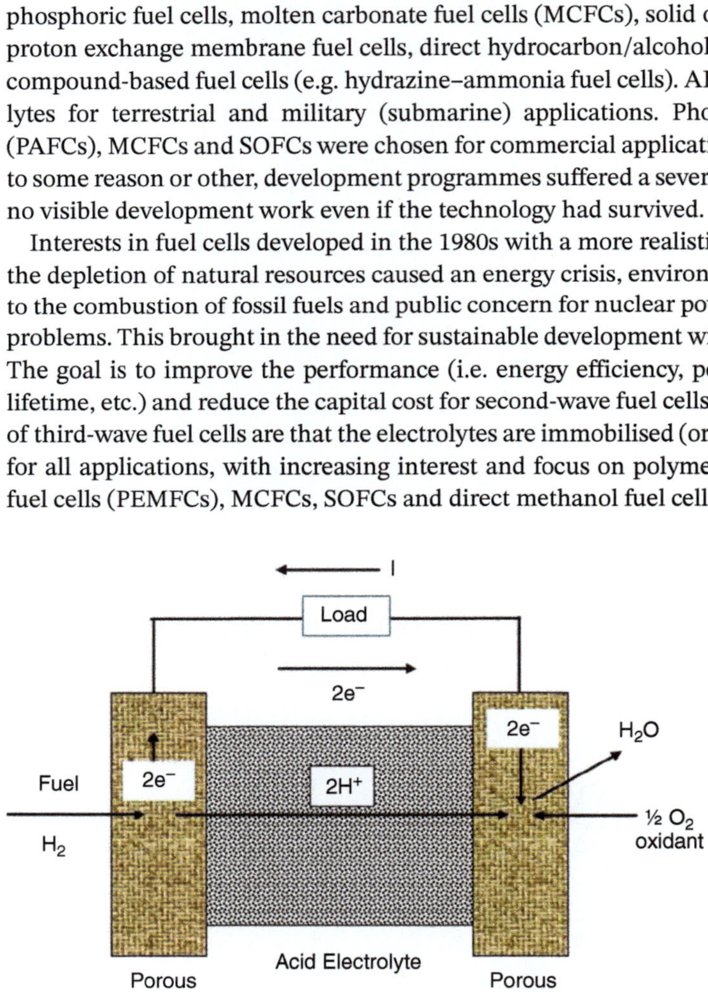

Figure 6.66 Schematic of a typical acid electrolyte fuel cell.

6.27.1 Types of Fuel Cells

Based on the electrolytes, fuel cells can be classified as PEMFCs, DMFCs, AFCs, PAFCs, MCFCs, SOFCs and reversible fuel cells (RFCs).

All types of fuel cells are being intensively developed around the world, including AFCs, PAFCs, MCFCs, SOFCs and PEMFCs. Only DMFCs survive in the family of direct hydrocarbon/alcohol fuel cells due to the slow rate of electrochemical reactions leading to poor performance. Except for DMFCs, all other fuel cells are being designed to operate on hydrogen as the fuel, and carbonaceous fuels are being reformed to hydrogen-rich gas mixtures first before being delivered to fuel cells. The classification is based on the type of electrochemical reaction, the catalysts, the fuel required and the temperature level. All these are important to decide the most suitable fuel cell required for a particular application because each fuel cell has its own merits and limitations for a specific application.

Fuel cells are primarily classified by the kind of electrolyte they employ. These characteristics, in turn, affect the applications for which these cells are most suitable. There are currently several types of fuel cells under development, each with its own advantages, limitations and potential applications. Table 6.3 shows some of the operating characteristics and relevant status of the fuel cells [111].

The maximum energy density of hydrogen is one of the most prominent features of its use as fuel in fuel cells. The other relevant important property of hydrogen is that it is highly reactive when appropriate catalysts are used. Its production source and storage methods, as discussed elsewhere in the book, also favour hydrogen.

Figure 6.67 shows membranes for middle- and high-temperature PEMFCs, and Figure 6.68 shows AFCs with mobile electrolytes and supported gas-diffusion electrodes. Figure 6.69 shows a schematic diagram and operating principles of PAFC units, whereas similar

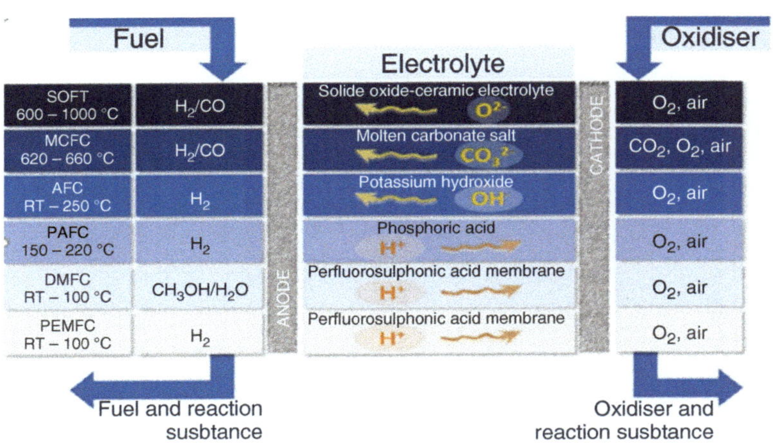

Figure 6.67 Membranes for middle- and high-temperature polymer electrolyte membrane fuel cells [112]/MDPI/CC BY 4.0.

Table 6.3 Comparison of fuel cell technologies [111].

Fuel cell type	Common electrolyte	Operating temperature	Typical stack size	Electrical efficiency (LHV)	Applications	Advantages	Challenges
Polymer electrolyte membrane (PEM)	Perfluorosulphonic acid	<120 °C	<1–100 kW	60% direct H_2;[i] 40% reformed fuel[ii]	• Backup power • Portable power • Distributed generation • Transportation • Specialty vehicles	• Solid electrolyte reduces corrosion and electrolyte management problems • Low temperature • Quick start-up and load following	• Expensive catalysts • Sensitive to fuel impurities
Alkaline (AFC)	Aqueous potassium hydroxide soaked in a porous matrix or alkaline polymer membrane	<100 °C	1–100 kW	60%[iii]	• Military • Space • Backup power • Transportation	• Wider range of stable materials allows lower cost components • Low temperature • Quick start-up	• Sensitive to CO_2 in fuel and air • Electrolyte management (aqueous) • Electrolyte conductivity (polymer)
Phosphoric acid (PAFC)	Phosphoric acid soaked in a porous matrix or imbibed in a polymer membrane	150–200 °C	5–400 kW, 100 kW module (liquid PAFC); <10 kW (polymer membrane)	40%[iv]	• Distributed generation	• Suitable for combined heat and power (CHP) • Increased tolerance to fuel impurities	• Expensive catalysts • Long start-up time • Sulphur sensitivity

Molten carbonate (MCFC)	Molten lithium, sodium, and/or potassium carbonates soaked in a porous matrix	600–700 °C	300 kW–3 MW, 300 kW module	50%[v]	• Electric utility • Distributed generation	• High efficiency • Fuel flexibility • Suitable for CHP • Hybrid/gas turbine cycle	• High-temperature corrosion and breakdown of cell components • Long start-up time • Low power density
Solid oxide (SOFC)	Yttria stabilised zirconia	500–1000 °C	1 kW to 2 MW	60%[vi]	• Auxiliary power • Electric utility • Distributed generation	• High efficiency • Fuel flexibility • Solid electrolyte • Suitable for CHP • Hybrid/gas turbine cycle	• High-temperature corrosion and breakdown of cell components • Long start-up time • Limited number of shutdowns

diagrams for other fuel cells, such as MCFCs, PEMs, DFMCs and SOFCs, are shown in Figures 6.70–6.73, respectively. Figure 6.74 shows the number of fuel cell vehicles in the world, and Figure 6.75 shows worldwide hydrogen refuelling stations. As evident from both

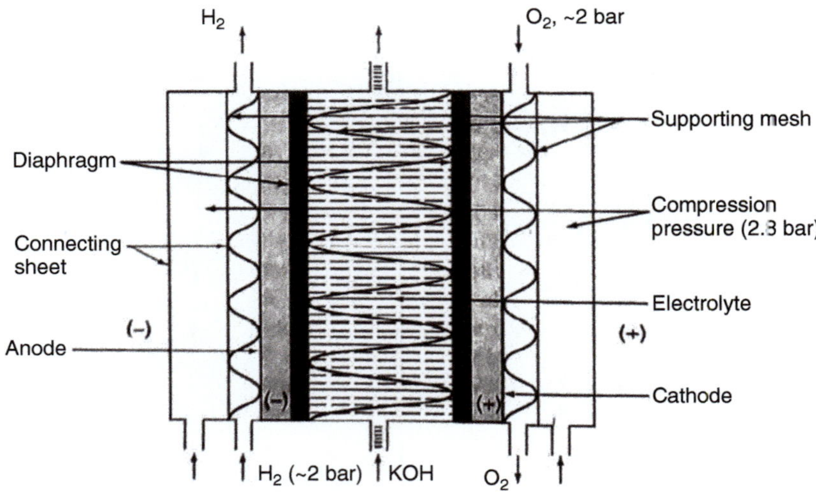

Figure 6.68 Alkaline fuel cells (AFCs) with mobile electrolyte and supported gas-diffusion electrodes.

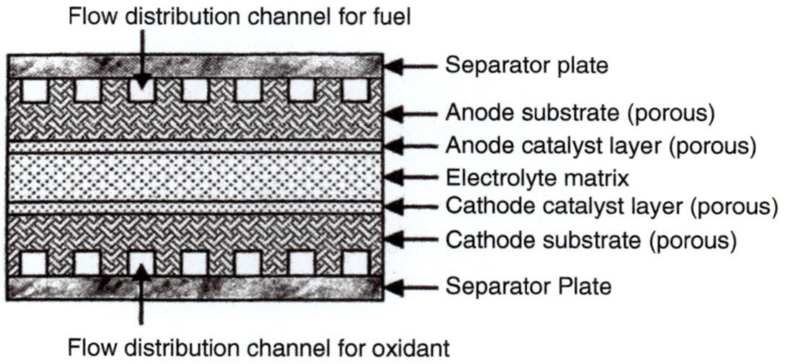

Figure 6.69 Schematic of a phosphoric acid fuel cell (PAFC) unit.

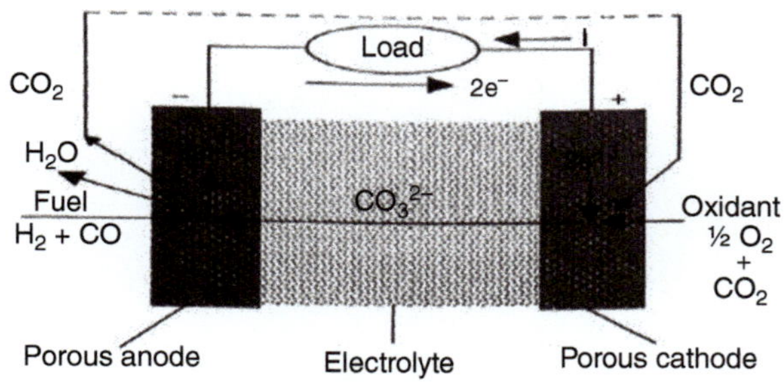

Figure 6.70 Schematic of an MCFC illustrating its operational principle.

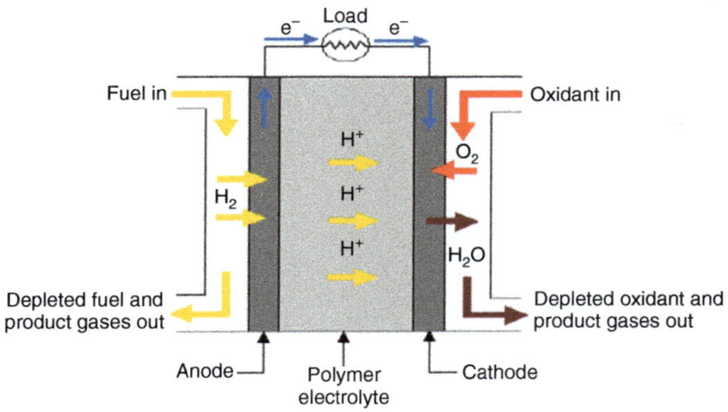

Figure 6.71 PEM fuel cell.

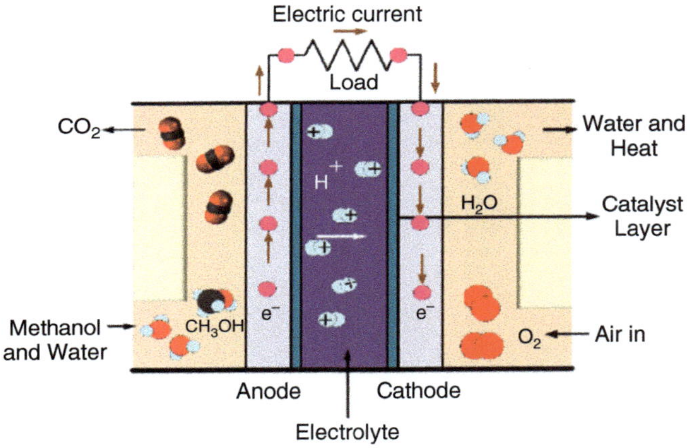

Figure 6.72 DFMC fuel cell.

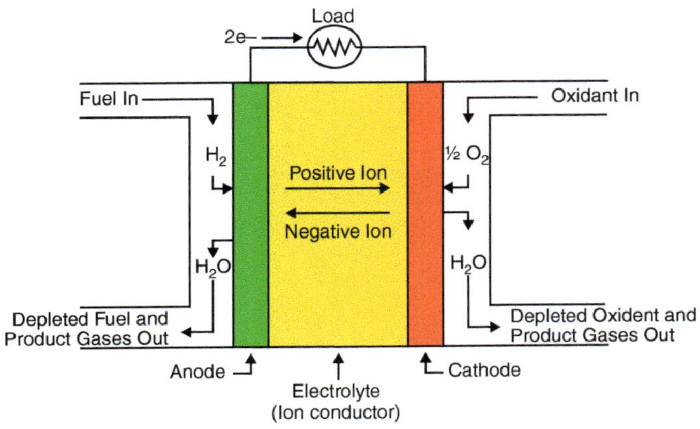

Figure 6.73 Solid oxide fuel cell.

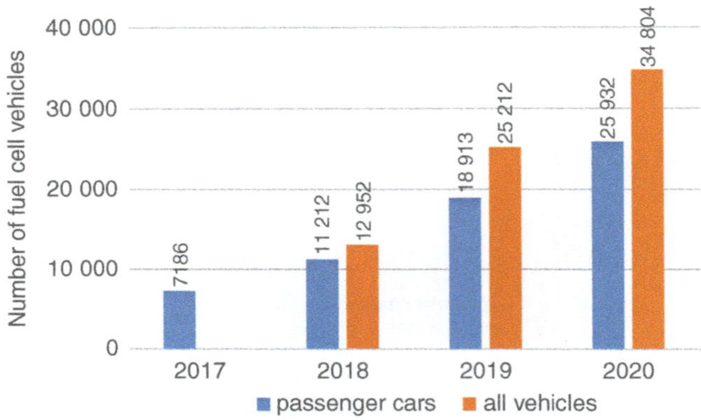

Figure 6.74 Number of fuel cell vehicles (worldwide). *Source:* Adapted from [113].

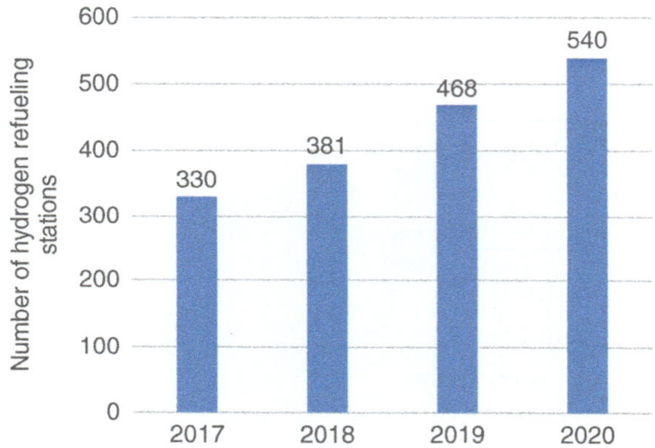

Figure 6.75 Increasing number of worldwide hydrogen refuelling stations. *Source:* Adapted from [114].

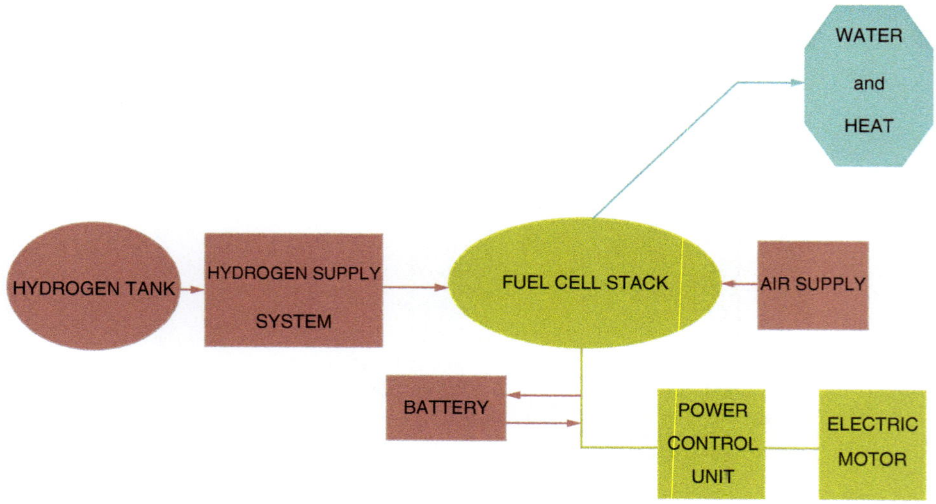

Figure 6.76 Block diagram of a fuel cell vehicle.

figures (Figures 6.74 and 6.75), the total number of vehicles and the total number of refuelling stations show an increasing trend. The numbers shown are until 2020, and the total numbers in both cases must have increased substantially by now. Figure 6.76 shows the schematic diagram of a fuel-cell vehicle.

Depending upon their application, all fuel cells have different features. Sazali et al. [115] discussed various fuel cells, indicating how their optimum efficiency is not limited by the Carnot efficiency of the thermal cycles. The paper also mentions recent developments in the technology of fuel cells. The analysis was performed with specific reference to modelling for PEMFCs, SOFCs and DMFCs. Modelling and optimisation provide ample guidelines for specific fuel cell system design, and necessary modifications can be carried out according to the desired designs. Among all these fuel cells, PEMFCs and PAFCs use hydrogen as fuel, and DMFCs use direct methanol (CH_3OH) as the fuel. Most other fuel cells use natural gas, biogas and coal gas. Thus, fuel cells are powered by either direct hydrogen or hydrogen-rich fuels.

6.27.2 Hydrogen Powertrains

There are several powertrains, such as conventional ICEs, battery electric vehicles (BEVs), plug-in hybrid vehicles (PHEVs) and fuel cell electric vehicles (FCEVs). PHEVs combine an electric motor with an internal combustion engine. They can operate on electric power alone for shorter distances and then switch to the gasoline engine when needed. Therefore, considering the present discussion, a comparative assessment can be made between BEVs and FCEVs. BEVs operate solely on electric power stored in on-board batteries. FCEVs use a fuel cell stack to convert hydrogen gas into electricity, which powers the electric motor. They do not rely on traditional fuels but use hydrogen stored in high-pressure tanks. Both BEVs and FECVs have zero tailpipe emission characteristics. BEVs have a lower overall carbon footprint if charged with clean energy. Similarly, for FECVs, the production and distribution of hydrogen can involve emissions if not sourced sustainably. Now, BEVs have a lower range compared to FCEVs and longer charging times compared to refuelling. FCEVs have quick refuelling times. The situation might change because the fast-charging infrastructure for BEVs and the limited hydrogen refuelling infrastructure for FCEVs are expanding. As far as performance is concerned, BEVs have the capability of instant torque, providing quick acceleration and smooth operation. However, it is limited by battery capacity for longer trips. FCEVs show comparable performance to conventional vehicles, with electric motor benefits. Last but not least, as far as energy source and sustainability are concerned, BEVs depend on the electricity grid and can be sustainable if powered by renewable sources. On the other hand, for FCEVs, theoretically, hydrogen has infinite source potential. However, the actual situation depends upon the economic viability of location-specific resources for sustainable production.

Low-temperature PEMFCs are suitable for automotive applications because of their high energy density, high efficiency and cold start capability with fast start-up times. At present, the capital and running costs of FCEVs are relatively higher than those of BEVs. A considerable fall in price is expected as the demand increases and the manufacturing volume rises [116]. The other advantages of FCEVs are that they have longer driving ranges and the time for refuelling is less than that of BEVs. The lifetime of batteries is much shorter than that of the hydrogen tank. There are a number of factors, such as climate, frequency of charging and discharging, deep discharge and overcharging, etc., which affect the lifetime of the battery. In contrast, hydrogen tanks can have fast and frequent refilling and deep discharging without any effect on the life period. As far as safety aspects are concerned, FCEVs are equally safe as BEVs or conventional IC engines. However, the safety considerations of hydrogen are different in terms of its combustion characteristics. Safety aspects of hydrogen energy have been described in Chapter 5 of the book. Compared to IC engines, FCEVs are quieter because they do not have any gear shifting mechanism. Vibration in running vehicles is much lower in FECVs. Both FCEVs and BEVs are zero-emission vehicles. FCEVs do not need any electricity network for refuelling [116].

6.27.3 Fuel Cell in the Transport Sector

FECVs have a very effective role in the decarbonisation of the transport sector, with a focus on cars, as cars constitute almost 50% of total vehicles in the world.

Several automobile manufacturers, such as Toyota, Honda and Hyundai, have developed FC vehicles. Many others have already built up the vehicles and are testing them in the pre-production phase. More recently, many others have also initiated efforts to build up FC vehicles. Selmi et al. [117] reviewed the technology of FCEVs and provided a picture on their recent global prospective.

Hydrogen possesses a very distinctive property in that it can be converted by electrochemical processes into electricity in a fuel cell. This process is much more efficient than the traditional route of converting conventional fuels to mechanical energy [113]. Thus, hydrogen-powered fuel cell vehicles, usually referred to as 'fuel cell electric vehicles' or 'FCEVs', are characteristically electric vehicles in which hydrogen is directly converted into electricity. BEVs have grown over the years for automotive applications. BEVs and FCEVs have almost similar components except for the propulsion system. One of the major constraints in BEVs is the weight of the batteries. In contrast, the advantage of FCEVs in automotive applications is that the amount of on-board hydrogen is much lighter than that in batteries. Moreover, it has been reported that FCEVs produce a much higher specific energy of 120 MJ kg^{-1} compared to 44 MJ kg^{-1} for gasoline [114]. Fuel cell vehicles present a decisive advantage in terms of refuelling times over battery electric options, as hydrogen can be refuelled in comparable time periods to conventional liquid fuels. FC vehicles are entering the transportation sector very rapidly. There have been close to 35000 FC vehicles, as shown in figure [118], and more than 600 hydrogen refuelling stations.

FCEVs are equipped with a hydrogen storage tank compressed at 700 bar. PEM fuel cell technology has been the most common type of fuel cell used in transportation applications. Hydrogen fuel cell vehicles are extremely sensitive to the purity of the fuel. It has been observed that in some models of FC vehicles, even 4 ppb (parts per billion) of hydrogen sulphide can destroy the fuel cell. Several vehicle manufacturers worldwide have already set deadlines for the carbon reduction targets and are developing FC vehicles to meet those targets. Advances in power electronics, including high reliability, high power density and improved efficiency taking into consideration the power, driving range, tank capacity, etc., are being introduced in FC vehicle models.

As evident from the block diagram above, the fuel cell stack is the central part of the overall system. Its function is to convert chemical energy into electrical energy that is subsequently given to the electric motor. The hydrogen supply system regulates the hydrogen supplied from the tank to the stack. Oxygen is supplied from the air supply system. The air supply system is a very important subsystem as it helps improve the output characteristics of the fuel cell. It consists of an air filter, humidifiers and an air compressor. As the names indicate, the air filter and humidifier maintain the quality, and the air compressor has a very important role considering the flow rate. The heat and water management system separates the water and recycles the heat to the cabin of the vehicle. The power control unit supplies power from the FC to the electric motor.

Oldenbroek et al. [119] modelled the national electricity, heating, cooling and transport systems of five countries, namely Denmark, Germany, Great Britain, France and Spain, for the year 2050. In the analysis, it has been assumed that only 50% of the passenger cars are grid-connected FCEVs. The analysis shows that it is possible to have renewable hydrogen that can be used as a fuel in the transport sector as well as energy storage for backup power plants. Several significant conclusions emerged from this analysis, which are applicable to all five countries. It is possible to produce an adequate amount of hydrogen to cater to the needs of FC vehicles. Appropriate electrolyser and facilities for hydrogen storage need to be established for this purpose. Several country-specific analyses have been carried out. It has been suggested that enough hydrogen can be produced during the summer and can be stored in underground salt caverns or empty gas fields to be used during the winter.

At this point in time, FECVs are expensive, but the TCO will definitely decrease as demand for these vehicles increase, leading to mass production. Improvement in fuel consumption and addition of a range extender will widen its application. Fuel-cell-based passenger vehicles have already been produced by only a few car manufacturers in some developed countries, mainly because of inadequate infrastructure. Recently, the majority of automobile manufacturers have identified FCEVs as a long-term solution, and many countries have strengthened and initiated their efforts in this direction. An extensive infrastructure for refuelling needs to be built to facilitate the rapid growth of FCEV population. However, in the initial phase, only a few refuelling stations might be established close to the location from which several types of return-to-base vehicles, such as delivery vans and passenger-sharing taxis, that need single and infrequent refuelling can operate.

6.27.4 Fuel Cell Buses and Trucks

Fuel cell buses have been in operation in several countries in Europe, North America, Japan, etc., and it is a mature technology with a proven record. Now there is a great interest and enthusiasm for FC buses as the best option for medium to long range as well as for suburban routes to long distance. South Korea, China, the Netherlands, Canada, France, Australia, India and many other countries have been testing and deploying fuel cell buses in various cities and regions as part of the activity to promote hydrogen-based mobility.

On-board hydrogen tanks carry around 40 kg of hydrogen in these buses. Fuel cells have great potential for use in trucks. It is possible for light goods vehicles to operate only within short distances using batteries and at times using range extender vehicles. However, for far-off distances with high energy requirements, long-haul fuel cell trucks are the appropriate non-carbon alternative. For this, the necessary refuelling network has to be built. Several reputed automobile manufacturers, such as Kenworth and Toyota, designed hydrogen-powered fuel cell trucks that reduce a large quantity of GHGs compared to baseline diesel trucks. Many other companies in different countries are also working on the development of hydrogen-fuelled FC trucks, keeping high efficiency and lower fuel consumption in mind. In many countries, two-wheeled vehicles are popular transport systems. The Intelligent Energy Emissions Neutral Vehicle (ENV) is a fuel-cell-based hydrogen-fuelled motorcycle that is equipped with a 6 kW, 48 V motor and 1 kW fuel cell. In the initial stage of development, it had already been tested with a top speed of 80 kmph and a range of 160 km. The range and speed would be further raised as the production continues. Suzuki also developed FC-based hydrogen-fuelled two-wheelers.

6.27.5 Off-road Transport Trains

There have been efforts to use hydrogen fuel cells in off-road transport trains. Recently, in 2022, a fleet of five hydrogen passenger trains, developed by the France-based company Alstom, started running on non-electrified zero-emission rail lines. Initially, these trains, called 'Coradia iLint', had exhaustive test demonstration in Germany, Austria, the Netherlands and France. It has been estimated that by replacing the diesel-powered trains with these hydrogen trains, it has been possible to reduce about 4000 tonnes of carbon dioxide out of the atmosphere each year in that locality. Diesel smoke has vanished off. This train has been designed to run at a speed of about 136 km h^{-1} though it usually runs at a speed of 80–120 km h^{-1}. The trains are fuelled from a hydrogen tank and a single tank supplies fuel to the train to cover about 1000 km. A train's hydrogen fuel cells generate electricity to power electric motors, which drive the train's wheels, similar to the way electric trains operate. Hydrogen is stored in tanks on the train and is combined with oxygen from the air in the fuel cells to produce electricity. Recently, the world's first hydrogen filling station for trains was built in Bremervörde, Germany, in cooperation with LINDE to facilitate refuelling the trains supplied by Alstom. The hydrogen refuelling station will have a capacity of around 1600 kg of hydrogen per day. Several European countries have withdrawn diesel from regular use in trains and have electrified the train lines. In such situations, particularly on rail lines that are not very frequently used, hydrogen could be a better option. Apart from Germany, some other European countries have expressed interest in using these hydrogen trains, and demonstrations have been made by Alstom on the respective railway tracks. In addition to all these factors, the production of green hydrogen is being rapidly followed, and in the course of time, this would make hydrogen more sustainable and FCEV trains might become less expensive than diesel from a TCO perspective. Piraino, Genovese and Fragiacomo [120] studied an advanced mobility concept by analysing both the energy performance and the economic viability of an integrated system, refuelling infrastructure for regional hydrogen trains. The study aimed at evaluating the relevant features of the system components for the hydrogen facility and the fuel cell hybrid train. A regional line of southern Italy was considered for detailed modelling studies. Two 180 kW fuel cells, in parallel, were considered both for passenger and freight locomotives for the analysis of the hybrid railways. Similarly, to analyse the hydrogen fuelling station, daily hydrogen consumption was assumed to be about 250 kg. To cater to this need, the station had a 670 kW PEM electrolyser, producing 260 kg of daily hydrogen that was stored at a pressure of 350 bar. The analysis stresses on two very significant parameters, namely the number of passengers and the freight weight, within a range of ±20%. Very encouraging results were obtained from the sensitivity analysis conducted for this study. For stringent demand scenarios, fuel cell systems chosen for the passenger trains showed an average efficiency of 50%, and the freight transport of the fuel cell stacks required an average efficiency of 47%. These high values of fuel cell efficiency beyond 47% and facility efficiency over 50% were always achieved. In addition, attractive financial indicators were achieved in terms of the levelised cost of hydrogen and the TCO. This study proves the feasibility of using hydrogen in transport systems in trains and broadly in heavy-duty transport. 'The investigated case study showed how these technologies can provide a sustainable and valid option to the mobility sector, both from a technical point of view, by guaranteeing important values of infrastructure energy efficiency and a reduced train hydrogen consumption, and from an economic point of view, with acceptable return-on-investment and payback period' [120]. Logan et al. [121] carried out studies to assess the capability of electric trains (ETSs) and hydrogen trains (HTs) to meet the Paris Agreement emission target. In this study, the projected levels of carbon dioxide emissions from ETs and HTs in Japan and the United Kingdom from 2020 and 2050 were estimated. The results obtained from this study were compared with conventional diesel-fuelled trains (DFTs) to evaluate the environmental compatibility of each fuel and to determine which was the best from an emission point of view. It was estimated that the process of electrolysis should be adopted to generate the large amount of hydrogen required to switch over to total hydrogen operation. However, looking at the problem from a realistic point of view, it was also concluded that electrolysis was neither possible nor economical for small-scale applications. As an alternative route, hydrogen production could be carried out either by SMR using natural gas or by coal gasification, and subsequently, CCS could be adopted to cater to the large demand. This will be helpful to reduce greenhouse gas emissions. However, there would be varying levels of emissions depending upon the mode of production. Capital costs and time would be required to develop appropriate infrastructure for large-scale hydrogen production with abundant renewable energy.

An analysis of kilograms of carbon dioxide per kilometre travelled by train type for Japan between 2020 and 2050 in five-year increments shows that, with HTs, the emissions produced per kilometre decrease by 9% from 6.7 kg CO_2 km^{-1} in 2020 to 6.1 kg CO_2 km^{-1} by 2050 [121]. This study also discusses the cost, and it has been observed that further research needs to be carried out to evolve an economically viable efficient system for the large-scale storage of hydrogen for commercial applications.

6.27.6 Stationary Power

The stationary power sector, such as transportation, is an important area where hydrogen has the scope of significantly contributing to energy–environment problems. Felseghi et al. [122] discussed the specific characteristics of hydrogen as a clean energy that can be used for stationary power and significantly contributes to air pollution problems. The researcher carried out a strength, weakness, opportunity and threat (SWOT) analysis with the major objective for identifying the possibilities and subsequently developing the strategies such that hydrogen energy can effectively penetrate the stationary power sector as a substitute for the conventional model. The potential of hydrogen energy and the wider scope of its utilisation have been highlighted, and at the same time, the possible threats

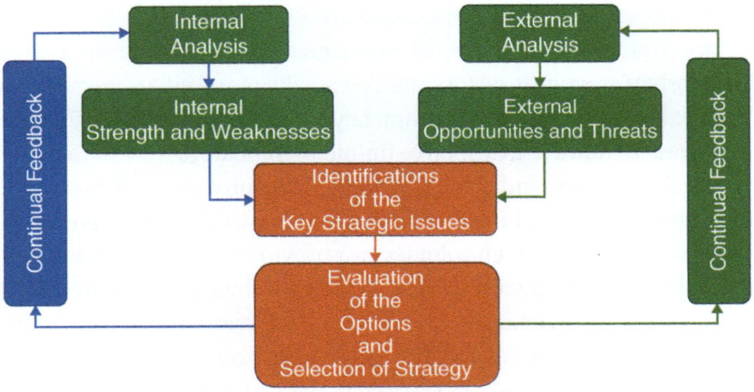

Figure 6.77 Schematic of the SWOT analysis.

from the external environment have been taken into consideration using a critical SWOT analysis. Possible options and implementation strategies for using hydrogen in the stationary power sector have been evaluated. The SWOT process adopted by the investigators is shown in Figure 6.77.

The significant factors that are considered in the SWOT analysis with respect to fuel cell technology are based on technological, environmental, social and economic factors [123] as shown in Figure 6.78.

There are several power-generating stations around the world that utilise hydrogen for stationary power generation. Some of these activities adopt various approaches to use hydrogen for stationary power generation, including hydrogen fuel cells, power-to-gas and integrated renewable energy systems. There are also several ongoing and upcoming projects of different magnitudes that evolve along with technology and energy policy. The Levenmouth community energy project, located in Fife, Scotland, has a 7 MW electrolyser that converts excess electricity from wind turbines into hydrogen, which is stored and used in a fuel cell to generate electricity when needed. The Western Australia Power-to-Gas Pilot

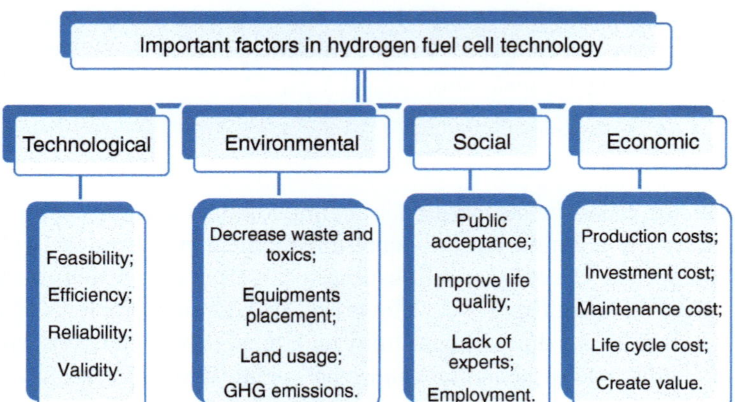

Figure 6.78 Factors relevant to the SWOT analysis of hydrogen and fuel cell technology. *Source:* Adapted from [123].

Plant uses excess renewable energy to produce hydrogen through electrolysis, which can then be injected into the natural gas grid or used in fuel cells to generate electricity. The Güssing renewable energy park in Austria has a long history of using renewable energy sources for power generation. The park includes a biomass gasification plant that produces hydrogen-rich gas, which is then used in ICEs and fuel cells to generate electricity and heat. The project Energiepark, Mainz, Germany, integrates a hydrogen electrolyser with a wind farm to demonstrate power-to-gas technology. The electrolyser produces hydrogen, which is combined with carbon dioxide to create synthetic natural gas (methane), that can be fed into the natural gas grid or used for power generation. In addition, the Hydrogen Park in Germany features a combined heat and power (CHP) plant that uses hydrogen fuel cells to generate electricity and heat for local buildings and industrial processes. The Hokkaido Wind Hydrogen Energy Demonstration Town, Japan, combines wind power with hydrogen.

6.27.6 Hydrogen in Gas Turbines

A closer look at the literature shows that several factors and selection of the appropriate components to perform satisfactorily under different operating conditions are important for the use of hydrogen in gas turbines. Nazari [124] and co-workers performed an exhaustive review and arrived at very useful conclusions. There has been a remarkable improvement in the emission characteristics of gas turbines by adding hydrogen to the conventional fuels in the combustor. The equivalence ratio has been observed to be a crucial parameter for flame stability. The oxides of nitrogen (NOx) from these gas turbines can be substantially reduced by altering the equivalence ratio and hydrogen dilution. Ditaranto and his team [125] have adopted EGR in which the oxygen-depleted flow is allowed to enter the combustor (Figure 6.79).

This recirculation decreases the reactivity of hydrogen [125], thereby decreasing the NOx emission level without affecting the flame stability. Such response characteristics leading to NOx control of hydrogen gas turbines are similar to those in ICEs, and researchers have reported that the injection of steam can eliminate irregular patterns of combustion during turbine operation. An overall comparative analysis of hydrogen-fuelled gas turbines with

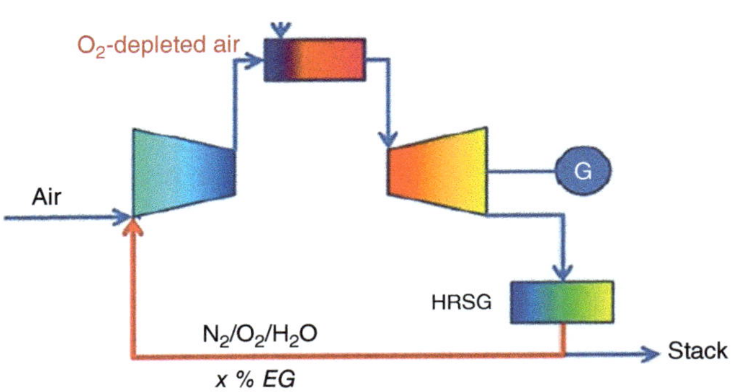

Figure 6.79 Exhaust gas recirculation (EGR) adopted for the gas turbine [125]/with permission of Elsevier.

conventional fuels shows that their exergy efficiency, similar that of fossil-fuelled gas turbines, is affected by the configuration and operating conditions. The system's exergy efficiency is affected by the addition of hydrogen to the base fuel. It has also been observed that hydrogen-fuelled gas turbines can be optimally integrated with other systems, such as solar PV cells. The performance characteristics of such integrated systems can be monitored by several factors. The present thrust on research on hydrogen gas turbines mainly involves studies related to the stability of flames and the emission characteristics of the system. These are, no doubt, very important aspects of investigation. However, further studies need to be carried out on turbines to obtain the specific advantages of hydrogen fuel. Perhaps the cycle analysis with hydrogen should go beyond the normal and recuperative Brayton cycle and involve the Rankine–Brayton cycle. Apart from exergy-economic analysis, important thermodynamic parameters, such as the pressure ratio, inlet temperature and optimal geometry of the components, etc., need to be considered. Such an exercise would be useful in providing design guidelines to optimise the cycle for hydrogen operation. The economic viability of the hydrogen production route and the choice of several hybrid technologies with renewable energy systems could be subsequently taken up for the design and development of hydrogen-specific gas turbines. At this point in time, it is perhaps worthwhile to consider the technical pros and cons of using hydrogen in the existing natural gas turbines. Some alterations are necessary in the existing hardware of combustor systems for efficient and safe handling of hydrogen. Apart from the hardware design, the operability and economic viability should be taken into account. Hydrogen combustion does not emit any noxious pollutants, such as CO, hydrocarbons, PM, etc., because it does not contain carbon atoms. Hydrogen combustion, at high temperature, produces oxides of nitrogen (NOx), which are regulated pollutants that contribute to acid rain and smog. However, several designs of combustors can be used at locations for applications where oxides of nitrogen do not pose any threat to the environment. It has been observed that diffusion combustors, which are designed on the basis of relatively older technology, generate higher levels of oxides of nitrogen. The positive feature of diffusion combustors is that they are stable and offer the flexibility of using several fuels. Presently, there are some diffusion combustors that operate as fuels with very high levels of hydrogen content, which are produced as by-products in several industrial processes of steel mills and petrochemical plants. In diffusion flames, there exists a stoichiometric region irrespective of overall stoichiometry. The primary zone of the diffusion combustor, because of its high temperature, generates more than 70 ppm NOx when burning natural gas and more than 100 ppm for liquid fuel.

Lean premixed combustors operate at a relatively lower temperature level, and the problem of NOx emission is inherently avoided. This is one of the significant reasons why the recent power plant installation uses this system even though they do not have the option to adopt flexible fuel for their operation. Presently, combined-cycle gas turbines have already been designed at a fixed temperature for specific applications where the NOx emission is limited to 2–25 ppm. It is expected that such premixed systems, designed for lower emission of nitrogen oxides, would tolerate some level of blending with hydrogen.

The heat transfer coefficients of the products of combustion in a hydrogen-operated gas turbine are higher than those of a natural gas turbine. Since the heat transfer to a rotating turbine is crucial to the peak temperature of the turbine, in a hydrogen-operated turbine, it becomes necessary to reduce the inlet valve temperature as the level of hydrogen increases.

This reduction in the inlet valve temperature, in turn, affects the overall cycle efficiency that would have been higher because of hydrogen fuel.

Considering the typical combustion characteristics of hydrogen, certain design features need to be closely analysed. Despite the glaring advantage of hydrogen combustion from improvements in performance and emission features, flashback has proven to be the major stumbling block in its widespread implementation. Modern premixed combustors, designed for operation with natural gas, need to be appropriately converted and retrofitted to take care of the higher flame speed of hydrogen. The other possible problem that could arise due to hydrogen is blow-off. It is quite likely that the flow velocity in a combustor could reach much higher values, about 150 km h^{-1}. Under such circumstances, it becomes very difficult to control the rapid flame from flying downstream. However, a blend of natural gas and hydrogen up to about 50% by volume could be tried in the combustors designed for natural gas. The other factor that needs to be closely attended to involves the problem of combustion instability. At present, low NOx combustors exhibit several degrees of oscillation depending upon the fuel used and the operating conditions. This poses a difficult problem to have a static design to ensure stable and reliable operation, particularly for situations where the fuel composition widely varies.

There have been studies on the production of oxides of nitrogen in the industrial gas turbines. The study takes into account the influence of flame stabilisation on the air intake hole diameter [126]. The efficient function of gas turbines is critical to the combustion chamber. As mentioned earlier, the production of oxides of nitrogen is related to the combustion process that takes place. The influence of flame stabilisation on the air intake diameter, the quantity of NOx produced and the exhaust gas temperature are all interrelated and affect the gas turbine efficiency. York and Zemansky investigated the hydrogen system for a heavy-duty gas turbine to assess its low NOx characteristics [127]. Tests were conducted with a specially designed 'premixing fuel injector' to avoid the undesirable combustion symptoms of flashback and at the same time ensure low emission. Stable and smooth operation could be achieved up to a temperature level of approximately 1850 K.

To assess the performance and combustion conditions, tests were carried out under 'full gas turbine conditions' and at several compositions of various fuels, such as hydrogen, methane and nitrogen. Long-duration full load tests were conducted at the target gas turbine temperature and pressure for 100 h with hydrogen comprising 90% by volume of the reactants. It was observed that the NOx emission level was drastically reduced to single digits in ppm with hydrogen–nitrogen fuel. The results are very important and provide important guidelines to achieve ultra-low emission by hydrogen combustion in gas turbines [127].

As in the case of transport and stationary power applications, the degree of hydrogen integration into gas turbines and specific projects are changing rapidly over time as technology advances and energy policies evolve. Some examples are given below. Therefore, there may be several more projects from pilot projects and research to full-scale commercial operations. Siemens Energy, the German energy company, has been actively working on hydrogen integration into gas turbines. They conducted tests with hydrogen blends in their SGT-400 industrial gas turbine. In Japan, Mitsubishi Hitachi Power Systems have been researching and developing hydrogen-operated and hydrogen-rich gas turbines. The Magnum power plant in Eemshaven in the Netherlands has been exploring the use of hydrogen in gas turbines for electricity generation. In the United States of America, The US Department of Energy and companies have been researching hydrogen integration into gas turbines. The National Renewable Energy Laboratory (NREL) has been working on projects related to hydrogen-enriched natural gas combustion in turbines. In South Korea, companies such as Doosan Heavy Industries have been involved in hydrogen-related turbine projects. Government agencies, companies and research institutes in Australia, the United Kingdom and many other countries are also working on the use of hydrogen blends in gas turbines.

6.27.7 Hydrogen for Maritime Applications: Ships, Submarines and Boats

Hydrogen fuel cells have been tested for marine applications, including ferries. Recently, the emissions in some areas of urban ports have already received attention. Local air quality can be improved in the ports with a single depot. Most ferries are designed for specific applications. The cost of hydrogen in marine applications is a point of concern because ferries consume about 200 kg of hydrogen per day Recently, in November 2022, two 200 kW fuel cell modules were installed by Ballard Power Systems of Canada for marine applications to be operated by the Norwegian company Norled. Ballard's FCwave™ module is a significant achievement in adopting zero emissions in marine applications. The ferry is 82.4 m long and has a capacity to accommodate about 300 passengers and 80 cars. The vessel is designed to attain a speed of 9 knots (one knot is equal to one nautical mile per hour or 1.852 km h^{-1}). There is a provision of 80 cbm (cubic metre) of hydrogen tank in the vessel for hydrogen storage. Green hydrogen production and PEM technology will be adopted. This is likely to reduce annual carbon emissions to the level of about 95%. DHYBRID, a German company, has developed a hydrogen fuel cell hybrid boat known as 'Alsterwasser', which operates on Hamburg's Alster Lake and demonstrates the clean and quiet features of hydrogen-powered maritime transportation. The boat's propulsion system utilises a hydrogen fuel cell along with batteries to provide efficient and environmentally friendly propulsion for passenger transport and events on the lake.

The transportation of water by hydrogen has not yet broken into the mainstream. The International Maritime Organisation (IMO) has expressed concern over the situation and is now in progress of revising the greenhouse gas strategy, and the shipping sector will adopt the guidelines in developing the infrastructure required to reach the zero-emission target. Recently, some of the largest green hydrogen producers along with some members of the global shipping sector have started collaborating to rapidly adopt green hydrogen-based fuels to achieve total decarbonisation of the maritime sector by 2050. Van Hoecke and team [128] have discussed the problem of using hydrogen for maritime applications. They have put more emphasis on the challenges for developing the appropriate hydrogen storage system for maritime application. This study suggests that the development of fuel-supply infrastructure and on-board safety are the major challenges at present. Hurdles for the development of hydrogen storage for the maritime industry are discussed in the development of hydrogen storage technologies. The most likely challenges will be the development of a new bunkering infrastructure and suitable monitoring of safety to ensure the safe operation of these hydrogen carriers on board the ship. The maritime sector had previously set a target of bringing down the CO_2 emission level by 2050 by at least 50% of the CO_2 level that existed in 2008. It is important that improved technologies be adopted for using hydrogen in ships, and the necessary infrastructure needs to be built to achieve this target. Liquefaction and regasification plants have to be installed. A well-defined supply chain has to be in place to facilitate distribution and storage of the fuel. In Denmark, a team of industries, including shipping companies and energy providers, aims at an ambitious target of producing more than 250 000 tonnes of hydrogen for a year to be used for all three major modes of the transportation sector, such as buses and trucks, maritime vessels and aeroplanes.

In recent years, the use of fuel cells in ferries has been successfully demonstrated at several locations of the world. Since fuel cells can operate without natural gas, methanol and diesel, it is possible to produce hydrogen from these fuels using reformers. Such technologies may be conveniently adopted in small ferries to be operated on hydrogen-powered fuel cells. Research needs to be carried out to optimise the production mechanism of hydrogen as well as to evaluate the economic viability of the fuel cell stack and powertrain. In the initial phase, the rapid use of hydrogen in maritime transport can be adopted by employing ferries and subsequently this technology could be upgraded for use of hydrogen in relatively large merchant vessels. However, since large merchant vessels, because of heavy freight shipping, require a more energy-dense fuel, ammonia derived from hydrogen can perhaps be more suitable for such applications. Problems related to the economically affordable production cost of hydrogen and appropriate refuelling infrastructure to store adequate amounts of hydrogen need to be addressed to ensure the rapid implementation of these technologies. H_2/O_2 fuel cells, particularly low-temperature fuel cells, have characteristics that make them ideal for powering submarines. They neither produce any emissions nor any waste products. They produce only water and thus can maintain zero buoyancy. In the absence of any moving parts, they operate quietly. They reject heat at low temperature. Fuel cell power plants are also used for surface ships and boats, both as main propulsion engines and auxiliary generators. Det Norske Veritas and Germanischer Lloyd (DNV GL) identified hydrogen as a potential alternative fuel for shipping [129] and developed strategies for its techno-economic evaluation. Among several alternatives being investigated, fuel cells and wind-assisted propulsion systems are found to be suitable for ship applications. Research and development activities are in progress, and it may take some time for fuel cell technology to mature to be used in ships. DNV's recent Maritime Forecast to 2050 has discussed technical aspects such as design features and the economic viability of prospective alternative fuels.

Ustolin et al. [13] provided a very exhaustive overview of the use of liquid hydrogen in maritime transportation systems. The paper has discussed several significant aspects such as environmental pollution (in terms of equivalent emission) during hydrogen production and other safety-related technical challenges for the typical combustion characteristics of hydrogen. The discussions very clearly emphasise the well-established fact that it is essential to minimise the carbon footprint to ensure and accelerate hydrogen's entry to any mode of transport sector, including maritime applications. Several aspects, such as appropriate infrastructure for bunkering, on-board storage mechanism and utilisation, safety, codes, regulations, etc., have been discussed. The reasons that might create difficulty in the use of LH2 technology in the maritime sector have been discussed and the possible solutions have bene highlighted [130]. The paper also highlights the potential of hydrogen being used as fuel in all modes of transport, such as rail, road, maritime and aviation. However, LH2 is preferred over compressed gas in the maritime sector, where large ships need fuel for long-range voyages. Apart from environment compatibility and safety constraints, it is also essential that the necessary infrastructure need to be of the size of the harbour built up. The hydrogen production cost and the cost of components for handling and storing liquid hydrogen need to be reduced. The overall system design for the liquid hydrogen system needs to be optimised considering the type of ship and the size of the harbour.

Serna and Tadeo [131] discussed the salient features of offshore electrolysis plants powered by the wave energy of sea. This is a proposal to have electricity from a system that is entirely isolated from an electrical grid. This integrated plant (Figure 6.80) consists of a PEM electrolyser and a reverse osmosis (RO) system with the capacity to produce an appropriate quality of water with adequate conductivity. The system also has a compression unit to store hydrogen. There are also batteries installed to store electricity for some time. The paper also discusses some relevant guidelines for the choice of location to install such plants for optimum performance.

The German Type 212 class and the Italian Tirado class are advanced designs of non-nuclear submarines. These designs have a diesel propulsion system and an additional air-independent propulsion (AIP) system using proton exchange membrane (PEM) compressed hydrogen fuel cells. The idea of the AIP system for submarines was thought of as an improvement on the age-old hydrogen–oxygen propulsion system mainly from a safety point of view with fire and explosion (Figure 6.81).

The distinctive feature of these submarines is that they can switch to the AIP system for silent slow cruising. These can stay submerged for up to three weeks without any vibration under extremely quiet conditions and are virtually undetectable.

Shifting the boats to hydrogen operation could offer significant advantages over existing conventional technologies. Operating a boat with a fuel cell could totally eliminate the CO_2 and other greenhouse gases that are responsible for sea pollution. A study by Sandia National Laboratories established that a boat can be designed, modified and developed to operate on hydrogen fuel cells. This is both technically and economically feasible now.

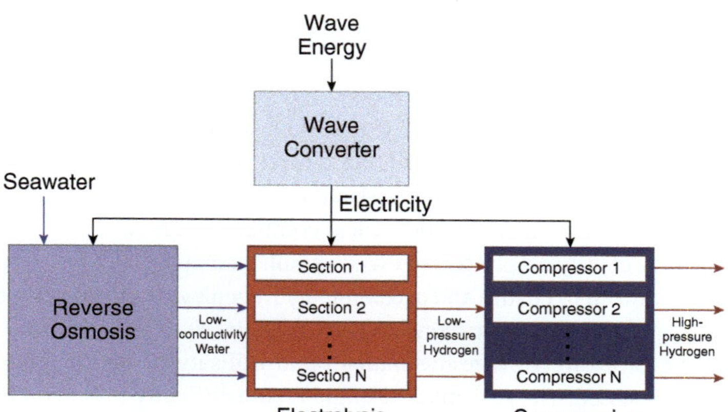

Figure 6.80 Electricity generation system from seawater [131]/with permission of Elsevier.

Figure 6.81 Hydrogen-powered submarines.

Sandia has, in this context, released a report specifying the technical details of such a system and, based on the strength of their research experience, built up the hydrogen-fuelled passenger ferry called 'Water-Go-Round'. As far as the most exciting exploration of Arctic Oceanography is concerned, hydrogen is perhaps the most ideal fuel as it does not have the waxing and freezing problems that in this case are the shortcomings of petroleum fuels.

Ballard's The FCwave™ fuel cell module is claimed to have proven the zero-emission features that are necessary for commercial application. There are several joint ventures that Ballard has with industries for the development of zero-emission fleets, such as megawatt fuel cells for large vessels with ABB, passenger ferry with Kongsberg Maritime and CMAL (HySEAS III), the M/F Hidle–FLAGSHIPS ferry in Norway with Norled and Westcon, the FLAGSHIPS push boat in France with ABB and LMG Marine, the ELEKTRA push boat with BEHALA and TU Berlin and the Hjelmeland ferry in Norway with Norled and Westcon (M/F Hydra). Nedstack Fuel Cell Technology, the Dutch company has been involved in various hydrogen fuel cell projects, including supplying fuel cell systems for ships. They have collaborated on projects such as the "Johanna" fuel cell boat, which is used for passenger transport. The HySeas III project is aims to develop a hydrogen fuel cell-powered ferry for use in Scotland's Orkney Islands. The project aimed to demonstrate the feasibility of using hydrogen fuel cells to decarbonise short sea shipping routes. The energy observer is a hydrogen-powered catamaran that embarked on a global journey to showcase the potential of renewable energy and hydrogen in maritime transport. It combines solar panels, wind turbines, and a hydrogen fuel cell system to generate and store energy for propulsion.

In addition, several international collaborative projects are also being discussed. Hydrogen boats have been developed and demonstrated by several researchers in a number of countries. Several research institutes and industries are putting up their best efforts to revolutionise the marine technology that would ensure the use of hydrogen in boats with higher efficiency. Bristol hydrogen boat was commissioned by the Bristol City Council in 2010 as part of its Green Capital initiatives. These boats were powered by a 12 kW fuel cell via two permanent magnet DC motors (Figure 6.82). The ferry carries 12 passengers and 2 crew members. The hydrogen fuel and refuelling station for the ferry are supplied by Air Products.

Figure 6.82 Hydrogen-powered boats. *Sources:* Amadea K/Wikimedia Commons and Bristol Hydrogen Boats.

6.27.8 Hydrogen in Aviation and Air Transport

Aircraft emission, although it is of no lesser consequence, has not received the attention it needs. Of course, aviation is one of the toughest sectors to decarbonise. The steps of using sustainable aviation fuels such as fuels produced from electricity (power-to-liquids or 'PtL'), hydrogen and battery electric aircraft are being considered. SAF from biomass has been successfully demonstrated. Biofuels are not totally emission free, and in that case, hydrogen is considered a suitable propulsion fuel. Hydrogen, if used for air propulsion, needs to be liquefied to supply the required range. Moreover, combustion turbines may be necessary because the fuel cells may not have the power during take-off. The current aircraft, using hydrogen, will be limited in the ranges to about 2500 km. Redesigning the aircraft to increase the range and install the appropriate storage technology without decreasing the number of seats might be a very time-consuming affair. There is another question that is related to climate benefit of using hydrogen. It is certain that water vapour coming out of the aircraft will be very low in quantity. However, water vapour at high altitudes causes radiative forcing, thereby contributing to the problem of global warming. Therefore, these are some of the points that need to be considered before any long-term solutions are evolved for aircraft propulsion. However, fuel cell buses and other ground vehicles can be operated near the airport to maintain the air quality. Another application area of fuel cells lies in unmanned aerial vehicles (UAVs) that can be used for both civilian and military purposes. These are more efficient and operate with much less vibration than their fossil fuel counterparts. Moreover, since fuel cell UAVs are much lighter than battery-powered UAVs, they can have a much longer range.

It is well known that NASA started using liquid hydrogen as a rocket fuel more than a half a century ago, as early as in the 1950s. Liquid hydrogen has some specific advantages as fuel for commercial subsonic and supersonic aircraft. One of its major advantages lies in its high energy content (142 MJ kg^{-1}), which is 2.8 times higher than that of conventional jet fuel. Hoelzen et al. [132] conducted a very useful study on the relevance of green hydrogen in aviation. The technology of hydrogen use in aircraft has been studied earlier by several researchers even though a number of questions need to be answered before widespread application. H_2 aircraft technology has been studied, but the critical factor based on a critical cost analysis related to infrastructure needs to be assessed. Hoelzen and team [132] carried out a cost analysis that shows the significance of hydrogen infrastructure essential for aviation. In this study, short- and a medium-range aircraft were modified for hydrogen propulsion. In this modelling study, in addition to the aircraft, the direct operating cost (DOC) related to the infrastructure of both aircraft was also evaluated. A detailed economic analysis carried out by the researchers clearly shows that low-cost green liquid hydrogen is the most desirable fuel for the aviation industry. Depending upon the quality of hydrogen, with the present level of technology, the total DOC of the system could vary over a wide range for short-range aircraft and 15–102% for medium-range aircraft.

Rondinell et al. [133] discussed various aspects concerned with the use of hydrogen as an aviation fuel. Specific design criteria and optimum operating conditions should be judiciously evaluated to meet the challenges related to safety. The environmental compatibility of the system is of high priority. Their studies were carried out on the use of liquid hydrogen as an aviation fuel. The aircraft configuration, critical dimensions, weight, and volume of the aircraft have been evaluated from an environmental point of view. This study concludes that cryogenic liquid hydrogen is the most sustainable aviation fuel from both environment and safety points of view. However, existing problems related to the economic viability of adequate amounts of hydrogen production and efficient storage and delivery systems need to be addressed.

Choi and Lee [134] analysed liquid hydrogen fuel in aviation. Researchers built four models of an aircraft with liquid hydrogen models. The analysis considered the expected fleet growth on the basis of the current picture. Liquid hydrogen costs were estimated considering the environmental regulations. According to this analysis, it has been observed that approximately 66–100% of all aircraft (from existing aircraft) need to be converted to liquid hydrogen aircraft in 2050. As far as economic viability is concerned, it is expected that the cost of liquid hydrogen would be rather low compared to the cost of jet fuel.

Regarding the launch of space vehicles starting with Ariane 1 to Ariane 6, it is well known that liquid hydrogen and liquid oxygen can be used for space vehicles. Ariane 5 used over 12 000 m^3 of liquid hydrogen as fuel. Of course, various difficulties, technical and non-technical in nature, have not been discussed here. The transition of the aviation industry to commercial flights on hydrogen is a long-term goal due to several technical, infrastructure and safety criteria. However, active research is underway to explore this possibility.

The MAHEPA (modular approach to hybrid-electric propulsion) consortium consisting of Pipistrel Vertical Solutions, Compact Dynamics, DLR, H2Fly, Politecnico di Milano, TU Delft, University of Maribor and University of Ulm has already developed the *HY4* hydrogen-powered, hybrid-electric aircraft. HY4 has flown at an altitude of 7230 feet (about 2203 m) and has demonstrated a 77-mile (about 124 km) journey between two airports. H2FLY has focused on converting existing regional aircraft to use hydrogen fuel cells as a power source. They retrofit a Dornier 228 aircraft with a hydrogen fuel cell system for propulsion. H2FLY has established itself as a leader in the development and advancement of hydrogen-electric propulsion systems for aircraft. The successful completion of Hy4's first flight demonstrates the feasibility and potential of hydrogen-electric propulsion systems for aircraft. ZeroAvia (United Kingdom and United states) is developing hydrogen-electric Piper M-class aircraft for hydrogen propulsion in aviation. On 19 January 2023, ZeroAvia took its maiden flight with its 19-seat Dornier 228 testbed aircraft. A full-size prototype hydrogen-electric powertrain is installed on the left wing of the aircraft. ZeroAvia is getting its activity accelerated to operate in commercial routes by 2025. In 2022, ZEROe demonstration Airbus launched the 'ZEROe Demonstrator' with the objective of testing hydrogen combustion technology on an A380 multimodal platform. Airbus ZEROe designed three aircraft for hydrogen propulsion. All three ZEROe concepts are hybrid-hydrogen aircraft. They are powered by hydrogen combustion through modified gas turbine engines. Liquid hydrogen is used as fuel for combustion with oxygen. Apart from traditional commercial aviation, hydrogen fuel cells are developed to create electrical power that complements the gas turbine. Such systems would lead to a highly efficient hybrid-electric propulsion system. It is expected that by conducting exhaustive ground and flight tests, appropriate mature technology could be developed to adopt regular hydrogen propulsion systems by 2025.

6.27.9 Hydrogen Use in the Domestic Sector

Hydrogen-operated gensets, both with SI and CI engines, have been discussed earlier. These gensets can be used to decentralise domestic power supply systems.

Broadly speaking, hydrogen can be used for space heating and cooling and for water heating in the same manner as natural gas. However, some modifications to the burner and fuel-supply system need to be made considering the properties of hydrogen. Since no other emissions are generated, these heaters are being safely used indoors.

Instead of flame combustion, hydrogen can be used in catalytic combustors to directly heat and humidify the air. These flame-free catalytic diffusion burners have the advantages of high-efficiency and pollution-free operation as there would be practically no NOx and CO_2. In these burners, hydrogen diffuses outwards and oxygen diffuses inwards. The reaction occurs over a platinum catalyst supported on a sintered metal disc/cylinder. Figure 6.83 shows the schematic diagram of a hydrogen catalytic burner and Figure 6.84 shows the picture of such a diffusion burner.

Hydrogen through either flame or catalytic burners can be used for cooking purposes. However, appropriate design features need to be installed to avoid backfire. Hydrogen in the compressed or hydride form can be used for domestic applications. Cooking stoves

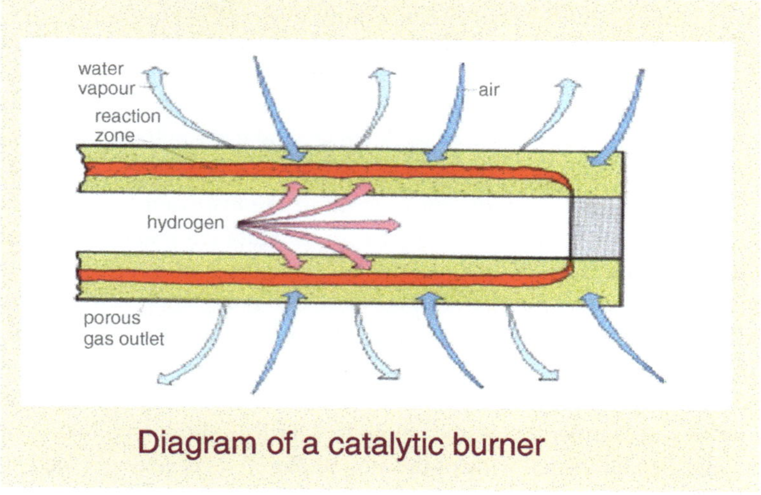

Figure 6.83 Catalytic burner.

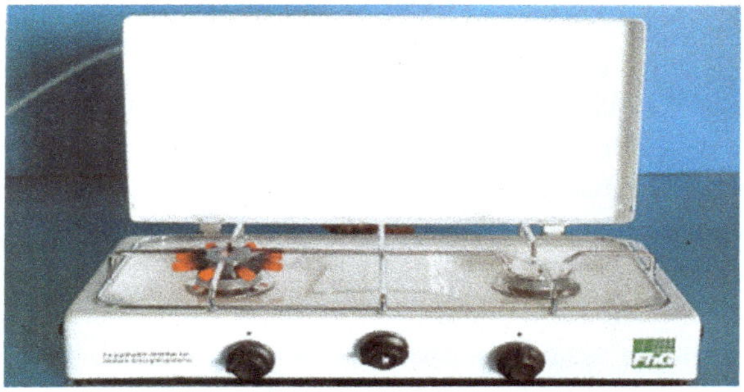

Figure 6.84 Catalytic hydrogen as a diffusion burner.

have also been developed using hydrogen as part of research pursuits in several institutes, and the MNRE of the Government of India had sponsored projects on this topic. A typical hydrogen stove thus developed is shown in Figure 6.85.

Flashback control of the flame into the fuel–air mixture is one of the major design criteria for building up these stoves. The tendency of flashback has been found to be avoided with natural gas and propane if the fuel velocity is adjusted to a range higher than the normal flame velocity. However, this technique does not work for hydrogen because of its high flame velocity. It is observed that the frequency of the backfire is substantially reduced if hydrogen is prevented from mixing with air before the burner port. Such a design configuration has been found to have lower NOx emissions.

Apart from domestic cooking stoves, the design and development of hydrogen as burners for industrial application have also been investigated. However, these projects have thus far been restricted to limited research. Kozhukhova, du Preez and Bessarabov [135] discussed the application of catalyst materials for safe domestic application. A large number of catalysts have been investigated for catalytic hydrogen combustion (CHC). Since Pt has been found to be most suitable for hydrogen-related reaction, the major difficulties for the catalyst support material have been due to several reasons, such as low thermal conductivity, cost of catalyst supports, degradation of supports in the severe thermal environment at the higher temperature level, etc. The prospective application of hydrogen in domestic appliances, such as catalytic burner and cooker, has been discussed. The successful use of Pt/SiC has been demonstrated in high-temperature CHC-based cookers.

Figure 6.85 A typical hydrogen stove. *Source:* cetechamerica.

Concluding Remarks

Considering the infinite source potential and planet-friendly characteristics of hydrogen, it is emerging as an undisputed candidate for the decarbonised world. The use of hydrogen in most of the key applications from small domestic appliances to large-scale industries and in land, air and water transport systems has been successfully demonstrated. Demand for hydrogen is growing rapidly. Although the major part of the demand still continues from the traditional uses in refineries and some industries, many new areas of applications have been successfully achieved and need to be scaled up. Specific safety aspects needed for handling hydrogen are important; and necessary steps and appropriate training should be given to the persons involved in dealing with hydrogen in any applications.

Successful demonstration of the contribution of pure hydrogen in the direct reduction of iron (DRI) has indicated the start of new steel projects using the DRI technique. Existing vehicles based on ICEs have been developed for total hydrogen operation with fuel injection systems and safety features installed at appropriate locations. These vehicles have shown improved thermal efficiency and practically zero-emission characteristics. In the transition phase, until adequate infrastructure for hydrogen production is built, the existing vehicles can run supplemented with other hydrocarbon fuels, which would be useful to reduce emissions. The first fleet of hydrogen fuel cell trains has already been operated successfully, and now many countries have started plans for using hydrogen fuel cells in trains, buses and other modes of road transport. Several pilot projects have been planned to use hydrogen for power generation. Many of these systems have been developed on a modest pilot scale and are likely to evolve into prototypes. Policies have been formulated, time-bound goals have been fixed and active implementation is going on in many countries to meet climate pledges of net zero emissions by 2050. International collaboration in research, development and marketing has started in several areas of hydrogen energy technology. Many institutes and laboratories in different countries have joint projects related to hydrogen energy. Recently, liquefied hydrogen from Australia was shipped to Japan in February 2022. Thus, the emerging trend of advancement in technological innovation, market dynamics, regulatory support and public acceptance ensures the widespread application of hydrogen energy in several carbon-neutral energy systems for a greener and more resilient future.

Abbreviations

ABB	Asia Brown Boveri
AFC	Alkaline Fuel Cell
AIP	Air Independent Propulsion
ATDC	After Top Dead Centre
BASF	Badische Anilin- und Sodafabrik
BFL	Backfire Limit
BMEP	Brake Mean Effective Pressure
BSFC	Brake Specific Fuel Consumption
BTDC	Before Top Dead Centre

BTE	Brake Thermal Efficiency
BTL	Biomass-To-Liquid
CC	Combustion Chamber
CCS	Carbon Capture Storage
CHC	Catalytic Hydrogen Combustion
CI	Compression Ignition
CMAL	Caledonian Maritime Assets Limited
CMI	Continuous Manifold Ignition
CNG	Compressed Natural Gas
CO	Carbon Monoxide
CR	Compression Ratio
CTL	Coal-To-Liquid
DAP	Diammonium Phosphate
DC	Direct Current
DISI	Direct Injection Spark Ignition
DME	Dimethyl Ether
DNV GL	Det Norske Veritas, Germanischer Lloyd
DOC	Direct Operating Cost
DRI	Direct Reduction of Iron
DRP	Direct Reduction Process
EGR	Exhaust Gas Recirculation
FCEV	Fuel Cell Electric Vehicle
FCH JU	Fuel Cell Hydrogen Joint Undertaking
FSN	Filter Smoke Number
GHG	Green House Gas
GO	Graphite Oxide
GTL	Gas-To-Liquid
HC	Hydrocarbon
HCCI	Homogeneous Charge Compression Ignition Engine
HCNG	Hydrogen Compressed Natural Gas
HDR	Hydrogen Donor Diluent Reduction
HRR	Heat Release Rate
HYBRIT	Hydrogen Breakthrough Ironmaking Technology
ICE	Internal Combustion Engine
ICI	Imperial Chemical Industries
IEA	International Energy Agency
IGCC	Integrated Gasification Combined Cycle
IIT	Indian Institute of Technology
IMO	International Maritime Organization

IPA	Intermountain Power Agency
IVC	Intake Valve Closure
LHV	Low Heating Value
LNG	Liquified Natural Gas
LPDI	Low Pressure Direct Cylinder Injection
LPG	Liquified Petroleum Gas
MCFC	Molten Carbonate Fuel Cell
MDPI	Multidisciplinary Digital Publishing Institute
MJ	MilliJoules
MNRE	Ministry of New & Renewable Energy
MON	Motor Octane Number
MPE	Minimum Pulse Energy
MW	Mega Watts
NASA	National Aeronautics and Space Administration
NOx	Oxides of Nitrogen
PAFC	Phosphoric Acid Fuel Cell
PCCI	Premixed Charge Combustion Ignition
PEM	Polymer Electrolyser Membrane
PFI	Port Fuel Injection
PV	Photovoltaic
RFC	Reversible Fuel Cell
RO	Reverse Osmosis
RON	Research Octane Number
SAF	Sustainable Aviation Fuel
SCR	Selective Catalytic Reduction
SI Engine	Spark Ignition Engine
SMR	Small Modular Reactor
SOFC	Solid Oxide Fuel Cell
SWOT	Strength Weakness Opportunity Threat (Analysis)
TCO	Total Cost of Ownership
TDC	Top Dead Centre
THC	Total Hydrocarbons
TKE	Turbulent Kinetic Energy
TMI	Timed Manifold Injection
ULSD	Ultra Low Sulphur Diesel
UNIDO	United Nations Industrial Development Organization

References

1 Zhang, B., Zhang, S.-X., Yao, R. et al. (2021). Progress and prospects of hydrogen production: opportunities and challenges. *Journal of Electronic Science and Technology* 19 (2): 100080.

2 Okonkwo, E.C. et al. Sustainable hydrogen roadmap: a holistic review and decision-making methodology for production, utilisation and exportation using Qatar as a case study, September 2021. *International Journal of Hydrogen Energy* 46 (10–11): 35525–35549.

3 Dalena, F., Senatore, A., Marino, A. et al. (2018). Chapter 1 – Methanol production and applications: an overview. In: *Methanol, Science and Engineering* (ed. A. Basile and F. Dalena), 3–28. ISBN: 978-0-444-63903-5.

4 Yu, K.M.K., Curcic, I., Gabriel, J., and Tsang, S.C.E. (2008). Recent advances in CO_2 capture and utilization. *ChemSusChem* 1: 893–899.

5 Ren, M., Zhang, Y., Wang, X., and Qiu, H. (2022). Catalytic hydrogenation of CO_2 to methanol: a review. *Catalysts* 12: 403.

6 Åhmana, M., Olssonb, O., Vogla, V. et al. (2018). Hydrogen steelmaking for a low-carbon economy: A joint LU-SEI working paper for the HYBRIT project. *EESS Report no 109 SEI Working Paper WP 2018-07 September*

7 Öhman, A., Karakaya, E., and Urban, F. (2022). Enabling the transition to a fossil-free steel sector: the conditions for technology transfer for hydrogen-based steelmaking in Europe. *Energy Research & Social Science* 84: 102384.

8 Wang, Y.Q., Liu, Y.H., Wang, S. et al. (2020). Hydrogen agronomy: research progress and prospects. *Journal of Zhejiang University. Science. B* 21 (11): 841–855.

9 Russell, G., Nenov, A., Kisher, H., and Hancock, J.T. (2021). Molecular hydrogen as medicine: an assessment of administration methods. *Hydrogen* 2 (4): 444–460.

10 Fydrych, D. and Łabanowski, J. (2015). An experimental study of high-hydrogen welding processes. *Revista de Metalurgia* 51 (4): 10–3989.

11 Balestri, M., Benelli, G., Donatini, F. et al. (2007). Enel's Fusina hydrogen-fed power generation plant. *2007 International Conference on Clean Electrical Power*. Capri, Italy, pp. 456-463

12 (2021). News, World's largest hydrogen fuel cell power plant. *Fuel Cells Bulletin* 2021 (11): 5–6. Available online 1 December 2021, Version of Record 1 December 2021. https://doi.org/10.1016/S1464-2859(21)00602-7.

13 Khalili, S., Rantanen, E., Bogdanov, D., and Breyer, C. (2019). Global transportation demand development with impacts on the energy demand and greenhouse gas emissions in a climate-constrained world. *Energies* 12 (20): 3870.

14 IEA. Global CO_2 Emissions from Transport by Subsector, 2000-2030. IEA, Paris.

15 Erren, R.A. and Hastings Campbell, W. (1933). Hydrogen: a commercial fuel for internal combustion engines and other purposes. *Journal of the Institute of Fuel* VI: 277–291.

16 Cecil, W. (1822). On the application of hydrogen gas to produce a moving power in machinery, with a description of an engine which is moved by the pressure of atmosphere upon a vacuum caused by explosions of hydrogen gas. *Cambridge Philosophical Society* L: 217.

17 Hoffmann, P. (1981). *The Forever Fuel, the Story of Hydrogen*. Boulder, Colorado: Westview Press.

18 Weil, K.H. (1972). The Hydrogen I.C. Engine—its Origin and Future in Emerging Energy–Transportation–Environment System. SAE 729212.

19 Oehmichen, M. (1942). *Wasserstoff als Motoreibmittel, Deutsche Kraft Fahart-Forschung Heft 68*. Berlin: VDI, Verlag GmbH.

20 Verhelst, S. and Wallner, T. (2009). Hydrogen-fueled internal combustion engines. *Progress in Energy and Combustion Science* 35: 490–527.

21 Das, L.M. (1990). Hydrogen engines: a view of the past and look into the future. *International Journal of Hydrogen Energy* 15 (6): 425–433.

22 White, C.M., Steeper, R.R., and Lutz, A.E. (2006). The hydrogen-fueled internal combustion engine: a technical review. *International Journal of Hydrogen Energy* 31: 1292–1305.

23 Furuhama, S. (1981). State of the art and future trend of Hydrogen Fuelled Engines. *JSAE Review* 4.

24 Escher, W.J.D. (1975). *Hydrogen-Fueled Internal Combustion Engine. A Technical Survey of Contemporary US Projects, ETA Report PR-51*.

25 Verhelst, S. (2014). Recent progress in the use of hydrogen as a fuel for internal combustion engines. *International Journal of Hydrogen Energy* 39: 1071–1085.

26 Homan, H.S., De Boer, P.C.T., and McLean, W.J. (1983). The effect of fuel injection on NOX emissions and undesirable combustion for hydrogen-fuelled piston engines. *International Journal of Hydrogen Energy* 8: 131–146.

27 Finegold, J.G. and Van Vorst, W.D. (1974). Engine performance with gasoline and hydrogen. Paper Presented at the THEME Conference at Miami.

28 King, R.O. and Rand, M. (1955). The oxidation, decomposition, ignition and detonation of fuel vapours and gases XXVII, the hydrogen engine. *Canadian Journal of Technology* 33: 445–469.

29 King, R.O., Wallace, W.A., and Mahapatra, B. (1948). The oxidation, ignition, and detonation of fuel vapors and gases V. The hydrogen engine and detonation of the end gas by the igniting effect of carbon nuclei formed by pyrolysis of lubricating oil vapor. *Canadian Journal of Research F26*: 264.

30 King, R.O., Hayes, S.V., Allan, A.B. et al. (1958). The hydrogen engine: combustion knock and related flame velocity. *Transactions of the Engineering Institute of Canada (EIC)* 2 (4): 143–148.

31 Das, L.M. (1990). Fuel induction techniques for hydrogen operated engine. *International Journal of Hydrogen Energy* 15: 833–842.

32 Sirens, R. and Verhelst, S. (2001). Influence of injection parameters on the efficiency and power output of a hydrogen fuelled ENGINE, *ICE Technical Conference, 57-64*. https://doi.org/10.1115/ICES2001-106.

33 McCarley, C.A. (1981). A study of factors influencing thermally induced backfire in hydrogen-fuelled engines and methods of backfire control. IECEC.

34 Varde, K.S. and Frame, G.A. (1984). Development of a high pressure injection for SI engines and results of engine behaviour. *5th World Hydrogen Energy Conference*.

35 Varde, K.S. and Frame, G.A. (1982). A study of combustion and engine performance using electronically hydrogen fuel injection. Paper presented at World Hydrogen Energy Conference-IV, Pasadena, California.

36 Mathur, H.B. and Das, L.M. (1991). Performance characteristics of hydrogen fuelled SI engine using timed manifold injection. *International Journal of Hydrogen Energy* 16: 115–127.

37 Homan, H.S. (1978). An experimental study of reciprocating internal combustion engine operated on hydrogen. PhD Thesis, Cornell University.

38 Furuhama, S. and Azuma, H. (1978). Hydrogen injection engine. *Proceedings of World Hydrogen Energy Conference 4.*

39 Furuham, S. et al. (1981). A LH2 engine fuel system on-board cold GH2 injection into two-stroke engine with LH2-pump. *ASME Paper No 81-HT.*

40 Martorano, L. and Dini, D. (1983). Hydrogen injection in two-stroke reciprocating gas engine. *International Journal of Hydrogen Energy* 8: 935–938.

41 Das, L.M. (1991). Exhaust emission characterization of hydrogen-operated engine system: nature of pollutants and their control techniques. *Intrnl. Journal of Hydrogen Energy* 16 (11): 765–775.

42 Murray, R.G. and Schoeppel, R.J. (1971). Emission and performance characteristics of an air-breathing hydrogen-fuelled internal combustion engine. *Proceedings of Sixth Intersociety Energy Conversion Engineering Conference (IECEC), Paper No. 719019.*

43 Wei, H., Zhu, T., Shu, G. et al. (2012). Gasoline engine exhaust gas recirculation – a review. *Applied Energy* 99: 533–544.

44 Zheng, M., Reader, G.T., and Hawley, J.G. (2004). Diesel engine exhaust gas recirculation—a review on advanced and novel concepts. *Energy Conversion and Management* 45: 883–900.

45 Jayakrishnan, K. (2018). Development of a hydrogen fuelled multi-cylinder vehicular engine with H2-SCR aftertreatment system. Ph.D. Thesis, Indian Institute of Technology Delhi.

46 Bhattacharyya, S. and Das, R.K. (1999). Catalytic control of automotive NO_x: a review. *International Journal of Energy Research* 23: 351–369.

47 Roy, S., Hegde, M.S., and Madras, G. (2009). Catalysis for NO_x abatement. *Applied Energy* 86: 2283–2297.

48 Krishnanunni, J., Bhankar, H., Chugh, D. et al. (2020). Strategies for stable operation of a H_2–ICE at high equivalence ratios with use of unburnt and externally injected hydrogen for reducing NO_x emissions. *International Journal of Hydrogen Energy* 45: 22125–22137.

49 Koebel, M., Elsener, M., and Kleemann, M. (2000). Urea-SCR: a promising technique to reduce NOx emissions from automotive diesel engines. *Catalysis Today* 59: 335–345.

50 Al-Baghdadi, M.A.R.S. (2004). *Effect of compression ratio, equivalence ratio and engine speed on the performance and emission characteristics of a spark ignition engine using hydrogen as a fuel, Renewable Energy* 29: 2245–2260.

51 Subramanian, V., Mallikarjuna, J.M., and Ramesh, A. (2007). Intake charge dilution effects on control of nitric oxide emission in a hydrogen fueled SI engine. *International Journal of Hydrogen Energy* 32: 2043–2056.

52 Das, L.M. (1996). Hydrogen oxygen reaction mechanism and its implication to hydrogen engine combustion. *International Journal of Hydrogen Energy* 21 (8): 703–715.

53 Mathur, H.B. and Das, L.M. (1991). The complete reference can be given as Combustion Related Studies on a Timed Manifold Injection Hydrogen Engine. In: *Energy and*

Environmental Progress I- Hydrogen Energy and Power Generation, vol. D (ed. T.N. Veziroglu), 223–240. Nova Science Publishers, New York.

54 Bihari, B., Gupta, S.B., Sekar, R.R. et al. (2005, January). Development of advanced laser ignition system for stationary natural gas reciprocating engines. *Internal Combustion Engine Division Fall Technical Conference* 47365: 601–608.

55 Liguori, S., Kian, K., Buggy, N. et al. (2020). Opportunities and challenges of low-carbon hydrogen via metallic membranes. *Progress in Energy and Combustion Science* 80: 100851.

56 Onorati, A., Payri, R., Vaglieco, B.M. et al. (2022). The role of hydrogen for future internal combustion engines. *International Journal of Engine Research* 23 (4): 529–540.

57 Pavel, N., Bärwinkel, M., Heinz, P. et al. (2018). Laser ignition-spark plug development and application in reciprocating engines. *Progress in Quantum Electronics* 58: 1–32.

58 Shukla, A., Vaghasia, J., and Mistry, M. (2022). Effect of laser ignition on combustion and performance of internal combustion engine: a review. *Energy Conversion and Management: X* 13: 100166.

59 Srivastava, D.K., Weinrotter, M., Iskra, K. et al. (2009). Characterisation of laser ignition in hydrogen–air mixtures in a combustion bomb. *International Journal of Hydrogen Energy* 34 (5): 2475–2482.

60 Alberti, A., Munafò, A., Koll, M. et al. (2019). Laser-induced non-equilibrium plasma kernel dynamics. *Journal of Physics D: Applied Physics* 53 (2): 025201.

61 Weinrotter, M., Ast, G., Kopecek, H., and Wintner, E. (2005). An extensive comparison of laser-induced plasma ignition and conventional spark plug ignition of lean methane-air mixtures under engine-like conditions (No. 2005-01-0248). SAE Technical Paper.

62 Agarwal, A.K., Singh, A.P., and Pal, A. (2017). Effect of laser parameters and compression ratio on particulate emissions from a laser ignited hydrogen engine. *International Journal of Hydrogen Energy* 42 (15): 10622–10635.

63 Ma, J.X., Alexander, D.R., and Poulain, D.E. (1998). Laser spark ignition and combustion characteristics of methane-air mixtures. *Combustion and Flame* 112 (4): 492–506.

64 Pal, A. and Agarwal, A.K. (2016). Effect of laser pulse energy on laser ignition of port fuel injected hydrogen engine. *International Journal of Hydrogen Energy* 41 (1): 675–682.

65 Dale, J.D., Smy, P.R., and Clements, R.M. (1978). Internal combustion engine-an experimental study. *SAE Transactions* 780329: 1539–1548. https://doi.org/10.4271/780329.

66 Natkin, R.J., Tang, X., Boyer, B. et al. (2003). Hydrogen IC engine boosting performance and NOx study. *SAE Transactions*, Technical Paper 2003-01-0631, SAE 2003 World Congress & Exhibition 865–875.

67 BMW (2009). *BMW Hydrogen Engine Reaches Top Level Efficiency*. Munich, Germany: BMW.

68 Matthias, N.S., Wallner, T., and Scarcelli, R. (2012). A hydrogen direct injection engine concept that exceeds US DOE light-duty efficiency targets. *SAE International Journal of Engines* 5 (3): 838–849.

69 Ikegami, M., Miwa, K., Shioji, M., and Esaki, M. (1980). A study on hydrogen-fuelled diesel combustion. *Bulletin of the JSME* 23 (181): 1187–1193.

70 Ikegami, M., Miwa, K., and Shioji, M. (1982). A study of hydrogen-fuelled compression ignition engine. *International Journal of Hydrogen Energy* 7 (4): 341–353.

71 Welch, A.B. and Wallace, J.S. (1986). Performance characteristics of a hydrogen fuelled diesel engine with ignition assistance. *Final Report, NRCC, Report No. DSS, Contract File No.24SU.31155-2-2664. Serial No. ISU 82-00340.*

72 Szwaja, S. and Grab-Rogalinski, K. (2009). Hydrogen combustion in a compression ignition diesel engines. *International Journal of Hydrogen Energy* 34: 4413–4421.

73 Mathur, H.B., Das, L.M., and Patro, T.N. (1993). Hydrogen fuelled diesel engines: performance improvement through charge dilution techniques. *International Journal of Hydrogen Energy* 18: 421–431.

74 Patro, T.N. et al. (1990). Utilization of Hydrogen fuel in small utility CI engine and its effect on Engine Performance, emission and combustion characteristics. Ph D Thesis, Indian Institute of Technology, Delhi, New Delhi, India.

75 Mathur, H.B., Das, L.M., and Patro, T.N. (1992). Hydrogen fuel utilization in CI engine powered end utility systems. *International Journal of Hydrogen Energy* 17: 369–374.

76 Patro, T.N. (1993). Combustion study of hydrogen fueled DI diesel engine simplified het release analysis. *International Journal of Hydrogen Energy* 18: 231–241.

77 Frantzis, C., Zannis, T., Savva, P.G., and Yfantis, E.A. (2022). A Review on experimental studies investigating the effect of hydrogen supplementation in CI diesel engines—the case of HYMAR. *Energies* 15: 5709.

78 Das, L.M. and Polly, M.2005, January). Experimental evaluation of a Hydrogen Added Natural Gas (HANG) Operated SI Engine. *SAE Paper No 2005-26-29.*

79 Sandhu, S.S., Babu, M.K.G., and Das, L.M. Effect of hydrogen supplementation on the performance, combustion and emission characteristics of a natural gas fuelled S.I. engine. *International Journal of Alternative Propulsion* 2 (2): 181–196.

80 Sandhu, S.S. et al. (2013). Investigations of emission characteristics and thermal efficiency in a spark-ignition engine fuelled with natural gas–hydrogen blends. *International Journal of Low-Carbon Technologies* 8 (1): 7–13. https://doi.org/10.1093/ijlct/ctr038.

81 Mathai, R., Malhotra, R.K., Subramanian, K.A., and Das, L.M. (2012). Comparative evaluation of performance, emission, lubricant and deposit characteristics of spark ignition engine fuelled with CNG and 18% hydrogen-CNG. *International Journal of Hydrogen Energy* 37: 6893–6900.

82 Athawe, M. (2013). Experimental investigation on utilization of hydrogen and CNG in a port injected SI engine. PhD Thesis, Indian Institute of Technology, Delhi, India.

83 Duk Sang, K. (2017). Development of hydrogen-compressed natural gas blend engine for heavy duty vehicles. *International Journal of Automotive Technology* 18 (6): 1061–1066.

84 Mehra, R., Duan, H., Juknelevicius, R., and Ma, F. (2017). Progress in hydrogen enriched compressed natural gas (HCNG) internal combustion engines – a comprehensive review. *Renewable and Sustainable Energy Review* 80: 1458–1498.

85 Mathai, R. (2012). Hydrogen and CNG as fuel for IC engines durability studies of a spark ignition engine for hydrogen blended CNG fuel. PhD Thesis, Indian Institute of Technology, Delhi, India.

86 Lynch, F. E. (1975). Backfire control techniques for hydrogen-fueled internal combustion engines. In: Hydrogen energy. *Proceedings of the Hydrogen Economy Miami Energy Conference, Miami Beach,* Fla., March 18–20, 1974. Part B. (A75-44751 22-44). New York: Plenum Press, pp. 717–726.

87 Mohammed, M.K., Balla, H.H., Al-Dulaimi, Z.M.H. et al. (2021). Effect of ethanol-gasoline blends on SI engine performance and emissions. *Case Studies in Thermal Engineering* 25: 100891.

88 Parthsarathy, M., Joshua, I., Lalvani, R. et al. (2015). Effect of hydrogen on ethanol–biodiesel blend on performance and emission characteristics of a direct injection diesel engine. *Ecotoxicology and Environmental Safety* 134 (Pt 2).

89 Al-Baghdadi, M.A.S. (2003). Hydrogen–ethanol blending as an alternative fuel of spark ignition engines. *Renewable Energy* 28 (9): 1471–1478. http://www.elsevier.com/locate/renene.

90 Yamin, J.A., Sheet, E.A.E., and Rida, K.S. (2021). Relative change in SI engine performance using hydrogen and alcohol as fuel supplements to gasoline. *Advances in Science and Technology Research Journal (ASTRJ)* 22 (6): 144–155.

91 Jeon, J. and Bae, C. (2013). The effects of hydrogen addition on engine power and emission in DME premixed charge compression ignition engine. *International Journal of Hydrogen Energy* 38 (1): 265–273.

92 Choudhuri, A. and Gollahalli, S.R. (2000). Combustion characteristics of hydrogen–propane mixtures. *International Journal of Hydrogen Energy* 25 (5): 451–462.

93 Sevrouk, K.L. et al. (2016). Numerical and experimental analysis of propane–hydrogen mixture ignition in air. *Journal of Physics: Conference Series* 774: 0120.

94 Karagöz, Y. and Köten, H. (2019). Effect of different levels of hydrogen + LPG addition on emissions and performance of a compression ignition engine. *Journal of Thermal Engineering* 5 (2, Special Issue 9): 58–69. Yildiz Technical University Press, Istanbul, Turkey.

95 Verma, S., Kumar, K., Das, L.M., and Kaushik, S.C. (2021). Effect of hydrogen enrichment strategy on performance and emission features of biodiesel–biogas dual fuel engine using simulation and experimental analyses. *Journal of Energy Resources Technology* 143 (9): 1–18.

96 Verma, S. (2019). Comparative assessment of an IC engine using alternative fuels based on energy and exergy. PhD Thesis, Indian Institute of Technology, Delhi, India

97 Alias, M.S.M et al. (2019). A review of hydrogen addition in an HCCI engine fueled with biofuels. *AIP Conference Proceedings 2059, 020045*

98 Murugesan, P. et al. (2022). Role of hydrogen in improving performance and emission characteristics of homogeneous charge compression ignition engine fueled with graphite oxide nanoparticle-added microalgae biodiesel/diesel blends. *International Journal of Hydrogen Energy* 47 (88): 37617–37634.

99 Ambaprasad, G. and Sharma, K. (2021). Prospects of hydrogen in HCCI engines-a review. *nternational Journal of Applied Engineering Research* 16 (7): 538–551. ISSN 0973-4562.

100 Ibrahim, M.M. and Ramesh, A. Investigations on the effects of intake temperature and charge dilution in a hydrogen fueled HCCI engine. *International Journal of Hydrogen Energy* 39 (26): 14097–14108.

101 Ibrahim, M.M. and Ramesh, A. Experimental investigations on a hydrogen diesel homogeneous charge compression ignition engine with exhaust gas recirculation. *International Journal of Hydrogen Energy* 38 (24): 10116–10125, 2013.

102 Jeon, J. et al. (2011). Effect of hydrogen and DME injection in homogeneous charge compression ignition engine with DME second injection strategy. *Institution of Mechanical Engineers – Internal Combustion Engines: Improving Performance, Fuel Economy and Emissions*, 157–166. https://doi.org/10.1533/9780857095060.5.157.

103 Furuhama, S., Hiruma, M., and Enomoto, Y. (1978). Development of a liquid hydrogen car. *International Journal of Hydrogen Energy* 8: 61–81.

104 Furuhama, S. and Fukuma, T. (1986). High output power hydrogen engine with high pressure fuel injection, hot surface ignition and turbocharging. *International Journal of Hydrogen Energy* 11: 399–407.

105 Finegold, J.G., Lynch, F.E., et al., (1973). The UCLA hydrogen car: design, construction and performance. *SAE Paper 730507*, 1626–163.

106 Lynch, F.E. (1983). Parallel induction: a simple fuel control method for hydrogen engine. *International Journal of Hydrogen Energy* 8 (9): 721–730.

107 Gopalakrishnan, R., Throop, M., Richardson, A., and Lapetz, J., (2007). Engineering the ford H2 IC engine powered E-450 shuttle bus. *SAE Technical Paper 2007-01-4095.*

108 Lapetz, J., et al., (2006). The design, development, validation and delivery of the Ford H2 ICE E-450 shuttle bus. In: *1st (Int) Symposium on Hydrogen Internal Combustion Engines*, pp. 20–33.

109 Wallner, T. et al. (2008). Fuel economy and emission evaluation of a BMW hydrogen 7 monofuel demonstration vehicle. *International Journal of Hydrogen Energy* 33: 7607–7618.

110 BMW. (2006). BMW Hydrogen 7 Rescue Guidelines.

111 https://www.energy.gov/eere/fuelcells/comparison-fuel-cell-technologies

112 Ebrahimi, M. et al. (2021). A review on ionic liquids-based membranes for middle and high temperature polymer electrolyte membrane fuel cells (PEM FCs). *International Journal of Molecular Sciences* 22: 5430.

113 Arshad, A., Ali, H.M., Habib, A. et al. (2019). Energy and exergy analysis of fuel cells: a review. *Thermal Science and Engineering Progress* 9: 308–321.

114 Bethoux, O. (2020). Hydrogen fuel cell road vehicles: state of the art and perspectives. *Energies* 13: 5843.

115 Sazali, N., Wan Salleh, W.N., Jamaludin, A.S., and Mhd Razali, M.N. (2020). New perspectives on fuel cell technology: a brief review. *Membranes* 10: 99.

116 Brandon, N.P. and Kurban, Z. (2017). Clean energy and the hydrogen economy. *Philosophical Transactions of the Royal Society A* 375: 20160400.

117 Selmi, T., Khadhraoui, A., and Cherif, A. (2022). Fuel cell–based electric vehicles technologies and challenges. *Environmental Science and Pollution Research* 29: 78121–78131.

118 Samsun, R.C., Antoni, L., Rex, M., and Stolten, D. (2021). Deployment Status of Fuel Cells in Road Transport: 2021 Update Energy & Environment, Band/Volume 542.

119 Oldenbroek, V., Wijtzes, S., Blok, K., and van Wijk, A.J.M. (2021). Fuel cell electric vehicles and hydrogen balancing 100 percent renewable and integrated national transportation and energy systems. *Energy Conversion and Management: X* 9: 100077.

120 Piraino, F., Genovese, M., and Fragiacomo, P. (2021). Towards a new mobility concept for regional trains and hydrogen infrastructure. *Energy Conversion and Management 228: 113650.*

121 Logan, K.G., Nelson, J.D., McLellan, B.C., and Hastings, A. (2015). Japan and the UK: Emission predictions of electric and hydrogen trains to 2050. *Transportation Research Interdisciplinary Perspectives* 10 (2021): 100344.

122 Felseghi, R.-A., Carcadea, E., Raboaca, M.S. et al. (2019). Hydrogen fuel cell technology for the sustainable future of stationary applications. *Energies* 12: 4593. https://doi .org/10.3390/en12234593 . http://www.mdpi.com/journal/energies.

123 Maier, D., Maftei, M., MaieR, A., and Bitan, G.E. (2019). A review of product innovation management literature in the context of organization sustainable development. *Amfiteatru Economic* 21: 816–829.

124 Nazari, M.A., Alavi, M.F., Salem, M., and Assad, M.E.H. (2022). Utilization of hydrogen in gas turbines: a comprehensive review. *International Journal of Low-Carbon Technologies* 17: 513–519. https://doi.org/10.1093/ijlct/ctac025. Published: 21 March 2022.

125 Ditaranto, M., Li, H., and Hu, Y. (2014). Evaluation of a pre-combustion capture cycle based on hydrogen fired gas turbine with exhaust gas recirculation (EGR). *Energy Procedia* 63: 1972–1975.

126 Crismaru, I.V. and Dragomir-Stanciu, D. (2015). The influence of the flame stabilization air intake holes diameter on the no production in industrial gas turbine. *Procedia Technology* 19: 730–736.

127 York, W.D., Ziminsky, W.S., and Yilmaz, E. (2013). Development and testing of a low NOx hydrogen combustion system for heavy-duty gas turbines. *Journal of Engineering for Gas Turbines and Power*, Paper no GTP-12-1258 135 (2): 022001. (8 pages).

128 Van Hoecke, L., Laffineur, L., Campe, R. et al. (2021). Challenges in the use of hydrogen for maritime applications. *Energy Environment Science* 14: 815–843.

129 Frithiof, N. (2021). Five lessons to learn on hydrogen as ship fuel, *Maritime Impact: Our Expertise in Stories* (15 July 2021).

130 Ustolin, F., Campari, A., and Taccani, R. (2022). An extensive review of liquid hydrogen in transportation with focus on the maritime sector. *Journal of Marine Science and Engineering* 10: 1222. https://doi.org/10.3390/jmse10091222.

131 Serna, Á. and Tadeo, F. (2014). Offshore hydrogen production from wave energy. *International Journal of Hydrogen Energy* 39 (3): 1549–1557.

132 Hoelzen, J., Silberhorn, D., Zill, T. et al. (2022). Hydrogen-powered aviation and its reliance on green hydrogen infrastructure: review and research gap. *International Journal of Hydrogen Energy* 47: 3108–3130.

133 Rondinelli, S., Sabatini, R., and Gardi, A. (2017). Challenges and benefits offered by liquid hydrogen fuels in commercial aviation. *International Journal of Sustainable Aviation* 3 (3): 200–216.

134 Choi, Y. and Lee, J. (2022). Estimation of liquid hydrogen fuels in aviation. *Aerospace* 9: 564.

135 Kozhukhova, A.E., du Preez, S.P., and Bessarabov, D.G. (2021). Catalytic hydrogen combustion for domestic and safety applications: a critical review of catalyst materials and technologies. *Energies* 14: 4897.

136 Das, L.M. (2016). Hydrogen fuelled Internal Combustion Engine (Chapter 7). In: *Compendium of Hydrogen Energy, Volume 3*. Woodhead Publishing Series.

7

Life Cycle Sustainability Assessment, Durability and Material Compatibility

I don't want to protect the environment. I want to create a world where the environment doesn't need protection.

– Anonymous

CHAPTER MENU

7.1 Introduction

Since hydrogen is considered one of the most prospective candidates to ensure a sustainable future, it is worth considering its life cycle sustainability assessment (LCSA). An

Hydrogen Energy: Production, Safety, Storage and Applications, First Edition. L. M. Das.

integrated LCSA refers to the evaluation of all environmental, social and economic negative impacts and benefits in decision-making processes towards more sustainable products throughout their life cycle [1]. An overview of the life cycle analysis (LCA) with various phases is provided in Figure 7.1.

The word sustainability stems from the Latin 'sub-tenere', assimilated 'sustinere' (to hold up). However, as per the implication and significance as laid out by the World Commission on Environment and Development (WCED, Brundtland Commission) of the United Nations (UN) in 1987, 'Sustainable development is development that meets the needs of the present without compromising the ability of future generations to meet their own needs' [2]. In the present context, it means human sustainability on planet Earth. The overall analysis of LCSA consists of several components, such as life cycle assessment (LCA), life cycle costing (LCC) and social life cycle assessment (SLCA) [2]. Thus, it is essential to identify relevant impact indicators by which a meaningful environmental profile can be constructed. Considering the above components, LCC methodology takes into account the cash flow analysis of the product over its entire useful lifespan. LCC assessment methods provide economic indicators from the point of view of life in three ways [3]: conventional LCC, environmental LCC and social LCC. LCC is perhaps the only systematic financial cost accounting that has been followed for a long time.

In general, conventional LCC involves investment, operating and maintenance costs. The overall economic benefits during its life are also evaluated in the LCC analysis. Although hydrogen combustion does not emit any major pollutants (particularly in fuel cells), its production in many ways has a negative impact on the environment. A clear picture of the actual environmental burden due to hydrogen energy can be estimated from resource consumption, energy requirements and emissions from the point of view of the

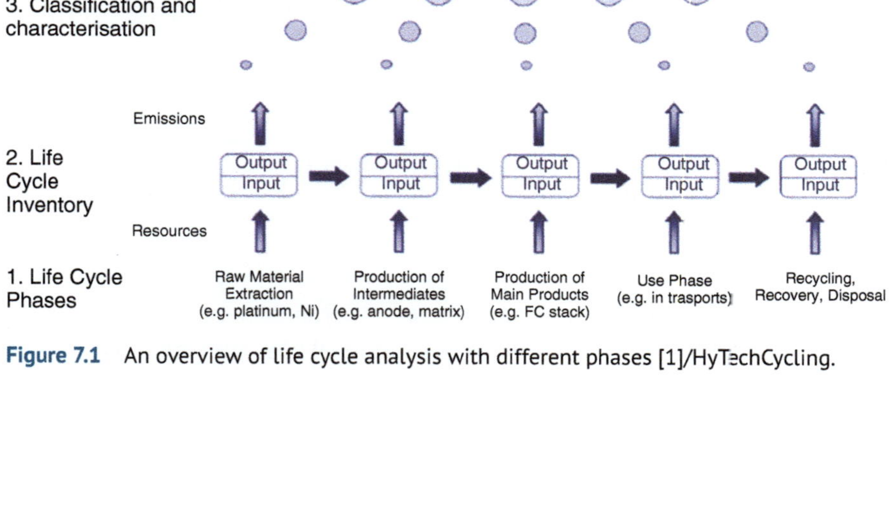

Figure 7.1 An overview of life cycle analysis with different phases [1]/HyTechCycling.

life cycle. Thus, environmental LCC evaluates the economic aspects of environment-related issues, such as damage due to global warming. The economic analysis of ecological LCC differs from that of the conventional LCC system. For the purpose of analysis, systems, boundaries and functional units (FUs) need to be identified. Studies have been conducted to calculate the cost of damage due to environmental degradation considering a wide range of pollutants responsible for disrupting the climate and affecting human health [4, 5]. Social LCA is a very broad area that includes all societal and environmental externalities, and it is very difficult to evaluate these in terms of cost. Numerous factors are taken into consideration, such as social welfare, job quality, lifestyle and various differences from different perspectives of social life. As a result, the analysis of social LCC methods is extremely heterogeneous and highly dependent on a particular society, making it difficult to formulate multiple versions. In recent years, several countries have adopted LCA strategies based on their environmental experience.

7.2 Life Cycle Analysis

Life cycle assessments play a very significant role in life cycle sustainability assessments. LCA consists of various stages of the product from raw material to ultimate disposal. This complete LCA is often referred to as a 'cradle-to-grave' analysis. LCA was initially developed in the 1960s, but it had no long-term impact. Since then, the situation has changed, and LCA has become an essential part of a product's life and its impact on the environment. Broadly, the life cycle consists of selection of appropriate raw materials, transportation and processing, manufacturing and assembly and the life cycle of the product and waste management. Figure 7.1 shows a broad schematic depiction of the life cycle process, and Figure 7.2 shows the phases of a LCA.

LCA attracted attention of global communities and researchers in the 1970s and progressed to environmental burden analysis. In the next two decades, during the 1980s and 1990s, 'full-fledged life cycle impact assessment (LCIA) and LCC models' were introduced [6]. Social and environmental aspects were introduced later, and a very rigid analysis has already been developed.

Environmental LCA has significantly developed over the past three decades. While LCA developed merely from energy analysis to a comprehensive environmental burden analysis in the 1970s, full-fledged life cycle impact assessment (LCIA) and LCC models were introduced in the 1980s and 1990s, and social LCA and particularly consequential LCA were derived in the first decade of the twenty-first century. There are several recent

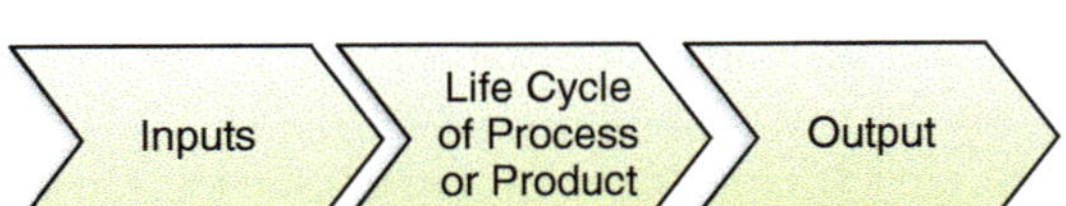

Figure 7.2 Stages of a life cycle process.

developments aimed at broadening traditional methods of LCA to include a more comprehensive life cycle sustainability analysis.

7.2.1 Stages of Life Cycle Assessment

Figure 7.3 highlights the main steps and applications of LCA. In the first stage, the goal and scope of the study must be very clearly identified. In this case, the functional unit must be defined as well as the system and the boundary. An appropriately chosen functional unit is the quantifiable unit of a product or service. FU is an evaluation of the studied system compared to any reference system. The inventory data are related to FU. A set of rational assumptions and limitations must be considered. A very critical aspect of this stage is related to impact categories and the methods to be used. When more than one product or function is involved, this state of assessment is more relevant for allocating environmental burdens.

It is pertinent to note that manufacturing consists of the manufacturing of each element of the product and their assembly. After the product has been packaged and transported, it must then be delivered/distributed to its final destination. During the entire life of the product, several factors should be considered, such as energy consumption and emissions, as well as end-of-life (EoL) treatment and final disposal, including the possibility of reusing the product. LCA is a rational, cradle-to-grave assessment of technologies, allowing an evaluation of their environmental impact. However, economic and social implications do not come under the purview of this analysis. For a decision to be made, this needs to be combined with other tools.

7.2.2 Life Cycle Inventory (LCI)

As part of the second stage, inventory analysis is a systematic methodology used in the field of LCA to quantify and assess the environmental impacts of a product, process or service over its entire life cycle. This phase involves the collection and quantification of data related to inputs, outputs and emissions throughout the life cycle of the product or process being assessed. As part of this analysis, all elementary flows, such as mass flow, energy flow (renewable and

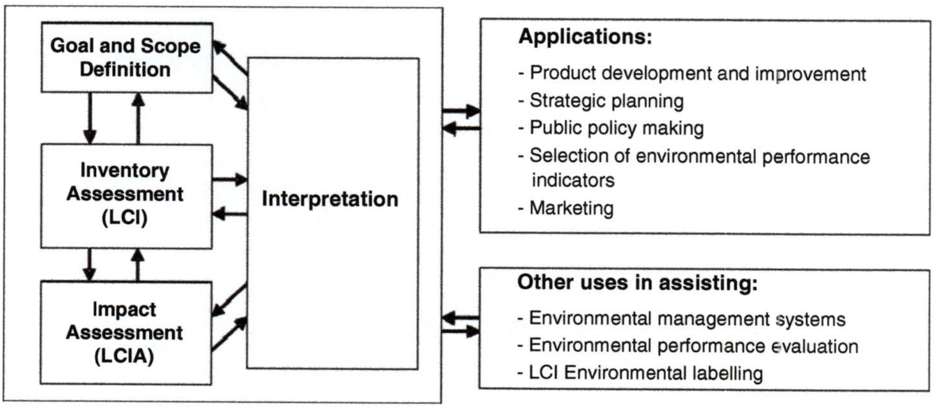

Figure 7.3 Main steps and applications of life cycle assessment.

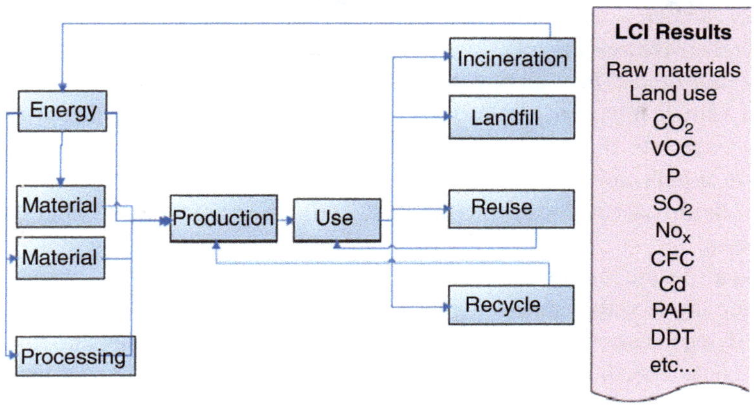

Figure 7.4 Flows of information needed for a life cycle inventory [7]/UNEP.

non-renewable emissions) and species to air, water and land, are considered. In the elementary flow diagram (Figure 7.4), the inventory list shows the grouping of products.

7.2.3 Life Cycle Impact Assessment (LCIA)

LCIA is the third phase of the LCA methodology. It considers the cumulative burdens stemming from the entire product life cycle, including the environmental aspects and impact of the product system. These steps are very clearly depicted in Figure 7.5. In this

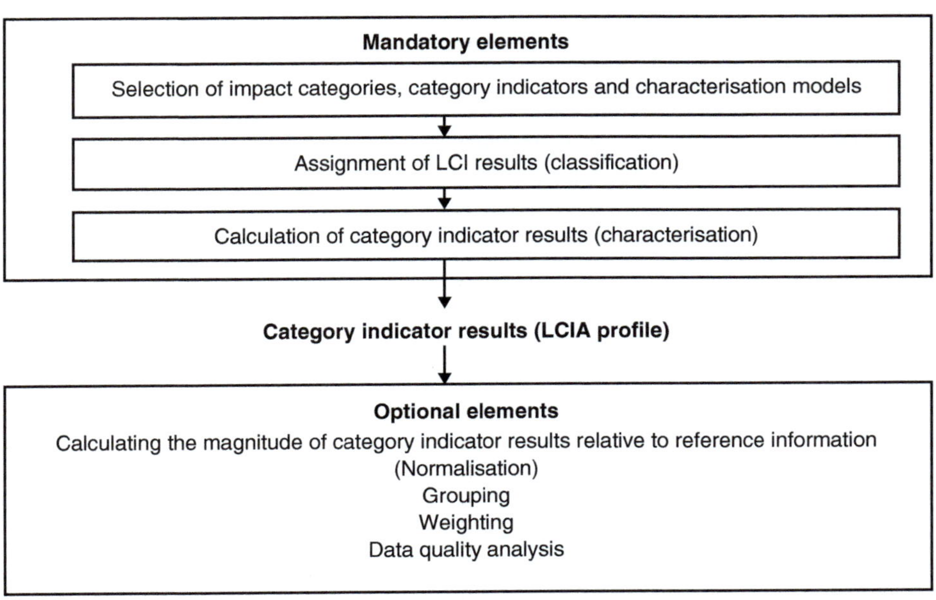

Figure 7.5 Elements of the life cycle impact assessment (LCIA) phase, International Standard Organization [8].

phase, all mandatory elements and certain other additional options are selected, which are essential for a comprehensive assessment. This life cycle impact assessment is based on the concept of category indicators. Then, all optional elements of environmental burdens are grouped together in a single indicator. The LCIA process converts the data obtained at the life cycle inventory stage into more meaningful indicators by using the following three steps: (i) the selection of indicators and characterisation models, (ii) classification to associate inventory data with impact categories and (iii) characterisation to provide the values of the category indicators.

Each category has a specific link to a specific environmental mechanism. For example, infrared radiative forcing is responsible for global warming, whereas proton release is responsible for acidifying emissions. The selection of the particular optional element, however, depends on the objective, the goal and its significance according to specific situations.

This is a very significant phase in which the results of LCIA analysis are translated into environmental impacts. Such an assessment is carried out by impact assessment methods, at the midpoint or endpoint level [9]. Evaluation involves identifying information, quantifying it and then assessing it for its environmental compatibility. Figure 7.6 shows how the relevant set of information can be processed to indicate the categories of impact and the characteristics of damage. The broad indicator descriptors used are in the *International Reference Life Cycle Data System (ILCD) Handbook* (EC-JRC, 2011) [9] and the life cycle impact method [10]. At present, several countries have formulated country-specific integrated product policies (IPPs). The goal is to assess an environmentally friendly product in the most cost-effective manner and its performance throughout its life cycle. It consists of the chain from the extraction of natural resources through their design, manufacture, assembly, marketing, distribution, sale and use until final disposal as waste.

In the IPCC's Fifth Assessment Report [11], midpoint characterisation factors were included in the most significant characterisation models. The relevance of some of these

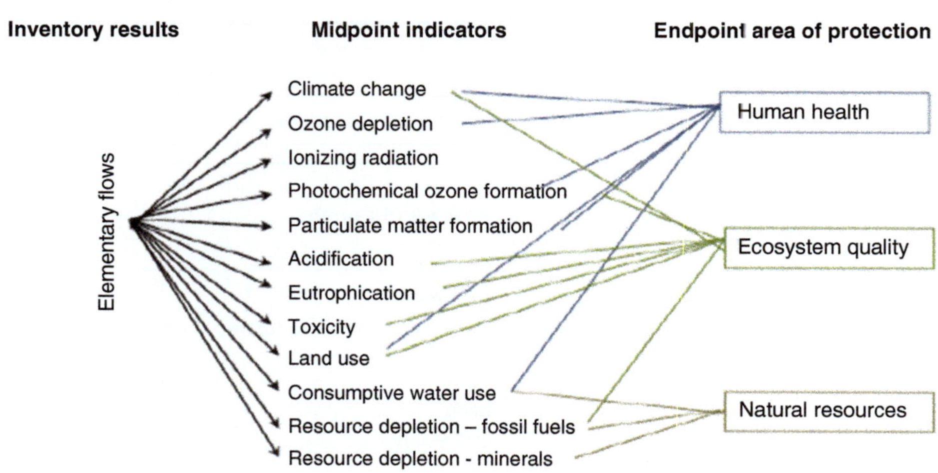

Figure 7.6 Framework of midpoint and endpoint indicators for LCA.

significant midpoints, such as global warming potential (GWP) (which describes climate change), acidification, eutrophication, etc., is discussed later in this chapter.

7.2.4 Fourth and Last Phase of LCIA Life Cycle

In this final phase, the results of the previous phases are interpreted and analysed. The findings are often compared to benchmarks, standards or other reference points to draw conclusions and make recommendations. The goal is to communicate the significance of the environmental impacts and identify opportunities for improvement. This is a very crucial step in the chain as it evaluates the information received from the earlier stages of LCA and/or LCIA. The conclusions derived from the entire set of analyses in this phase should take into account several key environment-related factors according to ISO 14040. In the critical review, all input data are evaluated for consistency, sensitivity, completeness and limitations.

7.3 Technical Review

Research articles published between 2004 and 2021 found that most LCAs for hydrogen production considered electrolysis from renewable energy, and the fuel cell was the most widely studied equipment with regard to electricity- and hydrogen-based LCAs. Most often, compressed storage is used for hydrogen storage. Only a few studies have been carried out on hydrogen or hydrogen-blended natural gas in gas turbines or combined cycles. However, there have been significant discrepancies depending on the main focus on different factors, such as identifying the system boundary and the method of analysis chosen. In the absence of relevant data and uncertainty in the approach of analysis at present, LCA studies with respect to hydrogen energy power generation have not matured and can be considered in the low technology readiness level (TRL), particularly in the absence of adequate relevant data and information with regard to material, energy input and outputs. Information or guidelines to understand the cause-and-effect relationships throughout the entire supply chain are not yet available. A useful approach to the analysis can be derived by identifying the categories of impact and determining the impact assessment method. LCA of alternative energy/fuels is also complicated because it is difficult to estimate indirect land-use change (ILUC) emissions, which are highly uncertain. Studies have taken into account several types of emission sources and sinks that are closely linked to biomass and soil carbon stocks [12]. The study relating to Europe provides rational quantitative information on several aspects, including the amount of land converted by the additional biofuel demand and subsequent greenhouse gas (GHG) emissions, as well as the effects related to this land conversion [12].

Al-Breiki and Bicer [13] analysed liquified natural gas (LNG), ammonia, methanol, DME and hydrogen to assess their environmental characteristics. Considering production and transport, it has been found that methanol and DME are less harmful than LNG. From the point of view of utilisation, however, emission perspectives are different. It has been observed that methanol emits 95.8 g of CO_2 eq. MJ^{-1} of methanol and DME emits 93.7 g of

CO_2 eq. MJ^{-1} of DME if the full life cycle of both DME and methanol is considered from natural gas reforming. These values are 30 and 27% higher compared to the full life cycle of LNG [13]. Using liquid hydrogen to generate electricity can reduce GHG emissions according to data relating to hydrogen. An amount of 9 kg of CO_2 eq. kg^{-1} of L-H_2 is emitted if liquid hydrogen is produced by steam reforming with carbon capture and storage (CCS). However, the emission level is reduced by about 40% if renewable energy is used instead of natural gas reforming.

Most of the reported literature deals with cradle-to-gate boundaries, whereas studies on cradle-to-grave analysis are rare. This is probably due to hydrogen's use in mobility. LCI studies have primarily been analysed using data and information available in literature, and in LCIA analysis. Global warming has been a significant factor. Harmonisation protocols are a very effective tool for a rational assessment of the environmental sustainability of hydrogen energy. This is usually performed by comparing the life cycle indicators against a conventional reference system. One of the major objectives of the harmonised protocol is to avoid any misleading 'confusion in terms of carbon footprint and cumulative energy demand' [14]. Quite often, hydrogen production by steam methane reforming (SMR) has been selected as the reference system because of its widespread application. The literature shows that GWP (as an indicator of carbon footprint) and cumulative non-renewable energy demand have been taken as indicators for LCA.

Figure 7.7 shows a schematic diagram of a harmonised hydrogen system. This system must be used in order to arrive at a definitive conclusion in terms of carbon footprint and cumulative energy demand. In addition to assessing the GWP, researchers have also carried out studies related to the acidification potential (AP) of hydrogen energy production systems. A comparison was made between hydrogen produced using SMR systems and renewable energy sources. A very significant result from this study indicates that, in the case of the hydrogen energy system, there is no correlation between acidification and carbon footprint [14].

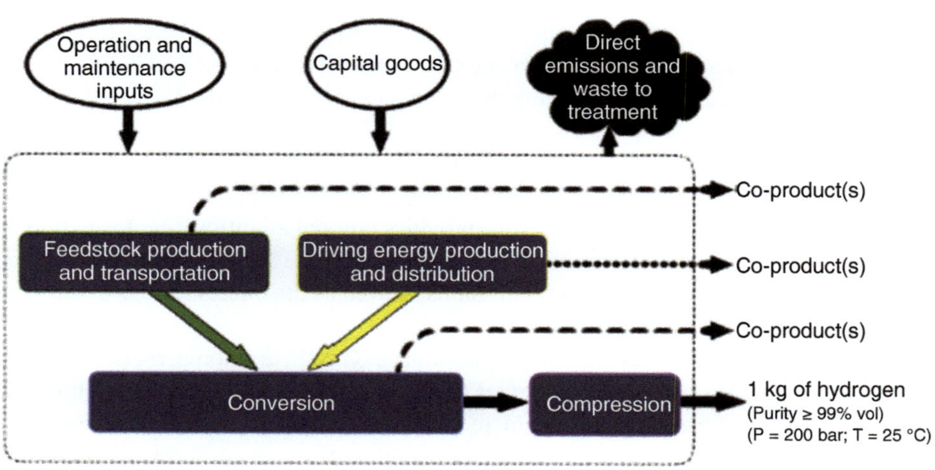

Figure 7.7 Schematic diagram of a harmonised hydrogen energy system [14]/with permission of Elsevier.

7.4 Life Cycle Assessment of Hydrogen Production

While sifting through the available literature with respect to LCA studies of different aspects of hydrogen energy, it is observed that the analysis using electrolytic production has received adequate attention and is widely studied. As far as impact analysis is concerned, most researchers have studied GWP. AP receives the second rank. Many other significant problems, such as toxicity, have not been adequately addressed. Antonio and co-researchers extensively reviewed the LCA of hydrogen energy systems [15]. The authors reviewed 97 scientific papers published up to 2015. The study looked at 509 original cases related to hydrogen systems. Based on this extensive literature survey, hydrogen production methods were classified into three broad groups: thermochemical, electrochemical and biological. Their study indicated that steam reformation and gasification were the most researched processes within the thermochemical category, whereas electrolysis was the most common process within the electrochemical category. Among the biological categories, fermentation was the most studied topic. As shown in the published literature on the topic of LCA of hydrogen energy systems, the studies are mostly comparative.

There is a glaring difference in the methodical choices. This difference in the approach to the analysis does not provide adequate guidelines or information to arrive at a well-defined interpretation. The findings of this study have been used to harmonise the results of the life cycle global warming impact of hydrogen production. Hydrogen production for this analysis was considered by the following three routes: (i) thermochemical, (ii) electrochemical and (iii) biological. Figure 7.8 presents the hydrogen production routes on which harmonisation studies were carried out.

The results of this study are highly relevant at this time when the 21st session of the Conference of the Parties (COP) is proposing to reduce GHG emissions so that the planet's average temperature level may not exceed 2 °C above pre-industrial levels. Original and harmonised results of the GWP of renewable energy with respect to the harmonised GWP of the SMR process were compared. The validity of several processes for producing hydrogen from renewable energy sources was examined. It was concluded that the

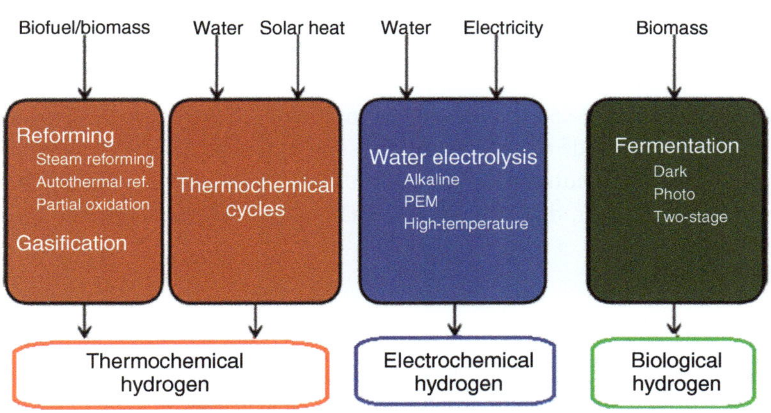

Figure 7.8 Harmonised studies on production routes for hydrogen [15]/with permission of Elsevier.

harmonisation protocol is more consistent with the thermochemical and biological categories than with the electrochemical category. A team of researchers extensively studied the application potential of the harmonisation protocol. As discussed earlier, a harmonisation protocol has already been developed that takes global warming into account. Apart from the use of GWP as a measure of carbon footprint, they also analysed the theme of a harmonisation protocol for cumulative non-renewable energy demand [16]. The results of the harmonised analysis for conventional hydrogen and 20 case studies of renewable hydrogen were compared. In this analysis, it has been observed that the cumulative energy demand shows a well-defined correlation.

Cetinkaya et al. [17] performed a comprehensive analysis of the subject taking into consideration the well-known processes of hydrogen production, such as steam reforming of natural gas, coal gasification, water electrolysis via wind and solar electrolysis, and thermochemical water splitting with a Cu–Cl cycle. A comparative evaluation of the energy equivalents of each method as well as the level of emissions equivalent to carbon dioxide emissions has also been carried out. As far as hydrogen production methods are concerned, this study shows that processes such as natural gas steam reforming, coal gasification and thermochemical water splitting with the Cu–Cl cycle methods are more advantageous than routes of production through renewable energy. From the point of view of emissions, thermochemical water splitting with the Cu–Cl cycle is the most efficient. In a study on the Life Cycle Assessment of Renewable Hydrogen Production via Wind/Electrolysis by Spath and Mann of NREL [18], researchers discussed in detail the impact on the environment of the production of hydrogen using wind/electrolysis. Wind and solar electrolysis have been observed to have advantages over other methods.

7.5 LCA-based Emissions

In any LCS study, greenhouse gas emissions have been observed to have a substantial impact on the environment. A detailed quantification of the effect of GHGs is therefore vital to any LCA of sustainability. Such quantification is carried out by the infrared radiative forcing values of different GHG species relative to CO_2 in terms of CO_2-equivalents (CO_2-eq).

In addition to oxides of nitrogen (NOx), volatile organic compounds (VOCs), oxides of sulphur and particulate matter PM2.5 are also relevant to general LCA studies. GWP analysis must take into account the mode of transportation, such as truck transport or pipeline transportation. Acidification and eutrophication must also be analysed from a global perspective. However, in this chapter, dealing with the LCA of hydrogen energy, the more relevant ones are described below.

7.5.1 Global Warming Potential

Global warming is caused by the release of greenhouse gases that disrupt the climate and is commonly referred to as 'climate change'. The release of aerosols or black carbons, depending on their intensity and location, will often change the environment and thus the climate, as is well known from atmospheric science. Hydrocarbons can also contribute to

Table 7.1 % Contribution to global warming potential (GWP).

GWP	% Contribution to GWP		
(gCO_2 equivalent/kg of H_2)	CO_2	CH_4	N_2O
970	97.9	0.6	1.5

global warming at times. Strictly speaking, the effects of global warming are also very mildly countered through the uptake of carbon dioxide into biomass and soil. However, these aspects have not been accounted for in the LCA in most available literature, and the impacts of greenhouse gases have not been considered. Climate change has already resulted in many severe consequences, such as the melting of the polar ice caps and a rise in sea levels, as well as unanticipated floods and draughts due to radical changes in rainfall patterns, rise in cyclones and associated weather events, loss of biodiversity and drastic disruption in the ecosystem. Anthropogenic activities involving fossil fuel combustion have been the major source of GHG emissions. The main areas are transport, agriculture, electricity, heat generation, several types of industrial activities, mining, etc.

The process of estimating the life cycle of GHGs is quite complex because the calculation of GHG intensity depends upon a number of parameters and methods adopted for calculations. GWP is expressed as a sum of the emissions made by CO_2, CH_4 and N_2O, and this is usually expressed as CO_2 equivalence. According to the Fifth Assessment Report of the IPCC, the GWPs of CH_4 and N_2O are 28 and 265 times higher than those of CO_2, respectively, over a 100-year timeframe. Table 7.1 shows the GWP (g CO_2-equivalent/kg of H_2). The major contributors to global warming are CO_2, CH_4 and N_2O.

A large amount of location-specific data and the validity of the assumptions adopted for analysis play a significant role. In Figure 7.9 [18], the life cycle GWP is shown in terms of carbon dioxide equivalents. The study results indicate that emissions from the operation of plants are extremely low and almost negligible compared to those caused by the manufacturing and construction of turbines. It has also been observed that an increase in wind turbine efficiency reduces resource consumption and emissions [18].

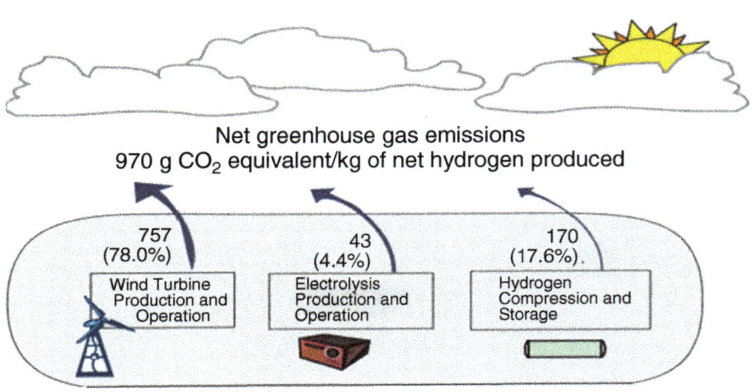

Figure 7.9 Life cycle global warming potential [18]/US Department of Energy/Public Domain.

Bhandari et al. [19] summarised 21 papers the team reviewed in the area of LCA studies on hydrogen production technologies. Most of the studies, according to the researchers, focused on the electrolysis process, and almost all analyses considered the impact of global warming. It has been observed that the impact of GWP on electrolysers is relatively much smaller, and the major point of concern in electrolytic hydrogen production was primarily due to electricity supply [19]. In these LCA analyses, AP has also been considered by a number of researchers though the problem of toxicity potential has hardly been studied. From an LCA standpoint, the method of generating electricity by hydropower and wind power was compared to electrolysis. In Figure 7.10, it is evident that in some cases the values of GWP widely differ from one study to another. This has been shown by extended red lines in the process. It is also clear from the figure that wind electrolysis is the most effective option, followed by hydro-electrolysis. The GWP values for grid-based electricity, taken from the German-based Union for the Coordination of Transmission of Electricity (UCTE) [16], are almost 30 times that of wind electrolysis. It is the electrolysis pathway of grid power that can vary from clean to very dirty, and this is what determines generation on the grid.

The share of each stage in the process of electrolysis in a wind turbine is shown in Figure 7.11. Wind power production and operation have a negligible contribution to GWP (4%), while wind turbine production and operation have the major contribution. This is primarily due to the high level of fossil fuel resources in the grid electricity mix.

Figure 7.12 provides a comparative depiction of GWP in several hydrogen production technologies, both electrolytic and non-electrolytic routes. As shown earlier in Figure 7.9, different values obtained from different studies for the same method are shown as extended red lines in the tip [19]. There is a high contribution to global warming from conventional processes, such as coal gasification and SMR. Considering the strong points in favour of wind electrolysis, it is briefly discussed here. In a wind electrolysis system, the electrolyser

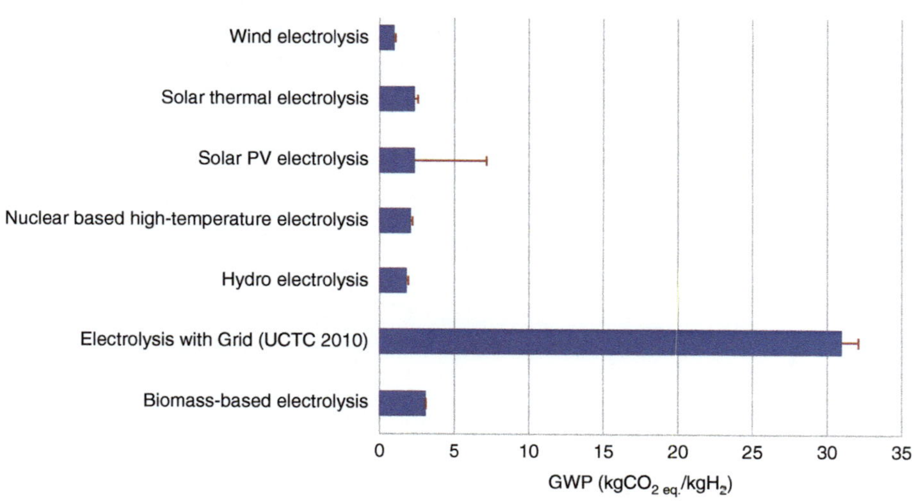

Figure 7.10 Global warming potential of various methods of hydrogen production [19]/with permission of Elsevier.

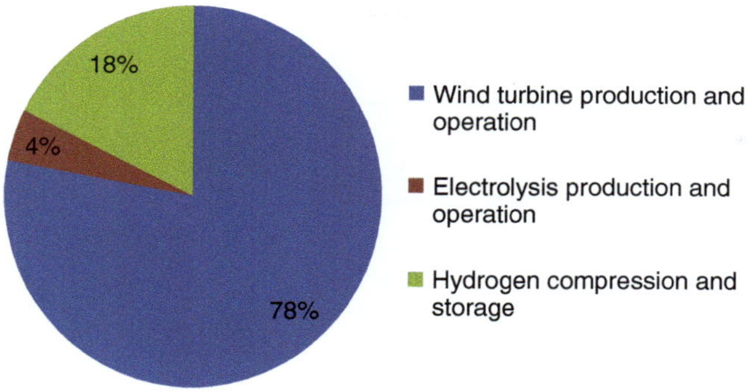

Figure 7.11 Share of GWP in wind electrolysis [19]/with permission of Elsevier.

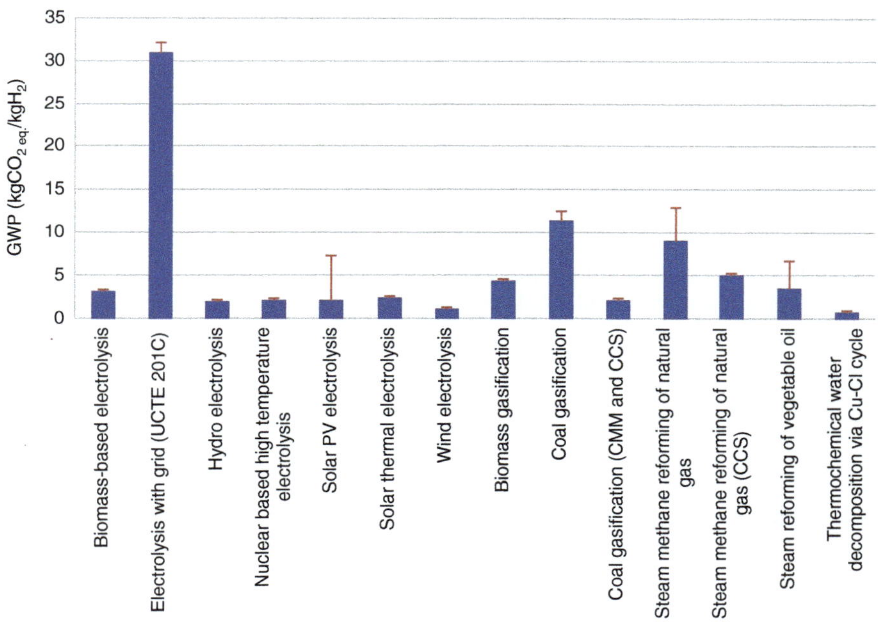

Figure 7.12 Values of GWP in various hydrogen production technologies. *Source:* Adapted from [16].

is located at a central plant, and hydrogen is the only product from the electrolyser. There is no decentralised production with a refuelling station as shown in Figure 7.13.

7.5.2 Acidification

When, due to some industrial activities as well as from vehicles and power plants, oxides of nitrogen (NOx), sulphur oxides (SOx), sulphuric acid and ammonia are emitted into the air, they react with moisture in the atmosphere and form acidic compounds, such as nitric acid, sulphuric acid, etc. These acids are subsequently deposited on land and water.

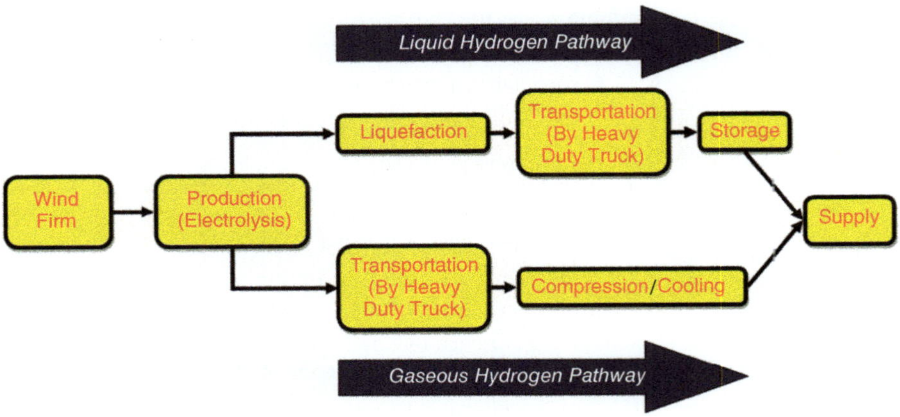

Figure 7.13 Wind electrolysis for liquid and gaseous hydrogen pathways.

Acidification is a problem of grave environmental concern. Several types of acidification should be critically analysed for LCA in a more precise analysis. It must be considered that ocean acidification is a result of the uptake of carbon dioxide (CO_2) from the atmosphere, which lowers the ocean pH value. When acidic compounds are deposited on land, they make the soil acidic by lowering its pH. Acidification causes acid corrosion of buildings, monuments and other materials, destroying their aesthetic value. Acidification of agricultural land is caused by the use of ammonia-based synthetic fertilisers. In Figure 7.14, we display the acidification values for different hydrogen production processes [19]. Nuclear, wind and hydro-powered systems exhibit a much lower AP than solar PV and SMR. Biomass-based systems (whether biomass gasification or biomass-based electrolysis) pose the direst scenario.

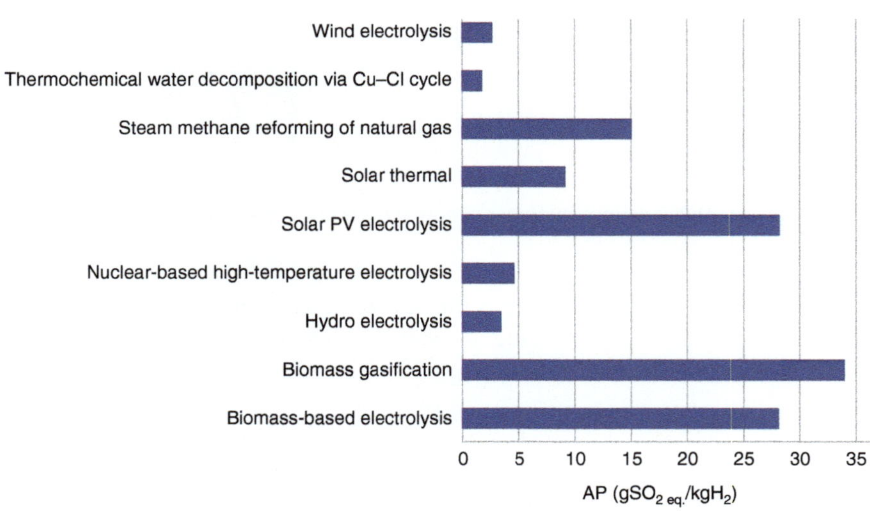

Figure 7.14 Average AP values in various hydrogen production technologies [19]/with permission of Elsevier.

Biomass-based processes are so high primarily because of the level of emissions during the production of biomass itself (which involves plantation, the use of fertilisers, etc.). Similar to global warming, acidification has not been a serious threat largely due to certain activities, such as manufacturing and construction, lasting only a short time. These activities leave a slender impact in its wake.

There is a strong correlation between GWP and AP, as shown in Figure 7.15 [14], for renewable electricity obtained from hydropower, solar and wind power. Life cycle implications need to be evaluated for acidification as well as any fossil-energy-consuming process. It is commonly known that NOx and SOx, produced by fossil fuels, ammonia and several other acidic gases, contribute to acidification. In areas of widespread industrial activities, this can become a problem of concern by developing acid rain. It is well known that there is a global concern for (CO_2) emissions into Earth's atmosphere, and the concern has mainly centred around its effect on climate. It will perhaps be relevant to mention here that CO_2 is also a major culprit causing ocean acidification. CO_2 produced in the atmosphere due to several industrial activities falls into the ocean. CO_2 products dissolve in seawater and decrease the pH value and concentration of carbonate ions. Recently, it has also been observed that shellfish face difficulty building shells and that corals have difficulty forming reefs. Both are made of carbonate compounds whose concentration is affected by acidification. The acidity of the ocean has increased by over 30% since the Industrial Revolution. Although the world's oceans are all connected, the effects of ocean acidification can unfold differently in various regions. The expression of ocean acidification can vary, especially in coastal waters, where additional drivers, including both human-caused and natural processes such as nutrient runoff and biological productivity, vary greatly over time and space. Local jurisdictions, nations and businesses seeking to

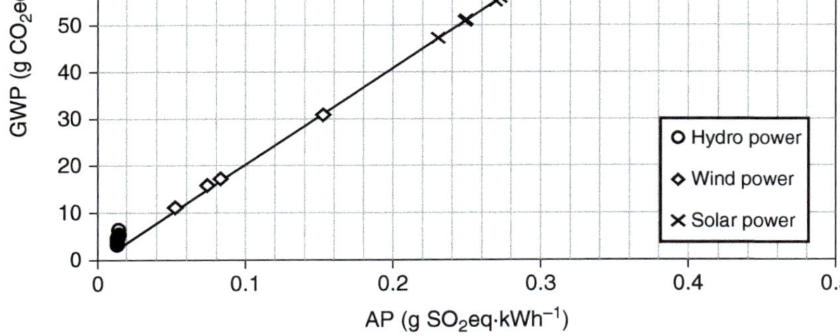

Figure 7.15 Relationship between GWP and acidification potential [14]/with permission of Elsevier.

implement adaptation and mitigation strategies need an accurate understanding, on global and regional scales, of how ocean acidification progresses.

Ocean acidification has disrupted marine ecosystems in several places, making it a topic of concern for the World Meteorological Organization on a global scale. The effects of ocean acidification are different in different regions. Near the coast, where humans and nature both contribute to the problem, its magnitude increases. Fertilisers and chemicals eventually end up in the oceans. Chemical runoff and biological productivity vary over time and space. Moreover, fish and corals are also destroyed. As far as LCA with respect to acidification is concerned, no uniform model has yet emerged. Some researchers believe that acidification is a suitable candidate for harmonisation.

A redesigned concept, IMPACT 2002+ LCIA, has been developed that 'proposes a feasible implementation of a combined midpoint/damage approach, linking all types of life cycle inventory results' [20]. This method considers 'elementary flows and other interventions through 14 midpoint categories that lead to four damage categories'. The novel aspect of this method is that it provides very significant information for a comparative evaluation of human toxicity and ecotoxicity [20].

Ocean acidification is not standardised in terms of how it is assessed over time. Time series observations of ocean chemistry must be collected in order to understand how the ocean changes. A comparative assessment can be made of the pattern of data variation in the ocean and coastal systems for the sake of standardisation. Scientists from the international community working with NOAA's Ocean Acidification Program and the Global Ocean Acidification Observing Network (GOA-ON) of NOAA (National Oceanic and Atmospheric Administration) [21] had a meeting in February 2020. The expert members of the team were experienced and had the relevant data with them, which were collected from time series stations at oceans all over the world. Consolidated and most recent CO_2 (fugacity of carbon dioxide) values for the global surface oceans and coastal seas were reported. There have been an additional 4.6 million CO_2 values reported relative to the previous version, which extends the record since 2011 [22]. 'The Surface Ocean CO_2 Atlas (SOCAT) is a synthesis of quality-controlled fCO_2 (fugacity of carbon dioxide) values for the global surface oceans and coastal seas with regular updates' [22]. Several versions of SOCAT (versions 1–4) have millions of relevant data, including annual surface water fCO_2 values. A large number of activities are being carried out by the group involved in The Global Ocean Data Analysis Project (GLODAP).

'The Global Ocean Data Analysis Project (GLODAP) is a synthesis effort providing regular compilations of surface-to-bottom ocean biogeochemical data, with an emphasis on seawater inorganic carbon chemistry and related variables determined through chemical analysis of water samples' [23]. This project involves the regular compilation of surface-to-bottom ocean biogeochemical data. In particular, it emphasises seawater inorganic carbon chemistry and related variables determined by chemical analysis of water samples [23]. Extremely relevant data have been collected and analysed in GLODAPv2.2019. Global water samples of more than 1.1 million have been collected through 840 cruises. Parameters such as salinity, oxygen, nitrate, silicate, phosphate, dissolved inorganic carbon, total acidity, pH, CFC-11, CFC-12, CFC-113 and CCl4 have been evaluated for quality control [23]. With additional relevant input data from several other scientists, the team started developing best practices so that a uniform approach could be adopted for an accurate global analysis to evolve long-term acidification trends. Since then, there have been several efforts at different levels, and uniform data quality standards based on observations can easily be accessed.

7.5.3 Eutrophication

Like acidification, the impact of eutrophication needs to be taken into account for an LCA. Eutrophication can be divided into two parts. Natural eutrophication and anthropogenic eutrophication. Nature plays a significant role in many activities, such as being affected by environmental events whereas anthropogenic eutrophication is due to human activities. The effects of anthropogenic eutrophication, at present, are much more severe than natural eutrophication. Eutrophication, caused by human activities, such as agricultural cropping, food processing and disposal of wastes, particularly with high organic compounds, wastewater treatment, etc., contributes to eutrophication. The release of nitrogen compounds during fuel combustion and volatilisation of ammonia from fertiliser, manure, etc., are also responsible for causing eutrophication. Figure 7.16 illustrates eutrophication. Although nitrogen and phosphorus are essential for regulating the ecosystem, water and soil are sometimes enriched by these nutrients. This causes an imbalance in the composition of species in the ecosystem.

Eutrophic waters reduce the biodiversity of the water and increase its toxicity. Excessive growth of algae and plankton in the water obstructs the flow of sunlight from reaching the bottom of the aquatic body. This causes dissolved oxygen to be depleted in the water. This leads to hypoxic conditions and the death of many types of larger life forms, such as fish.

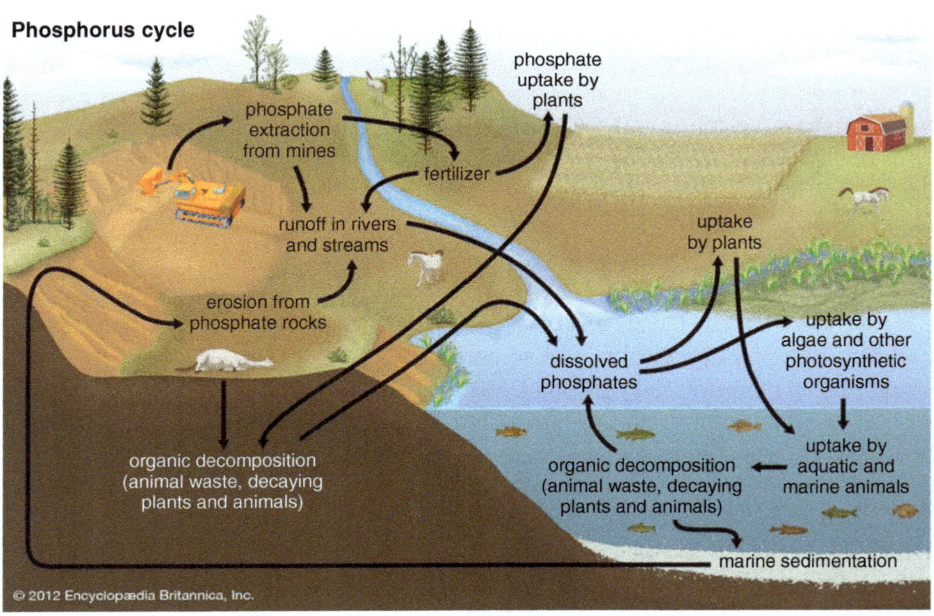

Figure 7.16 Process of eutrophication.

Interestingly, the situation also encourages other species that are adapted to such environments to relocate. Researchers have observed that crucian carp, even without oxygen, can survive for months.

There are certain factors, such as the disposal of wastes with high levels of organic content. Furthermore, fossil fuels are needed for transportation and electricity generation. These fuels release nitrogen compounds into the atmosphere and contribute to eutrophication. Eutrophication impact assessment is vital for LCA because the formation of greenhouse gases, such as nitrous oxide (N_2O), contributes to global warming, which is a major factor in climate change. The primary cause of eutrophication in terrestrial ecosystems is nitrogen oxides transported by automobiles and power plants. The abundance of phosphorus in water has been found to be one of the primary limiting factors that adversely affects the growth of plants. With agricultural activities taking place, there is a risk of phosphorus adhering to the soil, causing wastes, sewage disposal and detergents containing phosphorus to increase soil eutrophication. Until now, no international standard has been developed to assess the life cycle effects of eutrophication.

7.6 Comparative Assessment of the Hydrogen Production Process

Some efforts have been made here to have a comparative evaluation of the three major routes of hydrogen production, namely Steam methane reforming, gasification and electrolysis (Figure 7.17). Considering the increasing interest in gaseous fuels, and since natural gas, with methane as its main constituent, has already been used in several energy sectors, some studies have been conducted with biomethane and hydrogen as the most prospective representative gaseous fuel feedstock [24]. In the analysis, the most likely technological pathways have been identified, which are likely to be adopted. The production of biomethane from wastewater sludge, manure, landfill gas (LFG) and silage maize and hydrogen from natural gas combined with CCS, coal with CCS, biomass gasification, renewable electricity, grid electricity, wastewater sludge biomethane, manure biomethane and LFG biomethane have been adopted. The GREET (Greenhouse gases, Regulated Emissions, and Energy Transportation) model was used to determine the GHG intensity of various hydrogen pathways. Total well-to-wheel (WTW) GHG emissions were assessed, which included emissions from feedstock extraction, fuel production and combustion. During fuel combustion, there are no GHG emissions. As far as the analysis of CCS was concerned, energy use and associated emissions for the process of carbon capture were assessed. For this analysis, three hydrogen production routes are considered: SMR, gasification and electrolysis.

While the study was carried out without the European Union in the background, it provided a broad picture and relevant qualitative information.

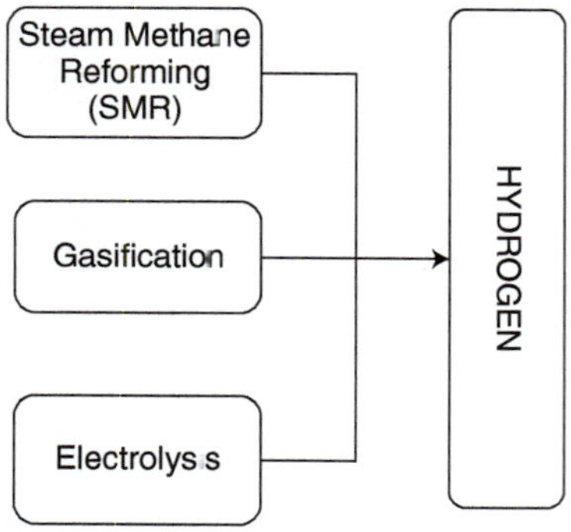

Figure 7.17 Three major routes of hydrogen production.

Because natural gas and coal produce high upstream GHG emissions, recovering and storing carbon dioxide (by CCS) from the produced syngas was considered in this study to ensure the production of low-carbon hydrogen. The process of electricity uses electricity to split water into hydrogen and oxygen. Whether electricity is supplied from the grid or renewable sources is very relevant to the intensity of GHG emissions in this context. Detailed diagrams of each of the three processes of production with various stages of operation in sequence are shown in Figure 7.18 [24].

Looking at the process of SMR, the feedstock could be natural gas (where methane is the major constituent) or biomethane. In the next stage, methane reacts with steam and produces hydrogen and carbon monoxide. As evident from the diagram, methane then passes through the water–gas shift, and syngas is generated. Approximately 60% of the total carbon dioxide produced at the SMR plant is in the syngas output [24]. Next, the hydrogen is separated from the remaining impurities by pressure swing absorption (PSA). Thus, the tail contains about 60% of the total amount of CO_2 generated. As evident from the figure, impurities in the tail gas are processed in the reformer, thereby causing all carbon to end up in the flue gas. It can also be seen in the flow diagram that there are three different streams of carbon dioxide. Therefore, a more detailed study could be conducted by installing three separate carbon capture units. However, in the analysis of the ICCT white paper, a detailed analysis has been carried out of the use of CCS for natural gas SMR. This is a very significant area of study considering its potential practical industrial applications. Carbon monoxide is also generated when carbonaceous materials (such as coal, biomass, forest residues, agricultural residues and crops) are gasified. Depending on the availability of local resources, the feedstocks for hydrogen production are passed through a water–gas shift. However, in the sequence, it is essential to install an acid gas removal unit to carry away the hydrogen sulphide that is formed. The next series of activities are as usual. After the capture unit separates carbon dioxide, PSA separates hydrogen. In contrast to the earlier two processes, SMR and gasification, electrolysis produces hydrogen of very high purity. Grid electricity and electricity from renewable sources are the most promising candidates. Figures 7.19 and 7.20 show how hydrogen is produced by electrolysis and gasification.

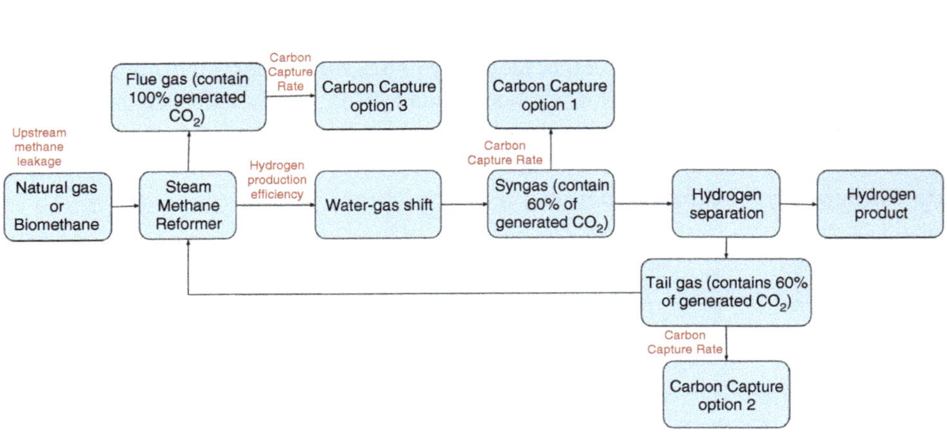

Figure 7.18 Sequence of hydrogen production by steam methane reforming [24]/International Council on Clean Transportation.

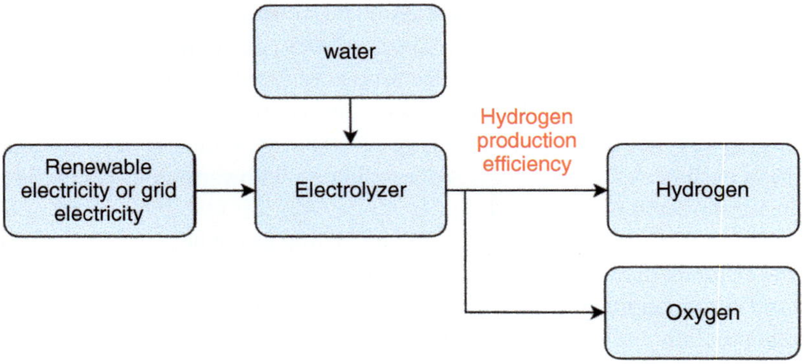

Figure 7.19 Hydrogen production from electrolysis [24]/International Council on Clean Transportation.

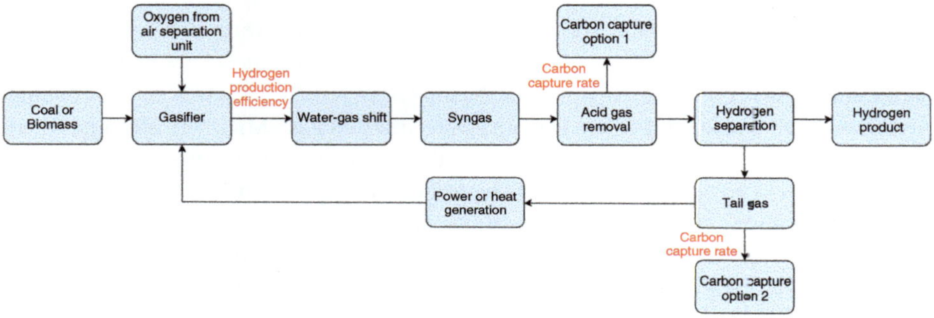

Figure 7.20 Stages of gasification of coal or biomass for hydrogen production [24]/International Council on Clean Transportation.

In contrast to the two methods of SMR and gasification, electrolysis yields hydrogen of very high purity. As indicated in the set of three figures, key parameters for natural gas SMR and CCS are hydrogen production efficiency, carbon capture rate and upstream methane leakage rate. Hydrogen production efficiency and carbon capture rate are key parameters for coal gasification with CCS. Carbonaceous feedstocks, such as coal and biomass, can be used to produce syngas. This gas mostly contains hydrogen and carbon dioxide, as well as a few other products. Even though syngas can be used to produce synthetic fuels, it is possible to extract hydrogen from syngas.

7.7 Climate Target Criteria: Carbon Capture

Hydrogen production efficiency is the key parameter for biomass gasification and electrolysis. It is not possible to meet the climate target either through hydrogen production from coal with CCS or by grid electricity as these two pathways have the highest GHG intensity. Even though it is possible, by this technique, to achieve 99.9% carbon dioxide (CO_2)

capture from hydrogen from fossil fuels, this is not achievable with the present level of technology when 55% of the total CO_2 generated during hydrogen production can be captured with the current industrial practices. There are several factors that are critical to the production efficiency, the size of the plant and its operating temperature and pressure. The energy content also varies for different feedstocks used for the SMR process [25–28]. It has been reported in the literature that the production efficiency of hydrogen from SMR could be in the range of 56.6–5%. Among the above three routes of hydrogen production, SMR is technologically most mature and economically viable at the present time. This is why it is the most popular method and globally about 50% of the hydrogen is produced from natural SMR. Of course, depending upon the local resource potential and the demand, the method of hydrogen production varies. A closer look at the situations arising from the fossil-based hydrogen production route with CCS indicates that it does not substantially contribute to improving the climate. Normally, an SMR plant does not capture all CO_2 emissions. Combining CCS with fossil-based hydrogen pathways is unlikely to be beneficial to the climate in the near future. SMR plants capture only about half of the total CO_2 generated though an ideal system can capture almost more than 99%, which is not economically viable. Inherent leakages of some amount of CO_2 cannot be avoided during transportation and storage locations as well as during CO_2 injection. Carbon capture is an energy-intensive process. An extra amount of energy is required at several stages of the process, such as while driving the carbon capture unit, and in compressing the CO_2 that is captured. It has been estimated that the overall efficiency of an SMR plant with carbon capture is about 14% less than that without carbon capture. Some studies have reported that the process of producing hydrogen from auto thermal reforming (ATR) is more effective in capturing carbon than the SMR process. This could be because CO_2 is present in ATR in a single concentrated stream and is so easy to capture. On the other hand, there are three streams in the SMR process. However, ATR is not extensively adopted for hydrogen production as SMR. The present situation also does not show any trend of expansion in favour of ATR. There are other constraints in the ATR process contributing to the benefits of climate. In fact, the demand for electricity for the ATR process is more than that required for SMR. This would further be enhanced if carbon capture is installed in the plant [29–32].

A model was developed using the sorption-enhanced reforming (SER) process mainly because of the intrinsic features of the SER process in CO_2 capture which is of immense importance to the environment [33]. In this study, a comprehensive comparison was carried out with conventional hydrogen production processes. It was observed that a higher natural gas to hydrogen conversion efficiency was possible with the SER process.

7.8 Review of Hydrogen Transport Modes and Delivery Methods

Tayarani and Ramji [32] studied the LCA of gaseous hydrogen from three processes, such as SMR, solar electrolysis and biomass-based production, at a particular location and carried it to a different refuelling location. Researchers have conducted an LCA of three methods of hydrogen production and transportation in trucks and pipelines. Apart from evaluating the GWP, the impact of noxious pollutants, such as nitrogen oxide, VOCs and particulate matter PM2.5, has been evaluated. Analysis of GWP shows that the truck transportation pathway is more energy and ecologically intensive than pipeline transportation. Solar electrolysis through pipelines was ecologically less intensive than most of the other processes that produce pollutants. The system–boundary diagram with the truck pathway and the pipeline pathway used for the analysis is shown in Figure 7.21.

Very useful results with regard to several pollutants, such as carbon dioxide, volatile organic compound, oxides of nitrogen (NOx) and particulate matter with a size of 2.5 μm (PM2.5), were obtained from this analysis as shown in Figures 7.22–7.24.

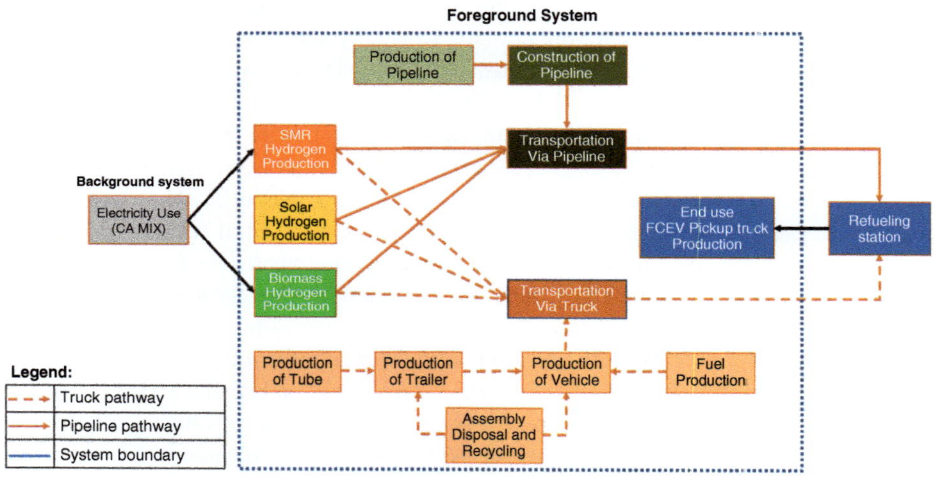

Figure 7.21 System–boundary diagram with the truck pathway and the pipeline pathway [32]/MDPI/Licensed under CC BY 4.0.

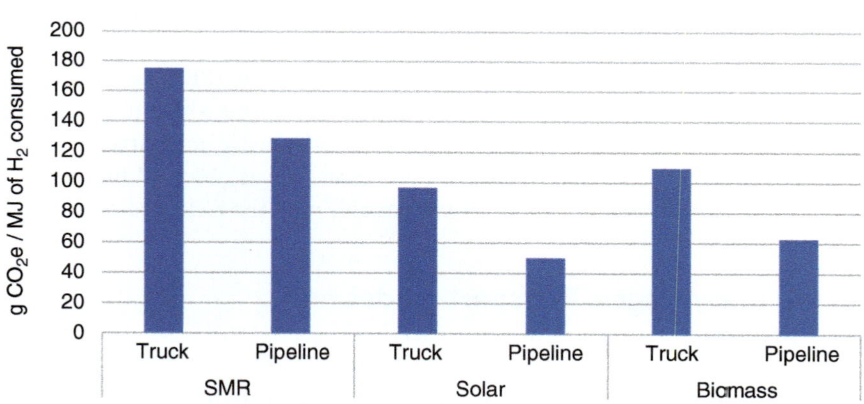

Figure 7.22 LCA-based emission (CO_{2e}) across different hydrogen production and delivery pathways [32]/MDPI/Licensed under CC BY 4.0.

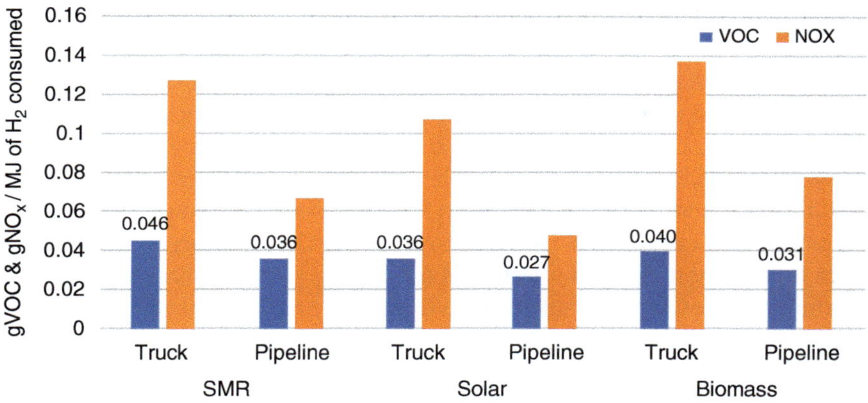

Figure 7.23 LCA-based VOC and NOx across different hydrogen production and delivery pathways [32]/MDPI/Licensed under CC BY 4.0.

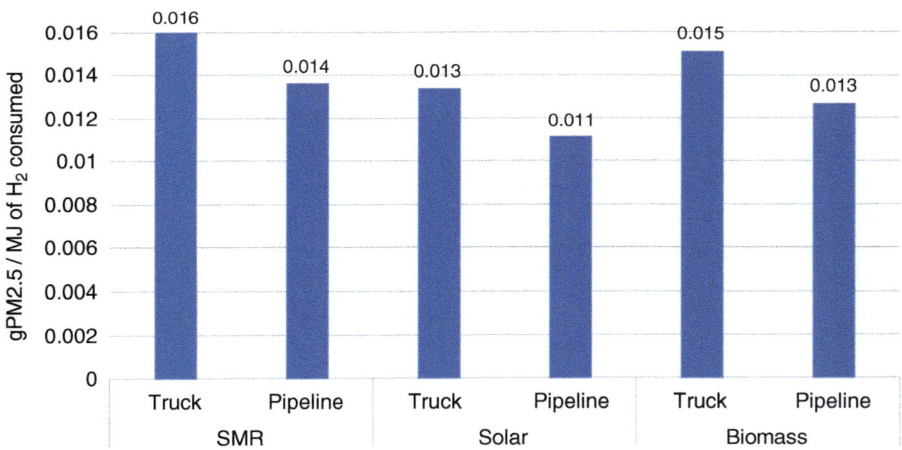

Figure 7.24 LCA-based PM2.5 across different hydrogen production and delivery pathways [32]/MDPI/Licensed under CC BY 4.0.

The exact picture of the level of pollutants and the corresponding pathways for the best/worst case scenario are glaringly reflected in the above graphs. In terms of the emission of carbon dioxide, biomass through pipelines seems to be the best option. As far as VOC and oxides of nitrogen are concerned, biomass via trucks is the worst case for oxides of nitrogen, and SMR via truck is the worst case for VOC emission [32].

Based on the analysis, solar electrolysis hydrogen production through pipeline transportation is the best option considering CO_2 eq. It was estimated to produce 50.298 gCO_2 eq./1 MJ H_2. In the case of PM2.5, SMR and biomass via trucks produce the highest PM2.5. As has been indicated, this set of results has not been arrived at from a definite location.

However, they do provide a relatively sound qualitative indication of hydrogen fuel transportation pathways.

Hamedani and the team of researchers [33] carried out their study on LCA from cradle-to-gate of hydrogen production in a small power plant. The power plant was operated by Agro-Industrial Residue Gasification. This was a small system (1 MWth). The configuration of the system consisted of a 'catalytic candle indirectly heated steam gasifier coupled with zinc oxide (ZnO) guard bed, water–gas shift (WGS) and pressure swing absorber (PSA) reactors'. It has been observed that the production of biomass has the highest impact. GWP and AP were the two most significant factors of concern from an environmental point of view. It was also observed from sensitivity analysis that an 'increase in hydrogen production efficiency does not necessarily result in a decrease in environmental impacts' [33]. It is important to understand that the environmental effects of hydrogen production are not totally dependent on the rate of production because it has often been observed that by-products have greater relevance in this context.

Koroneos et al. [34] discussed the LCA of hydrogen fuel production processes with a view to assessing the environmental burden due to hydrogen production processes. The use of photovoltaic energy has been found to be the worst offender for environmental pollution. This is because the manufacturing process of the photovoltaic modules significantly contributes to degradation of the environment. Moreover, the overall efficiency of the photovoltaic system is very low.

SMR of natural gas is responsible for very high equivalent emissions of CO_2 and SO_2. The production of hydrogen through wind, hydropower and solar thermal energy is a desirable alternative as they are extremely environmentally friendly. The level of emission is relatively much lower. Based on the comparison of different processes from an LCA perspective, renewable energy sources are the most environmentally friendly. Further studies are essential for optimizing the renewable energy source route to maximize energy efficiency, so that a major part of its available energy could be fruitfully utilized.

Cho et al. [35] utilised the emission data to evaluate the environmental impact assessment of hydrogen production via SMR. SMR is the most widely adopted industrial process of hydrogen production. The emission level of pollutants in hydrogen production depends upon the process condition and the location of the equipment installed in the industrial facility to control pollution. The researchers here used 'facility level real emissions data from multiple US EPA databases' and made some effort to assess the impact of hydrogen production. In their analysis, they considered 12 midpoint and 14 endpoint impacts for 33 hydrogen production facilities using SMR for hydrogen production. Global warming is very damaging, according to the analysis. It was observed that CO_2 emissions were the major contributors to global warming. The global warming impact was '94.6% of the endpoint impacts on human health' [35], whereas its contribution was 98.3% of the total endpoint impacts on terrestrial ecosystems. Thus, the LCA research results of almost all investigators have identified GWP as a major undesirable outcome of the hydrogen production process.

7.9 LCA for the Hydrogen Power Generation and Transport Sector

Rinawati et al. [36] conducted an exhaustive survey of the literature during the period 2004–2021 on the LCA of hydrogen-based power generation. This is an area in which sufficient relevant information is not available in the literature. Most LCA analyses have focused on electrolysis using renewable energy and compressed gas storage as the mode of analysis. In terms of hydrogen-based power LCA, the fuel cell was perhaps the only conversion equipment subjected to LCA. Fuel cells have been effectively utilised in a wide range of applications that require electricity. As they are also used for both stationary and portable generation units, they can be used for several domestic appliances, transportation and power generation. The character of LCA depends upon several parameters, such as the identification of system boundaries, FUs and, last but not least, methods adopted for LCIA. Furthermore, hydrogen can be produced and stored, and power can be generated in several different ways. Currently, the results of LCA studies conducted by different researchers are quite different. These discrepancies tend to produce an inaccurate picture considering the environmental impact of a project. Perhaps more tangible data may be obtained by adopting a harmonisation procedure for LCA results garnered through a number of studies. Thus, it is important to critically analyse the results of the different studies and optimise the parameters to establish a standard reference baseline.

The earlier study was related to power generation. Rinawati [37] and the team have carried out an exhaustive survey on the LCA of hydrogen for road transport applications. The researcher made an intensive review of more than 70 papers during the period 2000–2021 from the point of view of technical features and the methods adopted for analysis. The review involved about 350 cases of hydrogen being used for automobile applications. As part of this study, different segment approaches were taken, such as well-to-tank (WTT) and WTW, and ultimately, a comprehensive LCA was performed. WTT analysis mainly uses mass or energy as its functional unit. For total LCA as well as for WTW analysis, the vehicle distance was considered as the functional unit. In the transportation sector, fuel-cell-based hydrogen passenger cars were the most common. As far as hydrogen production is concerned, four methods, the thermochemical, electrochemical, thermal–electrochemical and biochemical routes, were considered [37]. From an environmental point of view, the analysis includes the identification of greenhouse gases and the determination of GWP and the subsequent effect on climate change. The analysis also took into account energy consumption in the transport sector, regulated emissions, etc. This was a very detailed and useful study. Its findings could provide useful guidelines for optimising the conditions for increased efficiency and reduced emissions, as well as for the widespread use of hydrogen in the automotive sector. Particularly for emerging energy systems, such as hydrogen transport, conducting LCA at an early stage of development is important in order to identify the most promising options as well as environmentally significant criteria. In an LCA of a vehicle, it is important to know the fuel cycle and the vehicle cycle. Figure 7.25 shows the vehicle cycle that consists of the chain from raw material to vehicle recycling and the fuel cycle is from feed extraction to vehicle operation [38].

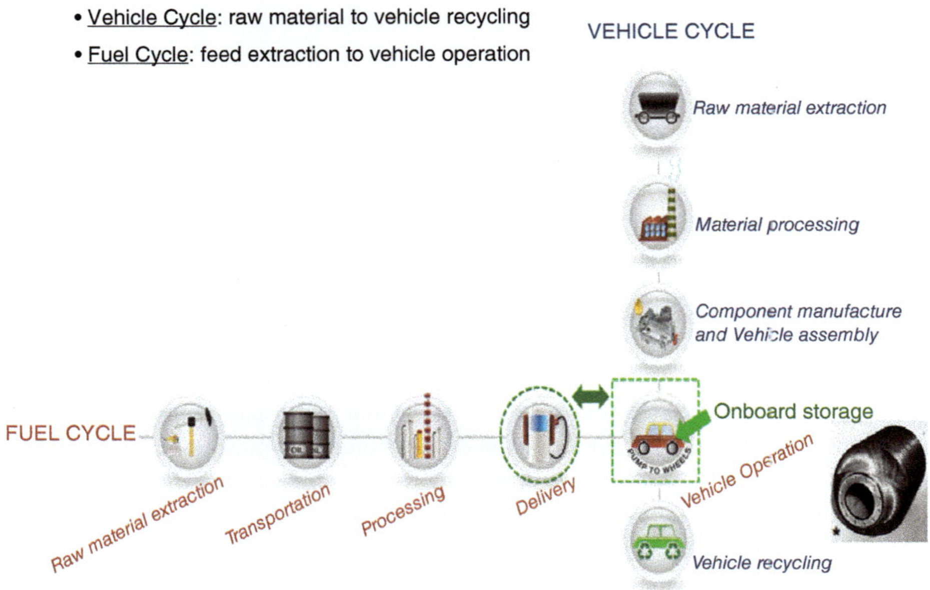

Figure 7.25 Stages for the fuel cycle and the vehicle cycle in sequence.

7.10 Analysis of Hydrogen Storage

In addition to production and utilisation in the transport sectors, LCA related to hydrogen storage is a crucial area from the implementation of technology point of view. Elgowainy et al. have summarised their results and suggested that on-board H_2 storage contributes 15–23% to the vehicle manufacturing cycle as depicted in Figure 7.26 [38].

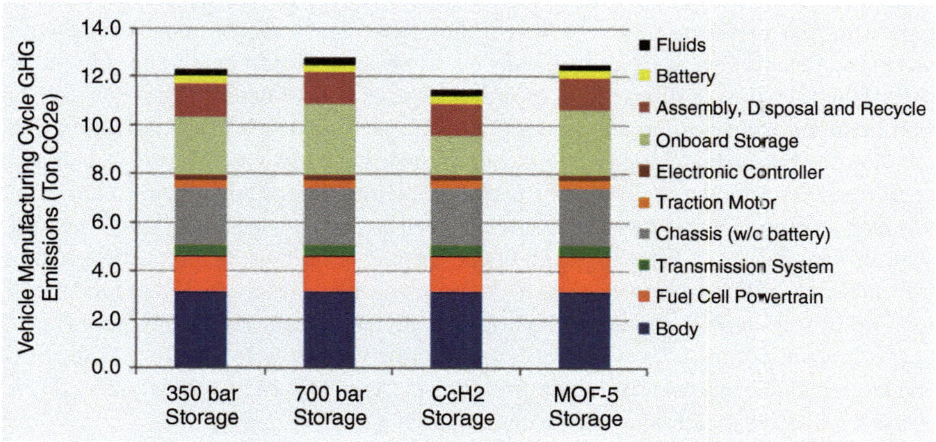

Figure 7.26 On-board H_2 storage contribution to vehicle manufacturing cycle [38]/US Department of Energy/Public Domain.

Simon and Bauer analysed several significant factors related to the LCA of conventional vehicular fuels, such as gasoline, diesel and natural gas [39]. To evolve a rational comparative picture, the LCA results for hydrogen production and relevant data for conventional fuels used for internal combustion engines (ICEs), such as gasoline, diesel and natural gas, have been generated. All data are shown according to the Euro 3 standard. Figure 7.27 shows the total GHG emissions for the production and combustion of gasoline, diesel and natural gas, which are used as conventional fuels for ICEs. The quantities of the fuels have been taken such that their energy content becomes equivalent to 1 kg of hydrogen. Figures 7.28 and 7.29 show the cumulative quantity of both classes of particulate matter for

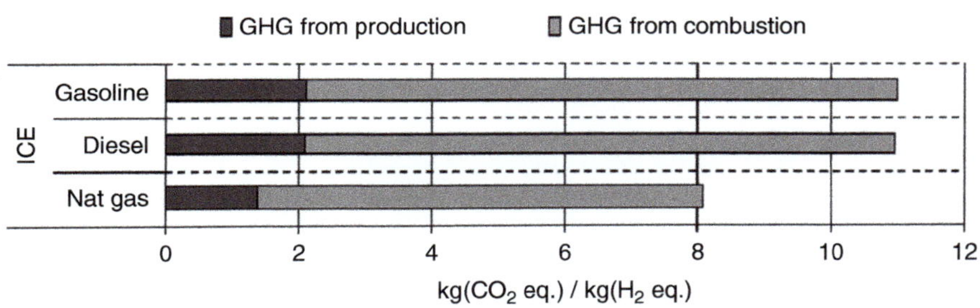

Figure 7.27 Total greenhouse gas (GHG) emissions from the production and combustion of 1 kg (H_2) equivalent quantities of gasoline, diesel and natural gas [39]/Cambridge University Press.

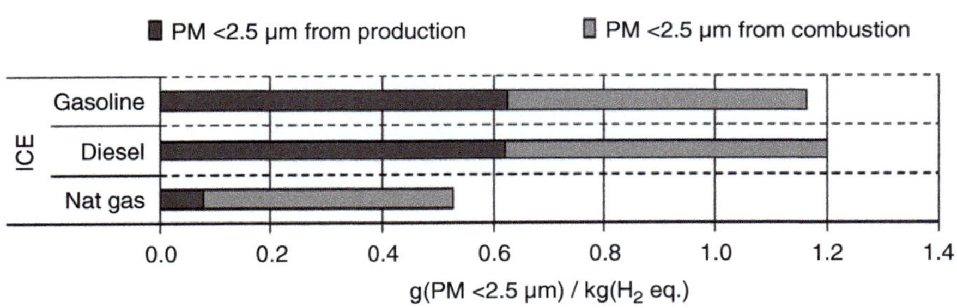

Figure 7.28 Emissions of PM less than 2.5 μm in diameter from the production and combustion of 1 kg (H_2) equivalent quantities of gasoline, diesel and natural gas [39]/Cambridge University Press.

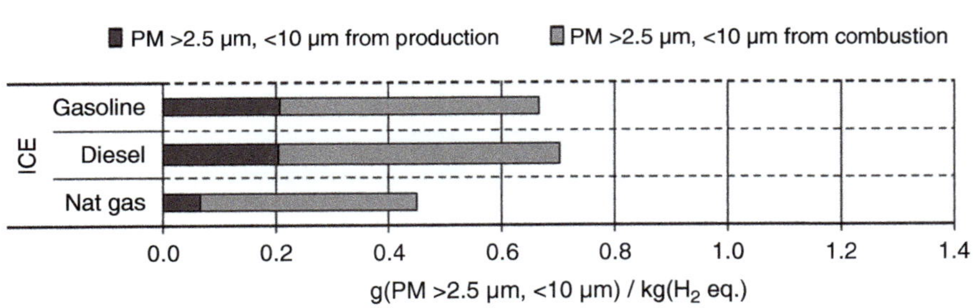

Figure 7.29 Emissions of PM between 2.5 and 10 μm in diameter from the production and combustion of 1 kg (H_2) equivalent quantities of gasoline, diesel and natural gas [39]/Cambridge University Press.

the three fuels (gasoline, diesel and natural gas) during production and combustion in an ICE. Figure 7.30 shows the non-methane volatile organic compounds (NMVOC) from production as well as combustion. It is slightly higher for diesel compared to gasoline. As shown in the figure, the value is substantially lower in the case of natural gas.

The formation of NOx is highly dependent upon the temperature of combustion. As evident from Figure 7.31, the level of NOx is higher in diesel combustion compared to gasoline, and it is much lower for natural gas.

The emission of oxides of sulphur (SO_2) has been a problem of concern for fossil fuels, particularly diesel and gasoline. This is why low-sulphur diesel and ultra-low-sulphur diesel have been developed, and ultimately, it has been decided that sulphur must be removed from fossil fuels. Figure 7.32 shows a comparative depiction of SO_2 emissions during the production and combustion of all three vehicular fuels.

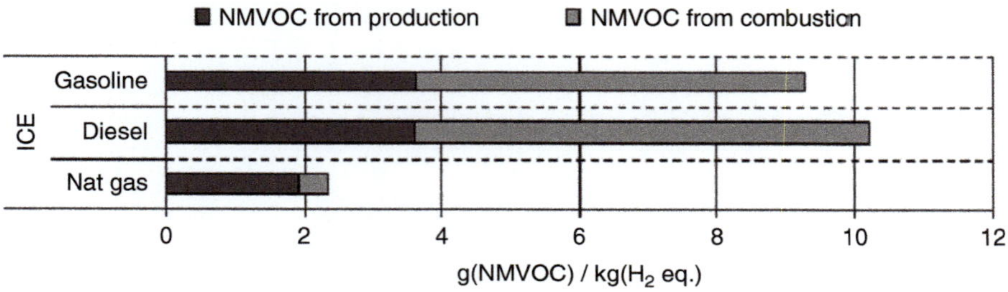

Figure 7.30 Emissions of non-methane volatile organic compounds (NMVOC) from the production and combustion of 1 kg (H_2) equivalent quantities of the ICE fuels [39]/Cambridge University Press.

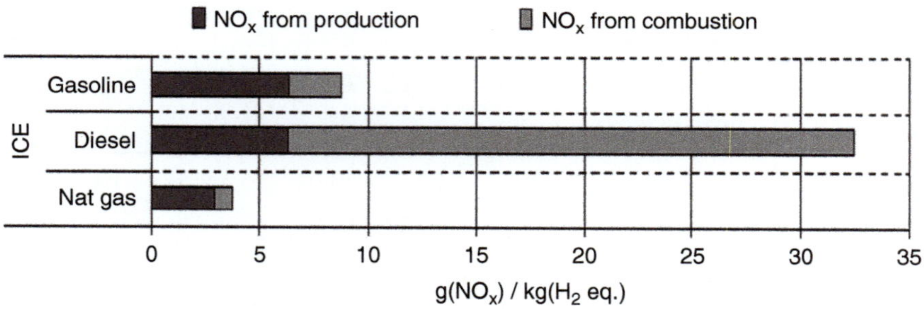

Figure 7.31 Emissions of nitrogen oxides (NOx) from the production and combustion of 1 kg (H_2) equivalent quantities of the ICE fuels [39]/Cambridge University Press.

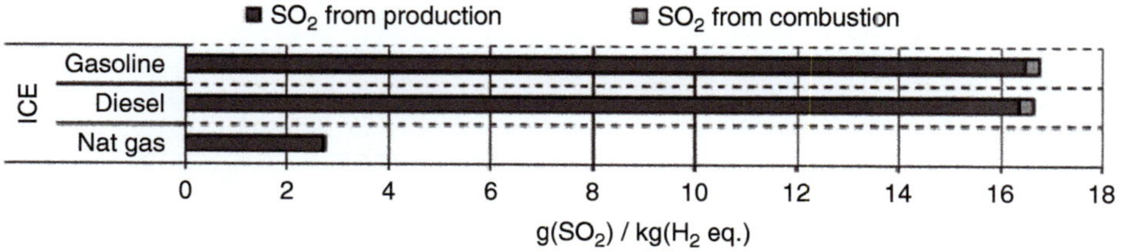

Figure 7.32 Emissions of sulphur dioxide (SO_2) from the production and combustion of 1 kg (H_2) equivalent quantities of the ICE fuels.

Antonio and co-workers [14–16] examined and identified existing and novel EoL technologies for recovering valuable materials from hydrogen-based devices. Systems such as stacks of proton exchange membrane (PEM) fuel cells, solid oxide fuel cells (SOFCs), water electrolysers, and alkaline water electrolysers have been studied, although specific attention has been paid to PEMs and alkaline electrolysers. The current level of knowledge does not provide definite information about specific fuel cell strategies. From the perspective of sustainability, a rational analysis of the strengths, weaknesses, opportunities and threats will provide guidelines for preparing a road map of the life cycle of fuel cells and hydrogen devices. Specifically, this study analyses short-term, medium-term and possible long-term applications. The short-term applications of fuel cells are likely to be limited to a niche market. The use of fuel cells is gradually increasing, and the hydrogen economy is expected to continue to grow rapidly. However, the widespread application of fuel cells is expected in the mid-term within a developing hydrogen economy. EoL strategies are very significant because of the rapid penetration of fuel cells into several areas, such as the transport, power, agriculture and domestic sectors.

There are many hazardous materials that cause damage to the environment. The paper discusses in great detail the processes through which these are recovered. This study shows that it is feasible to recover some valuable materials and components of fuel cell hydrogen devices by judicious application of EoL technologies. This is a step forward towards achieving a sustainable hydrogen economy. At present, the application of EoL is restricted to the recovery of valuable materials from stacks of proton exchange membrane fuel cells (PEMFCs), alkaline water electrolysis (AWE) and SOFCs. The activities mainly focus on the recovery of platinum group metals (PGMs). Moreover, conventional hydrometallurgical and pyro-hydrometallurgical pathways are now being used. PGMs are expensive materials and are used as major catalysts in PEM stacks. Research activities are in progress to tackle several aspects of these problems. Suitable non-noble metals (as substitutes for platinum) are being identified, and their economic viability and long-term stability are being investigated. Efforts are being made to recover other materials from stacks apart from PGMs. In this context, it is imperative to stress that appropriate EoL technology must be developed. This will allow us to implement mature technology more safely and economically.

Dhanushkodi et al. [40] reviewed and analysed various informatics tools and computer-based technology designed to collect, organise, analyse, interpret and present data and information on fuel cell technology study aimed at both the vehicle cycle and the fuel cycle of fuel cells with the objective of assessing the environmental impact at various stages. Inventory and impact analyses of the life cycle, such as vehicle material production and vehicle assembly–PEMFC production process, production and distribution of feedstock, fuel production and an assessment of the environmental factor, were carried out in this study. It has been observed that the production of fuel stacks has a greater environmental impact than the use of stacks in vehicles. In addition, the study shows that the environmental impact is mainly due to the materials used for catalysts and materials and energy used for flow field plates. Recycling materials for stack generation might be a useful approach. A comparative assessment of hydrogen-fuelled PEMFC cars with respect to conventional gasoline-fuelled vehicles is provided. Efficiency and greenhouse gas emission characteristics have been analysed from several routes of hydrogen production.

These routes include natural gas reformation, electrolysis of water and the use of renewable energy, such as solar and wind, biomass gasification, etc. Performance and emissions have been found to be better with fuel cells than with conventional engines. It is, however, essential to establish conditions for producing hydrogen from renewable resources at a relatively low cost.

Usai et al. [41] reviewed the LCA for fuel cell systems for light-duty vehicles and discussed its likely impact in the future. Vehicles operating on alternative fuels have been subjected to several research and development activities in the past. Fuel cell electric vehicles have received a lot of attention primarily because of their potential to drastically reduce greenhouse gases. A major obstacle to the widespread implementation of fuel cells has been the absence of an appropriate fuel infrastructure. Moreover, the production cost is abnormally high compared to the cost of Li-ion batteries.

Table 7.2 shows the list of components and sub-components that were analysed in this study.

Life cycle assessments must be more precise and rigid as technological development progresses. This should be performed with a high emphasis on the environmental aspects. In their study on the LCA of the current commercially available cell, the researchers observed that these systems need to be substantially modified/redesigned to ensure environmental compatibility. At present, approximately 5 tonnes CO_2 eq. is produced in an 80 kW fuel cell system with two-stage tanks with a capacity of 5 kg of hydrogen. In this context, the production of the tanks and the catalysts ought to be investigated closely and appropriate alternative steps deemed necessary to be adopted to enhance the environment's performance. Similarly, the function of some auxiliaries of the fuel cell needs to be closely examined. The use of renewable energy and appropriate replacements for platinum can significantly improve the environment.

Simon and Bauer [42] carried out some studies on the LCA of hydrogen production as well as on the use of fuel cells in vehicles. The major objective of their study was to concentrate on the quantities of specific materials used in fuel cells and perform a detailed inventory analysis with regard to production and the EOL to evaluate the current environmental

Table 7.2 Primary and sub-component analysis.

Primary components	Sub-components	
Fuel cell stack	Membrane	Gaskets
	Bipolar plates	Membrane electrode assembly
	Catalyst ink	Stack compression bands
	End plates	Stack housing
	Current collectors	Coolant
H_2 tanks	Tanks (carbon fibre and resin)	Tank auxiliaries (boss, foam and glass fibre)
	Air management	Fuel management
	Water management	Heat management
	Electronics (controller, sensors, wiring)	

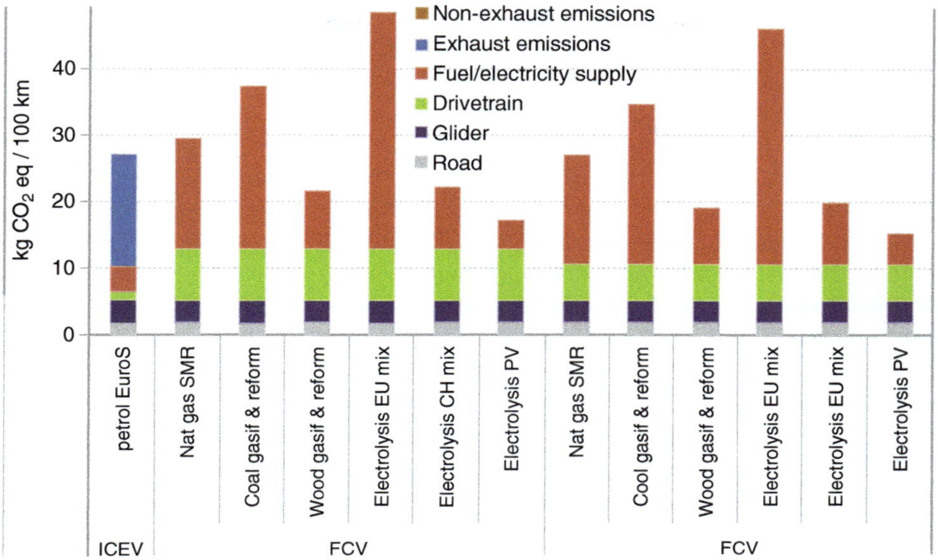

Figure 7.33 Life cycle GHG emissions from an FC vehicle using hydrogen from various sources. *Source:* Adapted from [42].

performance and assess the future. The authors carried out a sensitivity analysis and LCIA to determine the effects of significant fuel cell parameters. Membrane electrode assembly (MEA) plays a very significant role in the performance of fuel cells. In this study, it was observed that about '37 kg CO_2 eq./kW net system power in 2010 almost 50% is due to production of the MEA' [42]. This was due to the use of platinum catalyst material along with its extraction process. Therefore, if the catalyst loading is reduced, the situation will improve. Moreover, the sensitivity analysis indicates that it is necessary to adopt radical changes rather than gradual modifications in the design of fuel cells to reduce GHG emissions. This is because it depends on the source from which hydrogen is produced. Figure 7.33 shows that the LCA of a fuel cell system is highly dependent on the material and the route of hydrogen production. The emissions of SO_2 and particulate matter are almost the same as those from an IC engine. Therefore, as far as emission characteristics are concerned, a fuel cell vehicle operating on hydrogen produced from fossil fuels is no better than petroleum-fuelled IC engine vehicles from a life cycle perspective. However, a fuel cell using hydrogen from non-fossil sources can result in a substantial reduction in GHGs.

7.11 Durability Studies Related to Hydrogen Energy Utilisation

At this point in time, hydrogen's potential to provide a perpetual solution to both problems of fossil fuel shortage and environmental degradation has been clearly understood. Hydrogen can be used in several industries, including the transport sector and the power sector.

The word durability is derived from the Latin 'durabilis', which means 'enduring'. Therefore, durability tests of hydrogen-fuelled systems need to be studied to assess their potential for long-term applications. However, endurance and durability tests on hydrogen use have not been widely investigated.

Considering the application of hydrogen in vehicles, it is very significant that the challenges of safety must be addressed. This is whether stored in the form of compressed gas at high pressure, as a cryogenic liquid, or in metal hydride. When switching over to fuel cell vehicles, the major challenge is ensuring the safe transfer of power from the fuel cells to motors through high-voltage buses. Thus, a detailed exhaustive work plan was viewed from the perspective of hydrogen's potential to provide a solution to the safety-related problems for hydrogen vehicles. The study of hydrogen fuel containers involves three major tasks: (i) risk assessment studies to quantify potentially unsafe conditions, (ii) identifying and developing the tests required to provide an effective solution to the problems and (iii) evaluating the procedure that demonstrates that hydrogen-operated vehicles are as safe as conventional petroleum-fuelled vehicles [43]. Testing conditions used to assess the safety of the hydrogen containers included high and low temperatures during pneumatic pressure cycling, static hold time at high-pressure conditions and the number of pneumatic cycles. Hydrogen containers need to be tested under extreme temperature conditions to ensure that they can handle the stresses caused by thermal expansion and contraction. Cycling between high and low temperatures helps identify potential weaknesses in the container's structural integrity and material properties. This testing simulates real-world scenarios, such as rapid temperature changes during transportation or storage. Similarly, static hold time at high-pressure conditions needs to be carried out to assess how well the hydrogen container maintains pressure stability over time. It provides relevant information on container's sealing mechanisms and resistance to leakage or rupture under sustained high-pressure conditions. Repeated pneumatic cycling (filling and emptying) of the hydrogen container provides insights into its durability and fatigue resistance. This testing helps ensure that the container can withstand numerous loading cycles without developing cracks, leaks or other structural defects. A high cycle count demonstrates the container's ability to endure long-term usage without compromising safety. Considering the high diffusivity of hydrogen, it is quite likely that hydrogen continues to diffuse through the liners of the gas container. The rate and amount of hydrogen leakage are expected to increase during a number of pneumatic cycles. Cycle operation of a number of cycles, over a period of time, might cause blistering and cracking, thereby leading to excessive leakage. Physical damage is more likely to occur in low-temperature cycles when the rate of permeation is higher. In this context, wide variations in maximum and minimum temperatures during fuelling and normal vehicle operating conditions are crucial factors.

The test results in Table 7.3 show that the low baseline permeation values for the containers were substantially increased after 50 cycles. Pneumatic cycles at elevated temperatures resulted in visible damage, and the low-temperature ones did not exhibit any signs of damage during visual inspection. The researchers compared the normal interface between the liner end and the boss with the interface between the liner end and the boss. This was performed by cycling the container between +50 and +65 °C. Damage at 65 °C was visibly more severe than the damage with 50 °C cycle operation. Furthermore, two types of four containers and one type of three containers were tested to determine the effect of

Table 7.3 Temperature with respect to permeation rate. Adapted from [43].

Temperature	Permeation rate (cc h^{-1})				Visual inspection
	Baseline (before cycling)	After 50 cycles	After 100 cycles	After 200 cycles	
+65 °C	1.7	201	302	279	The liner pulled away from the end boss at the dome, with significantly more separation on the +65 °C container.
+50 °C	8.5	109	133	123	The liner blistered.
−25 °C	1.4	87	83	100	No damage was evident during the visual inspection.
−40 °C	4.6	78	84	108	

high-pressure static hold duration. There was no damage observed in type 4 containers with high-pressure static hold periods of 30, 100 and 500 h. Tests with type 3 containers showed no increase from the baseline value. Even at an increased hold time of 500 h, there was no evidence of permeation, deterioration, cracking, pitting or any other damage to the liner [43]. Typically, hydrogen fuel containers have a maximum life of 1833 fuelling cycles. Durability tests were carried out on different containers for 500, 1000, and 1833 pneumatic cycles. There was no damage sustained by any of the containers, and there was no visible difference among cylinders subjected to 500, 1000, and 1833 cycles.

Durability tests were conducted on hydrogen-fuelled ICEs, the findings of which have been reported [44]. The engine was instrumented to generate test data. Lubrication, bearings and wear surfaces were appropriately selected after the tests. Engine dynamometer tests were conducted for 1000 h under maximum power and maximum torque conditions. Data were also collected from a fleet of eight full-size hydrogen-operated pickup trucks. Long-term operation for nearly three years included intermittent maintenance, such as the replacement of the spark plug and the injector as required. Performance and emission characteristics were evaluated. Several times during the entire test period, emission characteristics were evaluated at the Argonne National Laboratory (ANL), and the concentration of pollutants was low. Lubricating oil analysis showed that there was no problem of concern due to wear in the engines. Studies on material compatibility were carried out after disassembling the engine, and there were no signs of wear. However, there were carbon deposits on the face of both inlet and exhaust valves as well as on the piston crowns. These are caused by excessive engine oil blow resulting from high intake charge pressure. However, a closer analysis after an impact test showed a small amount of fracture, which might be the result of embrittlement. The results of the performance, emissions and long-term durability tests show that hydrogen can be effectively used in engines and vehicles with improved performance and reduced emissions, even without an aftertreatment device.

Tests on an engine dynamometer for 1000 h provide very useful information. Most of the major parts of the engine were in reasonable condition. Only the spark plug was affected by corrosion, and the injector did not operate satisfactorily over the entire period. However, these components can be checked and intermittently replaced. Guidelines for the selection of appropriate spark plugs and injectors for long-term operation can also be chalked out during these long-term tests. Valve reliability was one of the key findings after the long-term durability test was completed. Wear was clearly visible on both intake and exhaust valves, and the intake valve had a more severe recession. The data from the operation of vehicles provided ample useful information. Spark plugs and fuel injectors as well as issues related to valves displayed similar problems as discussed above. Oil was changed regularly as per the maintenance schedule, and thus, oil analysis showed that the engine was in sound condition and without any major wear. However, the oil sample analysis in a couple of vehicles showed water in it. This could be due to the frequent use of start–stop conditions during driving, which results in the formation of excess water. It was observed that valve faces, tops of the piston and intake and exhaust ports had heavy black deposits. This might be due to the oil being pulled into the combustion temperature.

The Southwest Research Institute (SWRI) has recently initiated some activities to evaluate the components of compressed hydrogen gas tanks. The appropriate high-pressure test rig has been designed and developed for higher levels of pressure of more than 1360 atm. A wide range of temperatures between −4 and about 115 °C can be monitored during the tests. Such stringent and broad pressure ranges have been selected for tests such that the valves can operate without any failure beyond their normal range of operation [45].

At present, the potential of platinum as a catalyst in hydrogen fuel cells has been well established. However, considering its high cost, its commercial application is still a concern. On the other hand, the application of hydrogen fuel cells is rapidly penetrating several industries, including transport. Therefore, suitable and economically affordable alternatives are being sought. Platinum, considering its properties such as corrosion resistance and a higher level of conductivity, has attracted the attention of researchers as a support material for hydrogen fuel cells. Recently, researchers from the Queen Mary University of London and University College London (UCL), UK, produced graphene and developed hydrogen fuel cell catalysts. The technique adopted by the researchers was successful in developing 'high-quality graphene decorated with platinum nanoparticles in a one-pot synthesis'. The durability of graphene was also successfully tested over many cycles during a short period of time for the 'accelerated stress tests' recommended by the US Department of Energy (DoE). Graphene-based catalysts achieved 30% lower losses than conventional commercial catalysts during the same period of tests, which is very encouraging [46].

With the objective of achieving carbon-neutral goals, most automobile manufacturers focus on electric vehicles. Toyota, in addition to other alternative fuels, is working on hydrogen-powered automobiles.

Toyota developed a hydrogen-fuelled racing car (Figure 7.34) that was field tested at the Fuji International Speedway event in Oyama, Shizuoka Prefecture, on 21 May 2021 for 24 h endurance tests. In subsequent endurance tests, there were improvements in output by 24% and driving range by around 30%. In addition to work on the development of on-board liquid hydrogen, it is also working on solving problems related to the use of gaseous hydrogen in commercial vehicles and passenger cars. Currently, the company is developing liquid hydrogen cars that could be commercially viable. A difficult challenge is adapting the technology to hold hydrogen in the liquid form at −253 °C. Liquid hydrogen helps increase mileage and requires smaller hydrogen fuel stations [47].

Figure 7.34 Hydrogen-fuelled racing car [47]. *Source:* BrokenGearbox/Wikimedia Common.

7.12 Material Compatibility with Hydrogen Application

Many materials are affected when exposed to hydrogen gas. Temperature, pressure and mechanical loading all influence the nature and magnitude of deterioration. Embrittlement due to hydrogen is one of the major problems that can cause the material to lose its tensile strength and reduce its ductility. The result is often a reduction in load-bearing capacity, accelerating fatigue crack growth and brittle failure. However, a detailed discussion of the problem of embrittlement is made elsewhere in the book. The failure of several materials caused by hydrogen energy must be addressed clearly. The study of the tribological behaviour of some of the machines and components, such as piston rings and seal bearings, which are affected by the hydrogen environment, needs to be conducted from the point of view of LCA because of the situation leading to the physisorption of hydrogen molecules and subsequent chemisorption of dissociated atoms on metal surfaces. It is essential to evaluate the effect of hydrogen on the friction and wear characteristics of some materials. Although the primary objective of storing hydrogen at high pressure is to achieve high specific energy, there are still several issues associated with hydrogen storage systems. Metals fail due to the high rate of hydrogen uptake. Permeation of hydrogen in rubber and polymer seals is a typical problem often encountered in several applications.

Material selection is a crucial part of the reliable and safe design of any hydrogen system, including applications in transportation, stationary power generation, industrial heating, chemical processes, etc. Selection of the material to operate in its most extreme operating conditions is an essential criterion for applications where the material is exposed to hydrogen for a long time. Material selection for hydrogen applications needs to be considered from several angles, including its compatibility with the service conditions and exposure to hydrogen environments. Additionally, its mechanical properties need to be taken into account. The mode of failure can be determined by some preliminary testing or by consulting published literature. This information can provide some guidelines in this respect.

7.13 Ductility

Ductility refers to a material's ability to undergo plastic deformation (change in shape) under tensile stress without fracturing. It is a very crucial property in the context of material selection for use in hydrogen systems. These materials can accommodate local stress concentrations. They can be curved and reshaped without breaking. In contrast, brittle materials have only a small amount of elongation at fracture. Brittle materials do not show significant permanent elongation, and they fail suddenly when they are exposed to tensile stress.

7.13.1 Temperature Effect on Ductility

Temperature affects ductility. Most of the materials clearly show a distinct trend in the ductile to brittle transition temperature phase. The effect of a decrease in temperature is much milder in face-centred cubic (FCC) materials, such as austenitic stainless steels. However, this is particularly pronounced in body-centred cubic (BCC) materials, such as ferritic steels. Ferritic steels used in hydrogen applications, especially at low temperatures, can become more susceptible to embrittlement. As the temperature decreases, the energy available for plastic deformation also decreases. This can further exacerbate the brittleness induced by hydrogen, making ductility a critical factor in maintaining the structural integrity of the material. This characteristic needs to be carefully considered, especially for hydrogen applications at low temperatures. Metals, such as austenitic stainless steels, aluminium and its alloys, and copper and its alloys, work satisfactorily at low temperatures. Austenitic stainless steels have an FCC crystal structure, which provides them good ductility and toughness even at low temperatures. Aluminium and its alloys are lightweight and possess good mechanical properties at both ambient and low temperatures. They maintain their strength and ductility at low temperatures, making them suitable for applications in extreme cold. Copper and its alloys, such as brass and bronze, also exhibit good mechanical properties at low temperatures. They retain their ductility and thermal conductivity and thus are suitable for use in systems that operate at low temperatures. These are broad guidelines. Nevertheless, it is very important that design features, application areas and risk analyses are considered from a safety perspective. Chemical composition and strength are key factors in selecting the appropriate material. These materials are commonly chosen on the basis of their chemical composition and strength. Most commonly, aluminium alloys and copper alloys, low-alloy ferritic steels, austenitic stainless steels and C–Mn ferritic steels are suitable. High-strength ferritic and martensitic steels, titanium alloys and ductile cast irons are usually avoided. Polymers are not suitable for hydrogen applications as they are prone to catching fire. Hydrogen embrittlement is a phenomenon where the presence of hydrogen atoms in the metal's lattice structure can lead to reduced ductility and increased brittleness. This makes the material more prone to sudden and catastrophic failure, especially under tensile stress. The phenomenon of embrittlement caused by hydrogen is one of the serious problems related to hydrogen's effect on materials. Hydrogen-induced cracking is a problem that occurs when atomic hydrogen diffuses in the material and recombines with H_2 leading to high pressure, which induces cracks. A detailed discussion on embrittlement and other forms of hydrogen damage is provided in Chapter 8, and hence these phenomena are not discussed here. There have been scenarios when hydrogen has reacted with metals and formed metallic hydrides when this was not the objective. Specific reactions that occur between hydrogen and metals depend on different factors, such as the type of metal, the presence of impurities or catalysts, temperature, pressure and the duration of exposure. Sometimes, trace impurities in materials or environments can catalyse reactions between hydrogen and metals, leading to the formation of unexpected hydrides. When metals are exposed to a combination of hydrogen and corrosive or oxidising environments, complex chemical reactions can occur resulting in the formation of hydrides, oxides and other compounds on the metal surface. Similarly, because of the catalytic effect and the effect of pressure and heat, hydrocarbons and several other compounds are also formed.

7.13.2 Choice of Fire-resistant Material

Considering the typical physio-chemical properties of hydrogen, such as minimum ignition energy and high flame speed, it is essential that depending upon the specific application the system needs to be leakproof and the components need to be fire-resistant. In addition, in the case of a minor leak of a joint, the flame will spread rapidly if accidentally ignited. In such a case, a fire-resistant component would be able to tolerate the heat from the fire for at least some time. Therefore, choosing fire-resistant materials for hydrogen applications involves a careful consideration of several factors to ensure safety and performance. Broadly speaking, ideal materials should possess as many as possible characteristics as described hereafter. (i) The material should have high thermal stability to withstand elevated temperatures. (ii) It should be resistant to ignition and combustion in the presence of hydrogen flames. (iii) Chemical inertness to hydrogen and other potentially reactive species is a desirable feature. (iv) Good insulation properties of the material help to minimise heat transfer and prevent the spread of fire. However, it is not normally possible to obtain some material that should meet all these requirements without any modification. Moreover, there are several other additional requirements that also depend upon specific applications. Care must be taken to see that any coatings and paintings used should be compatible with fire. Moreover, the design of the system where hydrogen is used should have provisions for controlled venting and pressure relief to prevent overpressure explosion in the case of fire. Aluminium and copper alloys joined by brazing or soldering are damaged in hydrogen fires. It is imperative to select the proper material for different components and subsystems depending upon their location and exposure to the hydrogen system. Some non-metallic sealing materials, such as valve stem seals and polyethylene terephthalate (popularly known as 'PETE'), have failed in hydrogen fire cases. Various grades of steel, such as 316L, Inconel alloys, such as Inconel 625 and ceramics, have been found to be suitable as fire-resistant materials for hydrogen applications. Specialised fire-resistant coatings can be applied to structural materials to enhance their fire resistance. Some polymers, such as polyimides, are designed to be fire-resistant and can be suitable for specific hydrogen-related applications. Mineral wool is an insulating material that can provide effective fire resistance by limiting heat transfer.

7.13.3 Materials for Liquid Hydrogen Service

A material's fracture toughness at cold temperatures is a very significant parameter for liquid hydrogen service. Material selection for liquid hydrogen service must meet a range of requirements. The material chosen for liquid hydrogen should perform well under low-temperature conditions. A Charpy impact test is usually carried out to evaluate the impact energy and impact strength of the metal. It provides information related to the tough/brittle transition of the metal based on temperature. Some metals change from ductile to brittle in low-temperature conditions under certain circumstances. Such a situation can occur at a temperature significantly higher than cryogenic temperatures. Cryogenic compatibility, thermal expansion, permeability, ductility at low temperatures, corrosion resistance and thermal conductivity are some of the important criteria for selecting the material for using liquid hydrogen. For cryogenic tanks and pipelines, insulating materials should be selected to minimise heat transfer from the environment. Welding and any other joints should be leak free and mechanically robust at cryogenic temperatures. Specialised sealants and gaskets designed for cryogenic conditions are used to ensure leak-tight joints. In some situations, insulating materials, such as vacuum-insulated panels or multilayer insulation, are used to minimise heat transfer. Considering its low conductivity, polyurethane foam provides good insulation in cryogenic tanks and pipelines. Usually – aluminium alloys, such as 5083 and 5456, austenitic stainless steel, such as 304L and 316L, titanium and some other materials are used in systems handling liquid hydrogen because of their excellent corrosion resistance and good mechanical stability at cryogenic temperatures. Depending upon specific applications in cryogenic systems, materials such as niobium–titanium (Nb–Ti) and niobium–tin (Nb_3Sn) are also found to be suitable. Some composite materials, such as carbon-fibre-reinforced composites, can offer a balance of low thermal conductivity and high strength for cryogenic applications.

Hydrogen-compatible materials, polymers and metals with wider application potential for hydrogen infrastructure and storage are receiving a lot of attention. The Hydrogen Materials Compatibility Consortium (H-Mat), composed of Sandia National Laboratories (SNL, lead), Pacific Northwest National Laboratory (PNNL, lead), Oak Ridge National Laboratory (ORNL), Savannah River National Laboratory (SRNL) and ANL, is a framework for cross-cutting early-stage research and development (R&D) on hydrogen materials compatibility. H-Mat was launched in 2018 by the US Department of Energy's Hydrogen and Fuel Cell Technologies Office within the Office of Energy Efficiency and Renewable Energy [48]. R&D activities with well-defined objectives are being pursued in areas that have high practical relevance, such as atomic-scale to micro- and macro-scale modelling of hydrogen-induced degradation of polymers, experimental discovery of the mechanisms of hydrogen-induced degradation in polymers, development of hydrogen-resistant polymeric formulations, quantification of hydrogen effects on microstructural length scales in austenitic stainless steels and identification of hydrogen-resistant, high-strength ferritic steel microstructures. Clarification of the mechanisms of hydrogen degradation in high-strength aluminium alloys and development of a framework to quantify damage and crack nucleation in hydrogen environments are some of the areas that are being widely investigated [48].

The behaviour (e.g. fatigue life and resistance to buckling) of composite materials, austenitic steels and welded materials in cryogenic conditions (30 K) is being evaluated.

These R&D activities pursued in collaboration with industries and academia are of immense practical relevance.

Concluding Remarks

In most of the published literature, studies and investigations on LCA have focused on the environmental profile. Obviously, this aspect of the study is very relevant for a process or product analysis. However, the socio-economic profile is a very important consideration because a totally clear picture can emerge, and solutions can evolve only after critically analysing LCA, LCC and SLCA. There have been a few studies that have addressed the socio-economic impact. Most often a single parameter may not provide a solution. The rational integration of all of these into a critical thinking process provides extremely useful information and data. This is because it can be used to quantify the individuals responsible for the entire impact of a project. Adopting a promising pattern in practice is essential. Hydrogen system durability involves selecting materials that are resistant to embrittlement and permeation, designing robust components, considering operational and environmental conditions, and implementing appropriate maintenance and inspection practices. Durability assessments also provide information about material coatings, insulation and corrosion protection to extend the lifespan of hydrogen systems. It is important that appropriate materials need to be selected to safely encounter hydrogen's unique properties and potential interactions. Material compatibility in hydrogen systems needs to take into account hydrogen's small size, reactivity, purity, corrosion resistance, embrittlement, permeation, chemical reactions and compatibility with other gases from a performance and safety point of view. In summary, a comprehensive understanding of LCA, durability and material compatibility is needed to develop and deploy reliable technologies for the transition to a safe and sustainable no-carbon energy future.

Abbreviations

LCA	Life Cycle Assessment
LCSA	Life Cycle Sustainability Assessment
WCED	World Commission on Environment And Development
UN	United Nations
LCC	Life Cycle Costing
FU	Functional Unit
LCIA	Life Cycle Impact Assessment
ISO	International Standards Organization
ILCD	International Reference Life Cycle Data System
IPP	Integrated Product Policies

TRL	Technology Readiness Level
ILUC	Indirect Land-Use Change
LNG	Liquified Natural Gas
DME	Dimethyl Ether
GHG	Greenhouse Gas
CCS	Carbon Capture and Storage
LCI	Life Cycle Impact
SMR	Steam Methane Reforming
GWP	Global Warming Potential
COP	Conference of the Parties
NREL	National Renewable Energy Laboratory
IPCC	Intergovernmental Panel on Climate Change
PV	Photovoltaic
GOA-ON	Global Ocean Acidification Observing Network
NOAA	National Oceanic and Atmospheric Administration
SOCAT	Surface Ocean CO_2 Atlas
GLODAP	Global Ocean Data Analysis Project
CFC	Chlorofluorocarbons
LFG	Landfill Gas
GREET	Greenhouse Gases, Regulated Emissions and Energy Transportation
PSA	Pressure Swing Absorption
ICCT	International Council on Clean Transportation
ATR	Autothermal Reforming
SER	Sorption-Enhanced Reforming
VOC	Volatile Organic Compound
WGS	Water–Gas Shift
PSA	Pressure Swing Absorber
WTT	Well-to-Tank
WTW	Well-to-Wheel
NMVOCs	Non-Methane Volatile Organic Compounds
ICE	Internal Combustion Engine
PEM	Proton Exchange Membrane
PEMFCs	Proton Exchange Membrane Fuel Cells

AWE	Alkaline Water Electrolysis
SOFCs	Solid Oxide Fuel Cells
EoL	End-of-Life
MEA	Membrane Electrode Assembly
FC	Fuel Cell
UCL	University College London
FCC	Face-Centred Cubic
BCC	Body-Centred Cubic
PETE	Polyethylene Terephthalate
SNL	Sandia National Laboratories
ORNL	Oak Ridge National Laboratory
SRNL	Savannah River National Laboratory
ANL	Argonne National Laboratory
SLCA	Sustainable Life Cycle Assessment
UCTE	Union for the Coordination of Transmission of Electricity

References

1 Masoni, P. and Zamagni, A. (2011). *FC HY-Guide; Guidance Document for Preparing LCAs on Fuel Cells and H_2 Technologies.*

2 Klöpffer, W. (2008). Life cycle sustainability assessment of products. *International Journal of Life Cycle Assessment* 13 (2): 89–95.

3 Hunkeler, D., Lichtenvort, K., and Rebitzer, G. (2008). *Environmental Life Cycle Costing*. New York: CRC Press/SETAC Books.

4 Costa, D., Quinteiro, P., and Dias, A.C. (2019). A systematic review of life cycle sustainability assessment: current state, methodological challenges, and implementation issues. *Science of the Total Environment* 686: 774–787. https://doi.org/10.1016/j.scitotenv.2019.05.435.

5 Bickel, P. and Friedrich, R. (2005). *ExternE – Externalities of Energy, Methodology 2005 Update*. Luxembourg: European Commission. ISBN 92-79-00423-9.

6 Guinee, J.B., Heijungs, R., and Huppe, G. (2011). Life cycle assessment: past, present, and future. *Environmental Science & Technology* 45: 90–96.

7 UNEP. Towards a Life Cycle Sustainability Assessment: Making informed choices on products, UNEP/SETAC Life Cycle Initiative (1972-2012).

8 ISO 14040. Environmental Management–Life Cycle Assessment–Principles and Framework. London, UK: British Standards Institution (2006).

9 EC-JRC (2011). *International Reference Life Cycle Data System (ILCD) Handbook – Recommendations for Life Cycle Impact Assessment in the European Context*, 1e. Luxemburg: European Commission-Joint Research Centre – Institute for Environment and Sustainability.

10 Verones, F., Henderson, A.D., Laurent, A. et al. (2016). LCIA framework and modelling guidance [TF 1 Crosscutting issues]. In: *Global guidance for life cycle impact assessment indicators*, vol. 1, 40–57. UNEP https://backend.orbit.dtu.dk/ws/portalfiles/portal/131994451/Global_guidance_LCIA_v.1.1.pdf.

11 Myhre, G., Shindell, D., Bréon, F.M. et al. Anthropogenic and natural radiative forcing. In: *Climate Change 2013: The Physical Science Basis. Contribution of Working Group I to the Fifth Assessment Report of the Intergovernmental Panel on Climate Change* (ed. T.F. Stocker, D. Qin, G.K. Plattner, et al.), 659–740.

12 Valin, H., Peters, D., van den Berg, M. et al. (2015). *The Land Use Change Impact of Biofuels Consumed in the EU: Quantification of Area and Greenhouse Gas Impacts*. International Institute for Applied Systems Analysis.

13 Al-Breiki, M. and Bicer, Y. (2021). Comparative life cycle assessment of sustainable energy carriers including production, storage, overseas transport and utilization. *Journal of Cleaner Production* 279 (10): 123481.

14 Valente, A., Iribarren, D., and Dufour, J. (2019). Harmonising methodological choices in life cycle assessment of hydrogen: a focus on acidification and renewable hydrogen. *International Journal of Hydrogen Energy* 44: 19426–19433.

15 Valente, A., Iribarren, D., and Dufour, J. (2017). Harmonised life-cycle global warming impact of renewable hydrogen. *Journal of Cleaner Production* 149: 762–772.

16 Valente, A., Iribarren, D., and Du, J. (2018). Harmonising the cumulative energy demand of renewable hydrogen for robust comparative life-cycle studies. *Journal of Cleaner Production* 175: 384–393.

17 Cetinkaya, E., Dincer, I., and Naterer, G.F. (2012). Life cycle assessment of various hydrogen production methods. *International Journal of Hydrogen Energy* 37 (3): 2071–2080.

18 Spath, P.L. Mann, M.K. (2004, February) Life cycle assessment of renewable hydrogen production via wind/electrolysis milestone completion report. *NREL/MP-560-35404, National Renewable Energy Laboratory Contract No. DE-AC36-99-GO10337*.

19 Bhandari, R., Trudewind, C.A., and Zapp, P. (2014). Life cycle assessment of hydrogen production via electrolysis – a review. *Journal of Cleaner Production* 85: 151–163.

20 Jolliet, O., Margni, M., Charles, R. et al. (2003). IMPACT 2002+: a new life cycle impact assessment methodology. *The International Journal of Life Cycle Assessment* 8: 324–330. https://doi.org/10.1007/BF02978505.

21 Gruber, N., Clement, D., Carter, B. et al. (2019). The oceanic sink for anthropogenic CO_2 from 1994 to 2007. *Science* 363 (6432): 1193–1199. https://doi.org/10.1126/science.aau5153.

22 Bakker, D.C.E., Pfeil, B., Landa, C.S. et al. (2016). A multi-decade record of high-quality fCO_2 data in version 3 of the Surface Ocean CO_2 Atlas (SOCAT). *Earth System Science Data* 8: 383–413. https://doi.org/10.5194/essd-2016-15.

23 Olsen, A., Lange, N., Key, R.M. et al. (2019). GLODAPv2.2019—an update of GLODAPv2. *Earth System Science Data* 11: 1437–1461. https://doi.org/10.5194/essd-11-1437-2019.

24 Zhou, Y., Swidler, D., Searle, S., and Baldino, C. (2021). Life-cycle greenhouse gas emissions of biomethane and hydrogen pathways in the European Union. *White Paper, October 2021 of ICCT (International Council on Clean Transportation)*.

25 Contadini, J.F., Moore, R.M., and Mokhtarian, P.L. (2002). Life cycle assessment of fuel cell vehicles a methodology example of input data treatment for future technologies. *The International Journal of Life Cycle Assessment* 7 (2): 73–84. https://doi.org/10.1007/BF02978850.

26 Mueller-Langer, F., Tzimas, E., Kaltschmitt, M., and Peteves, S. (2007). Techno-economic assessment of hydrogen production processes for the hydrogen economy for the short and medium term. *International Journal of Hydrogen Energy* 32 (16): 3797–3810. https://doi.org/10.1016/j.ijhydene.2007.05.027.

27 Liu, K., Song, C., and Subramani, V. (2010). *Hydrogen and Syngas Production and Purification Technologies*. Wiley.

28 Patwardhan, K., Rajyalakshmi, S., and Balaramakrishna, P. (2013). Optimised hydrogen production by steam reforming: Part 2. https://www.digitalrefining.com/article/1000841/optimised-hydrogen-production-by-steam-reforming-part-2

29 Kurokawa, H., Shirasaki, Y., and Yasuda, I. (2010). Demonstration of highly efficient distributed hydrogen production from natural gas with CO_2 capture. https://www.osti.gov/etdeweb/biblio/21400910

30 Keipi, T., Tolvanen, H., and Konttinen, J. (2018). Economic analysis of hydrogen production by methane thermal decomposition: comparison to competing technologies. *Energy Conversion and Management* 159: 264–273. https://doi.org/10.1016/j.enconman.2017.12.063.

31 Salkuyeh, Y., Saville, B.A., and MacLean, H.L. (2017). Techno-economic analysis and life cycle assessment of hydrogen production from natural gas using current and emerging technologies. *International Journal of Hydrogen Energy* 42 (30): 18894–18909. https://doi.org/10.1016/j.ijhydene.2017.05.219.

32 Tayarani, H. and Ramji, A. (2022). Cycle assessment of hydrogen transportation pathways via pipelines and truck trailers: implications as a low carbon fuel. *Sustainability* 14: 12510. https://doi.org/10.3390/su141912510.

33 Hamedani, S.R., Villarini, M., Colantoni, A. et al. (2018). Life cycle performance of hydrogen production via agro-industrial residue gasification—a small scale power plant study. *Energies* 11: 675. https://doi.org/10.3390/en11030675.

34 Koroneos, C., Dompros, A., Roumbas, G., and Moussiopoulo, N. (2004). Life cycle assessment of hydrogen production processes. *International Journal of Hydrogen Energy* 29 (14): 1443–1450.

35 Cho, H.H., Strezov, V., and Evans, T.J. (2022). Environmental impact assessment of hydrogen production via steam methane reforming based on emissions data. *Energy Reports* 8: 13585–13595.

36 Rinawati, D.I., Keeley, A.R., Takeda, S., and Managi, S. Life-cycle assessment of hydrogen utilization in power generation: a systematic review of technological and methodological choices. *Frontiers in Sustainability* https://doi.org/10.3389/frsus.2022.920876.

37 Rinawati, D.I., Ryota, A., Takeda, K.S., and Managi, S. (2022). A systematic review of life cycle assessment of hydrogen for road transport use. *Progress in Energy* 4: 012001. https://doi.org/10.1088/2516-1083/ac34e9.

38 Elgowainy, A., Reddi, K., Wang, M. (2013). Life-cycle analysis of hydrogen onboard storage options. *Argonne National Laboratory, DOE Fuel Cell Technologies Program Annual Merit Review and Peer Evaluation Meeting*. Meeting 14 May, 2012.

39 Simons, A. and Bauer, C. (2011). Life cycle assessment of hydrogen production. In: *Transition to Hydrogen*, https://doi.org/10.1017/CBO9781139018036.006. Cambridge University Press.

40 Dhanushkodi, S.R. et al. (2008). Life cycle analysis of fuel cell technology. *Journal of Environmental Informatics* 11 (1): 36–44.

41 Usai, L. et al. (2021). Life cycle assessment of fuel cell systems for light duty vehicles, current state-of-the-art and future impacts. *Journal of Cleaner Production* 280: 125086.

42 Simons, A. and Bauer, C. (2015). A life-cycle perspective on automotive fuel cells. *Applied Energy* https://doi.org/10.1016/j.apenergy.2015.02.049 (This paper is included in the Special Issue of Clean Transport edited by Prof. Anthony Roskilly, Dr. Roberto Palacin and Prof. Yan).

43 McDougall, M. and Stephens, D. (2013, August). Cumulative fuel system life cycle and durability testing of hydrogen containers. *Report No. DOT HS 811 832*. Washington, DC: National Highway Traffic Safety Administration (NHSTA).

44 Beauregard, G.P. (2011). Findings of hydrogen internal combustion engine durability. *Final Technical/Scientific Report. Project Period: 3/7/07 – 12/31/2010*. Phoenix: Electric Transportation Engineering Corporation.

45 SWRI, Test Durability of Fuel Tank Valves for Hydrogen-Powered Vehicles, un22, 202 Ref. http://www.swri.org/press-release/swri-tests-durability-fuel-tank-valves-hydrogen-powered-vehicles.

46 *Innovation News Network*, Environment 24 July 2020.

47 *Ainichi Japan Newspaper*, 5 January, 2023 (5NAGOYA, Kyodo).

48 H-Mat: Hydrogen Materials Consortium. US Department of Energy's H-Mat consortium is a good starting point: https://www.energy.gov/eere/fuelcells/h-mat-hydrogen-materials-consortium

8

Hydrogen-induced Damage (HTHA, Embrittlement and Blistering)

'The two most common elements on earth are hydrogen and stupidity'.

– Harlan Ellison

CHAPTER MENU

Hydrogen Energy: Production, Safety, Storage and Applications, First Edition. L. M. Das.

8.1 Introduction

In hydrogen environments, materials are subject to a variety of degradation phenomena. A close review of the literature indicates that there are several terms referring to the characteristics of damage occurring in the hydrogen environment over a period of time. The discussions in this chapter refer to the umbrella term 'hydrogen-induced damage' which covers a broad category of damage rather than a single category. These include high-temperature hydrogen attack (HTHA), blistering and embrittlement. There was no consistent terminology to distinguish among these phenomena. The characteristic difference among these forms of damage evolved through several transitional terminologies, such as low-temperature hydrogen, embrittlement, hydrogen-induced blistering, cracking due to hydride formation (in certain alloys), cracking due to internal hydrogen, hydrogen attack (HA), hot hydrogen attack (HHA), etc. The causes of these forms of damage are not clearly understood. Terms such as 'hydrogen embrittlement' (HE) and 'hydrogen-induced cracking' (HIC) were liberally interchanged. According to the *ASM Handbook*, HIC occurs when molecular hydrogen pressure builds up within a crack. In the case of HE, the metal loses its ductility, and the load-bearing capacity decreases because of the absorption of hydrogen atoms or molecules by the metal. This leads to cracking and sometimes even catastrophic brittle failures. The available literature clearly shows that there have been several types of damage and failures in situations where hydrogen is used. The basic causes and characteristics of failure are not well understood.

The nature of any form of the above damage could be internal or external. HE, among others, has been responsible for the deterioration of the mechanical properties of the material by reducing its ductility and making it structurally brittle. The extent of damage caused by embrittlement depends upon the microstructure of the material and the amount of hydrogen absorbed. Although a large number of experimental and theoretical studies have been carried out on the subject, it still lacks a complete scientific explanation perhaps because the complexity of its analysis depends on a host of parameters, such as mechanical-, material- and environment-based parameters related to loading, stress, crystal structure, etc.

Several terms – such as 'hydrogen-enhanced cracking', 'hydrogen environment-assisted cracking' and, in some cases, even 'hydrogen-induced cracking' – have been loosely used in place of HE. This creates a lot of confusion. These terms hide the basic characteristics of embrittlement in which hydrogen atoms, present within the metal lattice, cause a loss of ductility. Furthermore, HIC occurs when gaseous hydrogen precipitates in a typical sour environment (such as H_2S), leading to cracking precipitation. The hydrogen blistering process occurs when hydrogen ions are reduced to hydrogen atoms and atomic hydrogen, which diffuse through metals and accumulate in void spaces. Being highly reactive, these hydrogen atoms combine in the void or trap and form molecules. Because molecular hydrogen cannot diffuse, this accumulated gas builds up pressure in the void. This builds up pressure, and sometimes, due to this high internal pressure, visible distortions in the form of blisters are formed.

In this chapter, we discuss both phenomena (HIC and HE) in greater detail to avoid confusion. In addition to HTHA, blistering is also explained.

8.2 HTHA, HA and HHA

'High-temperature hydrogen attack' has been used differently in literature in different phases. To understand the minor differences between these terms and their connotations, a brief description of each is given below. An extensive review paper by Poorhaydari [1] covers in detail the research and development activities for over a century on the topic of HTHA. 'Hydrogen attack' was the most prevalent term in the twentieth century for irreversible high-temperature hydrogen damage. The term was also used occasionally for low-temperature HE in aqueous corrosive environments [2, 3], which was confusing. 'High-temperature, high-pressure hydrogen attack' is a transitional term between HA and HTHA that was used [4] to exclude low-temperature HE.

'Hot hydrogen attack' is another transitional term that was coined towards the end of the twentieth century to exclude low-temperature HE [5, 6].

8.3 High-temperature Hydrogen Attack

'HTHA' has been the most prevalent term with very distinctive characteristics in the twenty-first century for irreversible high-temperature hydrogen damage. Reference to the full term can be traced back to the 1990s [7], and the full acronym appears in the 5th edition of the American Petroleum Institute (API) RP 1941 in 1997, although other transitional terms, such as 'high hydrogen attack', were also used in the 1990s [8]. In this chapter, the term 'HTHA' is used to maintain consistency. HTHA is a phenomenon that mainly occurs on steel exposed to a hydrogen environment at an elevated temperature of more than 204 °C. This is a very different type of problem in characteristics from embrittlement or any other problem encountered at lower temperatures. Here, apart from temperature and pressure, the contents of chromium and molybdenum also play important roles. HTHA usually occurs in refineries, several chemical industries and sometimes at facilities with high-pressure steam boilers. With an increase in temperature and pressure, the possibility of HA increases. A higher temperature causes the dissociation of hydrogen that dissolves in the steel and then reacts with carbon in the steel, resulting in the formation of methane. This ultimately leads to either surface decarbonisation or internal decarbonisation. Surface decarbonisation occurs on the surface and draws carbon from the material. On the other hand, internal decarbonisation causes the penetration of atomic hydrogen into the material and forms methane with carbon. Usually, surface decarbonisation is not a concern because this only decreases the hardness of the material and increases its ductility. Thus, a particular application can be selected on the basis of these characteristics [9]. Shewmon carried out extensive research on the problems of HTHA and the consequent mechanism of methane production after the diffusion of hydrogen into metal takes place. These three phases in the sequence include incubation, rapid attack and saturation. During incubation, 'sub-micron methane-filled bubbles nucleate, and they continue to grow on either ferrite grain boundaries or on or near carbides'. There is no visible change in any mechanical properties during this phase, but the material undergoes a little expansion. There occurs a drastic visible change in the second phase. Small methane bubbles 'link up along the grain boundary' of steel [9]. This results in a reduction in the density and affects several other mechanical properties. The cavities become larger, and the material rapidly expands. The final phase is the saturation phase when the expansion rate gradually slows down to the zero level. During this phase, the long, narrow fissures interconnect and methane starts venting to the surface. This occurs when extensive carburisation takes place. Internal decarburisation however poses a serious threat in many respects. The formation of methane and the subsequent formation of voids drastically affect the mechanical properties and make the material extremely unsuitable for a variety of applications. The increasing content of Cr, Mo and V significantly increases the binding of carbon and greatly reduces the occurrence of HA. In 1949, Nelson collected a large amount of experimental data from different steels and, after rationalising the large amount of test data, prepared the well-known curves referred to as the 'Nelson curve'. Boundaries were marked in these curves, indicating pressure, temperature and chemical composition (Cr and Mo), under which HA would not take place. Industries used to successfully refer to these curves for several decades. These curves have been updated several times repeatedly due to a series of failures occurring unexpectedly within the industry. The American Petroleum Institute revised it and published it in the API-recommended practice 941 [10]. Several more recent experimental studies and the analysis of some incidents during operation have shown that some other important parameters need to be considered for a more accurate analysis [11, 12], and Nelson curves do not provide adequate guidelines and may not be relevant for the type of steels currently used in a high-temperature hydrogen environment.

8.4 Factors Affecting Hydrogen Attack

There are several factors that are responsible for HTHA to occur and its rate of attack. These factors are temperature, pressure, time of exposure, chemical composition, heat treatment, stress, microstructure, composition, microstructure and environment.

8.4.1 Temperature

It was observed that a limiting temperature plays a crucial role in the onset of the HTHA phenomenon. This depends upon a number of parameters, such as exposure time, composition of the material and pressure. Fletcher and Elsea [13] reported 'apparent permissible service temperatures' of several materials up to a pressure of about 0.7 MPa. The respective relevant temperature levels for the materials are approximately 2200 °C for carbon steel, 3300 °C for C–0.5 Mo and 1Cr–0.5Mo steel, 4000 °C for 2Cr–0.5Mo steel, 5400 °C for 3Cr–0.5Mo steel and 6100 °C for 6Cr–0.5Mo steel [13]. The critical temperature range is a very significant governing factor, which is determined by the corresponding chemical reaction. A normal situation with no symptoms of an attack can show signs of a severe attack if it is increased to more than 250 °C. On the other hand, raising the pressure level from

1000 to 2000 atm only moderately increases the extent of damage. Decarburisation occurs at the surface when carbon diffusion exceeds 6000 °C.

8.4.2 Pressure

Pressure, like temperature, is a very significant parameter with respect to HA. This is because HTHA is not a temperature-sensitive compound. As reported in the literature, different researchers have suggested different values of the threshold pressure of hydrogen for the onset of HTHA ranging from 0.7 to 3 MPa. It is possible that a wide difference stems from the tests and analyses of different researchers using their own studies and tests. Prager's observations [5] indicate that the rupture life of hydrogen (when the life drops) is approximately at the time when the pressure level of hydrogen is doubled, which closely agrees with the relationship provided by the following equation [14, 15]:

$$PCH_4 = P2H_2/K$$

8.4.3 Exposure Time

There has not been any systematic study that provides adequate technical information on the time of exposure in connection with the onset or severity of HTHA. Of course, it is well known that by increasing exposure time, it is possible to decrease both limiting temperature and pressure. Nonetheless, no well-defined curve for the relationship between temperature and pressure has been prepared that can provide adequate guidance. It is essential that *T–P* curves be generated after the test and analysis, and results are available for different materials for different durations of exposure time.

8.4.4 Stress

Stress is a significant parameter that promotes HTHA. The chances of HA increase with an increase in mechanical stress. Then, the total pressure (hydrogen partial pressure, additional stress during operation and residual stress from heat treatment) should be taken into account to estimate the level of mechanical stress. Residual stress increases material strength and also promotes HTHA. The prevention of HAs requires some stress relief treatment (heat treatment) at or above 650 °C. There are well-defined standards and procedures for such treatments. Sometimes, it has been observed that, under high-stress conditions, brittleness occurs before decarburisation during the HA phenomenon. The temperature at such times is relatively moderate. Sakai and Asami made some fascinating observation in their work about the effect of stress: (i) incubation time vanishes with applied stress and (ii) decarbonisation occurs faster with specimens under stress [16]. The void growth rate, in a quenched-and-tempered 2.25Cr–1Mo steel subjected to a pressure of 19.6 MPa and a temperature range of 525–550 °C, increases up to three to four times when the external stress exceeds 1–2 MPa [17]. It has also been estimated that an applied stress of 150 MPa is equivalent to increasing the hydrogen pressure by approximately 5 MPa for 2.25Cr–1Mo steel at 454 °C [18].

8.4.5 Composition of the Material

The availability of carbon and its level of solubility in the matrix is critical to the phenomenon of HTHA. Increasing the content of Cr, Mo and V increases the resistance to HTHA. Elements such as tungsten, zirconium, chromium, molybdenum, vanadium, titanium and niobium bind carbon in the form of stable metallic carbide with higher thermal stability and thus offer resistance to HA. At the same time, some situations, such as high carbon content in the material, the presence of excess Mn and the addition of aluminium and Ni, are also detrimental and need to be avoided. It has also been reported that the presence of silicon in steel influences the formation and relative stability of carbides [19]. Researchers have already observed that microalloyed steels, for example, samples with 0.02–0.14% Nb, show a relatively lower tendency for HTHA than carbon steel [20]. A review of the literature shows that a wide range of research and development activities related to HTHA have been carried out involving materials. Efforts are made to highlight some significant outcomes of research activities of some researchers on several types of materials. The stability of carbides, composed of carbon and specific metals at elevated temperatures, needs to be closely analysed to assess their interaction with hydrogen under a situation that could lead to HTHA. Cementite (Fe_3C), composed of 93% iron and 7% carbon, is brittle and is not a stable carbide at elevated temperatures. It dissolves rather easily within an interaction with atomic hydrogen. Similarly, C–0.5Mo steel, which belongs to a series of carbon–molybdenum low-alloy steels, is highly susceptible to HTHA. Depending upon the content of chromium, there are a series of carbides (M2C, M3C, M6C, M7C3 and M23C6), where 'M' stands for metallic elements and 'C' represents the interstitial elements. All five types have their stability conditions. Stability conditions of M3C, M2C, M7C3, M23C6 and M6C are presented in an increasing order. Incidentally, 2.25Cr–1Mo steels have all five types of elements in their composition [19].

Van der Burg et al. considered a model that was built on a combination of two different models. One was the equilibrium reaction model for the hydrogen-to-methane reaction, and the other was based on the phenomenon of growth of cavities due to creep and diffusion. This approach of analysis, to study the HA in 2.25Cr–1Mo steels, included the computation of Nelson curves on several types of carbides under different levels of stress conditions. The analysis was able to identify significant parameters that determine the lifetimes of hydrogen attack [21].

It is possible to enhance the carbide stability and increase the HTHA resistance of 2.25Cr–1Mo steel by reducing silicon to a maximum of 0.15% by weight and, at the same time, by adding a maximum of 0.15.% by weight of vanadium [22]. There are some other observations on the effect of materials on HTHA. Vanadium shows higher resistance to HTHA in Cr–Mo steels compared to Ti or Nb [23]. Woodward [24] used in situ scanning transmission electron microscopy (STEM) to observe the chemical reaction of steel Fe_3C carbides with high-temperature hydrogen. It is well known that sometimes hydrogen in steel severely affects the ductility of the material. Johnson [25], more than a century and half ago, described the possibility of the overall performance deterioration of steel by hydrogen leading to sudden failure.

To assess the short-term effect of the possible microstructural changes that could occur within a highly susceptible material at the onset of HTHA, researchers put a high carbon

(0.78%) eutectoid steel exposed at a pressure of 1 atm for a brief period in a high-temperature furnace. The long-term microstructural changes caused by HTHA were studied by the microstructural analysis of a sample of C–0.5 Mo steel that was operated in a hydrogen reboiler in the industry for 53 years. The final series of tests was carried out using a novel in situ STEM to observe in real time the actual microstructural changes made by HTHA. This made it possible to analyse and assess the effect of both short- and long-term effects on materials.

8.5 Hydrogen Embrittlement Phenomenon

HE and HAC differ from HTHA, blistering and other forms of damage differs from HTHA, blistering and other forms of damage. Embrittlement is a process that results in the degradation of the mechanical properties of materials in a hydrogen environment. HE is not restricted to steels only while it is a well-known phenomenon in high-strength steels, HE can affect a wide range of metallic and non-metallic materials. It occurs when hydrogen atoms penetrate the material's structure and cause reduction in its mechanical properties, making it brittle and more prone to cracking or failure. This phenomenon was reported by Johnson [5], in 1875, who immersed a piece of iron in different acids and observed 'some remarkable changes produced in iron and steel by the action of hydrogen and acids'. He observed that 'after a few minutes' immersion (half a minute will sometimes suffice) in strong hydrochloric or dilute sulphuric acid – a piece breaks after being bent once on itself, while before immersion it would bear bending on itself and back again two or three times before breaking'. There was also a very useful observation that the metal suffers from an 'extraordinary decrease in toughness and breaking strain'. Interestingly 'it is not permanent, but only temporary in characteristics, for with the passage of time the metal slowly regains its original toughness and strength'. Johnson's observations literally accelerated active embrittlement research. Since then both experimental and theoretical analyses have been carried out. Nevertheless, no robust scientific explanation for the complexity of HE is available. Barrera et al. [26] discussed the impact of hydrogen causing several types of damage to metals. The paper mainly emphasises that the problem of embrittlement analysis has occurred from the atomistic to the continuum scale. The results of both theoretical analysis and experimental characterisation of the problem have been correlated. The paper also identifies the macroscopic effects which affect the mechanical properties that are responsible for embrittlement. Strategies for material development to avoid embrittlement have also been highlighted [26]. Dwivedi and co-researcher have made an exhaustive review on the problem of HE in different materials [27]. The mechanism of embrittlement, different causes of embrittlement which manifest different effects in different forms, embrittlement models and techniques and tools to measure HE along with preventive measures are discussed. Various embrittlement models are discussed later in the chapter.

As far as materials are concerned, high-strength steels, particularly those with martensitic microstructures and higher carbon content, and various types of metal alloys, including stainless steels, aluminium alloys and titanium alloys, are susceptible to embrittlement. Tungsten, molybdenum, copper, nickel and their alloys experience HE, especially at elevated temperatures. Materials used for hydrogen storage, such as certain intermetallic compounds and metal hydrides, are also susceptible to HE due to their primary function of absorbing and releasing hydrogen. However, not a very common phenomenon, polymers (plastics) and ceramics can experience HE if they are exposed to a hydrogen environment. HE depends upon a series of factors related to mechanical environmental and material-related functions. Loading on the systems, stress and strain, fugacity and thermodynamics of hydrogen, gas source, crystal structure and phase stability are some of the significant factors that govern the mechanism of embrittlement. Embrittlement often leads to losses in tensile strength, ductility and fracture toughness. It also decreases the energy of cohesion and is also responsible for the accelerated development of fatigue cracks. The interaction of hydrogen with metals occurs in a sequence when diatomic hydrogen (HH), which is in a state of continuous motion, is absorbed on the metal surface and dissociates into monoatomic hydrogen (H). This monoatomic hydrogen is absorbed in the metal and diffuses through the interstitial spacing in the metal. It interacts with metallurgical properties, such as grain boundaries. Thus, there is a possibility of hydrogen causing a loss of mechanical properties, such as fatigue and fracture resistance. This phenomenon of HE, unlike other hydrogen degradation mechanisms, occurs at a relatively low temperature of less than 200 °C. The main causes of failure caused by HE include the material, the hydrogen environment and the tensile stress on the component. High-strength steels with a tensile strength beyond 1000 MPa are extremely vulnerable to HE when hydrogen atoms migrate to the zone of highest stress. As shown in Figure 8.1, when the crack is formed, hydrogen continues to move towards the tip of the crack. The crack continues to grow, leading to failure. There are some industrial activities and processes, such as acid pickling and electroplating, where there is a high possibility of exposure to hydrogen. Embrittlement becomes a concern when steel is exposed to acids or if there is any possibility of corrosion.

8.5.1 External and Internal Embrittlement

There are two different types of embrittlement observed in hydrogen: 'internal hydrogen embrittlement' and 'external hydrogen embrittlement'. The phenomenon of internal embrittlement was first observed more than a century ago during the manufacturing of steel. At that time, it was not clear how much hydrogen needed to be introduced into the steel-making process. There were situations when excessive hydrogen was absorbed into the bulk of a material, typically a metal, during its manufacturing, processing or service life. These hydrogen atoms can diffuse into the material's lattice structure and occupy interstitial sites, causing a weakening of the material's bonds and reducing its ductility and toughness due to internal embrittlement. Later, as the process of manufacturing of steel continued over the years, the problem was encountered frequently, and upon observation and analysis, the causes responsible for internal embrittlement were understood. Causes, effects and preventive and curative steps of internal

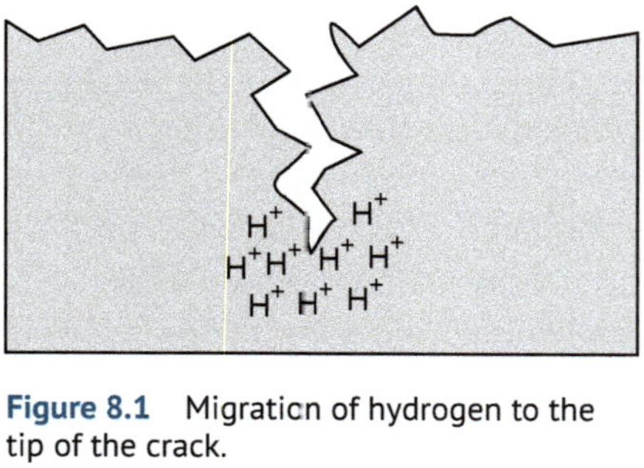

Figure 8.1 Migration of hydrogen to the tip of the crack.

embrittlement are clearly understood until now. Internal embrittlement no longer occurs. The term 'embrittlement' now refers to 'external embrittlement' that occurs due to the penetration of hydrogen into pressurised steel tanks. This affects the mechanical properties of the material and sometimes, under certain conditions, leads to the bursting of pressurised tanks. When hydrogen gas penetrates the surface of pressurised steel tanks, it can be absorbed into the material's microstructure and causes embrittlement. This embrittlement weakens the material's ability to deform plastically before fracturing, which reduces its toughness and ductility. As a result, the material becomes more prone to sudden and brittle failures, even under lower levels of applied stress. Pressurised tanks are used in various industries for storing gases or liquids under high pressures. If the steel material used to construct these tanks becomes embrittled due to the presence of hydrogen, it can lead to catastrophic failures, such as tank ruptures or leaks. These failures can have serious safety implications as they can result in explosions fires, or releases of hazardous materials.

8.5.2 Embrittlement Index

Apart from variables such as pressure, temperature, strain rate and strength of alloys, the embrittlement index is another factor that defines the sensitivity of HE. EI is the ratio of the relative reduction in the value of any mechanical property of hydrogen to its original value in an inert environment:

$$\mathrm{EI} = \frac{X - X_{\mathrm{H}}}{X}$$

where X stands for the mechanical property (such as ultimate tensile strength, elongation, etc.). Figure 8.2 shows the pattern of variation of EI with respect to pressure, temperature, hardness and strain rate.

As shown in the graphs in Figure 8.2, susceptibility to HE takes an upwards trend along with an increase in pressure right from the beginning. However, there is a point beyond which the rate of increase is extremely low. This indicates that the point of saturation has been reached and that any amount of additional hydrogen would have only a negligible effect. The graph related to temperature demonstrates that susceptibility to HE continues to increase slowly until it reaches the maximum value at room temperature and then decreases. At lower temperatures, hydrogen diffusion is slow, but at higher temperatures, hydrogen atoms are very energetic. The strain rate is closely linked to diffusion. The graph shows that there was adequate time for hydrogen to diffuse, whereas during rapid impact loading, the scope for diffusion was less. The hardness curve shows that the embrittlement index increases with hardness. High-strength alloys are closely linked to the strength of the material.

8.5.3 Characteristics of Hydrogen Embrittlement

It has already been discussed that the degree of deterioration depends upon the specific properties of a particular metal. This includes the temperature and pressure of hydrogen as well as the mechanical loading conditions. Embrittlement results in a loss of strength, ductility and fracture toughness. Strain rate sensitivity and susceptibility to delayed failure are

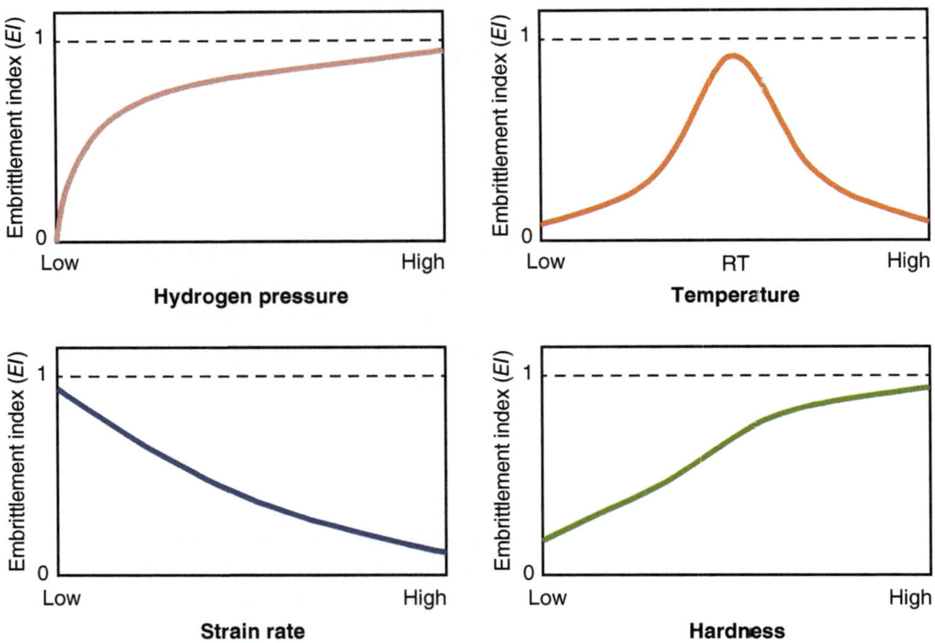

Figure 8.2 Embrittlement index with different parameters.

some of the significant characteristics manifested by HE. Due to all these conditions, sometimes the material fails catastrophically below its desired strength and stress. This makes the material brittle, and crack growth occurs due to dislocation emission. Thus, it is necessary to understand the fracture mechanics of crack growth in a hydrogen environment. Hydrogen can also be produced by corrosion reactions. HIC often occurs in welded materials. Sometimes, this is also referred to as 'cold cracking' because such cracking takes place at a relatively low temperature (such as near room temperature). Hydrogen-assisted cracking (HAC) results in subcritical growth in the material which occurs above a threshold level that is below the yield stress. However, cracking leading to fracture is initiated at stress between the yield stress and the ultimate stress. HAC is caused by diffused hydrogen content during welding. HAC can be reduced by heating electrodes before welding. It has also been found that heating after welding has been useful for reducing the chances of the occurrence of HAC. Several techniques such as gas chromatography, the glycerine technique and the mercury method are available to determine the content of diffusible hydrogen.

There are several industrial processes and common manufacturing processes, such as casting, machining, welding, carbonising, heat treatments, etc. In some of these situations, hydrogen may dissolve in the component, thereby leading to embrittlement. In some other situations, corrosion dissolves hydrogen into the metal. Hydrogen released from moisture (in coatings of welding electrodes, damp welding rods, etc.) is found to cause embrittlement. In some chemical or electrochemical processes, hydrogen atoms are generated at the surface by the reduction of hydrogen ions or water. They dissolve in the metal and thus become a source of hydrogen that could lead to embrittlement. Embrittlement has been a problem in the industry where the chemical reactions of sulphide stress cracking involve hydrogen sulphide.

8.6 Mechanisms of Embrittlement

HE has proven to be a serious bottleneck for the widespread use of hydrogen in a number of engineering applications, such as pipelines, high-pressure tanks and the use of steel in the automotive, chemical and many other industries. Challenges have been encountered in the storage, transport and utilisation of hydrogen. A review of published literature on contemporary research and developmental activities describing the multifaceted nature of HE has clearly shown that there is no single integrated theory that could consider all types of HE mechanisms. There have been several mechanisms explaining the phenomenon of HE over the past few decades. However, only some models have survived, which have taken into account the updated findings and adopted advanced calculations to develop more realistic models. Embrittlement is highly dependent on the hydrogen concentration at a specific location. In such circumstances, stress-enhanced diffusion takes place, thereby initiating cracks. These cracks continue to occur until the hydrogen concentration reaches a critical point. The growth of the crack stops only when the stress level substantially decreases. In general, there are several models built on different hypotheses and analysed field observations that explain the phenomenon of embrittlement. Quite often, the explanations only mention a few of the many factors that are responsible for resulting in embrittlement. There are situations when internal stress is developed due to the formation of new hydrides when the mechanical properties change. These are brittle and have a lower density than pure metal. This often leads to failure.

It has also been observed in many cases that the non-uniform distribution of hydrogen in metals under stress leads to situations where plasticity is much higher in certain locations, thereby giving rise to local microscopic deformation that ultimately results in failure. In some cases, there is a loss in atomic bonding strength in the presence of hydrogen, which leads to a well-pronounced effect of lattice decohesion. Embrittlement takes place and fracture occurs when stress surpasses the level of cohesive stress. These basic conditions and the migration behaviour of hydrogen in a particular metal are useful in assessing the susceptibility of metals to HE.

8.7 Embrittlement Models

There are a number of research papers discussing the primary causes of embrittlement. There are still a few debates and doubts. A broad review of the literature shows that the dominant mechanisms of HE are the hydrogen-enhanced decohesion (HEDE) mechanism, adsorption-induced dislocation emission (AIDE) and hydrogen-enhanced localised plasticity (HELP) model and hydrogen-enhanced strain-induced vacancy (HESIV) formation mechanism. In addition to these mechanisms, there are other phenomena that describe the tangible effects of corrosion and material degradation due to the impact of hydrogen. Most embrittlement problems can be explained by mechanisms that involve plasticity- and decohesion-based mechanisms. Among these models, HELP, HESIV and AIDE fall into the former category, whereas, as the term suggests, 'HEDE' is better explained by the decohesion mechanism. While the characteristics of a specific damage mechanism may differ from others, it is quite common that hydrogen-enhanced plasticity and decohesion simultaneously occur in metals [28–36] as well as the influence of dominant parameters, such as material chemistry, microstructure temperature, strain rate, hydrogen concentration, etc. There have been instances when plasticity becomes a prerequisite for HEDE to occur.

8.7.1 Hydrogen-enhanced Decohesion

The decohesion mechanism is one of the earliest mechanisms that was identified. It is based on the idea that hydrogen, when dissolved in steel or other metals, reduces the interatomic cohesive strength, thereby reducing cleavage surface energy. In a decohesion mechanism, there is a critical concentration of hydrogen atoms at which brittle fracture occurs. In this case, with a reduction in the decohesion mechanism, a critical concentration of hydrogen atoms accumulates in the grain boundaries, and the situation leads to premature brittle fracture, which is intergranular (IG) or quasi-cleavage (QC) in characteristics. Figure 8.3 shows the HEDE mechanism in gaseous hydrogen.

According to the figure, the situation develops and the maximum stress is established. Once the hydrogen concentration rises beyond the critical value, a crack forms along the cleavage plane. Computer simulation studies have also been carried out to analyse and support the HEDE mechanism [37, 38]. There are a few experimental techniques available to study the related kinetics on the basis of hydrogen, reducing cohesive force. Therefore, sometimes questions are raised about the validity of these analyses. Nevertheless, there is no dispute about the fact that the major cause of HE in high-strength steel is due to the HEDE phenomenon. This occurs because the high stress level at the crack tips causes high concentrations of hydrogen there [39]. Hydrogen (H) interacts with metallurgical features causing embrittlement and crack propagation.

8.7.2 Hydrogen-enhanced Local Plasticity Model

HELP model analysis is based on the hypothesis that hydrogen, dissolved in the metal, enhances dissociation motion. Beachem [40] observed this behaviour in the IG and QC fractures using a microscope. There have been a large number of studies on several aspects of

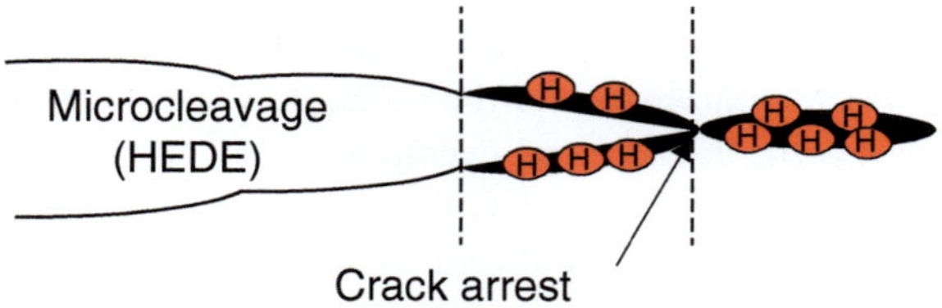

Figure 8.3 HEDE mechanism in gaseous hydrogen.

material failure because of hydrogen. Intergranular failure of nickel and iron, problems of hydrogen-enhanced fatigue, hydrogen-induced fracture structure, hydrogen-enhanced local plasticity, influence of hydrogen on dislocation, problems related to formation and behaviour of QC fracture in embrittlement on steel, role of H-H interaction, formation of local hydrides, etc. are some of the material-related problems caused by hydrogen in different situations [41–48]. Elaborate discussions on these aspects have not been provided here. Locally ductile processes cause failure, in a sense. It is possible that local dislocation mobility occurs at a relatively low stress level due to hydrogen. In these situations, it is the dislocation that is responsible for plastic deformation in a metal lattice. Local plastic deformation due to embrittlement leads to fracture. Transmission electron microscopy (TEM) is used to observe hydrogen-enhancing plastic events introduced to the column of the TEM, and this trend is reversed when hydrogen is removed. In fact, parameters such as microstructure, stress intensity of the crack tip and hydrogen concentration are crucial to the characteristics of fracture. The magnitude of plastic deformation is responsible for the development of dislocation structures beneath the fracture surfaces into dislocation walls or bands. Lin and co-researchers have proposed a model in which both hydrogen-enhanced plasticity and HEDE are analysed [49]. As far as loss of ductility and the suppression of dimples are concerned, the simulation results of this analysis are in close agreement with the experimental observations.

8.7.3 Adsorption-induced Dislocation Emission

In this mechanism, hydrogen is absorbed in the crack tip where the surface concentration occurs. A result of this absorption is the generation of dislocations within the material and their further motion. Consequently, the situation leads to dislocation injection and formation of microvoids that facilitate crack growth [50, 51]. Thus, AIDE is a combined effect of HEDE and HELP. A HEDE mechanism thins the intermetallic bond first, followed by a HELP mechanism for the formation of microvoids in the next step.

It is critical to know the mechanism of crack growth in an environment where the effects of embrittlement are more prominent. This is compared to a situation with no effect of embrittlement. In an environment with no embrittlement, dislocations mainly occur in the plastic zone at the crack tip. As a result, most of them contribute to microvoid formation and crack blunting due to their location. It is possible that a very small number of them, which intersect the crack tip, can advance the crack when hydrogen is adsorbed, embrittlement is evident and dislocations are generated at the crack tip. These developments lead to a situation where the strain in the plastic zone would be reduced because of the lower stress growth of the cracks. The microvoids will also be smaller due to the small amplitude of the strain. The AIDE mechanism with crack growth is shown in Figure 8.4. It has also been observed that the growth rate of cracks in hydrogen gas or in hydrogen-charged metals tends to be much slower in a liquid–metal environment.

8.7.4 Hydrogen-enhanced Strain-induced Vacancy Formation

Due to their strong influence on many mechanical properties of metals, defects such as vacancies and dislocations are critical to embrittlement. In essence, the title HESIV

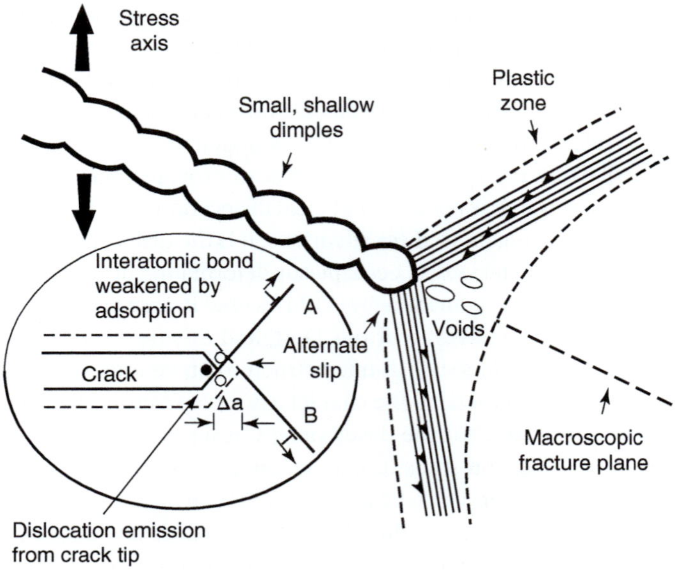

Figure 8.4 AIDE mechanism showing crack growth.

suggests the sequence of stages involved. Here, hydrogen first enhances vacancy formation and then stabilises them. Thus, their mobility is reduced. Vacancies accumulate and form clusters. Eventually, they coalesce into larger voids and microcracks. This phenomenon of microvoid coalescence causes premature failure of the material. Nagumo [52] provides a clear idea of the HESIV mechanism. A closer analysis of the above sequence of events shows that a large number of vacancies are the main cause of HESIV. Hydrogen has only an indirect role to play. It is also not clear by what mechanism hydrogen is able to enhance vacancy formation. The density of the material and the susceptibility of the material to embrittlement have been the foundation for building this model.

There have been studies related to the dominance of lattice defects causing the failure of metals due to hydrogen, and it has been observed that strained specimens absorb more hydrogen [53]. Thermal description spectrometry has been adopted to evaluate the hydrogen trapping characteristics of iron and steel [53, 54]. There have been several investigations into the effect of hydrogen on various materials with respect to embrittlement characteristics. In pure iron, hydrogen significantly contributes to enhanced vacancy formation. Researchers have also observed that increasing the hydrogen absorption capacity in medium carbon martensitic steel through strain-induced vacancies results in the glaring effect of embrittlement. Thermal desorption spectroscopy of tritium has been investigated to evaluate the effect of deformation in iron. It has been observed that 'the high density of point-like defects may play an essential role in the fracture process in the ductile–brittle transition region associated with plastic deformation' [55]. Researchers have also observed that increasing the hydrogen absorption capacity in medium carbon martensitic steel through strain-induced vacancies results in the glaring effect of embrittlement [56].

It has been discussed earlier that corrosion is the cause of HE. It is believed that the bolts of the San Francisco–Oakland Bay Bridge had cracked due to HE. The zinc-plated rods caused the reaction of zinc with water, which resulted in the introduction of hydrogen into the steel [57]. Lattice defects have also been found to cause embrittlement.

There have been studies related to the dominance of lattice defects causing the failure of metals due to hydrogen.

8.7.5 Hydride-induced Embrittlement

Metal hydrides are formed when hydrogen atoms diffuse into metals and form brittle metal hydrides. These metal hydrides (MHx) affect the mechanical properties of the metal and initiate cracking. This phenomenon is sometimes referred to as 'hydrogen cracking'. In most cases, hydrogen is obtained through casting, welding, heat treatment or corrosion. To prevent hydride-induced embrittlement, appropriate precautionary and preventive measures can be taken, such as controlling temperature, reducing stress and reducing corrosion.

As evident from the discussion above, in the hydride-driven embrittlement process, hydrides first nucleate in the stress field of a crack and subsequently develop into larger sizes. This takes place through the nucleation and growth of newly formed hydrides in the stress field and not through the growth of individual hydrides. Thus, hydride-induced embrittlement is a second-phase mechanism that involves the nucleation and growth of an extensive hydride field ahead of a crack. The auto-catalytic process of hydride nucleation and growth together with the brittle nature of embrittlement usually features in metals such as magnesium, vanadium, niobium, zirconium, tantalum and titanium, and their alloys are susceptible to this phenomenon in a hydrogen environment. Figure 8.5 shows the effect of hydride-induced embrittlement on a metal.

8.7.6 Hydrogen-enhanced Macroscopic Plasticity (HEMP)

Hydrogen-enhanced macroscopic plasticity is also sometimes referred to as hydrogen-enhanced macroscopic ductility. Hydrogen, when found in large concentrations, has been observed to affect the mechanical properties of steel. This weakens the strong arrangement

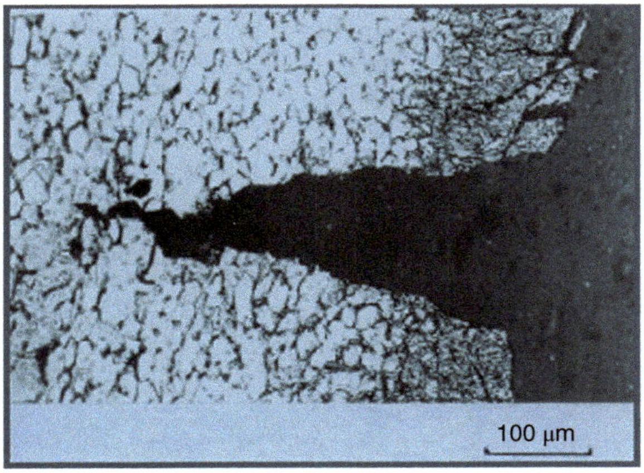

Figure 8.5 Hydride-induced embrittlement.

of the materials. The yield strength of steel is reduced by hydrogen diffusion. The material becomes soft and easy to bend. This phenomenon resulting in the loss of yield strength due to hydrogen is referred to as HEMP [27, 58].

8.7.7 Hydrogen-assisted Microfracture Mode (HAM)

The hydrogen-assisted/charged microfracture mode occurs in materials mainly due to the heavy hydrogen concentration at dislocations. Hydrogen charging decreases the ductility of the material. The mode of fracture also changes in this process. The flexibility of the material is decreased due to hydrogen, thereby changing the crack mode from cup and cone to fragile mode [27, 59]. This type of microfracture caused by hydrogen is referred to as hydrogen-assisted microfracture.

8.7.8 Decohesive Hydrogen Fracture (DHF) and Mixed Fracture (MF)

DHF is a brittle fracture that originates from the influence of hydrogen. Decohesion occurs when a material's grains or phases separate along grain boundaries or other interfaces, often resulting in a brittle fracture. When hydrogen is present, it can exacerbate this process, making the material more susceptible to fracture. In other words, decohesion, which is a form of brittle fracture, is caused by hydrogen. Brittle fracture initiated by hydrogen can eventually lead to material failure under certain circumstances. This is essentially due to ductile microvoid coalescence (MVC), which has some fisheye fractures. These fisheyes are brittle and are surrounded by MVC. In certain studies, the final fractography reflects that hydrogen causes both brittle and ductile materials to break down, and this is referred to as mixed fracture [27].

8.7.9 Hydrogen-assisted Microvoid Coalescence (HDMC)

An amplification of the term 'HAMC' gives the broader indication that it is a phenomenon that occurs when hydrogen influences the behaviour of microvoids in a material, leading to their coalescence and eventual failure of the material. Microvoid coalescence is characteristically a ductile fracture mechanism in which the crack spreads through several stages. It is characterised as a flexible brake system. Hydrogen atoms are absorbed into the material's lattice or grain boundaries and weaken the material, thereby making it prone to deformation. This leads to the formation of microscopic voids or cavities within the material, quite often at sites of stress concentration. Subsequently, hydrogen atoms migrate to the surfaces, and then the microvoids coalesce or merge together under the influence of hydrogen. This process weakens the material further and leads to macroscopic cracks or fractures. Therefore, in the sequence of events, at first, void nucleation, void growth and void coalescence take place, followed by crack extension, shear fracture of existing ligaments, etc. Due to the impact of hydrogen, localised disengagement and plastic deformation take place [27, 58].

8.8 Sensitivity Criteria for Materials to HE

The analysis of the effect of HE on different metals can be carried out from several perspectives, such as strength, composition, microstructure, processing and environmental conditions, etc. There is not a single set of universal criteria that applies to all materials, but some common factors that can influence a material's sensitivity to HE. Materials with certain microstructures, such as those containing large amounts of brittle phases, fine grain structures or regions prone to stress concentration (such as grain boundaries), can be more sensitive to HE. There have been many instances when hydrogen has been responsible for degrading the mechanical behaviour of metallic materials that have subsequently failed. A closer analysis of the causes of some of these incidents reveals the situations that led to this phenomenon. Higher-strength materials tend to be more susceptible to HE because of their reduced ductility and increased susceptibility to crack propagation. The presence of hydrogen, combined with other factors, such as stress and temperature, plays a crucial role in embrittlement. Elevated temperatures and high levels of hydrogen can increase the likelihood of embrittlement. High-stress conditions, rapid loading rates and cyclic loading can enhance the likelihood of HE. In the 1960s, to increase the mechanical strength of steel cylinders, their weight was reduced. This shows an adverse effect leading to embrittlement. Figures 8.6 and 8.7 show the damaging effects of embrittlement on the cylinders [60].

The above accidents showed that the problem was limited to high-strength steel only, but it was subsequently observed that the welded steel could also be affected by embrittlement (Figure 8.8).

Field experience and close observations have revealed that a wide range of stainless-steel equipment is affected by embrittlement. This is particularly true for the springs and diaphragms of valves and compressors as well as cryogenic vessels, particularly when these are made of stainless steel. It has been observed that austenitic steel metals are relatively safe because they normally do not show any major symptoms of embrittlement during a brief period of operation. However, after working in a hydrogen environment for a prolonged period, embrittlement occurs. Hydrogen atoms, being small, are capable of permeating solid metals easily and causing embrittlement. HE occurs in many metals, such as

Figure 8.6 Effects of embrittlement on a hydrogen transport vessel in 1980 [60], Hervé Barthélémy (2007), International Conference on Hydrogen Safety.

Figure 8.7 Hydrogen cylinder bursts: intergranular crack [60], Hervé Barthélémy (2007), International Conference on Hydrogen Safety.

Figure 8.8 Devastating effect of embrittlement on a welded steel pressure vessel [60], Hervé Barthélémy (2007), International Conference on Hydrogen Safety.

steel, iron, nickel, titanium, cobalt and their alloys, whereas copper, aluminium and stainless steels are relatively less affected by HE.

8.9 Susceptibility of Materials to Hydrogen Embrittlement

The terms 'sensitivity criteria' and 'susceptibility' are closely related but have slightly different meanings when discussing a material's response to HE. Sensitivity refers to a set of conditions or factors that determine how responsive a specific material is to HE. Sensitivity criteria are more quantitative and specific. On the other hand, susceptibility refers to a material's general tendency or predisposition to undergo HE. This is more qualitative in characteristics.

It has been observed that the fracture toughness of steel decreases as strength increases, thereby enhancing the possibility of HE leading to fracture. There have been several studies on metals that are susceptible to HE as well as materials that are likely to be attacked by hot hydrogen. There are five steps through which hydrogen interacts with metals. As far as diatomic hydrogen is concerned, there are three mechanisms. HH is in constant motion, undergoes absorption at the surface of the metal and dissociates to monoatomic hydrogen (H). However, there are two mechanisms for monoatomic hydrogen. Monatomic hydrogen (H) absorbs into metals and diffuses through interstitial spaces. Alloys of nickel, aluminium and copper along with austenitic stainless steels are quite susceptible. As far as the effect of HE in steel is concerned, high-strength steels, with strengths above 1000 MPa, are most vulnerable to embrittlement. From a usage point of view, such high-strength steels are commonly used in many significant applications such as nuclear industries, aerospace applications, automobiles and high-pressure hydrogen storage tanks. Steel with an ultimate strength of less than 1000 MPa (on the Hardness Rockwell Scale [HRC]) or a hardness of less than 32 (on the HRC) does not tend to be embrittled. From a safety and quality control point of view, it is always desirable to ensure that the hardness does not fall in the range of 32–36 because many failures have been reported in this range. Hydrogen also affects the fatigue properties of steel. However, high-cycle fatigue is relatively less affected by HE compared to low-cycle fatigue.

With regard to the problem of embrittlement of copper alloys, it has been observed that copper alloys containing oxygen are more readily affected by embrittlement when exposed to hot hydrogen. For annealing in a hydrogen-containing furnace atmosphere or brazing with a reducing flame, hydrogen diffuses through copper and the subsequent reduction of Cu_2O produces water vapour. In steam embrittlement, the two metallic copper atoms together with water form bubbles at the interface between two grains (grain boundaries). Vanadium, titanium and nickel, because of their high hydrogen solubility, often absorb a large amount of hydrogen and form hydrides that induce subsequent brittle failure.

8.10 Evaluation and Measurement of HE

There are several techniques and tools for measuring HE. The linearly increasing stress test (LIST) is a technique used for HE measurement in materials. It consists of two methods: the constant load test and the constant extension test. In either of the cases, it is necessary to determine the threshold limit of stress [61].

Quite often, a practical difficulty faced in this test is that the component does not fail within the expected time period. This is because it takes too long for fractures to occur. A constant extension rate test (CERT) is another method, which is also called 'slow strain rate testing' (SSRT) [62]. Continuous elongation will be applied until the specimen fails this test. In this test, the specimen continues to extend even after reaching the threshold stress value. There is another test referred to as the slow rate tension test. This is a very common test used to assess the embrittlement effect on steel [63–66]. The embrittlement index and fractographic images showing the signs of embrittlement are the basis of this measurement. However, since it is a destructive test, its use is normally restricted to miniature samples where structural integrity is not a problem of serious concern.

8.10.1 Temperature Desorption Spectroscopy (TDS)

TDS, also referred to in the literature as temperature-programmed desorption (TPD), has been the most widely adopted technique for measuring failures caused by HE and some other phenomena [67–69]. The quality and quantity of diffusible hydrogen can be determined by the TDS technique. Steel traps collect hydrogen. When a significant amount of absorbed energy is reached, hydrogen absorbs and releases thermal energy. This critical energy level is referred to as the desorption activation energy, and the temperature at which hydrogen atoms are released is called the 'desorption temperature'. Quadrupole mass spectroscopy [70, 71] is used to measure the quantity of hydrogen that remains on the surface. TDS has been proven to be very effective and reliable. Even 0.1 mg kg^{-1} of desorbed hydrogen can be measured with this instrument [72, 73].

8.10.2 Hydrogen Permeation Test

The hydrogen permeation test is used to measure the amount of diffusible hydrogen in steel and assess the susceptibility of steel to HE. The hydrogen permeation test is normally carried out in a double-cell configuration consisting of an entry cell (charging cell) and an exit cell (oxidation cell) separated by a steel membrane [74, 75]. Hydrogen is first introduced to the charging cell and then to the oxidation cell through a membrane. Permeation testing is often combined with other techniques for effective implementation [76].

8.10.3 Microstructural Analysis

Microstructural analysis, using scanning electron microscopy (SEM) and TEM, has proven very effective in studying the internal microstructure of a material and assessing the effect of hydrogen on the internal microstructure and characterisation of the material. SEM is a relatively simple technique that provides useful information through a three-dimensional image that provides the scope to study fracture structure: microvoids, dimples and fisheyes. However, SEM has a functional limitation for such studies because of its small magnification and low resolution.

The capability of TEM, in contrast to that of SEM, is that it has very high magnification and resolution. It has been reported that TEM works satisfactorily at a million times magnification and nanometre resolution. TEM offers a wide scope to obtain very significant information on morphology as well as topography. Relevant crystallographic information can also be obtained through this microscope. It is, however, necessary for the sample to be electron transparent for TEM imaging and less than 100 nm thick for correct TEM microstructure analysis.

8.10.4 Hydrogen Microprint Technique (HMT) or Hydrogen Microprint and Silver Decoration Techniques

HMT was developed by Ovejero-García [77]. Recently, through an improved HMT, it is now possible to locate the points from which hydrogen atoms are emitted. Therefore, this improved technique is useful to identify the location of embrittlement where hydrogen accumulates [78]. Tests were carried out to assess the detection efficiency of the conventional HMT. It has been observed that in a controlled atmosphere with a humidity level of 80% or higher, it is possible to achieve a highly increased detection efficiency up to 40% with nickel plating of the steel surface [79].

It is used to identify the path through which hydrogen is diffusible in metal. Once this is understood, it becomes easy to identify the specific microstructure of the path and the impact of hydrogen. The HMT strategy has been successfully applied to a variety of materials such as low-carbon steel, high-strength steel and austenitic stainless steel [80–81]. It has also been used in notched steel and deformed steel to assess the distribution pattern of hydrogen in the stress field. During the process, the surface of the hydrogen-charged specimen is covered with a photographic emulsion containing silver bromide (AgBr) crystals. When hydrogen atoms come out of the metal, they react with the silver salt and reduce the silver ions in the crystals to metallic silver. After the reaction, a photographic solution ($Na_2S_2O_3$) was used to eliminate unreacted AgBr crystals. As a result, silver particles are present at this location, and the excess unreacted gel is removed. The particles can be observed by SEM and analysed with an energy-dispersive X-ray spectrometer [82].

8.11 Embrittlement Prevention

Appropriate methods to prevent embrittlement are crucial factors from a safety and a practical application point of view. All methods are based on the principle of minimising the contact between hydrogen and metal during fabrication. A very effective method of allowing hydrogen to diffuse before embrittlement occurs is to pre-heat and post-heat the metal. This technique has proven to be useful in the welding process, particularly for high-strength and low-alloy steels such as chromium, molybdenum and vanadium alloys [83]. Selection of the appropriate material is crucial. The choice of the right material depends on the application to build a robust and reliable system that simplifies the monitoring process. The metals or alloys need to be identified such that the basic properties are not greatly affected and the chances of embrittlement are greatly reduced. Pickling and electroplating increase the possibility of hydrogen absorption. Acid pickling is used to remove oxide scale from the surface of steel. Hydrogen is present in acid pickling baths. As far as possible, acid pickling should be avoided. Electroplating is usually performed to deposit zinc on nuts, bolts, screws, etc. in order to protect them from corrosion. During the process of electroplating, hydrogen is produced at the surface of the metal that is being coated. Therefore, as a preventive measure, the electroplated components should be baked at a temperature between around 190 and 220 °C within a few hours after electroplating. As a result, embrittlement can be avoided. Baking with respect to HE refers to a controlled heat treatment process that is applied to facilitate hydrogen diffusion into the metal. Baking involves heating the metal to an elevated temperature to accelerate the diffusion of hydrogen atoms within the metal. As the metal is heated, hydrogen atoms gain enough energy to overcome the trapping sites and move from the lattice. The outgassing of hydrogen during baking reduces the hydrogen concentration within the metal, which in turn reduces the potential for HE.

Broadly speaking, there are a number of factors that cause HE in a material. The primary factors are the material strength and surface quality. The surface condition takes into account if any metallic coatings have been heat treated. Parameters such as pressure, temperature, amount of hydrogen and the time of exposure are critical. Many factors determine how much the material is affected and how much embrittlement affects it. Through the application of a protective coating, HE was successfully avoided by mixing different metal alloys into the base material. In some cases, inhibitors have also been effective. The use of titanium, aluminium alloys and niobium and graphene coatings has been reported to be effective in preventing HE. In steel, cadmium and nickel plating, as well as cadmium and titanium plating, can provide a protective barrier against diffusion. In several specific situations, several materials and specific techniques have been successfully used to prevent embrittlement, such as Pt, Cu, electroless nickel, Au, Sn and some Sn and Pb alloy coatings. Several studies have shown that Cu is more effective than Ni at reducing hydrogen diffusion in iron [76–78]. Lead coatings have also been found to be very useful in hydrogen cracking on a variety of steels in a wide variety of environments.

8.12 Blistering

Blistering is an irreversible process and causes a permanent reduction in the material's mechanical strength. Hydrogen blistering mainly occurs in metals of low strength and in metals exposed to a corrosive environment. The origin of blisters was reported for the first time more than eight decades ago. The figure shows the distinctive features of hydrogen blistering and embrittlement.

Hydrogen blistering is one of the most common types of corrosion in the petroleum industry. A large number of corrosives, such as carbon dioxide, chlorides, acids, sulphur and sulphate-reducing bacteria, etc., cause blisters. Sometimes blistering also takes place during repairs when metals are subjected to high temperatures. Controlling the appropriate thermal environment and monitoring the temperature in a particular industrial operation are critical factors.

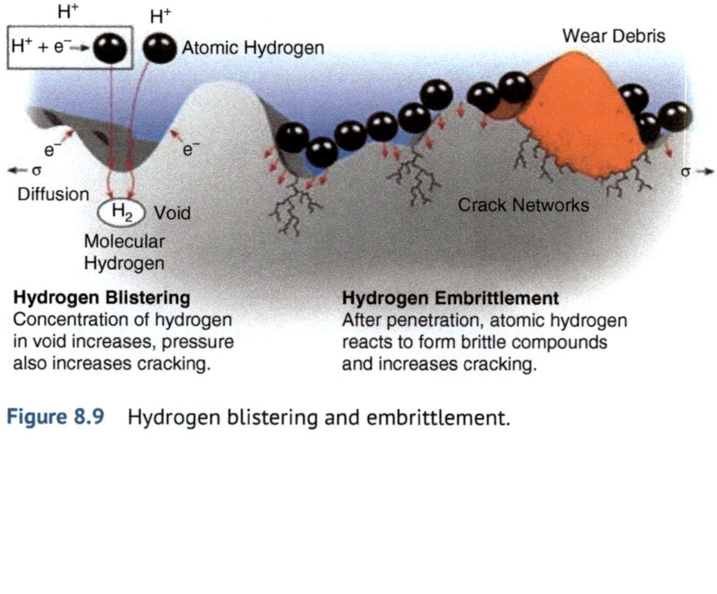

Figure 8.9 Hydrogen blistering and embrittlement.

8.12.1 Characteristics of Blisters

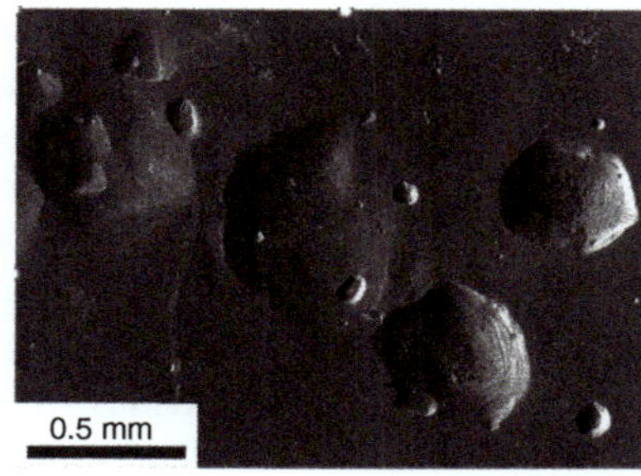

Figure 8.10 Blistering effect showing the feature of the blister [85] Elsevier B.V.

In one of the earlier research papers by Zapffe [84], several problems related to defects of steel and other effects associated with embrittlement were discussed. This paper briefly discusses and indicates the significance of the study on the effect of hydrogen-induced cracks, creep, flake formation, dislocation structure, etc., on which research activities have been carried out in subsequent years. Figure 8.9 shows the distortion on the steel surface where the blisters are circular in shape, and as evident from Figure 8.10, corrosion occurs on the surface, and therefore, these circular blisters have been formed. There is a distinct bulge with a thickness in the range of 5–20 μm on the edge of the blister, close to the corrosion environment [85].

In fact, it has been reported in the literature that the blister size varies over a very wide range starting from a few millimetres to metres [86, 87]. Researchers carried out experiments and observed that in the case of medium carbon steel with microstructures, cathodic hydrogen charging, without any external stress, causes hydrogen-induced blistering [88].

Since blisters are near the surface, cracks cause the blisters to curve towards the surface. Escobar et al. [89] observed that pure iron is more susceptible to cracks than high-strength ferrite steel. In pure iron, cracks develop inside the sample and then move faster to the surface releasing its internal pressure. Therefore, blister formation is quicker and more pronounced in pure iron than in high-strength steel. Griesche et al. [90] carried out some studies using neutron radiography and tomography on iron samples electrochemically charged with hydrogen. The investigations included both blisters (surface damage) and internal hydrogen-induced cracks. The samples in this study were kept until there was no hydrogen and all of them had escaped through blisters. A rapid fall of concentrated hydrogen started within minutes in blisters and near the surface. On the contrary, hydrogen remained for hours in cracks farther apart. After some time, the blisters and the cracks near the surface beneath the blisters did not contain hydrogen. However, the ones deeper inside the material, which did not have any blisters, contained hydrogen. Perhaps the blisters had openings on the surface through which hydrogen diffused into the surroundings. These observations are in close agreement with the research results of Hagi [91], who 'observed few blisters in this specimen'. Such a situation develops due to the 'Ease of swelling in the surface layer and suppression of crack growth due to release of precipitated hydrogen gas pressure' [91].

Many researchers have reported that similar conditions can lead to both the phenomena of blistering on the surface and stepwise cracking. In general, crack nucleation and propagation mechanisms are similar but different from the location-specific features of each phenomenon [92, 93]. Investigations on hydrogen-recharging time for blister and hydrogen-mediated cracking have been carried out by Li and co-investigators [94]. The increased charging time results in an increase in the height–width ratio in the case of hydrogen blisters and the length-to-depth ratio in the case of hydrogen-induced cracks. There is a change in hydrogen damage morphology in both cases. The critical role of grain–boundary cracking-induced blisters was also studied, and a crystallographic analysis was also carried out. These results provide helpful guidelines for the design of hydrogen-resistant structural steels through the control of grain boundaries. According to several research findings, current density and charging time play a very critical role in blister size and the rate of growth during electrochemical charging. Higher current densities and longer charging times result in an increased size of blisters formed at a faster rate [92, 95–97]. Li and the team [98] also studied several characteristics of blisters linked to surface roughness on pure iron. Researchers found that blister characteristics, such as the length of blisters and the width of blisters, increase with hydrogen-recharging time. This is true irrespective of whether the surface is polished or ground. However, the ground surface exhibits less sensitivity to blisters than the polished surface. There may be a higher density of dislocations along the surface and a higher residual stress on the surface. When large blisters are decorated with smaller blisters, dislocations and microvoids may lead to blistering [99]. This is a secondary stage because it is essential that large blisters be formed first to initiate hydrogen build-up in the material and subsequently form small blisters. As large blisters are necessary for the formation of secondary blisters, the phenomenon takes place after longer charging times.

Blister morphology of recrystallised and cold deformed ULC steel appears on the surface: (i) small dome-shaped blisters, (ii) large dome-forming blisters with small blisters on top of them and (iii) plateau-forming blisters [99]. It is evident from Figure 8.11 that the recrystallised material shows a distinct dome-shaped blister, whereas the cold deformed specimen shows three very distinct shapes of blisters. The first is a small dome-like shape with very clearly portrayed blisters, and the second is a relatively large dome-style structure with a number of small blisters. The third one is a plateau-like elevation with small blisters on it.

8.12.2 Preventive Measures Against Hydrogen Blistering

There are some specific steps that need to be followed to avoid blistering. In the first step, adequate care should be taken to ensure that impurities do not enter the metal at the

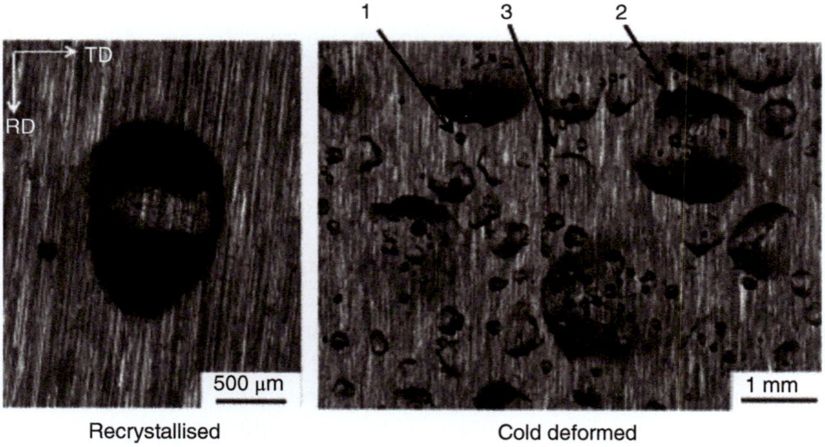

Figure 8.11 Blister morphology of recrystallised and cold deformed ultra-low-carbon (ULC) steel. *Source:* Laureys et al., 2017/with permission from Elsevier.

manufacturing stage. Choosing the right material for a specific application requires minimum voids and high corrosion resistance. The reduction of hydrogen ions and the formation of hydrogen molecules can be complex when phosphorus compounds, sulphide ions or arsenic compounds are present. Therefore, the concentration of atomic hydrogen increases, thereby creating a situation conducive to hydrogen blistering. It is desirable that these poisonous materials be removed. Killed steel, which is fully deoxidised before casting, is always preferred to rimmed steel. 'Killed steel', as its name implies, quietly solidifies in the mould without any gas being evolved. It possesses a very high degree of chemical homogeneity and no porosity. Therefore, many critical components are heat treated with it. Similarly, depending upon the suitability of the application, nickel-containing steels or nickel alloys should be used to avoid blistering as the rate of diffusion in these materials is extremely low. On many occasions, the use of several types of metallic, organic and inorganic coatings has proven effective in preventing blistering in steel. The coating should not allow hydrogen to penetrate. In addition, it should not have any chemical reactions with the ingredients inside the container. In several applications, depending upon the material, rubber, plastic coating and brick linings are being used. It is observed that claddings made of nickel-base alloys and stainless steel are also used for this purpose.

Concluding Remarks

There was no well-defined definition among various types of problems stemming from several forms of hydrogen attack. The present terminologies that distinguish HTHA, blistering and embrittlement have been adopted recently after a number of transitional terms. HTHA demonstrates itself in the following two forms: surface decarburisation (without fissuring) and internal decarburisation (with fissuring). HIC and blistering almost always occur at the same time. They are similar types of damage phenomena as far as the role of hydrogen is concerned. The major difference is that blisters are visible to the naked eye. As a result, they can easily be identified and treated accordingly. HE mostly occurs in steel and other materials. HE can be avoided if the concentration level of diffusible hydrogen is below the critical level. Selection of a suitable material is an important factor. The addition of titanium and aluminium alloys to the base material has proven effective in eliminating the problem of embrittlement. Zinc and nickel plating and some coating of graphene and niobium also help in avoiding embrittlement.

Abbreviations

AIDE	Adsorption-Induced Dislocation Emission
API	American Petroleum Institute
ASM	American Society for Metals
CERT	Constant Extension Rate Test
Cr	Chromium

DHF	Decohesive Hydrogen Fracture
EI	Embrittlement Index
Fe_3C	Cementite
HA	Hydrogen Attack
HAC	Hydrogen-Assisted Cracking
HAM	Hydrogen-Assisted Microfracture Mode
HDMC	Hydrogen-Assisted Microvoid Coalescence
HE	Hydrogen Embrittlement
HEDE	Hydrogen-Enhanced Decohesion
HELP	Hydrogen-Enhanced Local Plasticity
HEMP	Hydrogen-Enhanced Macroscopic Plasticity
HESIV	Hydrogen-Enhanced Strain-Induced Vacancy
HH	Diatomic Hydrogen
HHA	Hot Hydrogen Attack
HIC	Hydrogen-Induced Cracking
HMT	Hydrogen Microprint Technique
HRC	Hardness Rockwell Scale
HTHA	High-Temperature Hydrogen Attack
IG	Intergranular
LIST	Linearly Increasing Stress Test
MF	Mixed Fracture
MHx	Metal Hydride
Mn	Manganese
Mo	Molybdenum
MPa	Megapascal
MVC	Microvoid Coalescence
Ni	Nickel
QC	Quasi-Cleavage
SEM	Scanning Electron Microscopy
SSRT	Slow Strain Rate Testing
STEM	Scanning Transmission Electron Microscope
TDS	Temperature Desorption Spectroscopy
TEM	Transmission Electron Microscopes
TP curves	Total Product Curves
TPD	Temperature-Programmed Desorption
ULC	Ultra-Low Carbon
V	Vanadium

References

1 Poorhaydari, K. (2021). A comprehensive examination of high-temperature hydrogen attack—a review of over a century of investigations. *Journal of Materials Engineering and Performance* 30 (11): 7875–7908.

2 Effinger, R.T., Renquist, M.L., Wachter, A.G., and Wilson, J.G. (1951). Hydrogen attack of steel in refinery equipment. *American Petroleum Institute—Proceedings.*

3 Bonner, W.A., Burnham, H.D., Conradi, J.J., and Skei, T. (1953). Industry probes means of combating hydrogen attack on steel in refinery equipment. *Petroleum Processing* 8 (6): 878–888.

4 Sagues, A. (1979). Dilatometer for the in situ observation of high-temperature, high-pressure hydrogen attack. *The Review of Scientific Instruments* 50 (1): 48–51.

5 Prager, M. (2000). A Worldwide Look at Hot Hydrogen Attack, in American Society of Mechanical Engineers, Pressure Vessels and Piping Division (Publication) PVP, vol. 411.

6 Saugerud, O.T., Moore, D.E. and Odette, G. (1992). Model-based approach to evaluating the probability and critically of hot hydrogen attack in C-1/2 Mo equipment. *American Society of Mechanical Engineers, Pressure Vessels and Piping Division (Publication) PVP* 239: 46.

7 Cantwell, J.E. (1993). High temperature hydrogen attack: A 50 year overview, in corrosion 93. *The NACE Annual Conference and Corrosion Show, Paper No. 538,* Houston, Texas.

8 Kermad, A., Browne, R., Narayansingh, K., and Bigford, D. (1996). Hydrogen attack of 1Cr 1/2 Mo catalytic reforming reactors. *American Society of Mechanical Engineers, Pressure Vessels and Piping Division (Publication) PVP* 336.

9 Shewmon, P.G. (1976). Hydrogen attack of carbon steel. *Metallurgical Transactions A* 7: 279–286.

10 American Institute of Petroleum (2008). *Recommended Practice 941*, 7e.

11 Shih, H.M. A model for calculation of the Nelson curves for hydrogen attack. *Acta Metallurgica* 30: 537.

12 Parthasarathy, T.A. (1985). Mechanisms of hydrogen attack of carbon and 2-1/4Cr-1Mo steels. *Acta Metallurgica* 33: 1673.

13 Fletcher, E.E. and Elsa, A.R. (1964). *The Effects of High-Pressure High-Temperature Hydrogen on Steel.* Columbus, Ohio: Defense Metals Information Center.

14 Zapffe, C.A. (1944). Various, discussion on boiler embrittlement. *Transactions of the ASME* 66: 117.

15 Schuyten, J. (1947). Hydrogen attack on metals at high temperatures and pressures. *Corrosion and Material Protection* 4 (5): 13–18.

16 Yokogawa, K., Fukuyama, S., Kudo, K., and Shewmon, P.G. (1989). Effect of hydrogen attack on tensile and creep properties of low carbon. *International Journal of Pressure Vessels and Piping* 37: 365–385.

17 Sakai, T. and Asami, K. (1987). Effect of applied stress on hydrogen attack of 2 1/4 Cr-1Mo steel. *Journal of the Iron and Steel Institute of Japan* 73 (3): 551–557.

18 Sakai, T., Nomura, T., and Niccolls, E.H. (2000). The basis and application of the “Pw” parameter for high temperature hydrogen attack. *American Society of Mechanical Engineers, Pressure Vessels and Piping Division (Publication) PVP* 411: 45–56.

19 Imanaka, T. and Shimomura, J.I. (1984). Temper embrittlement and hydrogen attack on 2 1/4 Cr-1Mo steels in high pressure and high temperature hydrogen atmospheres. *5th International Conference on Pressure Vessel Technology*.

20 Sundararajan, G. and Shewmon, P. (1980). The hydrogen attack of HSLA steels. *Metallurgical Transactions A* 11A (3): 509–516.

21 Van der Burg, M., Van der Giessen, E., and Brouwer, R. (1996). Investigation of hydrogen attack in 2.25Cr-1Mo steels with a high-traxiality void growth model. *Acta Materialia* 44 (2): 505–518.

22 Shimomura, J. and Imanaka, T. (1985). Decomposition of carbides in 2–1/4Cr-1Mo steels during hydrogen attack. *Scripta Metallurgica* 19: 1507–1511.

23 Sakai, T., Asami, K., Kondo, N., and Hayashi, T. (1987). Effects of carbide forming elements on hydrogen attack and embrittlement in 2 1/4Cr-1Mo steel. *Tetsu-To-Hagane/Journal of the Iron and Steel Institute of Japan* 73 (2): 372–379.

24 Woodward, T.J. (2021). Techniques Understanding the High Temperature Hydrogen Attack of Steels Utilizing Novel In-Situ Analytical Transmission Electron Microscopy – A thesis submitted to The University of Manchester for the degree of Doctor of Philosophy in the Faculty of Science and Engineering.

25 Johnson, W.H. (1875). On some remarkable change produced in iron and steel by the action of hydrogen and acids. *Proceedings of the Royal Society of London* 23: 168–179.

26 Barrera, O., Bombac, D., Chen, Y. et al. (2018). Understanding and mitigating hydrogen embrittlement of steels. *Journal of Materials Science* 53 (9): 6251–6290.

27 Dwivedi, S.K. and Vishwakarma, M. (2018). Hydrogen embrittlement in different materials: a review. *International Journal of Hydrogen Energy* 43 (46): 21603–21616.

28 Lin, M., Yu, H., Ding, Y. et al. (2022). A predictive model unifying hydrogen enhanced plasticity and decohesion. *Scripta Materialia* 215 (1): 114707.

29 Dmytrakh, I.M., Leshchak, R.L., and Syrotyuk, A.M. (2015). Effect of hydrogen concentration on strain behaviour of pipeline steel. *International Journal of Hydrogen Energy* 40 (10): 4011–4018.

30 Djukic, M.B., Sijacki Zeravcic, V., Bakic, G. et al. (2014). Hydrogen embrittlement of low carbon structural steel. *Procedia Materials Science* 3: 1167–1172.

31 Wang, R. (2009). Effects of hydrogen on the fracture toughness of a X70 pipeline steel. *Corrosion Science* 51 (12): 2803–2810.

32 Koyama, M., Tasan, C.C., Akiyama, E. et al. (2014). Hydrogen-assisted decohesion and localized plasticity in dual-phase steel. *Acta Materialia* 70: 174–187.

33 Takakuwa, O., Ogawa, Y., Yamabe, J., and Matsunaga, H. (2019). Hydrogen-induced ductility loss of precipitation-strengthened Fe-Ni-Cr-based superalloy. *Materials Science and Engineering A* 739 (2): 335–342.

34 Krieger, W., Merzlikin, S.V., Bashir, A. et al. (2018). Spatially resolved localization and characterization of trapped hydrogen in zero to three dimensional defects inside ferritic steel. *Acta Materialia* 144: 235–244.

35 Nagao, A., Smith, C.D., Dadfarnia, M. et al. (2012). The role of hydrogen in hydrogen embrittlement fracture of lath martensitic steel. *Acta Materialia* 60 (13–14): 5182–5189.

36 Nagao, A., Dadfarnia, M., Somerday, B.P. et al. (2018). Hydrogen-enhanced-plasticity mediated decohesion for hydrogen-induced intergranular and "quasi-cleavage" fracture of lath martensitic steels. *Journal of the Mechanics and Physics of Solids* 112: 403–430.

37 Gavriljuk, V.G., Shanina, B.D., Shyvanyuk, V.N., and Teus, S.M. (2013). Hydrogen embrittlement of austenitic steels: electron approach. *Corrosion Reviews* 31: 33–50.

38 Geng, W., Wang, V., Li, J.-X. et al. (2018). Hydrogen trapping in carbon supersaturated alpha-iron and its decohesion effect in martensitic steel. *Scripta Materialia* 149: 79–83.

39 McMahon, C.J. (2001). Hydrogen-induced intergranular fracture of steels. *Engineering Fracture Mechanics* 68 (6): 773–788.

40 Beachem, C.D. (1972). A new model for hydrogen-assisted cracking (hydrogen "Embrittlement"). *Metallurgical Transactions* 3 (February): 441–455.

41 Martin, M.L., Somerday, B.P., Ritchie, R.O. et al. (2012). Hydrogen-induced intergranular failure in nickel revisited. *Acta Materialia* 60: 2739–2745.

42 Wang, S., Martin, M.L., Sofronis, P. et al. (2014). Hydrogen-induced intergranular failure of iron. *Acta Materialia* 69: 275–282.

43 Wang, S., Nagao, A., Sofronis, P., and Robertson, I.M. (2018). Hydrogen-modified dislocation structures in a cyclically deformed ferritic-pearlitic low carbon steel. *Acta Materialia* 144: 164–176.

44 Wang, S., Nagao, A., Edalati, K. et al. (2017). Influence of hydrogen on dislocation self-organization in Ni. *Acta Materialia* 135: 96–102.

45 Martin, M.L., Sofronis, P., Robertson, I.M. et al. (2013). Amicrostructural based understanding of hydrogen-enhanced fatigue of stainless steels. *International Journal of Fatigue* 57: 28–36.

46 Martin, M.L., Robertson, I.M., and Sofronis, P. (2011). Interpreting hydrogen-induced fracture surfaces in terms of deformation processes: anew approach. *Acta Materialia* 59 (9): 3680–3687.

47 Martin, M.L., Fenske, J.A., Liu, G.S. et al. (2011). On the formation and nature of quasi-cleavage fracture surfaces in hydrogen embrittled steels. *Acta Materialia* 59: 1601–1606.

48 von Pezold, J., Lymperakis, L., and Neugebeauer, J. (2011). Hydrogen-enhanced local plasticity at dilute bulk concentrations: the role of h–h interactions and the formation of local hydrides. *Acta Materialia* 59: 2969–2980.

49 Lin, M., Yu, H., Ding, Y. et al. (2022). A predictive model unifying hydrogen enhanced plasticity and decohesion Meichao. *Scripta Materialia* 215: 114707.

50 Lynch, S. (2012). Hydrogen embrittlement phenomena and mechanisms. *Corrosion Reviews* 30: 105–123.

51 Lynch, S.P. (1988). Environmentally assisted cracking: overview of evidence for an adsorption-induced localised-slip process. *Acta Metallurgica* 36 (10): 2639–2661.

52 Nagumo, M. (2001). Function of hydrogen in embrittlement of high-strength steels. *ISIJ International* 41 (6): 590–598.

53 Takai, K., Shoda, H., Suzuki, H., and Nagumo, M. (2008). Lattice defects dominating hydrogen-related failure of metals. *Acta Materialia* 56 (18): 5158–5167.

54 Takai, K., Yamauchi, G., Nakamura, M., and Nagumo, M. (1998). Hydrogen trapping characteristics of cold-drawn pure iron and eutectoid steel evaluated by thermal desorption spectrometry. *Journal of the Japan Institute of Metals* 62 (3): 267–275.

55 Nagumo, M., Ohta, K., and Saitoh, H. (1999). Deformation induced defects in iron revealed by thermal desorption spectroscopy of tritium. *Scripta Materialia* 40: 313–319.

56 Fuchigami, H., Minami, H., and Nagumok, M. (2006). Effect of grain size on the susceptibility of martensitic steel to hydrogen-related failure. *Philosophical Magazine Letters* 86: 21–29.

57 Chung, Y. (2014). Corrosion on the new eastern span of the San Francisco–Oakland Bay Bridge. *Materials Performance* 53 (11): 58.

58 Hirth, J.P. (1980). Effects of hydrogen on the properties of iron and steel. *Metallurgical Transactions A* 11: 861–890.

59 Sanchez, J., Lee, S.F., Martin-Rengel, M.A. et al. (2016). Measurement of hydrogen and embrittlement of high strength steels. *Engineering Failure Analysis* 59: 467–477.

60 Barthélémy, H. (2006). Compatibility of metallic materials with hydrogen – review of the present knowledge. *16th World Hydrogen Energy Conference (WHEC)*, Lyon, France (13–16 June 2006).

61 Baboian, R. (2005). *Corrosion Tests and Standards: Application and Interpretation*. ASTM International.

62 Parkins, R.N. (1979). Development of strain-rate testing and its implications. In: *Stress Corrosion Cracking-the Slow Strain Rate Technique* (ed. G.M. Ugiansky and J.H. Payer), 5, 25. ASTM International.

63 Zhang, T., Zhao, W., Li, T. et al. (2018). Comparison of hydrogen embrittlement susceptibility of three cathodic protected subsea pipeline steels from a point of view of hydrogen permeation. *Corrosion Science* 131: 104–115. https://doi.org/10.1016/j.corsci.2017.11.013.

64 Singh, D.K., Singh Raman, R.K., Maiti, S.K. et al. (2017). Investigation of role of alloy microstructure in hydrogen-assisted fracture of AISI 4340 steel using circumferentially notched cylindrical specimens. *Materials Science and Engineering A* 698: 191197. https://doi.org/10.1016/J.MSEA.2017.05.056.

65 Hu, Y., Dong, C., Luo, H. et al. (2017). Study on the hydrogen embrittlement of Aermet100 using hydrogen permeation and SSRT techniques. *Metallurgical and Materials Transactions A* 48: 4046–4057. https://doi.org/10.1007/s11661-017-4159-x.

66 Fan, Y.H., Zhang, B., Yi, H.L. et al. (2017). The role of reversed austenite in hydrogen embrittlement fracture of S41500 martensitic stainless-steel. *Acta Materialia* 139: 188–195. https://doi.org/10.1016/j.actamat.2017.08.011.

67 Wang, M., Akiyama, E., and Tsuzaki, K. (2007). Effect of hydrogen on the fracture behavior of high strength steel during slow strain rate test. *Corrosion Science* 49 (11): 4081–4097.

68 Dadfarnia, M., Sofronis, P., and Neeraj, T. (2011). Hydrogen interaction with multiple traps: can it be used to mitigate embrittlement? *International Journal of Hydrogen Energy* 36 (16): 10141–10148.

69 Fernández-Sousa, R., Betegón, C., and Martínez-Pañeda, E. (2020). Analysis of the influence of microstructural traps on hydrogen assisted fatigue. *Acta Materialia* 199: 253–263. https://doi.org/10.1016/j.actamat.2020.08.030.

70 Verbeken, K. (2012). Analysing hydrogen in metals: bulk thermal desorption spectroscopy (TDS) methods. In: *Gaseous Hydrogen Embrittlement of Materials in Energy Technologies Mechanisms, Modelling and Future Developments*, vol. 1, 27–55. Woodhead Publishing Series in Metals and Surface Engineering.

71 Venezuela, J., Tapia-Bastidas, C., Zhou, Q. et al. (2018). Determination of the equivalent hydrogen fugacity during electrochemical charging of 3.5 NiCrMoV steel. *Corrosion Science* 132: 90–106.

72 Bergers, K., Camisao de Souza, E., Thomas, I. et al. (2010). Determination of hydrogen in steel by thermal desorption mass spectrometry. *Steel Research International* 81 (7): 499–507.

73 Venezuela, J., Liu, Q., Zhang, M. et al. (2016). A review of hydrogen embrittlement of martensitic advanced high-strength steels. *Corrosion Reviews* 34 (3): 153–186.

74 Frappart, S., Feaugas, X., Creus, J. et al. (2012). Hydrogen solubility, diffusivity and trapping in a tempered FeCCr martensitic steel under various mechanical stress states. *Materials Science and Engineering A* 534: 384–393.

75 Zakroczymski, T. (2006). Adaptation of the electrochemical permeation technique for studying entry, transport and trapping of hydrogen in metals. *Electrochimica Acta* 51 (11): 2261–2266.

76 Figueroa, D. and Robinson, M.J. (2010). Hydrogen transport and embrittlement in 300 M and AerMet100 ultra high strength steels. *Corrosion Science* 52 (5): 1593–1596.

77 Ovejero-García, J. (1985). Hydrogen microprint technique in the study of hydrogen in steels. *Journal of Materials Science* 20: 2623–2629.

78 Ichitani, K., Kanno, M., and Kuramoto, S. (2003). Recent development in hydrogen microprint technique and its application to hydrogen embrittlement. *ISIJ International* 43 (4): 496–504.

79 Ichitani, K., Kuramoto, S., and Kanno, M. (2003). Quantitative evaluation of detection efficiency of the hydrogen microprint technique applied to steel. *Corrosion Science* 45 (6): 1227–1241.

80 Matsuda, S., Ichitani, K., and Kanno, M. (2008). Visualization of hydrogen diffusion path by a high sensitivity hydrogen microprint technique. *Environment Induced Cracking of Materials* 1: 239–248.

81 Nakatani, M., Fujihara, H., Sakihara, M., and Minoshima, K. (2011). Influence of irreversible hydrogen and stress cycle frequency on the fatigue crack growth property in high-strength steel and hydrogen visualization. *Procedia Engineering* 10: 2381–2386.

82 Ichitani, K. and Kanno, M. (2003). Visualization of hydrogen diffusion in steels by high sensitivity hydrogen microprint technique. *Science and Technology of Advanced Materials* 4: 545–551.

83 Choudhary, S., Vishwakarma, M., and Dwivedi, S.K. (2021). Evaluation and prevention of hydrogen embrittlement by NDT methods: a review material. *Proceedings* 6: 18.

84 Zapffe, C.A. and Sims, C.E. (1941). Hydrogen embrittlement internal stress and defects in steels. *Transactions of the American Institute of Mining and Metallurgical Engineers* 145: 225–259.

85 Isakov, M.G., Izotov, V.I., Karpel'ev, V.A., and Filippov, G.A. (2000). Kinetics of the damage formation in a low-carbon low-alloy steel upon hydrogenation. *The Physics of Metals and Metallography* 90: 302–308.

86 Skei, T., Wachter, A., Bonner, W.A., and Burnham, H.D. (1953). Hydrogen blistering of steel in hydrogen sulfide solutions. *Corrosion* 9: 163–172.

87 Nelson, G.A. and Effinger, R.T. (1955). Blistering and embrittlement of pressure vessel steels by hydrogen. *Welding Journal* 34: 12–21.

88 Ju, C.P. and Rigsbee, J.M. (1985). The role of microstructure for hydrogen-induced blistering and stepwise cracking in a plain medium carbon steel. *Materials Science and Engineering A* 74: 47–53.

89 Escobar, D.P., Miñambres, C., Duprez, L. et al. (2011). Internal and surface damage of multiphase steels and pure iron after electrochemical hydrogen charging. *Corrosion Science* 53: 3166–3176.

90 Griesche, A., Dabah, E., Kannengiesser, T. et al. (2014). Three-dimensional imaging of hydrogen blister in iron with neutron tomography. *Acta Materialia* 78: 14–22.

91 Hagi, H. (1997). Thermal evolution spectrum of hydrogen from low carbon steel charged by cathodic polarization. *Materials Transactions, JIM* 38 (11): 970–977.

92 Tiegel, M.C., Martin, M.L., Lehmberg, A.K. et al. (2016). Crack and blister initiation and growth in purified iron due to hydrogen loading. *Acta Materialia* 115: 24–34.

93 Dong, C.F., Li, X.G., Liu, Z.Y., and Zhang, Y. (2009). Hydrogen-induced cracking and healing behaviour of X70 steel. *Journal of Alloys and Compounds* 484: 966–972.

94 Li, X., Huang, W., Xubin, W. et al. (2021). Effect of hydrogen charging time on hydrogen blister and hydrogen-induced cracking of pure iron. *Corrosion Science* 181: 109200.

95 Pérez Escobar, D., Miñambres, C., Duprez, L. et al. (2000). Internal and surface damage of multiphase steels and pure iron after electrochemical hydrogen charging. *Corrosion Science* 53: 3166.

96 Dominguez-Aguilar, M.A., Olivares-Xometl, O., Likhanova, N.V., and Lopez-Fuentes, M. (2011). Cathodic induction of hydrogen blistering on carbon steel in alkaline sour environment. *Corrosion* 67: 850051.

97 Venegas, V., Caleyo, F., Hallen, J.M. et al. (2007). Role of crystallographic texture in hydrogen-induced cracking of low carbon steels for sour service piping. *Metallurgical and Materials Transactions A* 38: 1022.

98 Li, X., Wang, Y., and Wu, X. (2020). Effect of surface roughness on hydrogen-induced blister behaviour in pure iron. *Meta* 10 (6): 745.

99 Laureys, A., Van den Eeckhout, E., Petrov, R., and Verbeken, K. (2017). Effect of deformation and charging conditions on crack and blister formation during electrochemical hydrogen charging. *Acta Materialia* 127 (1): 192–202.

9

Path Forward

Climate change is like death; no one wants to talk about it.

– Amitav Ghosh

The technical aspects of hydrogen production, storage and utilisation have already been discussed. Discussions are focused here on ongoing efforts and the possible short- and long-term strategies with hydrogen as the building block of the clean energy economy.

The current energy consumption model is entirely dependent on fossil fuels. Today, about 83% of energy consumed comes from fossil fuels. Anthropogenic activities, related to fossil fuel combustion, are responsible for the rapid increase in atmospheric pollutant concentrations. As a result of extensive and irrational use of fossil fuels, climate change has been catastrophic, and biodiversity has been irreparably damaged. Vehicles and industrial activities are the major causes of urban air pollution that makes the air unbreathable. Noxious pollutants such as carbon monoxide (CO), unburnt hydrocarbons, oxides of nitrogen, tropospheric ozone, oxides of sulphur, smoke, particulates matter and a host of other pollutants have created severe health hazards and affected human beings, plants and animals. Buildings and structures and century-old monuments have been badly damaged by poor air quality. Absorption of terrestrial radiation by CO_2 and H_2O, global warming and greenhouse gases have caused multiple problems. Sea level rise, melting of the polar ice sheet, change in ocean current pattern, beach erosion, droughts, floods, forest fires and hurricanes have become concerns. A centuries-old seasonal cycle has been totally disrupted. Smog, acid rain, El Nino, etc., which were essentially local or regional in character, have now grown into major environment-related global problems. The amount of solar energy striking Earth every 80 min is nearly equivalent to annual global energy consumption. Solar energy is a limitless renewable resource available for another five billion years. Almost 71% of Earth is covered by water and approximately 96.5% comes from the oceans. Hydrogen can be derived from unlimited water sources in oceans using infinite solar energy. Interestingly, hydrogen upon combustion gives off water that is released back into the atmosphere from where it originated. Literally, an inexhaustible source of energy – an alchemist's dream! Fossil fuels are depleting alarmingly. Fossil fuel combustion results in environmental degradation, which threatens the very survival of life on Earth. Since the energy system is closely linked to environmental impact, it would be the most appropriate approach to evolve a single solution to this threat of global annihilation. Hydrogen is the

Hydrogen Energy: Production, Safety, Storage and Applications, First Edition. L. M. Das.

answer. The International Energy Agency (IEA) has suggested five key areas in 'comprehensive policy frameworks to facilitate hydrogen adoption across the entire energy system: (i) Establish targets and/or long-term policy signals. (ii) Support demand creation. (iii) Mitigate investment risks. (iv) Promote R&D, innovation, strategic demonstration projects and knowledge-sharing. (v) Harmonize standards and removing barriers.' [1].

There have been some very recent programs and collaborative projects on hydrogen production, storage and application. It is likely that the details given here might have changed in the meantime as these projects are in continuous process of development.

Recently, in August 2023, University of California, Santa Barbara has entered into a research agreement with New Hydrogen, Inc. Objective of this collaborative project is to produce hydrogen by a novel ground-breaking thermochemical approach, using heat instead of electricity. This is a prospective and economically viable approach. At present, it is possible to convert only about 20% of sunlight to electricity by using the solar panels. This expensive route of producing electricity can be avoided using solar heat directly by converting sunlight to heat with about 90% efficiency. The researchers are developing a molten liquid as the catalyst to split water continuously in a single redox chemical loop, to produce hydrogen and oxygen in separate chambers. The molten catalytic liquid is developed so that it can be reduced in one chamber, oxidized in another chamber, and can be continuously recycled and reused. Production of green hydrogen by the water-splitters at industrial temperatures of less than 1,000°C, depends on the cost of the catalysts used. Work is in progress to identify suitable substitutes to expensive rare earth materials, which are used as catalysts now.

Ionomr has developed a novel technology called the Shell Game Changer Accelerator, which is powered by NREL (GCxN). Ionomr worked with NREL and Shell to validate its Aemion+ anion-exchange materials (AEM) which would substantially reduce the capital cost of the equipment used in electrolysis for hydrogen production. In stability tests for over 800 hours, Aemion+ has demonstrated its potential as a high performing, eco-friendly and less expensive substitute to Iridium Titanium. This technology is highly promising to achieve green hydrogen economy.

Fabrum in New Zealand has been involved in green hydrogen production, storage, dispensing, and system integration for heavy transport, mining and aviation markets. Recently they have announced a $23M Series A financing led by London-headquartered AP Ventures with participation from Australia-based Fortescue Future Industries, Japan-based Obayashi Corporation and New Zealand-based K1W1. With this funding Fabrum plans to expand its activities – to meet the growing demand for its end-to-end hydrogen systems technologies across heavy transport, mining and aviation markets. H.W. Richardson (HWR), the heavy-transport company of New Zealand has been working on the development of dual-fuel hydrogen technology for heavy transport industry such as trucks. This is expected to result in 40% reduction in emission without any loss of power. HWR is planning their trucks to use hydrogen-refuelling station by Fabrum. The first green hydrogen production system is developed by Fabrum using a 1.1-Megawatt (MW).

Keeping in mind the temperature rising below 1.5 °C in the year 2100, IEA has estimated that about 530 Mt of hydrogen ought to be produced to achieve a net-zero scenario in 2050. An extensive infrastructure facility for electrolysis has to be developed for meeting a demand of 3585 gigawatts (GW) by 2050 [2]. Most hydrogen production routes are likely to switch to green hydrogen production by electrolysis. In the transition phase, for a few years before turning to total hydrogen energy, production of hydrogen from fossil fuels with carbon capture, utilisation and storage (CCUS) is gaining stupendous momentum. Projects on CCUS have begun to whittle down global greenhouse gas emissions. More than 60 industries, including the oil and gas sector, have already taken up CCUS projects that aim at capturing more than 40 million tonnes of CO_2. It is expected that global carbon capture capacity will grow astronomically during the next couple of decades until 2050. At one time, blue hydrogen was conceptualised as relatively 'clean' compared to grey hydrogen because blue hydrogen has a very high rate of carbon capture. There have been disputes about blue hydrogen. It has been reported that while carbon dioxide emissions are lower, fugitive methane emissions for blue hydrogen are higher than for grey hydrogen. This is because of the increased use of natural gas to power carbon capture [3]. In fact, there has not been enough experience on a large scale of storing carbon dioxide from carbon. Now the entire focus is on green hydrogen with a zero-emission target. Right availability of renewable energy is adequate to facilitate large-scale green hydrogen production. However, the trend is changing very rapidly. Global green hydrogen production by electrolysis rises fast every year. Large-scale electrolysis hydrogen production plants have been established in many countries. Green hydrogen is emerging as a potential game changer to ensure fast carbon neutrality. The International Renewable Energy Agency has reported the time period by which countries will have green hydrogen cheaper than blue hydrogen [4]. To meet such an enormity of needs in so many spheres of utilisation, sufficient amounts of green hydrogen must be produced. Policies have been framed to support hydrogen use in fuel cell vehicles. There could be some transient phases or alternative arrangements immediately adopted until enough hydrogen is produced. In addition, adequate appropriate infrastructure is to be built to switch over to total hydrogen. Of course, green hydrogen is the ultimate objective. In the meantime, research on some prospective future production processes such as solid oxide electrolyser cells (SOECs), anion exchange membranes (AEMs), electrified steam methane reforming (ESMR) and methane pyrolysis is needed.

Right now, hydrogen is a local business because of logistic problems and the cost. Most hydrogen used in commerce is transported through pipelines. Prospects of global hydrogen trade are likely to improve. Now, global hydrogen pipelines are about 5000 km in North America, Europe and Asia and they are getting expanded. Until a stable widespread hydrogen transfer system is established, during this transition phase, hydrogen could be sent with methane, methanol, ammonia, etc. for local industrial use. Handling and transporting this way would be relatively safer. Present storage facilities depend on how hydrogen will be used. In some locations, tanks or liquid dewars have been installed depending on commercial requirements. This is mostly for aerospace applications such as the Kennedy Space Centre in Florida where large storage vessels are kept. An efficient, large-scale hydrogen energy storage system needs to be established to ensure a reliable energy supply chain. In this context, underground seasonal hydrogen storage (USHS) is a highly promising technique as it gets rid of problems due to local and temporal variations in renewable energy production. Identifying the correct location for installing the USHS system requires a comprehensive geological survey. Other conditions such as the character of the subsurface structure, assessment of long-term safe storage capacity and possible biological reactions due to hydrogen injection need to be ascertained. USHS is a relatively new technology. In general, storage has mostly been adopted for specific applications under circumstances when production is hampered.

Gold Hydrogen Ltd. (GHY), an Australia-based company, as part of its project activities on the flagship Ramsay Project in South Australia, plans to start its drill testing for white (natural) hydrogen. It may be recalled that tests conducted in the Ramsay project, almost almost ninety years before, in 1931, had shown the presence of natural hydrogen (white hydrogen) in the well. This discovery did not get adequate attention, perhaps because hydrogen's potential was not realised at that time and the commercial values were not favourable. This natural hydrogen well is believed to be formed by the interaction of some rock formation and water. A more accurate study of the geological structure of the region would be useful to assess the amount of natural hydrogen in that region.

It is a fact that most hydrogen discoveries have been made by accident rather than by intentional effort. Interestingly, in the village of Bourakébougou in western Mali, some well-diggers attempted to create a water source by drilling a borehole to a depth of 108 meters, but it turned out to be a dry borehole, and they did not find any water at that depth. In the meantime, wind started blowing out of the hole. The wind exploded severely in the face of a driller who was smoking cigarette and peering into the hole, perhaps to assess if water level was visible. There was a huge fire, but there was no black smoke. Therefore, it was realised that the wind was in fact a stream of an odourless flammable gas. The gas well was plugged and shunned by the villagers for almost two decades until 2007. At a later phase, drillers on the haunt for fossil fuels accidentally discovered that spot and found that hundreds of feet below the arid earth lies a white hydrogen reserve. Today, the village of Bourakébougou, Mali has been powered by white hydrogen for more than a decade.

Researchers observe that hydrogen is generated continuously within the earth because of the reaction between water and rock. Hydrogen thus produced tries to percolate through earth's crust. In the process it sometimes accumulates in underground traps. U.S. Geological Survey (USGS), in one of their recent studies have found that there might be sources of natural hydrogen to meet the gradually increasing global demand. Until now, the well in Bourakébougou, Western Mali has been the only source of white hydrogen, which comes about to 5 tonnes per year.

Recently, as reported in September 2023, scientists in France have discovered white hydrogen in Lorraine region. On the basis of the gas data at 1100 metres below ground (14% hydrogen), it has been estimated that the Lorraine deposit could contain up to 46 million tonnes of white hydrogen. This is slightly more than half of the current global production of grey hydrogen. Boreholes drilled in Kangaroo Island and Yorke Peninsula in South Australia have been found to have hydrogen. British company Getech, using its Globe digital platform and additional artificial intelligence machine learning, can predict the location of natural hydrogen deposit sites. Many companies have accelerated their efforts to locate the untapped reservoirs of white hydrogen in several parts of the world.

Generating hydrogen from waste has started in some industries to some scale. The technology of generating hydrogen from waste varies widely depending on feedstock from different areas. Organic waste, agricultural wastes have been the common feedstock. Some processes have even used sewage and some even plastics that cannot be recycled. The strategy of producing hydrogen from waste cannot be green. With organic waste as feedstock while capturing any carbon emission, can be carbon-neutral. Strictly speaking, this cannot be labelled as "green" because no carbon capture technology is 100 percent effective. However, they are always "low-carbon" hydrogen. A major contribution of this technology is that it uses feedstock, which otherwise would have been disposed of in landfills or caused further pollution through several carbon-emitting disposal methods. Some industries have developed technologies, that are in the early stages of development. Contribution of waste-to-hydrogen technology is undeniable towards a cleaner environment. It holds the promising potential for improved carbon-neutral hydrogen production.

The Global Hydrogen Production Technologies (HyPT) Centre has established an international partnership among US, Australia, Canada, UK, Egypt, and Germany. The national leads of this newly established global centre has professors from Arizona State University (US), University of Adelaide (Australia), University of Toronto (Canada), and Cranfield University (UK). Major objective of the centre is to develop low-cost large-scale net-zero hydrogen. The centre aims at making net-zero hydrogen production more efficient and affordable with a target price of $1 per kilogram. The success of the research and development activities would be extremely helpful in the such as ammonia and steel production, cement, transportation, and energy storage. Now, in its initial phase, several projects, depending upon the expertise of the research team are being formulated. One group focuses on methane pyrolysis, a process that utilizes heat to break down natural gas into hydrogen gas and solid carbon particles and does not produce carbon dioxide. Similarly, other researchers with diverse expertise from several institutions in different countries work synergistically through knowledge sharing, towards a common goal to achieve sustainable hydrogen economy.

There are numerous types of steel, each type having its unique physical, chemical, mechanical, and magnetic properties. New steel alloys and specialized steel types and grades are developed to meet specific industrial and technological needs. However, steel industry is a significant contributor to global greenhouse gas emissions, accounting for approximately 8% of total emissions. Currently, every tonne of steel produced emits an average of 1.63 tonnes of CO_2. RINA Company's Hydra project, funded by the European Commission's Next Generation EU plan and supported by the Italian Ministry of Enterprises aims at 100% hydrogen fuelled steel production by offering near-zero carbon technology and capabilities. The plant, would have 30m high direct iron ore reduction (DRI) tower, which will use hydrogen as a reducing agent.

National Renewable Energy Laboratory (NREL) of Department of Energy (DOE) of United States and Honeywell have entered into a collaboration on Fuel Additives for Solid Hydrogen (FLASH) Carriers in Electric Aviation" project. Major objective of the project is to develop cartridge-based hydrogen storage solution for unscrewed aerial vehicles (UAVs). In The FLASH project, Honeywell's contribution involves testing the viability of fuel cartridge technology, developing prototype models, offering supply chain support, and conducting comprehensive assessments of fuel cell systems, FLASH can serve as a platform to mature the new hydrogen carrier technology developed at NREL as part of the Hydrogen Materials Advanced Research Consortium (HyMARC) project. The focus of the FLASH project is on solid materials with a high hydrogen capacity to facilitate quick release of hydrogen gas to be used in fuel cell. The material should be able to operate satisfactorily at a lower temperature of about 100°C. While internal combustion engines today can provide the required range for long-range drones, they come with trade-offs in terms of noise, vibration, and emissions, particularly carbon emissions. As a result, there is ongoing interest in developing more environmentally friendly and quieter propulsion systems for long-range drones, such as electric or hybrid systems, to mitigate these issues. Battery-powered electric Unmanned Aerial Vehicles (UAVs), also known as drones, are a suitable choice for reducing carbon emissions in short-range applications. They are environmentally friendly and can be used for tasks that do not require long-range travel or carrying heavy payloads. However, when it comes to applications that demand extended endurance and long-range flight capabilities, as well as the ability to carry substantial payloads, battery-powered UAVs may not be the ideal solution due to limitations in battery capacity and weight. On the other hand, the statement suggests that hydrogen technology can offer significant advantages for electric Vertical Take-Off and Landing (eVTOL) systems. Hydrogen can enhance the endurance and range of eVTOL vehicles. This is because hydrogen fuel cells can provide a more efficient and lightweight energy source compared to traditional batteries, which can be especially advantageous for longer-range and heavier payload missions.

A comparative assessment of current methods of hydrogen delivery shows that gaseous hydrogen is expensive for long distances, liquid tankers have a substantial energy loss and therefore pipelines seem to be suitable depending upon the demand and the distance. Even though liquefaction and storage of hydrogen has been used very successfully for space applications and chemicals, large-scale ship-based transport has not yet been commercialised. Hydrogen pipeline transmission is a mature technology, as the pipelines in the Rhine-Ruhr metropolitan area, installed in 1938, are still operational today. Carbon steel or stainless steel should be used to prevent embrittlement. Central production and distribution need to be optimised depending on demand and location. While sufficient practical experience has not been gained with this technique, geological salt caverns can be used to store hydrogen. Efforts need to be made to locate appropriate depleted fields and sites to establish more salt caverns. By doing this, hydrogen can be transferred over long distances and safely stored underground for a very long time without any loss of energy content. This can function as a strategic reserve catering to multiple applications. It has been reported that hydrogen-based fuels will account for about 13% of energy demand in 2050 [2]. Strategies, alliances and technology developments are moving forward across a wider geographical area to achieve net-zero emissions. Considering the growth pattern in industry and transport, it is quite likely that the majority of hydrogen produced will go there. Even though the technology of vehicle development (both internal combustion engine and fuel cell) has been developed, there are still some inadequacies, such as refuelling stations, on-board storage, etc., which need immediate attention. The cost of fuel cells and comparable driving ranges with electric vehicles (BEVs) are potential rivals for cars. In some transport applications such as rail transport, hydrogen has to face economic challenges because electrified rail transport has already been used in several countries for a long time. In power generation, a low-carbon hydrogen-based fuel such as ammonia along with hydrogen can be employed on short-term basis in gas turbines in the initial phase. CCUS may be considered for some applications, depending on the situation. Several projects and efficient collaboration among government, academia and industries have been initiated to evolve short-, medium- and long-term strategies to pave the path towards total decarbonisation of a wide range of sectors such as land transport, aviation, shipping, power sector, data centres, steel and several chemical industries.

As a transitory solution, hydrogen and compressed natural gas blends are used in gas distribution grids and vehicles. Quality regulations vary in different countries. In order to decide about blending in the gas network, it is necessary to analyse quality parameters, such as composition, calorific value and Wobbe index. Some experts suggest a combination of CO_2 with green hydrogen to produce energy-dense synthetic fuels will have applications in situations where hydrogen is not suitable.

Earlier the USA, European Union, India, Japan, Australia, China, Canada and some other countries had already started hydrogen energy activities which are getting accelerated now. More recently, many governments have been proactive in developing hydrogen infrastructure and expanding active research and development programmes on several aspects of the hydrogen energy system. It is within the last couple of years, nine countries have already adopted hydrogen energy strategies and another twenty-six countries have committed to have clean energy vector with hydrogen energy system. Several international projects have been initiated and some information from some of the programmes has been taken from a recent report ([4]) and given here. National programmes and international collaborations might have increased afterwards. Germany has already signed bilateral agreements with a number of countries, including Australia, Chile, Morocco, Namibia, Tunisia and Ukraine. Egypt and Zimbabwe already have installed over 100 megawatts (MW) of electrolysers. The hydrogen strategy of the European Union is to install 40 GW of renewable hydrogen electrolysers in the European Union by the year 2030. The National Hydrogen Strategy was adopted in Japan in 2017, and hydrogen-energy-related activities are continuing in full swing since then. Japan aims at rapid introduction of hydrogen energy technology in almost all energy sectors and has become the world's first hydrogen society. The Government of India has been pursuing research and development activities on hydrogen energy for more than three decades. The National Hydrogen Energy Road Map (NHREM) was initiated by the National Hydrogen Energy Board (NHEB), where activities on production, storage and utilisation of hydrogen are carried out with industry–academia collaboration. Recently, India launched its National Hydrogen Mission in August 2021 with the ambition of becoming 'a global hub for green hydrogen production and export'. The Government of India is considering making it mandatory for refineries and fertiliser plants to use green hydrogen. During COP26, the Government of Namibia has announced a green hydrogen project by HYPHEN. In 2020, Chile launched a green hydrogen strategy for 5 GW of electrolyser capacity by 2025. The project aims at a capacity of 25 GW by 2030. Morocco has adopted a National Hydrogen Commission in 2019 and has already prepared a green hydrogen road map in January 2021. HTEC in Canada had opened up a hydrogen retailing station in 2018. This was started with the objective of having a multi-station network to facilitate on-the-road operation of hydrogen electric vehicles. The number of stations has increased to five. These are some of the projects in different countries with a view to ensure fast penetration of clean hydrogen value chain, integrating the technologies, for a clean energy future. It is quite likely that by the time this book reaches the readers, things would have changed substantially with larger participation in hydrogen programmes in several spheres in many countries.

There are some technologies almost ready for commercialisation. Innovation in electrolyser manufacture; improved and economically viable routes of hydrogen production; use of carbon capture and storage technology; optimisation of hydrogen storage depending upon specific applications; and use of hydrogen in vehicles, fuel cells and several other industrial and domestic sectors – these are some of the important areas where the use of green hydrogen can be considered with adequate safety measures.

Given the 1.5 °C climate mitigation target, water electrolysis is becoming increasingly significant. Now, it is extremely critical to identify the optimum route for high-purity green hydrogen production along with its utilisation options. Lowering the capital cost of electrolysis, locating water resources, identifying renewable sources (solar, wind, geothermal, nuclear thermo-chemical water splitting, photo-electro-chemical electrolysis and biological methods) and implementing relevant conversion technology and environmental impact assessment are some of the major challenges to be investigated to move towards commercial readiness. Infrastructure must be built up for central-station and distributed hydrogen production.

As far as life cycle analysis is concerned, most of the research and analysis has been limited to hydrogen production. Many life cycle analyses present only an environmental profile. It is worthwhile to study socio-economic issues. There are practically no studies on cradle-to-grave analysis. Such a study is very relevant as it provides a holistic picture of hydrogen technology. This provides guidelines for robust decision-making towards a carbon neutral future. As far as storage is concerned, conventional storage methods need to be streamlined. Low energy efficiency of cryogenic storage and the need to store extremely large volumes of gaseous hydrogen are some of the inherent limitations of conventional storage. Reliability of composite vessels needs to be further improved for specific applications. Material compatibility of storage systems, regular inspection and maintenance, design of pressure vessels, embrittlement and blistering issues, and control of leakage at vulnerable points are some of the well-identified problems that require significant attention to be sorted out for regular applications.

A widespread investigation should be conducted into the storage in salt caverns, depleted natural gas fields and aquifers. In each case, the relevant properties of hydrogen should be considered to decide the mode and duration of storage. While considering storage in depleted natural gas fields, it is relevant to note that gas fields are widespread and not like salt caverns. In this case, properties such as high diffusivity, low viscosity and high compressibility factor of hydrogen ought to be assessed. The problem of embrittlement can be eliminated by limiting diffusible hydrogen content below the critical hydrogen concentration level for any application. Several other coating and plating techniques prevent diffusion or penetration into the materials. Eyebrows have been raised on safety issues with the entrance of hydrogen economy. Common man has to be appropriately educated and informed about how to handle hydrogen and stay near the hydrogen environment. Adequate ventilation and absence of unintended ignition source and propensity of leakage need to be scrupulously avoided.

Some of the technologies in the hydrogen value chain are yet to reach maturity for regular system design. This is essential for a technology to prove its potential for large-scale upgradation leading to commercialisation. Design and development of hydrogen-powered vehicles have technical feasibility with improved thermal efficiency and nearly zero-emission characteristics.

BMW AG, Sasol Ltd and Anglo American Platinum Ltd. (Amplats) are working on collaborative projects where BMW's fuel cell iX5 sport utility vehicle will be able to operate on green hydrogen delivered by Sasol. Amplats mines Platinum group metals by Amplats will be useful for fuel cell electric vehicles.

ZeroAvia, the aviation company has focussed on the development of hydrogen-electric propulsion systems. Use hydrogen fuel cells to generate electricity that powers the electric motors which turn the aircraft's propellers. The company aims developing such aircrafts targeting initially about 480km range in 9–19 seat capacity by the end of 2025. Apart from fuel cell propulsion system, there are plans to extend the range of activities to liquid hydrogen fuel storage and refuelling infrastructure.

Moog Aircraft in the UK, is leading a ground-breaking collaborative project to develop precision fuel controls for large commercial air craft with gas turbines. The members of this collaborative project include the University of Bath/IAAPS, Cranfield University, Baker Hughes Druck, Curtiss-Wright Corporation, and Carter Manufacturing. It has the technical support of GE Aerospace.

In October 2023, California Department of Transportation (Caltrans) signed a contract with Stadler Rail, Inc, as a result of which zero-emission, hydrogen intercity passenger trains will be used for transportation. These vehicles would be developed on Stadler's successful Fast Light Intercity and Regional Train (FLIRT) concept. These will be self-powered multiple-unit hydrogen fuel cell trains without a separate locomotive. Such innovative clean transportation projects will be helpful in building an expanded clean rail network.

Bus builder Iveco and carmaker Hyundai have signed a memorandum of understanding (MoU) to work together and complement each other's expertise and resources to build hydrogen fuel cell bus called E-WAY H2. The hydrogen fuel cell bus, which is designed, based on Hyundai's cutting-edge HTWO fuel cell system in conjunction with an Iveco 310-kilowatt electric motor for its power source, generated significant interest at an event on October 2023. This 12 meters long, low-floor city bus is expected to be a very desirable option for both urban and suburban routes.

Recently, in September 2023, India's first Hydrogen Fuel cell powered bus was launched in New Delhi. Tata Motors, country's largest vehicle manufacture and Indian Oil Corporation Ltd. (IOCL), the country's largest petroleum company contributed to this pioneering achievement in India towards greener mobility. The prominent features of the bus include 350-bar hydrogen storage system, 70-kilowatt fuel cell stack, electronic braking system and stability control for higher safety. 75 kg of hydrogen to be used by these buses are produced from renewable sources by Indian Oil Corporation (IOC). Intelligent transport system, new generation telematics enhance vehicle management, safety and efficiency by leveraging advanced technology and data-driven insights.

As decarbonisation strategies are adopted in a wide scale for transport sector, hydrogen fuel has increasingly garnered attention as a significant player in mitigating emissions within the trucking sector.

Nikola's flagship hydrogen truck model, the Nikola One, is a Class 8 heavy-duty truck designed for long-haul freight. It features a hydrogen fuel cell powertrain that generates electricity for propulsion. The truck also includes 320 kWh lithium batteries that power the fully electric drivetrain. Since peak torque is reached almost instantly as the foot touches the throttle, the truck can shoot up to very high speed within seconds. It is claimed that this truck has a range of 1200miles (close to 2000km) without any emission.

While Toyota is better known for its passenger car hydrogen fuel cell model, the Mirai, the company has also shown interest in developing hydrogen fuel cell trucks. Toyota launched its Class 8 hydrogen fuel cell truck venture in 2017 under the name Project Portal, and things keep moving. California Air Resources Board has already certified the company's hydrogen fuel cell powertrain kit, a soup-to-nuts system: a comprehensive and ready-to-install package that includes all the necessary components such as storage tanks, fuel cell stacks, batteries, electric motors etc. Similarly US truck manufacturer Paccar and Toyota have joined hands to develop Kentnworth and Peterbilt to bring hydrogen trucks to road. Toyota and VDL have brought the first fuel cell truck from VDL with Toyota technology as part of their collaboration which is planned to expand in different countries of Europe to ensure CO_2-neutral mobility by establishing a sustainable hydrogen infrastructure.

Hyundai Xcient Fuel Cell: Hyundai's hydrogen truck model, the Xcient Fuel Cell, is a Class 8 heavy-duty truck designed for commercial transportation. It is equipped with a hydrogen fuel cell system that provides power for the vehicle. In these trucks, ten hydrogen tanks are arranged two-up in a frame behind the truck rising to nearly 4 meters in height of the tractor's roof fairing. The tanks have a volume of about approximately 1,750 litres of compressed hydrogen at around 690bars, totalling a little less than 70 kilograms of fuel. It takes about half an hour for refilling. The Xcient Fuel Cell has a range of around 650 km on a single hydrogen fill, which makes it suitable for long-haul applications.

Daimler, the parent company of Mercedes-Benz, has developed the GenH2 Truck, a hydrogen-powered heavy-duty truck. The vehicle is equipped with fuel cell technology and features high-performance lithium-ion batteries for additional power and energy recuperation. The GenH2 Truck is designed for long-haul transportation and is expected to have a range of around 1000km on a single hydrogen tank. Fuel cell technology, established for a number of years, is now entering a relatively wider application area. High cost and limited durability have been major stumbling blocks for decades. Currently, several projects have been successfully carried out in many countries and several fuel cell vehicles have been designed and are operating on roads in many countries. By the year 2021, more than 40 000 fuel cell electric vehicles were in circulation around the world. There seems to be an accelerating demand as the figure has now gone to approximately 56 000 by February 2023. Use of fuel cells is now rapidly entering portable, stationary and different modes of transport applications. Since internal combustion engines can tolerate some level of impurities in hydrogen fuel, modified hydrogen engines can be used in the transport and power sectors in some locations until commercially available fuel cell vehicles are ready for widespread implementation all over the world. Similarly, the technology of transporting liquid hydrogen in vessels has been successfully demonstrated and it is expected that their number will increase.

Optimum hydrogen-operated gas turbines are yet to be developed even though such systems have been constructed and studied. Several industries are trying to blend hydrogen into existing natural gas plants and evaluating the performance and design parameters to build hydrogen-ready power plants. In long-term endurance tests with natural gas blending, useful information is gathered on performance and emission characteristics necessary to determine the optimum mix ratio. Data related to material compatibility, life cycle analysis and techno-economic feasibility can also be derived from these tests. Polymer-based distribution networks in existing grid infrastructure can tolerate ~20% hydrogen volume to avoid embrittlement. These tests would provide guidelines for adjustments and modifications necessary to ensure smooth hydrogen gas turbine operation.

It is technically feasible to use renewable hydrogen in several energy sectors without disturbing existing infrastructure. Depending on the specific situation, the mode of energy transition and sector integration can be worked out. Together with technological development, several important steps – such as international collaboration, thoughtful planning, stable and uniform long-term policy frameworks and mechanisms for fixing standards – are essential to accelerate the hydrogen energy transition to a greener future.

References

1 International Energy Agency (IEA), Global Hydrogen Review 2021 https://www.iea.org/reports/global-hydrogen-review-2021

2 *Renewable should overtake coal within five years to secure 1.5 C goal,* International Energy Agency (IEA) Report on Energy, May 2021

3 Robert W. Howarth, Mark Z. Jacobson, *How green is blue hydrogen?* Energy science and engineering Volume 9, Issue 10 October 2021 Pages 1676-1687

4 *Geopolitics of the Energy Transformation: The Hydrogen Factor,* IRENA, International Renewable Energy Agency, Abu Dhabi (2022).

Index

Hydrogen Energy: Production, Safety, Storage and Applications, First Edition. L. M. Das.

d

e

h

i

m

n

V

W

Z